Basic Concepts of Neuronal Function

A Multilevel, Self-Teaching Textbook

Don L. Jewett, M.D., D.Phil. (*Oxon.*)
Associate Professor of Orthopaedic Surgery, University of California, San Francisco, School of Medicine, San Francisco

Martin D. Rayner, Ph.D. (*Cantab.*)
Professor and Chairman, Department of Physiology, University of Hawaii, John A. Burns School of Medicine, Honolulu

BASIC CONCEPTS OF NEURONAL FUNCTION

A Multilevel Self-Teaching Textbook

LITTLE, BROWN AND COMPANY
BOSTON/TORONTO

First Edition

Library of Congress Catalog Card No. 82-82796

ISBN 0-316-46310-8

Printed in the United States of America

SEM

v

Contents

This preface is addressed to teachers who may want to use this book in their classes. Students are welcome to read this portion, which outlines the goals and background for the writing of this book, or they can proceed directly to Chapter 1.

PRINCIPLES IN A FOREST OF FACTS

It is our firm conviction that biology in general and neurophysiology in particular have reached milestones in their development. As any scientific field progresses from its initial descriptive phase to analytic maturity, typically a major change occurs in the teaching techniques required to introduce newcomers to the field. The descriptive phase requires a historical-anecdotal teaching method. The principles that underlie the field are not entirely clear, even to its teachers, and the historical approach exposes the excitement of the "chase," of the logical sequence by which the major innovators have made their successive contributions. In contrast, when the analytic phase is reached, the introductory course is made up of the generally accepted underlying principles of the field. At this point, the historical sequence may be abandoned in favor of a sequence designed to introduce the field with maximal clarity, with minimal ambiguity, and in the shortest possible succession of logical steps.

This book attempts to teach basic principles of neurophysiology. Even if one recognizes the superiority of teaching the basic principles, how is this to be accomplished in practice? We have found that the teaching process (as a closed loop) can clarify the principles underlying neurophysiology. It is our experience that if motivated, intelligent students do not understand an explanation, there must be a basic flaw in the teaching, i.e., in the textbook lectures given to the students. The unsatisfactory replies that we formerly gave to students' questions led us to write this book.

STUDENT LEARNING WITH A MULTILEVEL TEXTBOOK

Even with a logical sequence of principles, a textbook still must communicate well to many students in different curricula in a variety of fields in a multitude of schools. Usually this requirement is ignored, being generally directed to the students of the author, a solution that reflects only the limitations of the traditional textbook. Students are remarkably adaptable if given half a chance. With these factors in mind, we wrote this book in a multilevel format (see Chapter 1), to communicate to a variety of students (with different interests) within the confines of a single book. **Students interested in "just the basics" can stick to the "core" material (which is clearly labeled), venturing into more complex material only when available time and mood coincide. Those interested in greater depth will profit from reading and understanding the "didactic simplifications,"** especially if someday they will have to teach this material themselves! Thus, by clearly indicating the intended readership, paragraph by paragraph, we do neither student a disservice and may broaden the outlook of both.

Although we wrote this book in the context of medical school education, we are confident that it will also be useful in dental and pharmaceutical schools, and in graduate and undergraduate courses in physiology, zoology, physiological psychology, and bioengineering.

viii

The relationship of this book to other scientific writings is shown in the accompanying diagram. The goal of this multilevel textbook is to cover the basic part of the conceptual continuum more simply than the standard textbook does, without trying to supply *all* the more complex descriptive and experimental detail usually presented in a standard textbook. However, **this book offers sufficient knowledge for students to be able to get from standard works (and more advanced texts) whatever additional information they may want.** This book is more detailed than the popular scientific writings for the layperson (e.g., *Scientific American*) because the minimal knowledge of neurophysiology needed in medicine or in other specialized courses is greater than the knowledge that can be culled from lay sources.

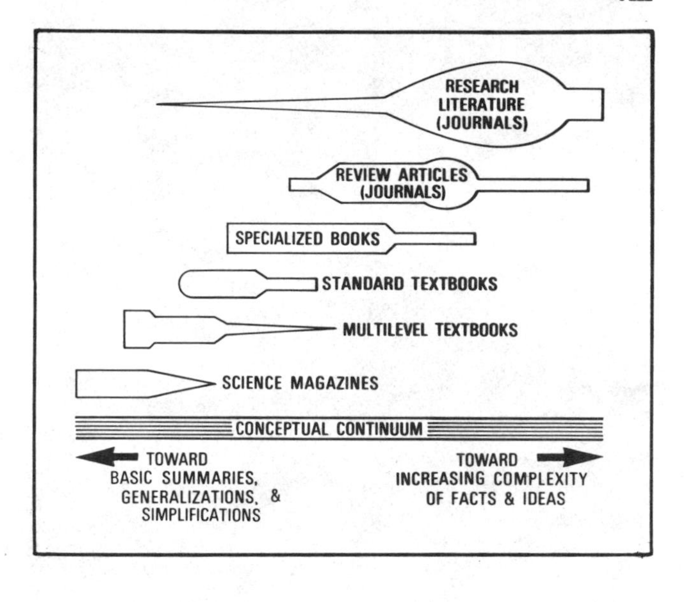

TESTED TEXTBOOKS

The role of multilevel textbooks in the medical curriculum has been described elsewhere [25]. The basis of any successful teaching material must rest on the experimental method; that is, the teaching ability of the material must be user-tested. In practice, we elicited student feedback with sentence-by-sentence criticisms, made the appropriate corrections to the text, and then repeated the process. On the basis of user testing in medical and pharmaceutical schools, as well as in undergraduate classes in neurophysiology and bioengineering, and graduate classes in physiology, we believe that the apparently large audience for which this book is intended can actually be served well by it.

EASE OF TEACHING

This book is written so as to ease the burden of teaching. We seek to serve a large range of teacher interests and backgrounds by making the book both adaptable to individual instructors' preferences and also as self-teaching as possible, thus freeing the teacher from the more mundane tasks of information transmission and (in our experience) raising the quality of questions posed by students. The first feature is provided in the following way: Each paragraph is numbered (with a small number to the left of the last line) sequentially from the top of the page. Thus, "page 305.2" refers to the second paragraph on page 305. In this way, the index can directly cite a given paragraph, and so can an instructor! Such numbering allows the instructor to modify the levels that we have assigned to the various parts of the book. It is a simple matter to give to students a list of the paragraphs that are to be included or excluded from a given level, keeping the basic definitions of level as described in Chapter 1. The instructor may say, "The following paragraphs are to be considered as second level: 305.2, 307.3. . . . The following are to considered third level: 305.4, 419.7. . . ." *Of course, only changes in emphasis need to be so listed. In a few minutes students can mark their copies of the text to correspond to the individual instructor's judgment as to the importance of each paragraph.* One should not underestimate how much students value explicit delineation of what the instructor considers important!

One word of caution to those who modify the levels available in the text: We have ensured that no material in the first or second level is dependent on material in the third or fourth level. Thus, material can be moved *to the left* (made "more important") without concern for whether parts of the learning sequence will be disrupted. However, if material is moved *to the right* (made "less important"), teachers should be aware that they may have demoted a section on which a later idea depends. (The students will point out such discrepancies, should they occur!) To minimize such occurrences, we tried to limit the amount of material making up the intellectual core (the first and second levels).

viii

ix

OTHER UNUSUAL FEATURES

The casual observer flipping through the book may notice a number of unusual gimmicks. We assure you that each and every gimmick has been tested and retained on the basis of student reactions rather than the authors' idiosyncrasies. These features are as follows: (1) The multilevel format, already described and discussed, is used. (2) Questions are interspersed throughout the text to encourage active student participation. (3) Either hints or answers to the questions are given in easy-to-find locations on nearby pages. (4) The book is printed the oblong way so that students can see more of the organization of the levels at one time. This format also means the line length is longer, there is space for illustrations and notes in the right-hand margin, and turning the pages to read the hints is easier. (5) The most important statements for review purposes are in boldface type. (6) At times a lighthearted approach is used (even at the expense of revealing the personalities of the authors) since students find such small diversions a welcome respite during long study hours. (7) Each paragraph is numbered, not only to allow flexibility to individual instructors, but also to permit the user of the index to find a specific reference quickly and accurately. (8) There are numerous illustrations, many of them original, to aid in visualizing concepts.

FROM MICROSCOPIC TO MACROSCOPIC

As you can see from the preceding description, this book is written for students, in contrast to many textbooks that seem to have been written for the authors' peers and have little regard for the "teachability" of the result. For this reason we took an approach that we did not ourselves experience as students: to move always from the microscopic to the macroscopic and from principles to specifics. While some instructors may find that this approach is not to their taste, we are confident that their students will like it. We trust that after one more generation this problem will lessen, as neurophysiology makes the transition to a field based on principles that must, of necessity, involve subcellular (molecular) events.

A WORD OF CAUTION

We ask that those who would like to evaluate this text do so by a thorough reading of the sections of interest. We doubt that a quick "flip-through" will allow a reader to appreciate the value of the text, given the number and types of innovations in both format and content. Even with a thorough reading of a section, those who are knowledgeable may think that "this feature is not necessary" because they cannot remember what it is like to encounter the material in ignorance! In other words, we are surer of the student's response than we are of the teacher's! But we take this to be a good sign, for a scientific field has not progressed unless students take a shorter time to learn the same material than their teachers did. The intellectual path in the newer material is shorter and, consequently, is foreign in some ways to readers who traversed the longer path some time ago.

We recommend strongly that those wishing to experience the full usefulness of our approach to membrane phenomena follow the "shorter version" described in the next section. The consistency of viewpoint, emphasizing the electrical-capacitative analogy, is best appreciated by seeing how readily it allows understanding of steady state and transient potentials and electrical, mechanical, and chemical transmission, as outlined.

ix

A SHORTER VERSION

This book is intended to show the most confirmed skeptic (student or teacher) that the approach we have taken to membrane potentials not only is consistent with more advanced approaches, but actually leads directly to established and successful concepts of membrane phenomena, such as the Hodgkin-Huxley equations. To achieve this completeness, we included several advanced chapters and sections that need not be read by those wishing to learn the basic physiology of membrane potentials. Students who are pressed for time will find that the following list of chapters presents a complete description of the basic concepts underlying membrane potentials, axonal conduction, generator and receptor potentials, and synaptic transmission: Chapters 3, 4, 6, 7, 9, 10, and 11 (a total of about 200 pages). Teachers wishing to evaluate this text will find this shorter version of value.

Those teachers who find the shorter version of value to their students may wish to *communicate with the first author*, giving him a rough estimate of the number of students who used this core material. Such information could be very useful in deciding if a separate, shorter book should be published.

WITH THANKS

We acknowledge, with thanks, the help we have received from the many students, both predoctoral and postdoctoral, who took the time to offer criticisms of previous versions; from Lloyd D. Partridge, Ph.D., for his thorough review and criticisms of an early version of the manuscript; from the typists, Leslie Williams, Victoria Stephens, Stephen Feinstein, and Nancy Kennelly, who spent long hours assembling complex materials; from Jean C. Lieberman, Ph.D., and Al Averbach, who did the editorial work; from Mark Mikulich, who did most of the work on the illustrations; from Dean Julius Krevans, University of California, San Francisco, School of Medicine, who gave us his support; and from the staff of Little, Brown and Company for bringing the project to completion. We give special thanks to our publisher for being receptive to a book in such an unusual format.

D. L. J.
M. D. R.

Basic Concepts of Neuronal Function

A Multilevel, Self-Teaching Textbook

1

Introduction to a Multilevel, Self-Teaching Method

WHY THIS BOOK?

As knowledge in neurophysiology has advanced, the complexity and difficulty of the standard textbook have increased in like measure. We have reached the point at which many "beginning" textbooks are too advanced for the novice! Even more important, **almost all textbooks offer more information than any student can assimilate** or even use! Under these circumstances, the student has no way to distinguish what is most important (i.e., should be learned) from what is less important (i.e., should be ignored at first). As a result, many students resort to memorizing relatively trivial (but attractive) details while failing to master the major concepts of the field.

The main purpose of this book is to make clear those ideas that are of fundamental importance in neurophysiology. That often these ideas are not readily absorbed from standard textbooks seems to us not a failure on the part of the student, but a defect in the standard textbook format, which does not give adequate clues to the relative importance of different sections of the subject matter. Hence, we wrote this book in a multilevel format.

MULTIPLE LEVELS

To distinguish clearly the importance of the material presented in this book, the text is divided into four "levels" of significance. These levels are marked by different amounts of indentation and different numbers of vertical lines. You should look **now** through the book to see the format.

The **first level,** indented 5 spaces from the left-hand margin and marked by **four** vertical lines, is used for the **most basic** and **most important generalizations,** which introduce or summarize the core material. These brief statements indicate what material immediately follows. On the first reading, you may not understand the vocabulary or concepts. However, **upon review, you will find that these summaries help you remember the main points.**

The **second level,** indented 10 spaces from the left-hand margin and marked by **three** vertical lines, is used for detailed exposition of the core material of neurophysiology, the material all students are expected to understand and master (and on which later material is based). Levels 1 and 2 together comprise about 50 percent of the book.

The **third level,** indented 15 spaces from the left-hand margin and marked by **two** vertical lines, contains material that only the more interested students need to learn. Here are found further ramifications of general principles, the 5 percent of exceptions to rules that are 95 percent correct, experimental verifications for some of the "facts" presented at the second level, and so on. This level makes up about 25 percent of the book.

The **fourth level,** indented 20 spaces from the left-hand margin and marked by **one** vertical line, presents obscure points that the authors enjoy, references to mathematical derivations, more detailed descriptions of experimental methods

3

of interest to graduate students in physiology, and so on. The fourth level makes up about 25 percent of the book.

How the multilevel arrangement will help you. (1) You can read only what is most appropriate for you at the time, **without having to read everything.** (We know that as an eager student, often you read everything in the textbook—but you may not always have time for that!) (2) If you are interested in something, you can find additional information by going to a higher level in this book, rather than searching in some other book. (3) You can **easily review the main points** before an exam.

An efficient way to use this book. (1) When you start to read a chapter, you may find it helpful to get an overview of the material to be presented by flipping through the chapter, **reading just the headings.** (2) Study the chapter, concentrating on the first and second levels, that is, the core material. The first-level generalizations will be clear to you only when you have mastered the second-level material. (3) If you are interested and have already mastered the second level, feel free to read into higher levels, **but that material is not necessary for understanding what follows in the succeeding first and second levels.** (4) Before the exam, review at least the main points in the first and second levels. **The first level is especially valuable for such reviewing.** Follow any changes in level that your instructor has given you.

BOLDFACE WORDS AND SENTENCES

A book can seem pretty dull if it contains only summarizing statements that you are supposed to memorize. Often it is hard for the reader to follow both the ideas and the language in such outlines. So this book contains **additional** comments, introductory statements, careful and detailed explanations, and so on, to make the book easier to read and follow. However, **the really crucial sentences are in boldface type** so that you will be sure to notice them on **first reading** and be able to **review** them easily later. You have probably underlined important parts of your previous books—feel free to underline the portions you want to emphasize—the boldface indicates what WE want to emphasize!

WRITING SPACE

On the right-hand side of many pages, there is space for you to make notes, comments, reminders, and so on. The more you work with a book, the more useful it is to you, both now and in the future, so don't hesitate to use this space.

FIGURES

Figures are located in the space on the right-hand side of the page. When a figure is repeated, the original figure number is shown after a slash mark; for example, Fig. 6-2/4-18 is a duplicate in Chap. 6 of the original Fig. 4-18. This numbering system also applies to equations. This system allows you to trace back in the text easily, should you wish to review the text that accompanied the original figure or equation.

3

4

PARAGRAPH NUMBERING

To the left of the vertical lines are small numbers at the end of each paragraph. These numbers permit the index to direct you to a specific paragraph! A reference in the index to 302.3 indicates the third paragraph on page 302.

REFERENCES

Students should remember that the "facts" presented in textbooks are merely conclusions based on experimental evidence, which may change when more accurate (or more complete) studies are made. Wherever facts are quoted that have not achieved widespread acceptance and hence are **not** yet in textbooks, we cite the source by a number enclosed in brackets, corresponding to the Bibliography at the end of the book. However, some of the references in this book are other textbooks. They are included to allow students the opportunity either to study an alternative explanation of a baffling fundamental concept or to obtain an introduction to the research literature through a standard textbook containing a more complete discussion of the experimental background material than we considered advisable here. These references are also given in brackets. Since not all the texts listed in the Bibliography will be readily available to the average student, we included multiple references for the most basic concepts. However, it would make difficult reading if every concept mentioned in every text were referenced in this book; while we tried to list the clearest alternative explanations, no slight of an uncited text is intended when referencing does not include it.

QUESTIONS

Interspersed throughout the text at all levels (except the first) are **questions.** There are different classes of questions, **indicated by how they are labeled.**

QUESTION: This heading indicates questions that follow the material just presented, and it will help you understand the material. **You should try to answer these as you read.** If you can answer these questions, then you know the basic material of that level.

Question: This heading indicates questions that are more difficult, require other knowledge, or involve more complex reasoning. (Note that it is not set in all capitals.) Feel proud if you answer them correctly, but don't worry if you don't.

EXAM QUESTION: This heading indicates samples of questions that you should be able to answer in an exam. They are taken from old exams or are like those that might be asked. They are placed at the ends of some sections and chapters, just as exams come after you have covered a number of topics. Use these questions to get a feel for how well you are remembering the material.

NOTE: **Since the questions are segregated according to level, you need concern yourself with only those questions in the level that you are reading or that you have mastered.**

NOTE ALSO: If there is a sequence of questions, we put them in order of increasing difficulty.

4

5

NOTE FINALLY: The questions in most textbooks are at the ends of the chapters, where they are usually ignored. The fact that questions in **this** book are **in the text** indicates that we mean them to be of **help to you in your learning.** Thus, many of them are quite simple—if you have understood what you have read, often you will be reassured by answering such questions easily. However, the harder questions may give you some insight into how the thinking of the neurophysiologist differs from that of normal people!

HINTS

What good are questions without answers? In this book we intend for you to learn most of the material without additional aid, so answers **are** provided—they are called hints because sometimes they do not give you the answer directly, but give an additional hint, so that you can work it out yourself. **It is very important that you try to get the answer yourself before looking at the hint.** Write down the answer **before looking.** Of course, you can cheat and peek at the answer before thinking about it; no one will know, but you won't really have had to chance to use your available knowledge, especially when it is less than you would like it to be! **Pulling together information into a usable form** is a most important part of your training. Let's put it another way: Neither scientific problems nor patients come with hints attached! There is a skill involved in actively using your own brains. It takes practice to learn to reason things out. If you miss the question, who will ever know? And you get to read the hint whether you're right or wrong!

Where to find hints. At the end of each question, there is a hint number with an arrow [e.g., "(Hint 5↓)"]. If the arrow points **down,** then the hint is at the bottom of the **next** two-page spread (see diagram). If a page number is also given, the hint is on the bottom of that page. Hint numbers with the arrow pointing **up** are at the bottom of the **preceding** two-page spread.

DON'T FORGET: **Try working on the answer before you lift the page to look at the hint!**

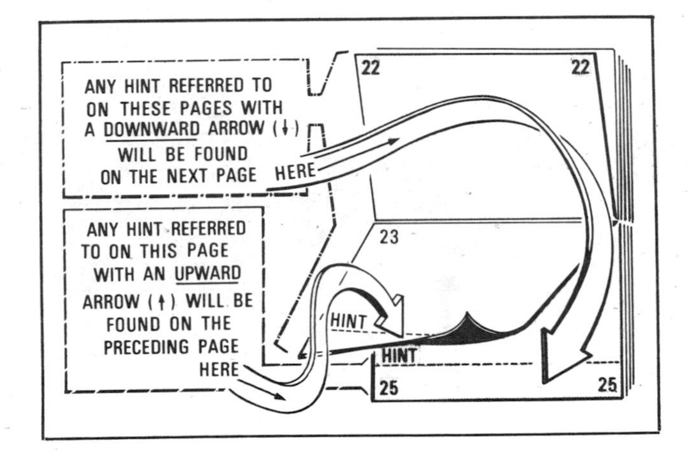

THE END OF THE BEGINNING

Soon you will be able to determine firsthand whether these gimmicks are as helpful to you as they have been to our previous students.

As teachers, we are pleased that by means of this book our efforts will reach many students; but we regret that by the nature of such long-distance, one-way communication, we do not have the advantages of direct interaction with our readers. Writing a book, like studying, can be a solitary occupation: authors, in the dark of night, endeavor to put their thoughts into clear phrases while students, in the dark of night, endeavor to grasp and retain what was written. We hope that you will let us know of both the failures and successes of our efforts, in the hope that the next edition will be improved, just as previous versions of this material have been improved by the students who took the time to tell us their reactions. We would be happy to hear from you by mail; or, if your path somehow crosses ours, please don't hesitate to meet us directly.

5

7

Neuroanatomic Background of Neurophysiology

INTRODUCTION

Since students may come to study neurophysiology from a variety of backgrounds, this chapter provides essential terminology, particularly in neuroanatomy. (Those who are familiar with these ideas may still find it of some interest to skim this chapter; the approach used here is different from the usual viewpoint of neuroanatomy.)

Classically, anatomy is the study of form, while physiology is the study of function. Of course, such a distinction implies that form and function **can** be considered separately. Historically, a biological science begins with the study of form, progresses to the study of function, and finally realizes that the two are largely inseparable.

Hence quite detailed anatomic descriptions often must be included in the body of the text, at the point where the interrelationship between form and function will be understood best. This chapter merely provides a general background in the concepts and terminology common to many chapters of the book. Students should consult textbooks of histology and neuroanatomy to appreciate the full range of anatomic knowledge. Here we skim off from that body of knowledge only those points having direct application to the main goal of this book: a description of the principles on which the nervous system operates, as understood at present.

The general approach of this book is to go **from** an understanding of **the parts** of the nervous system **to the interworking** of those parts; we hope to show that in many cases the behavior of large groups of cells can be inferred from an understanding of the functioning of single cells.

Thus, **we move from the microscopic to the macroscopic,** from a single cell to groups of cells, and from simple principles to the more complex interactions of several variables. Let us take the same approach now—let us treat anatomy by going from the cell to the tissue, to the organ, to the organ system. Hence we start with cells.

CELL THEORY

Our modern understanding of the nervous system rests on the idea that the nervous system consists of a complex organization of single cells.

Each cell is the progeny of another cell, and each is delineated from others by a boundary: the cell membrane. The cell membrane separates the internal and external environments of the cell. Outside the cell is the extracellular fluid, which, although it may still be inside the body, is a jumble of chemicals in dilute solution, in contrast to the intracellular fluid of the cytoplasm. Inside the cell are highly organized macromolecules that are concerned with the cell's replication, growth, and maintenance (the nucleus); with the cell's energy metabolism (the mitochondria); and with any special products that the cell may produce for extrusion or excretion (the endoplasmic reticulum and Golgi apparatus).

Table 2-1.

Dimensions Related to the Meter Standard		Dimensions in Common Neurophysiological Use	
1 mm (millimeter)	$= 10^{-3}$ m (meter)	1 cm (centimeter)	$= 10^{-2}$ m
			$= 10^{4}$ μm
1 μm (micrometer, formerly called micron)	$= 10^{-6}$ m		$= 10^{8}$ Å
		1 mm (millimeter)	$= 10^{-1}$ cm
			$= 10^{3}$ μm
			$= 10^{7}$ Å
1 nm (nanometer)	$= 10^{-9}$ m	1 μm	$= 10^{-4}$ cm
			$= 10^{-3}$ mm
			$= 10^{4}$ Å
		1 Å (angstrom)	$= 10^{-10}$ m
			$= 10^{-8}$ cm
			$= 10^{-7}$ mm
			$= 10^{-4}$ μm

Table 2-2. Units of Time in Common Neurophysiological Use

1 ms (millisecond) = 10^{-3} s (second)

1 μs (microsecond) = 10^{-6} s

1 Hz (hertz) = 1 cycle/s

Table 2-3. Units of Volume in Common Neurophysiological Use

1 mL (milliliter) = 10^{-3} L (liter)

1 μL (microliter) = 10^{-6} L

1 nL (nanoliter) = 10^{-9} L

(Note that often the milliliter is used as the standard of reference since 1 mL of H_2O has a volume of 1 cm^3.)

9

Since cells have dimensions on the order of 10 μm (Table 2-1), it is easy to understand the controversy in the last century as to whether cells are separate entities. The basic separation of individual cells could not be fully established before the invention of the electron microscope, which can clearly resolve the cell membrane, a structure with a thickness of about 8 to 10 nm (Table 2-1). Quantitative study of cells requires a familiarity with the standard units of dimension used in description of cells and their internal organelles. These units are given in Table 2-1. While it would be simple if all measurements were quoted in terms of the meter as a standard (see the left-hand side of Table 2-1), many measurements in neurophysiology are based on the **centimeter** as a standard (see the right-hand side of Table 2-1). In electron microscopy, it has been normal to use angstrom units. Fortunately, these standard units are readily interconverted, and you should familiarize yourself with the conversions in Table 2-1. (Similar conversions for units of time, volume, quantity, and concentration are shown in Tables 2-2, 2-3, 2-4, and 2-5.)

The use of prefixes in scientific discourse greatly simplifies communication. You are familiar with the following prefixes: milli = 10^{-3}, micro = 10^{-6}, nano = 10^{-9}, kilo = 10^{3}, and mega = 10^{6}. The use of these prefixes is not limited to traditional science but can add clarity to other measures:

1. The milli-Helen is that amount of beauty necessary to launch one ship.
2. If we define a *kluge* as "an ill-assorted collection of inappropriate devices forming a barely functioning whole" which occupies 1 m^3, then a megabuck is the estimated cost (to Congress) of a kluge, or the actual cost (after overruns) of a millikluge. Furthermore, a *jiffy* is the estimated time to complete a microkluge.

However, some care must be taken in analyzing older words; e.g., a microtome is **not** a small book.

QUESTION: What is the thickness of the 8- to 10-nm cell membrane in angstroms? (Hint 1↓)

QUESTION: Given that certain *pores* occur in a nerve membrane at a frequency of about 13 per square micrometer, how many of these pores would be found in 1 cm^2 of membrane surface? (Hint 2↓)

There was a raging controversy in biology at the turn of the century about protoplasmic continuity. Some, including Golgi, thought that nerve cells were connected by a diffuse fibrillary network forming an anastomosis. However, by using the Golgi silver technique Cajal was able to establish indirect evidence that both axons and dendrites had free nerve endings. Cajal and Golgi shared the

Table 2-4. Preferred Units of Quantity

The standard is either the **mole** (the gram equivalent of the molecular weight) for a compound or the **equivalent** for an ion (where this is the gram equivalent of the atomic or ionic weight divided by the valence).

1 mmol (millimole) = 10^{-3} mol (mole)

1 μmol (micromole) = 10^{-6} mol

1 nmol (nanomole) = 10^{-9} mol

1 pmol (picomole) = 10^{-12} mol

Similarly,

1 mEq (milliequivalent) = 10^{-3} Eq

and so on. Since a molecular (or equivalent) weight of *any* substance contains the same number of molecules (or ions), these units convey more information than the weight of substances in grams.

Table 2-5. Preferred Units of Concentration

The standard is the **molar** (or the **normal**) solution, which contains 1 mol (or 1 Eq) dissolved in 1 L of solvent. Thus

1 *M* (molar) = 1 mol/L (moles/liter)

1 m*M* (millimolar) = 1 mmol/L

and so on. Because of the potential for confusion between moles (a quantity) and molar (a concentration), we express concentration in terms of **moles per liter** (mol/L) or **equivalents per liter** (Eq/L) throughout this book.

9

Nobel prize in 1906, and Golgi devoted his acceptance speech to further attack on the neuron theory established by his fellow prize winner [3, p. 38]!

1

Although derived initially from a single cell, the **cells** that make up a given organism **undergo differentiation** (during embryonic and later development), **leading to specialization of the daughter cells.**

2

A group of cells that together fulfill some overall function is called a **tissue.** Masses of tissue that are observable to the naked eye are called **organs.** Finally, groups of organs that function together in achieving broad goals (as perceived by biologists) are called **organ systems,** for example, the circulatory, respiratory, and digestive systems.

3

This book, then, deals with the nervous system: a collection of organs and tissues whose cells are specialized to transmit information rapidly over considerable distances, to detect signals from the external environment (external to the animal's body), to detect internal states of the body, to analyze and store the information, and to control those aspects of the animal that allow it to interact with the external environment (behavior by means of muscle contractions, secretions of odors or enzymes, etc.).

4

(In more down-to-earth language: the nervous system's job is to organize the four Fs of an animal's behavior relative to its environment: Feeding, Fighting, Fleeing, and reproductive behavior.)

5

The preceding description does not clearly separate the functions of the endocrine system with regard to the four Fs. While the endocrine system plays an essential part in such activities, we do not discuss these ideas because this book deals with those parts of the nervous system below the head, for the most part. **Information flow in the nervous system is characterized by its rapidity and high selectivity, as contrasted with information flow in the endocrine system.**

6

EVOLUTION OF THE NERVOUS SYSTEM

Since multicellular organisms have evolved through a sequence, going from simple to more complex forms, it is probably easier to grasp the reasons for the complex nature of the mammalian nervous system by considering the steps through which it **might** have moved in an evolutionary sequence.

7

Although examples corresponding to each stage of the sequence that we present can be found among currently living phyla (which the zoologists among our readers will readily recognize), these living representatives do **not** lie in a continuous phylogenetic sequence. Nevertheless, such a sequence of increasing behavioral and anatomic complexity may well have taken place among our distant ancestors. The parallel

8

Fig. 2-1. This and other figures in this sequence show the hypothetical evolution of the nervous system.

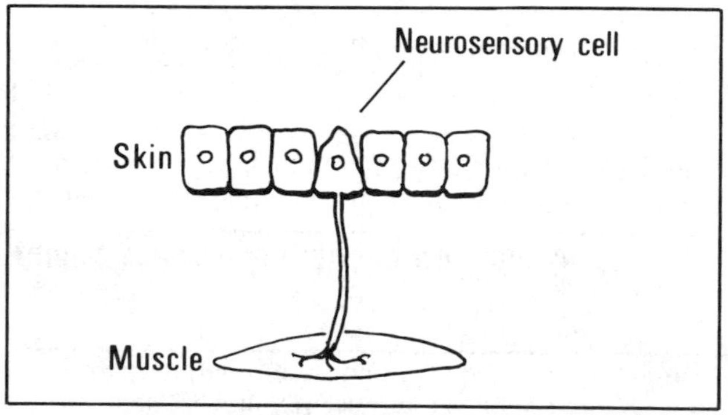

Neurosensory cell
Skin
Muscle

(Modified from E. L. House and B. Pansky, *A Functional Approach to Neuroanatomy* [2nd Ed.]. New York: McGraw-Hill, 1967.)

Fig. 2-2. Further development from Fig. 2-1.

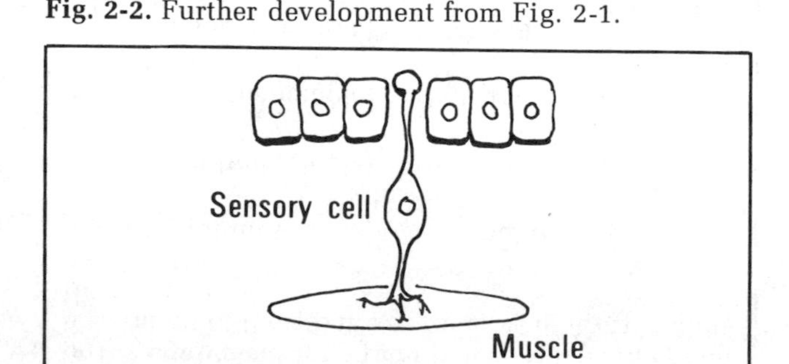

(Modified from E. L. House and B. Pansky, *A Functional Approach to Neuroanatomy* [2nd Ed.]. New York: McGraw-Hill, 1967.)

evolution of form and function shown here emphasizes the interdependence of these not-so-separate aspects of the nervous system.

By the time the degree of differentiation shown in Fig. 2-1 has been reached, the animal is already multicellular, with some cells specialized for external protection (skin) and others for movement (muscle). Increasing body size accentuates a persistent problem: the need for rapid transmission of information over distances greater than that of the average cell (say, 10 μm). Hence we see a further specialization in which processes from one of the former skin cells elongate and reach the underlying muscle layers, thus forming a *neurosensory* cell (Fig. 2-1).

Notice that the neurosensory cell of Fig. 2-1 is already a complex structure, with its outer portions acting as a sensory ending, a transmission line to propagate information, and, presumably, a neuroeffector region that stimulates the innervated muscle cell. Further specialization (Fig. 2-2) involves movement of the cell nucleus toward the middle of the animal or even separation of function (Fig. 2-3) into a transducer (receptor) cell (a specialized sensory cell responsive to different aspects of the environment) and a sensory neuron that receives and transmits information to effector organs.

As evolution proceeds, the behavior of the animal becomes more complex, with a given muscle able to respond to inputs from different areas and with any given receptor able to influence the contraction of more than one muscle (Fig. 2-4).

In Fig. 2-5, we see further developments: (1) **interneurons** that increase the flexibility of the connections; (2) the bringing together of most of the neurons into a region where they can easily interact: the **central nervous system** (CNS), which is protected (in mammals) by a bony framework; and (3) the grouping of the processes of the sensory and motor cells into peripheral nerves: the **peripheral nervous system** (PNS). (We return to the details of Fig. 2-5 in a few moments.)

Finally, as the behavior of the animal becomes still more complex, showing goal-seeking behavior, prolonged memory, and learning (rapid adaptation to new environments), there is a marked proliferation of the interneurons, greatly increasing the bulk of the CNS and leading to separation of parts of the CNS, which are now **interconnected by long nerve tracts.** The increase in the bulk of the forebrain is shown in Fig. 2-6.

The relative size of the parts of the CNS in humans is depicted in Fig. 2-7, which shows the skull and spine divided in the middle (sagittal) plane. The cerebral hemispheres of the

Fig. 2-3. Further development from Fig. 2-2.

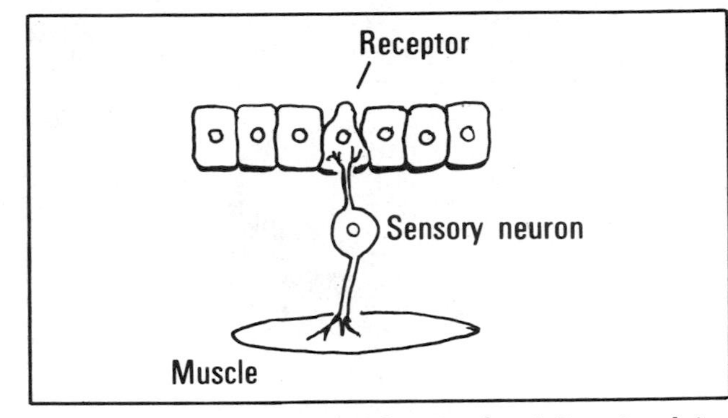

(Modified from E. L. House and B. Pansky, *A Functional Approach to Neuroanatomy* [2nd Ed.]. New York: McGraw-Hill, 1967.)

Fig. 2-4. Addition of extra interconnections for more complex responses. Small arrows show the direction of information flow.

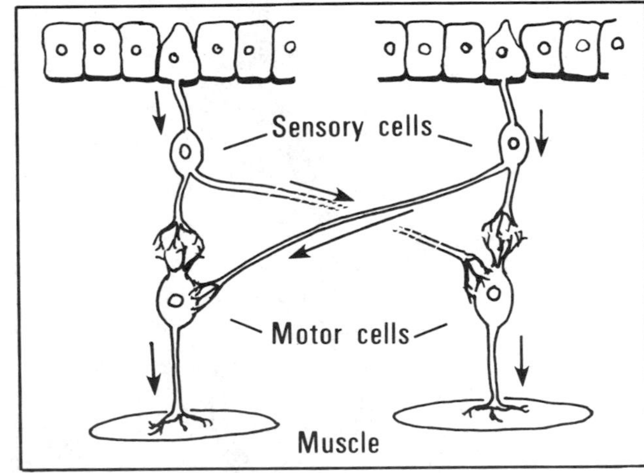

(Modified from E. L. House and B. Pansky, *A Functional Approach to Neuroanatomy* [2nd Ed.]. New York: McGraw-Hill, 1967.)

HINTS

1. 80 to 100 Å, right? One nanometer is 10^1 Å.
2. One micrometer is 10^{-4} cm, so there is 10^8 μm^2 in 1 cm^2. Hence there will be 13×10^8, or 1.3×10^9, pores per square centimeter.

12

Fig. 2-5. Further development, with interneurons intercalated between sensory and motor cells. Small arrows show the direction of information flow.

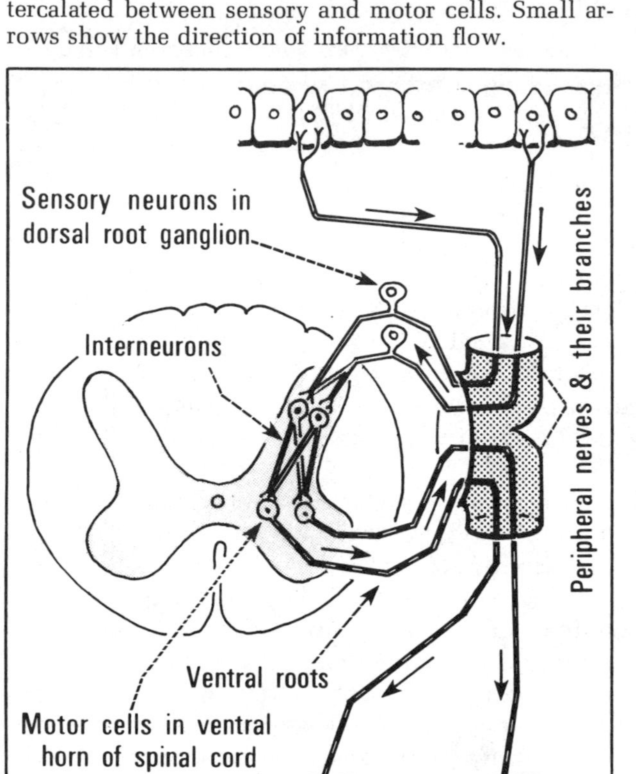

Fig. 2-6. Development of the brain during evolution. The great bulk represents multiplication of interneurons.

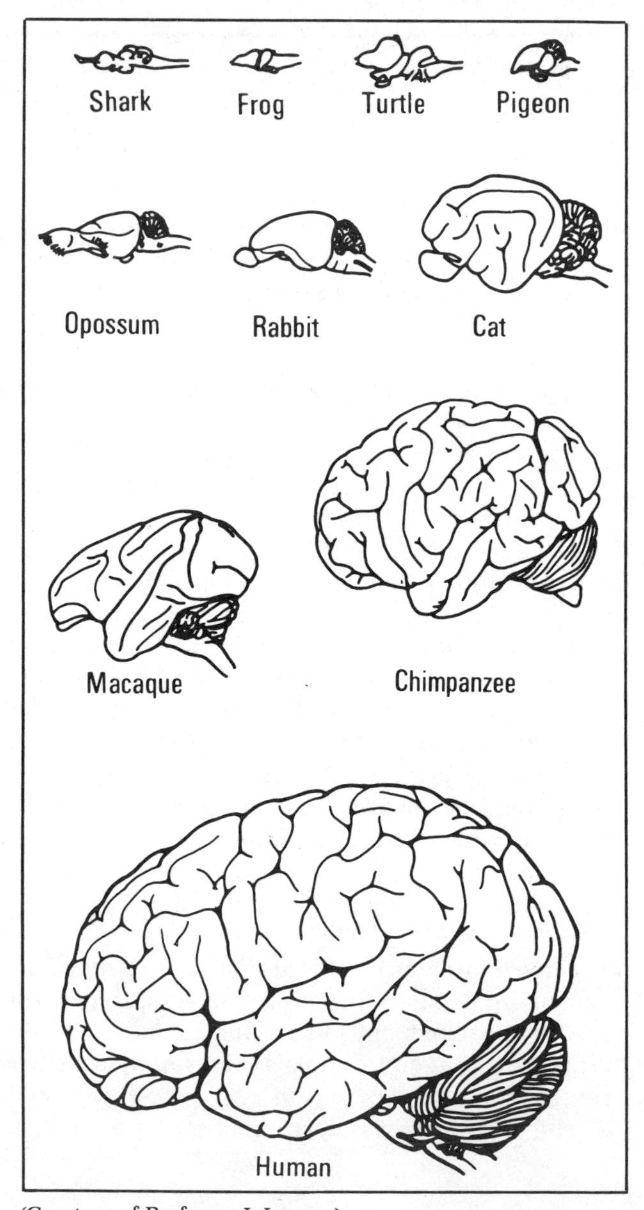

(Courtesy of Professor J. Jansen.)

Fig. 2-7. Representation of the brain and spinal cord to show relation of cord, spinal nerves, and vertebrae. Anterior and posterior roots are shown as single nerves emerging from each segment.

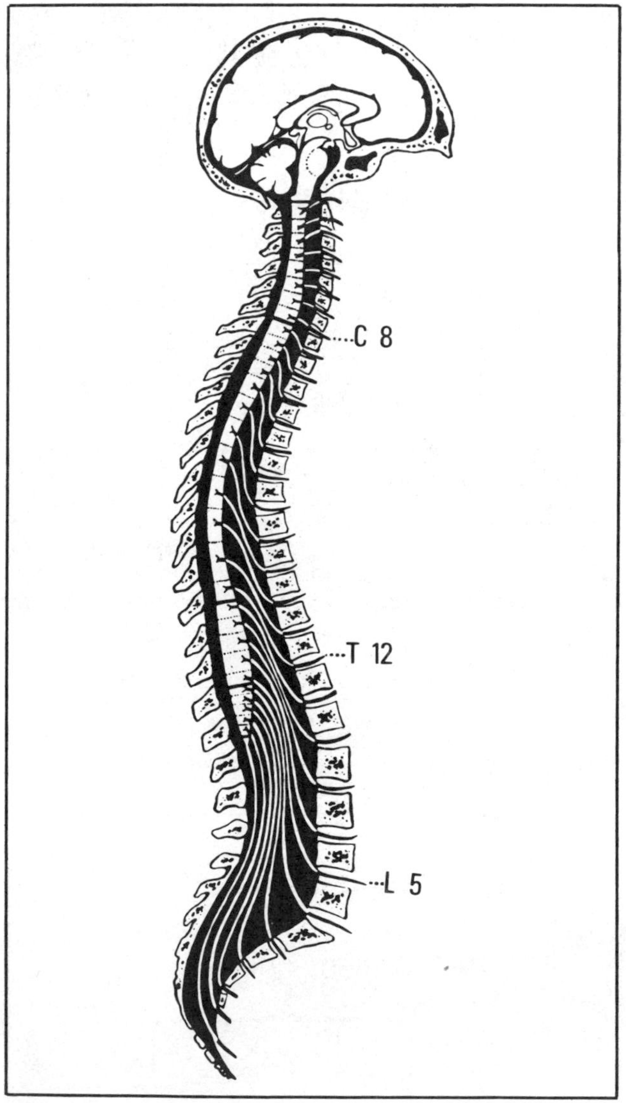

C 8

T 12

L 5

(From E. Gardner, D. J. Gray, and R. O'Rahilly, *Anatomy.* Philadelphia: Saunders, 1975.)

12

13

brain overlie the brainstem, which connects with the spinal cord. The nerve roots leave the spinal cord and pass between the bones of the spine to reach distant parts of the body as peripheral nerves. The CNS and PNS of *Homo sapiens* are shown in Fig. 2-8.

A detailed picture of the anatomy of a segment of the spinal cord is shown in Fig. 2-9. Note that the dorsal and ventral roots merge as they leave the spine, forming the peripheral nerves. Note also that the general plan in Fig. 2-5 was drawn to show the similarity to the anatomy in Fig. 2-9. In both Figs. 2-5 and 2-9, you should note the following anatomic structures: the dorsal and ventral roots, each of which normally conducts information in only one direction, and the peripheral nerves, with their many branches (shown diagrammatically) in which information is transmitted in both directions because the peripheral nerves contain processes of both sensory and motor neurons.

The gray matter of the spinal cord (Fig. 2-9) comprises the region of the cell bodies and interconnections between neurons. The white matter represents the processes of cells (axons) communicating with other parts of the CNS.

The *white matter* is white because of **myelin,** which surrounds the cell processes making up the tracts (as well as many of the cell processes in peripheral nerves). The function of myelin seems to be to speed transmission of information. Obviously, faster transmission was needed over the course of evolution as animals became larger and as the brain enlarged.

THE NEURON, PARTICULAR AND GENERALIZED

Let us now return to the microscopic anatomy of the neuron. With a large number of interneurons subserving many different functions, it should be of little surprise that **neurons come in a great variety of shapes and patterns.**

Some of the different shapes and patterns are shown in Fig. 2-10, which illustrates cells from the mammalian spinal cord, cerebellum, and cerebral cortex. Figure 2-11 shows some of the various cell types that occur in the mammalian retina alone.

In the face of such complexity, you will understand that **it is difficult to apply any uniform anatomic terminology** to such a wide variety of **structures.**

Figure 2-12 shows some of the terms applied to portions of motor and sensory neurons. Note that **the axons of these two neurons make up the peripheral nerves** (below the dotted line), **while they both also form part of the CNS** (above the dotted line).

In the next section of this chapter, we describe a simple physiological (i.e., functional) classification of the parts of a nerve cell. However, a number of commonly used anatomic terms must be clearly understood before you can venture on to the detailed descriptions of neuron function contained in later chapters. These terms are defined in Table 2-6.

Fig. 2-8. Dissection of the CNS and PNS of the human.

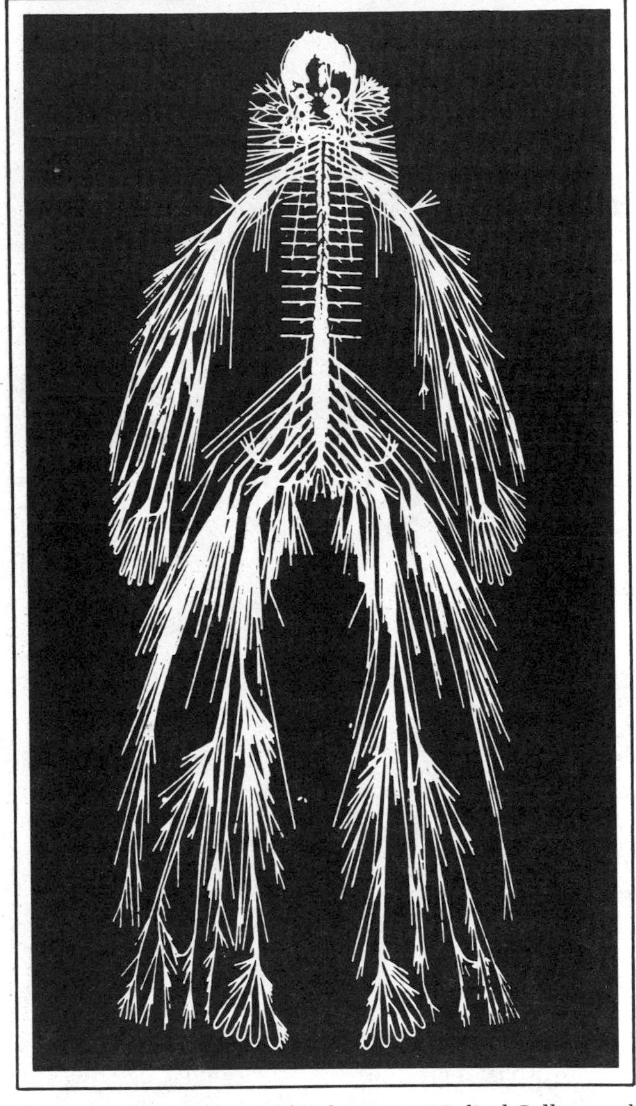

(Courtesy of the Archives of Hahnemann Medical College and Hospital of Philadelphia).

13

Fig. 2-9. Section of spinal cord.

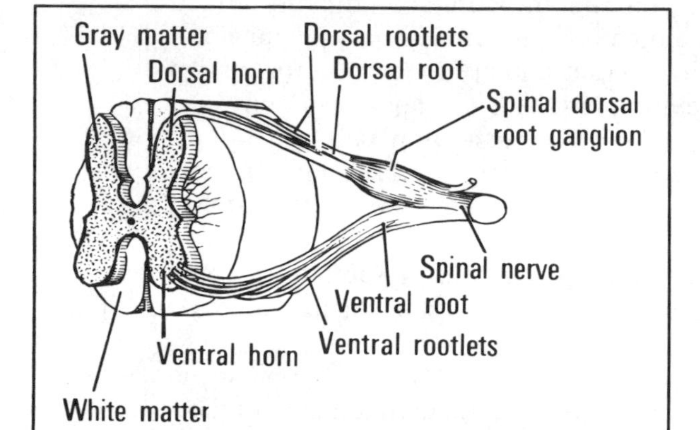

(Redrawn from Hirschfeld and Leveille, in C. M. Goss [Ed.], *Gray's Anatomy of the Human Body* [29th Ed.]. Philadelphia: Lea & Febiger, 1973.)

Fig. 2-10. Neurons from different parts of mammalian nervous system. From left to right: motor neuron from spinal cord, pyramidal cell from cerebral cortex, Purkinje cell from cerebellum, and interneuron from cerebral cortex.

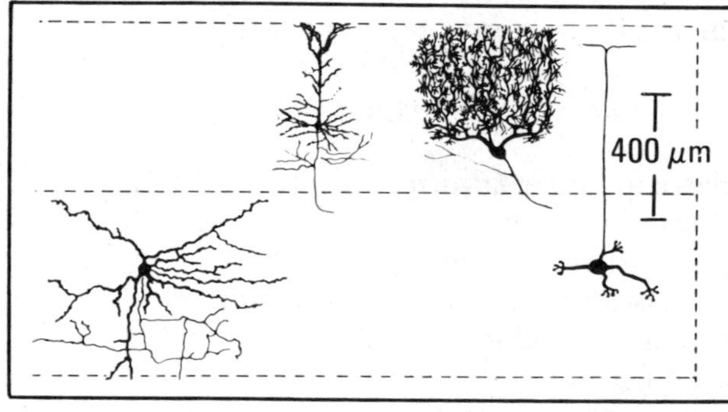

(From D. Bodian, Introductory survey of neurons, *Cold Spring Harbor Symp. Quant. Biol.* 17:1, 1952.)

Fig. 2-11. A few of the cell types found in mammalian retina (nuclei not shown).

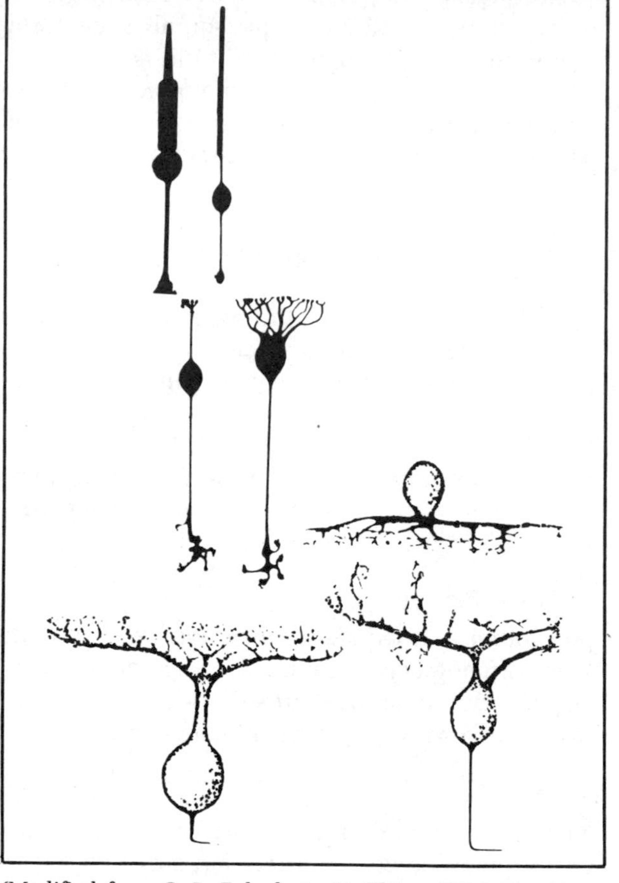

(Modified from S. L. Polyak, in H. Klüver [Ed.], *The Vertebrate Visual System*. Chicago: The University of Chicago Press. Copyright 1957 by The University of Chicago.)

Fig. 2-12. Anatomic labels for portions of sensory and motor neurons. Direction of information flow is CNS to PNS in the motor neuron and PNS to CNS in the sensory neuron. Note that axon length is not to scale (axons can be as long as 1 meter).

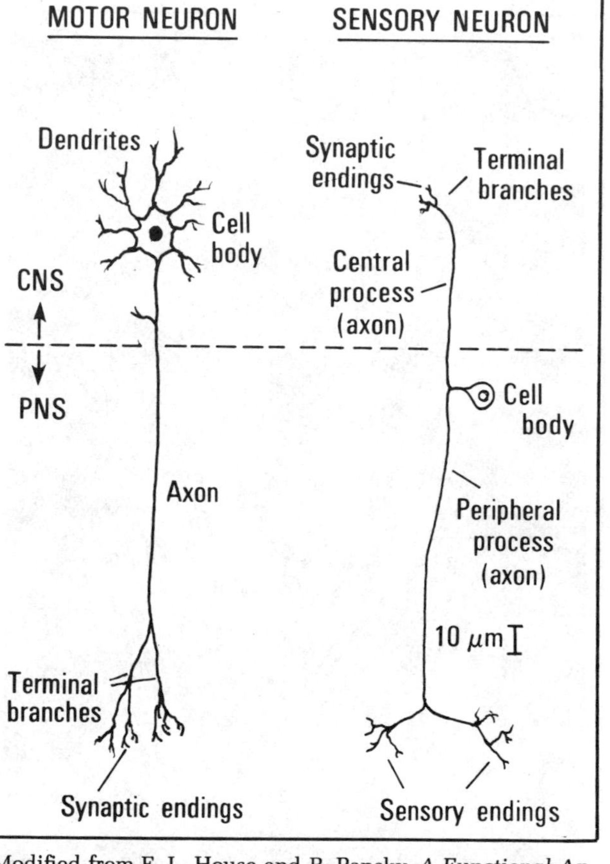

(Modified from E. L. House and B. Pansky, *A Functional Approach to Neuroanatomy* [2nd Ed.]. New York: McGraw-Hill, 1967.)

15

Table 2-6. Anatomic Terms in Common Neurophysiological Use

A. General descriptive terms

Term	Definition
Afferent	= centripetal transfer (toward the CNS), as of information, nerve impulses, etc.
Efferent	= centrifugal transfer, away from the CNS
Distal	= farther from the center of the body
Proximal	= closer to the center of the body
Anterior and posterior	= terms sometimes used by human anatomists (because of the human's erect position) for ventral and dorsal, respectively. Thus, sometimes the afferent dorsal root of a spinal nerve (see Fig. 2-9) may be called the posterior root. Similarly, the ventral horn of the gray matter of the spinal cord is frequently called the anterior horn.
Orthodromic	= conduction in the "usual" direction (i.e., centrifugal flow of nerve impulses in a motor nerve fiber)
Antidromic	= conduction in the "unusual" direction (i.e., centripetal flow of nerve impulses in a motor nerve fiber)

B. Cell types

Term	Definition
Neuron	= a cell of ectodermal origin specialized for transmission of information by means of changes in electrical potential across the cell membrane (or any cell autogenetically derived from a clearly neuronal precursor)
Sensory neuron	= the first nerve cell on the **input** side of neuronal processing; typically, directly responsive to stimulation from the external or internal environment
Motor neuron (motoneuron)	= the last nerve cell on the **output** side of neuronal processing; typically, directly innervating muscle cells
Interneuron	= any neuron lying between the sensory and motor neurons. (This definition is adhered to quite rigidly by invertebrate neurophysiologists. However, vertebrate neurophysiologists often use the term *interneuron* to imply small interneurons whose specific function has not yet been identified, referring to other, better characterized interneurons as, for example, upper motor neurons or second-order sensory neurons.)
Neuroglia	= cells (not neurons) specialized for a variety of supporting functions, few of which are yet well understood
Schwann cells	= PNS counterparts of the neuroglial cells of the CNS. (The functions of the Schwann cell are considered in greater detail in a later section of this chapter, p. 20.7)

15

16

Table 2-6. (Continued)

C. Parts of a neuron

Term	Definition
Cell body	= central part of the neuron containing the nucleus
Axon	= (1) neuroanatomic usage: a neuronal process carrying impulses *away from* the cell body; (2) neurophysiological usage: a neuronal process specialized for long-range, self-propagating information transfer, toward or away from the cell body or both (see Fig. 2-12). Axons do not contain Nissl substance.
Dendron (or dendrite)	= (1) neuronatomic usage: any neuronal process that carries information *toward* the cell body. (Thus the peripheral process of the sensory neuron shown in Fig. 2-12 would be called a dendron by the neuroanatomist or a sensory axon by the neurophysiologist.) (2) neurophysiological usage: a neuronal process involved in short-range information transfer (often toward the cell body) and/or information processing. The term is normally confined to those processes that appear to be relatively unspecialized extrusions of the cell body; frequently these contain Nissl substance.
Synapse	= region of functional contact between two nerve cells or between a nerve cell and an effector cell (e.g., a contractile or secretory cell). Typically, transmission is unidirectional and synapses are functionally polarized. Thus one of the apposed cell membranes may be called *presynaptic* and the other *postsynaptic*. Clearly the presynaptic membranes are the output regions (to other cells), while the post-synaptic areas provide the input to the cell (from other cells).
Sensory ending	= that part of a sensory neuron specialized to receive information from the cell's environment (rather than from another nerve cell). Sensory endings respond to either the external or the internal environment of the body.
Nissl substance	= dark-staining "particles" seen in the light microscope when basic dyes are used to stain nervous tissue. Nissl substance is not found in neuroglial cells—thus its presence differentiates neurons from neuroglia. Nissl substance is found in the cell body and dendrites, but not in the axons of nerve cells. Since the advent of the electron microscope, it has been shown that Nissl substance actually consists of parallel rows of interconnecting tubules of rough endoplasmic reticulum.
Axon hillock	= small area at the base of the axon that would be considered part of the cell body except that observations with the light microscope show no Nissl substance in this area. The physiological significance of the axon hillock is brought up in later chapters.

16

PHYSIOLOGICAL CLASSIFICATION OF PARTS OF A NEURON

In a textbook of neurophysiology, there seems no need to justify inclusion of a physiological (i.e., functional) classification of the parts of a neuron. We do not intend this classification to replace time-honored terms such as *axon* and *dendron*, but merely to clarify the basis of neuronal function.

1

Neurons can be considered as having four functional subdivisions:

1. **An input portion,** where information is received from other neurons, from other tissues, or from the environment
2. **A short-range transmission portion,** by which information is transmitted from the input portion to a location less than a maximum distance of about 5 mm
3. **A long-range transmission portion—the conductive portion,** by which information can be propagated more than a meter from the site of impulse initiation
4. **An output portion,** where the cell transmits information to another neuron, to muscle cells, to secretory cells, etc.

2

This classification is especially useful because, as you might suspect, **there are different cellular mechanisms for each separate function in the classification.** Figure 2-13 shows how the motor and sensory neurons of Fig. 2-12 can be labeled according to this functional classification.

3

Finally, Fig. 2-14 shows the sequence in which the various functional regions are discussed in this book: Chapter 4 (not shown) deals with the genesis of membrane potentials in all parts of a neuron. Chapters 6 and 7 cover the basic mechanisms of long-range transmission. Chapter 9 describes how sensory endings work. Chapter 10 deals with output-input interaction at the synapse between nerve and skeletal muscle cells. Chapter 11 considers the synapses of the CNS and some of the complex interactions that can take place during short-range transmission when many synapses are present within a confined membrane area. Finally, Chap. 13 brings together all these components in a description of the motor and sensory control of muscle function—the systems required to provide accurate control of movement and posture. Chapter 12 (not shown in Fig. 2-14) provides a background understanding of the control systems discussed in Chap. 13. Thus you can see that **this book is organized along the lines of the functional classification just described.**

4

Just for completeness, we include a description of the different types of neuronal functions in Table 2-7. You can't hope to understand the ideas and terminology at this point, but you may find it of interest if you review this part after reading the rest of the book.

5

Fig. 2-13. A functional classification of parts of sensory and motor neurons. Neurons are same as shown in Fig. 2-12.

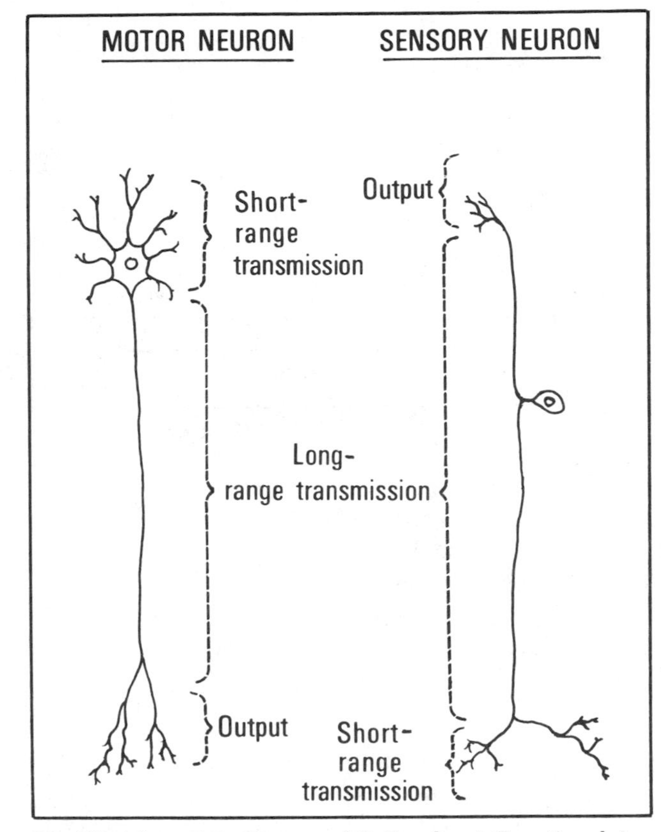

(Modified from E. L. House and B. Pansky, *A Functional Approach to Neuroanatomy* [2nd Ed.]. New York: McGraw-Hill, 1967.)

EXAM QUESTION: Match the cell structures in the left-hand column with the appropriate description or definition in the right-hand column.

1. Axon	A. Contains only the axons of motor neurons
2. Dendrite	B. A part of the *input* portion of a neuron
3. Peripheral nervous system	C. A neuron wholly within the CNS
4. Dorsal root	D. Contains the processes of sensory cells
5. Ventral root	E. A cell whose axon makes up peripheral nerves and transmits information from the CNS to muscle
6. Interneuron	F. That part of the nervous system lying outside the protection of the skull and spine
7. Motor cell	G. Composed of both sensory and motor cell processes (axons)
8. Peripheral nerve	H. The *long-range transmission* portion of a neuron
9. Synapse	I. Functional connection between two neurons

1 Mark your answers; then check Hint 3.↓

METABOLIC ASPECTS OF NEURON FUNCTION

Now, on the one hand, it is very obvious that living organisms are bags of water filled with complex organic molecules. On the other hand, often it is not clear what all those complex molecules do! 2

In other words, there can be no doubt that underlying all the functions and processes described in this book there must be chemical actions and reactions, but the state of the biochemical "art" usually does not allow us to paint a clear picture of the detailed workings of subcellular components. 3

As you will see in the coming chapters, in some places the details of cell structure, as shown at the level of the electron microscope, indicate the function of that part of the cell. But the specific functions of the cell nucleus in the physiological activities of the cell, such as a neuron, are far from clear. 4

Fig. 2-14. Sequence of topics covered in this book, showing chapter numbers. The neurons are as in Figs. 2-12 and 2-13.

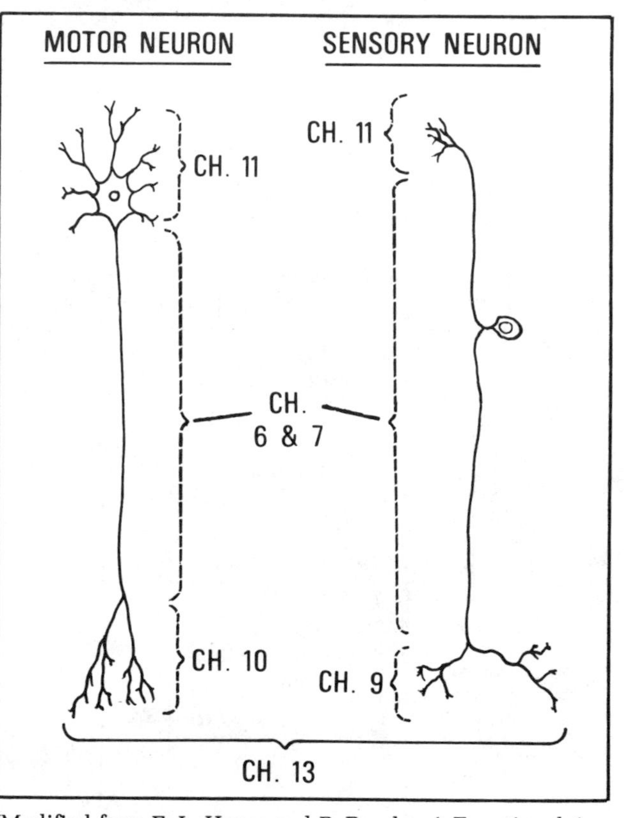

(Modified from E. L. House and B. Pansky, *A Functional Approach to Neuroanatomy* [2nd Ed.]. New York: McGraw-Hill, 1967.)

19

The cell nucleus seems to provide for the general metabolic needs of the cell, those that **sustain the cell structure and function** with regard to **synthesis of new constituents** and the **utilization of energy sources by the cell.** 1

There is no question that the nervous system requires considerable energy for continued operation. Most of this energy derives from the metabolism of glucose, which in turn requires a ready, steady supply of oxygen. (It is common knowledge that even brief periods of lack of oxygen have profound effects on the behavior of the nervous system.) The cell nucleus regulates the supply of enzymes for such metabolism. 2

In regions of the cell where one supposes energy to be needed, it is common to find mitochondria, which contain the necessary sequences of enzymes to break down glucose and release its chemical energy. 3

Such energy is often stored in high-energy phosphate bonds in the molecule adenosine triphosphate (ATP). This molecule is built up from adenosine diphosphate (ADP) and inorganic phosphate. Thus, **in regions where energy is required, ADP and ATP are often found.** 4

The extreme length of some neurons makes it extremely difficult to supply the distant parts of the cell with essential substances from the cell nucleus. It has been shown that whereas the enzymes required for the synthesis of the synaptic chemical transmitters can be put together only in the cell body, the transmitters themselves may be synthesized in the synaptic endings in close proximity to their point of release. 5

Thus one can readily imagine the logistic problems faced by the motor nerves to the muscles of the foot, which may be 10^6 μm long even though the internal diameter of the axon may be no more than 2 to 5 μm! At such great distances from the nucleus, a transport system dependent on only thermal agitation (Brownian movement) would require months to years to move a substance from the nucleus to the end of the axon [27, p. 7]! It is not difficult to perceive the need for a mechanism of "fast" *axonal transport* of intracellular substances, such as has recently been discovered [41]. The "fast transport" mechanism has been shown to be able to transport large molecules, enzymes, and even particulate matter at rates of 400 mm/day (with a range of 50 to 2000 mm/day). 6

It has been suggested that the mechanism of movement is similar to that in muscle (described in Chap. 10) in which *transport* filaments slide along the microtubules or neurofilaments known to be present in axons. **The transport process, like muscular contraction, requires ATP for energy and Ca^{2+} as well.** 7

Table 2-7. Types of Neuronal Function and the Mechanisms Underlying Them

Function	Mechanism
Input (excitation or inhibition)	Transduction (sensory endings) Chemical synapse Electrical synapse
Short-range transmission	Electrotonic spread Decremental conduction?
Long-range transmission	Action potentials (Coding by frequency and/or other patterns)
Output (excitation or inhibition)	Chemical synapse Electrical synapse

19

Interestingly, a *reverse* transport mechanism (i.e., one moving material from distal to proximal) has been shown to move material at about **half the rate of the proximal-distal mechanism.** Thus, there is a form of "conveyer belt" that moves material to and from the cell nucleus. The role that such a system could play in the maintenance of distant portions of the neuron is readily apparent.

1

Furthermore, such a two-way conveyer-belt system can also explain how the cell nucleus responds to damage to the distal parts of the axon (cell chromatolysis): it apparently produces increasing amounts of material. In this way, substances returning to the nucleus may provide a biochemical feedback (see Chap. 12) that controls nuclear production of materials needed in the periphery.

2

Finally, the *trophic substances* postulated as being secreted from the terminals of nerves in order to affect postsynaptic membranes (Chap. 13) may be complex molecules synthesized under the genetic control of the nucleus.

3

While the fast transport rate of 400 mm/day does not seem "fast" in ordinary terms, it is fast compared to the previously known rate of movement of materials in axons when a nerve fiber has been cut and the axon regenerates. **The rate of growth of the distal axon is 1 to 10 mm/day, when it is regenerating.**

4

The changes seen in a mammalian motor cell after transection of the axon and the subsequent regeneration are shown in Fig. 2-15. Note that **the portion of the axon distal to the cut degenerates when it is cut off from the nucleus (Wallerian degeneration).** The cell nucleus then undergoes a process of *chromatolysis*, during which the axon regenerates. The large volume of material that must be synthesized is indicated by the relatively large volume of the axon relative to that of the cell body (a ratio as high as 250:1).

5

In mammals, birds, and reptiles, regeneration of axons occurs in only the PNS. No functional regeneration in the CNS of these animals occurs after damage. The reasons are not clear, since regeneration can occur in the CNS of amphibia [40, pp. 134–136].

6

SCHWANN CELL SHEATH

The neurons of the peripheral nervous system lie embedded in a sheath of Schwann cells. However, the importance of this Schwann cell sheath in neuron function is little understood [8, pp. 1058–1061; 27, pp. 5–9; 45, p. 80; 63, p. 47]. Muscle fibers are not surrounded by Schwann cells.

7

In the simplest instance of the *unmyelinated* peripheral axon (see Fig. 2-16), one or more axons appear to be engulfed in a single Schwann cell. Only a rather narrow (about 100 Å) **mesaxon** remains to allow ion exchange between the extracellular fluid surrounding the Schwann cell and the periaxonal fluid in the space between the axon membrane and its surrounding Schwann cell membrane.

8

Fig. 2-15. Regenerative cycle of motor neurons (Rhesus sciatic nerve). Cycle of changes in cytoplasmic Nissl substance (ribonucleoprotein), correlated with axon amputation and axon regeneration. At height of reduction of Nissl substance (chromatolysis), the regeneration of disproportionately large axoplasmic volume is barely underway.

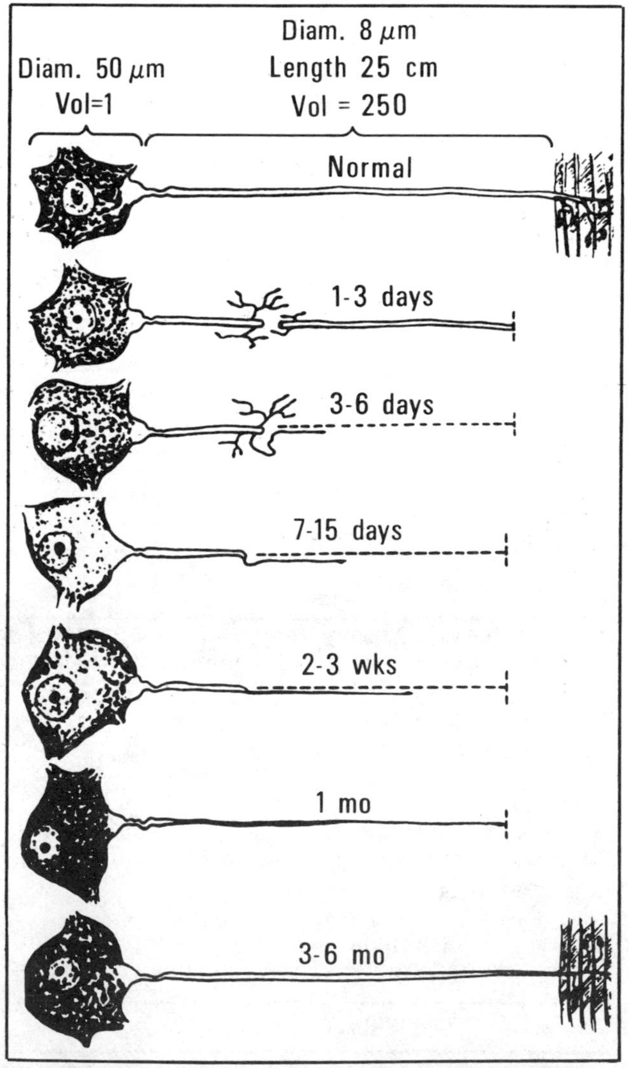

Diam. 50 μm
Vol=1
Diam. 8 μm
Length 25 cm
Vol = 250
Normal
1-3 days
3-6 days
7-15 days
2-3 wks
1 mo
3-6 mo

(From D. Bodian, Nucleic acid in nerve-cell regeneration, in *Symposium of the Society for Experimental Biology, No I: Nucleic Acid.* London: Cambridge University Press, 1947.)

Physiological evidence suggests the presence of a diffusion barrier separating the immediate periaxonal fluid from the true extracellular medium [45, p. 80]. Unfortunately, it is not yet clear whether this physiologically observed *Hodgkin-Frankenhaeuser space* corresponds to the anatomic *periaxonal space*. 1

In spite of our lack of knowledge about the function of the Schwann cell, two of its functions are well understood: its role in the formation of myelin and its role during nerve regeneration. 2

Myelin Formation

You learn in Chap. 7 that the most rapidly conducting nerve fibers are surrounded by a *myelin sheath*. These myelinated axons are seen in the light microscope to be surrounded by a sheath of material with a high affinity for lipophilic stains. This sheath is interrupted periodically along the length of the axon by *nodes of Ranvier*, in which the sheath appears to be thin or absent. This structure is shown diagrammatically in Fig. 2-17. 3

Since the advent of the electron microscope, it has been shown that this sheath is formed by superimposed layers of Schwann cell membrane. The myelin sheath is laid down by the Schwann cell "rolling up on itself," much as a window shade does, as shown in the sequence of drawings in Fig. 2-18. The final, fully developed sheath consists of many hundreds of layers of closely apposed double membranes between which almost all the Schwann cell cytoplasm has been squeezed out. (Nevertheless, a long and very narrow mesaxon sometimes can be traced that leads in from the extracellular fluid toward the periaxonal space.) 4

Fig. 2-16. Relation of unmyelinated fibers to Schwann cells. Drawing represents cross-section through single Schwann cell enveloping six unmyelinated fibers. Note that Schwann cell membrane is everywhere intact, but invaginates to surround nerve fibers.

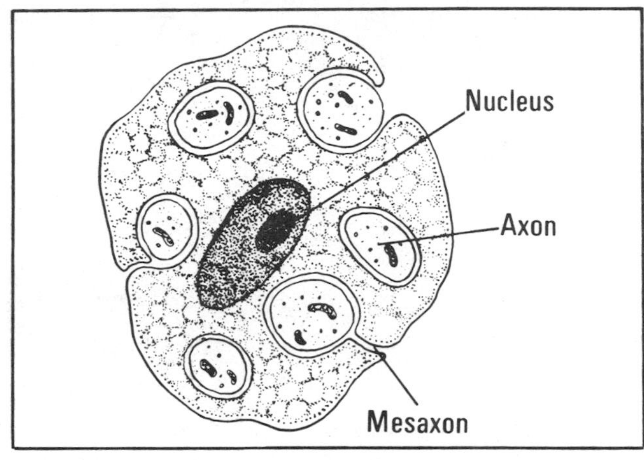

(Modified from G. M. Wyburn. *The Nervous System*. New York: Academic Press, 1960.)

HINTS

3. 1-H, 2-B, 3-F, 4-D, 5-A, 6-C, 7-E, 8-G, 9-I.

4. The answer is 2.5×10^{-13} mEq, based on

$$\text{Volume of periaxonal space} = 100\ \text{Å} \times 10\ \mu m \times 1\ mm$$
$$= 10^{-6}\ cm \times 10^{-3}\ cm \times 10^{-1}\ cm$$
$$= 10^{-10}\ cm^3$$
$$= 10^{-13}\ L$$
$$K^+\ \text{concentration} = 2.5\ mEq/L$$

5.
$$\text{Total number of } K^+ \text{ ions} = (6 \times 10^{20})(2.5 \times 10^{-13}) = 1.5 \times 10^8$$
$$\text{Total membrane area} = 10\ \mu m \times 1000\ \mu m = 10^4\ \mu m^2$$
$$\text{Number of } K^+ \text{ ions per } \mu m^2 = \frac{1.5 \times 10^8}{10^4} = 1.5 \times 10^4$$

Fig. 2-17. Nerve cell (spinal motor neuron of frog) with some of its central (synaptic) and peripheral (neuromuscular) contacts.

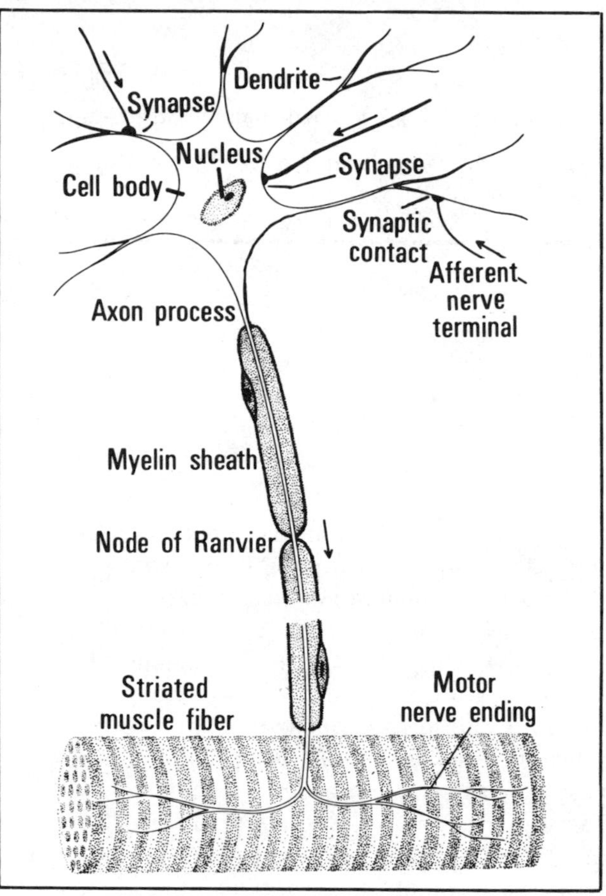

(From B. Katz, *Nerve, Muscle, and Synapse.* New York: McGraw-Hill, 1966.)

Fig. 2-18. *A:* cross-section of nerve fiber (axon) and Schwann cell; *B, C:* development of myelin sheath.

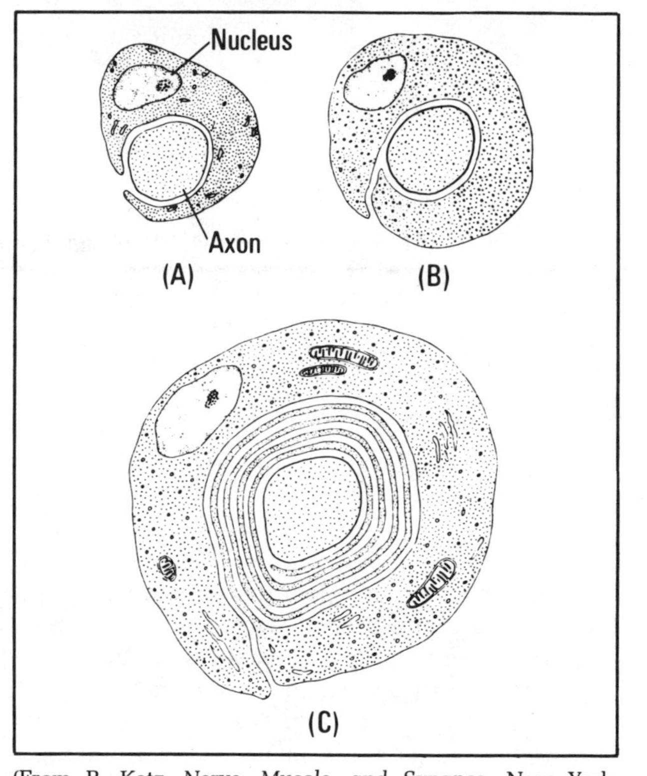

(From B. Katz, *Nerve, Muscle, and Synapse.* New York: McGraw-Hill, 1966.)

23

It used to be thought that the nerve membrane was exposed to the extracellular fluid at the node of Ranvier [27, p. 9]. Now, however, it is clear that the Schwann cell sheath is continuous at the nodes [65, p. 1158], although it may be essentially unrolled and have a relatively large mesaxon at those points. Even at the nerve-muscle synapse, the nerve tissue appears to be engulfed in a covering Schwann cell [39, p. 1205].

Nerve Regeneration

When the distal end of a cut axon degenerates, the complex myelin structure also breaks down. However, the Schwann cells themselves do not degenerate, but form *neurotubes*. These neurotubes act as guides for the regenerating axons. In this process, there does not seem to be any specificity between regenerating axon and neurotube: motor axons may grow down neurotubes formed by degeneration of sensory axons and vice versa. (Thus motor and sensory functions may not return completely even when the cut ends of a severed nerve are accurately apposed before the start of the regeneration process.) Myelin is finally reformed about the regenerated axons.

Many other potential functions of the Schwann cells and their CNS counterparts, the glial cells, have been proposed. In general, these functions have been ascribed to such *supporting* cells simply for lack of better explanation. Katz has aptly commented on the suggestion that neuroglia might be involved in memory: ". . . [A]t present the only feature which can really be said to be shared by memory and glia [neuroglia] is that very little is known about either of them" [27, p. 7].

EXAM QUESTION: *Choose the single correct answer from the alternatives provided.* If the periaxonal space of an internode 1 mm long is 100 Å across and the axon circumference is 10 μm, what quantity of K^+ ions (in milliequivalents) does this space contain if it is in equilibrium with an extracellular fluid containing 2.5 mEq/L of K^+ ions? (See Hint 4.↑)

(*a*) 2.5×10^{-10} mEq
(*b*) 2.5×10^{-13} mEq
(*c*) 2.5×10^{-15} mEq
(*d*) 2.5×10^{-16} mEq

EXAM QUESTION: Given that there are 6×10^{20} ions per milliequivalent, how many K^+ ions are distributed per square micrometer of axonal membrane area? (Hint 5↑)

23

25

Physical and Chemical Background of Membrane Potentials

26

1 In this chapter, we review all the basic ideas required to understand our treatment of membrane potentials and cell excitability (Chaps. 4 through 8).

2 For most readers, the majority of these concepts will be familiar ground. Our aim is to assist your recall and fill in any gaps in your background knowledge before you become frustrated for lack of some simple basic concept.

POLARITY AND MAGNITUDE OF CELL MEMBRANE POTENTIAL

3 The cell membrane potential is seemingly ubiquitous in living cells. It is present in not only nerve and muscle cells but all the other cells from which recordings have been made, including kidney and liver cells and even plant cells!

4 It is important to realize that **a membrane potential is always a potential difference;** i.e., the measurement is that of the voltage between two points—in this case, the inside of the cell with respect to the outside. It is impossible to measure the "potential" of a single point. (This obvious statement is necessary because it is surprisingly easy to lose sight of simple physical principles in a biological context!)

5 Although early extracellular measurements had indicated that the inside of the cell was negative with respect to the outside, a clear demonstration of the magnitude of the membrane potential was not possible until the development of the intracellular micropipette. Originally used to study plant cells, the micropipette was adapted by Graham, Ling, and Gerard to the study of animal tissue. Micropipettes are improbable devices made by heating glass tubing and then pulling it rapidly in a longitudinal direction so that the diameter becomes very small (as small as 0.5 μm!) while the ratio of wall thickness to lumens remains constant. These small electrodes can penetrate cell membranes, with the membrane sealing around the electrode tip after penetration so that the potential difference across the cell membrane can be recorded, apparently without significantly altering the properties of the impaled cell. Modern micropipettes can be used to study many cells, but they are still too large to penetrate the smaller dendrites and axons.

6 The **magnitude** of the cell membrane potential is **about 90 mV.** This value can vary somewhat from cell to cell, but is rarely over 100 mV.

7 QUESTION: What is 100 mV in volts? (Hint 1↓)

8 **Thus the cell membrane potential is described as being about −90 mV, where the sign indicates the polarity of the inside of the cell with respect to the outside.**

26

27

ESSENTIALS OF ELECTRICITY REVIEWED

1 Since you are going to spend some time learning how the −90-mV potential comes about, **the basic ideas of electricity must be clear in your mind.**

2 A potential difference between two points (such as exists across the membrane) **may cause a movement of charge.**

3 The unit of charge is the coulomb (C), where 1 C = 6 × 10^{18} electron charges; the flow of charge is called **current,** measured in amperes (A), where 1 A = 1 C/s passing a given point.

4 Note that **all currents** may be said to **travel in complete circuits.** Of the many analytic elements used in the description of electrical circuits, there are only three whose properties need to be understood in order to follow the ideas presented in this book: batteries, resistors, and capacitors. The following sections review these elements.

5 **Batteries.** A **battery** (labeled E in Fig. 3-1) may be defined as a device for storing potential chemical energy in such a way that the energy can become available to drive a current when the circuit between the battery *poles* is completed. In the ideal battery, the amount of chemical energy so stored is considered to be infinite, hence the potential should not be affected by the rate or duration of energy release. Thus, the battery symbol represents a *constant-voltage source* that is not affected by the actual amount of current flowing in the circuit.

6 **Resistors.** When a battery (as defined previously) is attached to a resistor (labeled R in Fig. 3-1), a current flows in the circuit (see Fig. 3-1). *Note:* Although it is well known that currents in metallic conductors are a result of electron movements, **throughout this book the direction of current flow is presented as the direction of movement of positive particles.** That is, current flows from the positive to the negative pole of the battery. (Later you learn the reasons for this definition in the section entitled "Ionic Currents in Solution.")

7 The relationship between the battery voltage, the *size* of the resistor, and the amount of current is given by **Ohm's law:**

$$I = \frac{E}{R} \qquad \text{Eq. 3-1}$$

where I = current, in amperes (A)
E = potential difference across the batttery, in volts (V)
R = resistance in ohms (Ω)

8 The earliest accepted definition of the volt was based on Ohm's law: 1 V was defined as the potential difference required to drive a current of 1 A through a column of mercury 106.3 cm long and 1 mm² in cross-section (the international

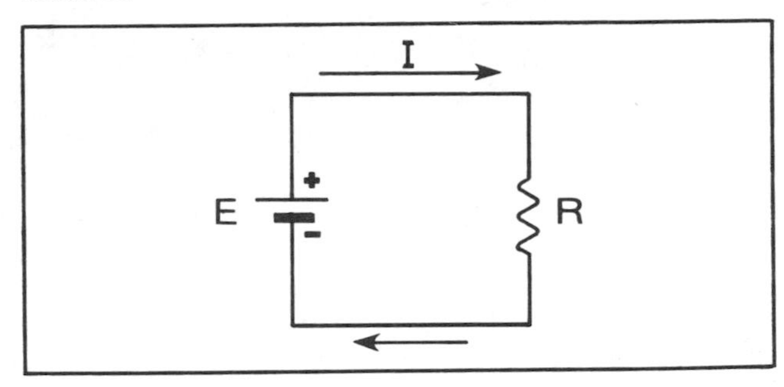

Fig. 3-1. Battery E passing a current I through a resistance R.

27

28

definition of a 1-Ω resistor). Subsequently, the volt has been defined in absolute units on the basis of the work that must be done to move a given amount of charge. A potential difference of 1 V will perform 1 joule (J) of work in movement of 1 C of charge. Voltage thus can be represented as *joules per coulomb* (J/C) or *dynes times centimeters per coulomb* (dyn·cm/C) when it is remembered that 1 J = 10^7 dyn·cm.

1

2

Note that **for constant *E*, *I* is inversely related to *R*.**

QUESTION: If a 6-V battery produces a 3-A current in the circuit of Fig. 3-1, what is the value of R in ohms? (Hint 3↓)

3

The resistance in ohms of a real object such as a piece of copper wire can be obtained by inserting it into a circuit such as that shown in Fig. 3-1 and measuring the voltage and current across the wire. However, if the length and diameter of the piece of wire are known, one can also derive a value, the **specific resistivity** ρ, which is a constant property of the material—in this case, pure copper. Once the specific resistivity is known, the resistance of any other piece of pure copper can be calculated from

$$R = \frac{\rho L}{A} \qquad \text{Eq. 3-2}$$

where R = resistance (Ω)
L = length of the material (cm)
A = cross-sectional area (cm^2)
ρ = specific resistivity (Ω·cm)

Notice that for any given value of ρ, R **increases** when L is increased and **decreases** when A is increased. In the study of cell membranes, currents passed across the membrane normally are measured as **current per unit membrane area** [for example, microamperes per square centimeter ($\mu A/cm^2$)]. When such *current densities* and the applied voltage are substituted into Ohm's law, R is given in units of ohms times square centimeters ($\Omega \cdot cm^2$). Often, this value is called the **specific resistance** of the membrane in question, and it represents the resistance of a 1-cm^2 membrane area.

4

Question: If the specific *resistance* of a cell membrane 100 Å thick is found to be 4×10^3 Ω, what is the specific resistivity of the membrane substance? (Given: 100 Å = 10^{-6} cm.) (See Hint 5.↓)

5

28

Resistance, as a concept, describes how effectively a circuit element "resists" flow of charge. In membrane physiology, often it is intuitively easier to invert this concept and think in terms of the *ease* with which charge can pass across a cell membrane. We thus define **conductance** g as the **reciprocal of resistance,** with units of "reciprocal ohms," or mho (the word *ohm* spelled backward).

1

Thus g = $1/R$ and Ohm's law can be written as

$$I = gE \qquad \text{Eq. 3-3}$$

Note that here, **for a constant** ***E*****, current varies directly with conductance;** e.g., **when** the variable **conductance** g of Fig. 3-2 **increases,** ***I*** **also increases.** We use this relationship repeatedly when dealing with membrane currents in subsequent chapters.

2

Clearly, we can also define a **specific conductivity** with units of mho per centimeter and a **specific membrane conductance** (the conductance of a 1-cm² membrane area) with units of mho per square centimeter.

3

Question: What is the specific conductance of a membrane 100 Å thick with specific resistivity of $4 \times 10^{-9}\ \Omega \cdot$cm? (See Hint 2. ↓)

4

A circuit involving two resistors in series with a battery is important in understanding much of what follows.

5

Such a *voltage-divider* circuit of two resistors in series is shown in Fig. 3-3. Since the current must be the same through the two resistors in series, it is easy to derive the following equation:

$$V_1 = E\frac{R_1}{R_1 + R_2} \qquad \text{Eq. 3-4}$$

This equation shows that V_1 is a fraction of the battery voltage E, as determined by the ratio of R_1 to the total resistance.

6

If you haven't the time to derive Eq. 3-4 yourself, you can find the proof in Hint 7.↓

7

Fig. 3-2. Battery passing a current through a variable conductance g. Note that the conductance symbol is a resistor. A conductance and a resistance are the same physical entities—they differ only in their units of measurement.

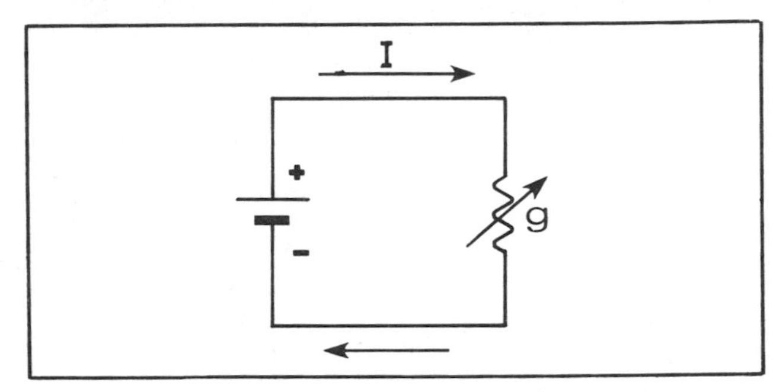

Fig. 3-3. Voltage-divider circuit of two resistors in series.

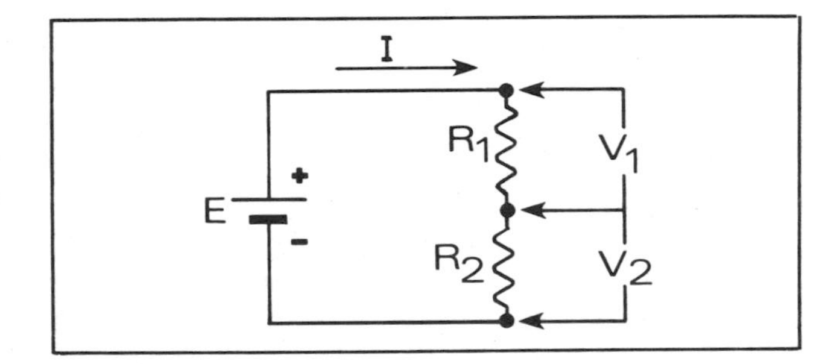

HINT

1. 0.1 V. If you missed this, go on to Hint 4. ↓

30

It is instructive to see **what happens to V_1 as R_1 is varied,** given that E and R_2 are constant. You will find this of interest since **an important part of the understanding of membrane potentials involves the change of a resistance in series with a battery.** Therefore, let's look carefully at what happens. If R_1 is very low, then V_1 is almost zero, as you can readily see from Eq. 3-4, where $R_1/(R_1 + R_2)$ is almost zero. This is the situation when the two points on either side of R_1 are connected by a very low-resistance path, such as a wire. This example verifies the simple fact that **there can be no difference in potential between two points connected by a low resistance** (high conductance).
1

If the resistance of R_1 is raised until $R_1 = R_2$, *then* you can see from Eq. 3-4 that $V_1 = E/2$ (the voltage of E is equally divided by the equal resistances). This is the most obvious example of the voltage-divider aspects of this circuit.
2

Finally, **if R_1 went to infinity** (as would occur if R_1 were an open switch), **then V_1 would go to E,** no matter the value of R_2.
3

The last example shows why a device to measure voltage should have as high an input resistance as possible (where R_1 would represent the input resistance of the measuring device and R_2 the output resistance of the voltage source). The higher the input resistance, the more accurate is the measurement of E by means of observing V_1.
4

When considering membrane potentials later in this chapter, you will encounter the idea of the **internal resistance of a battery.**
5

The circuit of Fig. 3-3 and Eq. 3-4 can help you to understand the idea of internal resistance of a battery. You must realize that **the battery symbol** as defined previously **cannot represent a real battery,** because if the battery of Fig. 3-1 were hooked up to a zero resistance (a short circuit), the battery would produce (by Ohm's law, Eq. 3-1) an infinite current! No real battery can do this. Therefore, the *circuit diagram* for a real battery connected to an external resistance must be diagrammed as shown in Fig. 3-4. A moment's glance should tell you that you are familiar with this circuit, for it is the same as that in Fig. 3-3 with the parts somewhat rearranged. The resistance R_2 is now the internal resistance of the real battery, whose external terminals are at points A and B.
6

Now you can see how the internal resistance (R_2) helps you to understand the real battery. If the R_1 resistance is made zero by shorting the battery, then the presence of R_2 still limits the current to $I = E/R_2$, rather than the *infinite* current of the "ideal" battery when short-circuited. **A good battery** (one able to deliver current into a low external resistance without significantly dropping its voltage) has **low internal resistance.** As the battery becomes discharged, the internal resistance increases to the point at which the voltage drops noticeably; e.g., a light connected to a discharged battery is dim.
7

QUESTION: If we assume that in Fig. 3-4 initially, $R_1 > R_2$, what happens to V_1 when R_2 increases to the point at which $R_1 = R_2$? (See Hint 6.↓)
8

Fig. 3-4. A "real" battery (represented by the dotted line) and an external resistance shown in symbols used in this book.

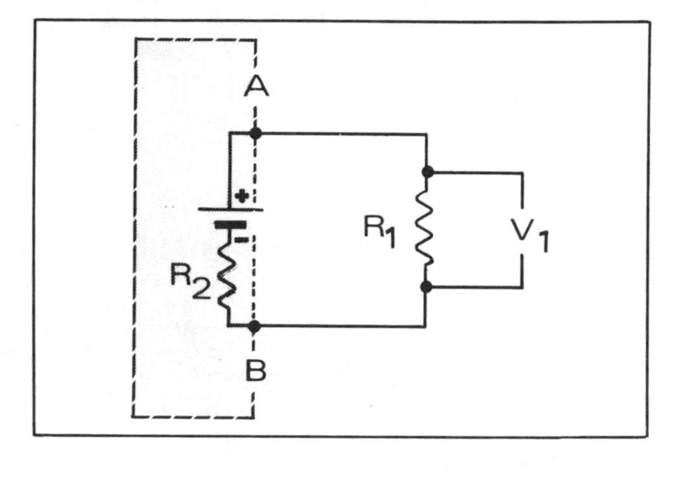

30

In practice, batteries differ not only in voltage and internal resistance but also in their physical size. A 12-V transistor battery is a small fraction of the bulk of a car battery. This difference reflects the total quantity of charge stored, as measured in ampere-hours. 1

You will see later that a small cell is like a small battery: It runs down more quickly if it cannot be recharged. 2

Capacitors. A capacitor (labeled C in Fig. 3-5) is a device that can "store" charges on its *plates*, with **the plates being closely spaced conductors separated by a nonconducting** (insulating) **medium.** 3

Across a given capacitor, the amount of charge asymmetry is directly proportional to the applied voltage. This is given by

$$Q = CV \qquad \text{Eq. 3-5}$$

where Q = *quantity* of charge, in coulombs
C = size of the capacitance, in farads
V = potential across the capacitor, in volts 4

Here C is just the constant of proportionality between Q and V, just as R is the constant of proportionality between E and I in Ohm's law. 5

Note that both V and E are symbols for voltage. With one exception (later identified), we adopt the convention that E represents an independent variable, whereas V is a dependent variable. Thus E represents the voltage of the *source* (e.g., a battery) while V is the voltage on a circuit element (e.g., a resistance or capacitance). Hence V in Eq. 3-5 implies that the voltage on the capacitor is the result of the movement of charge Q produced by the battery. 6

Fig. 3-5. Circuit that will charge the capacitor when the switch is closed.

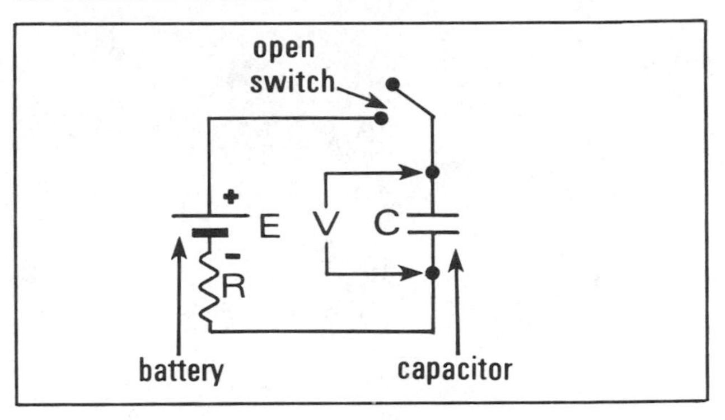

HINTS

2. When the specific resistivity is $4 \times 10^9\ \Omega\cdot\text{cm}$, then specific resistance is $4 \times 10^3\ \Omega\cdot\text{cm}^2$. Specific conductance is the reciprocal of specific resistance; therefore, the specific conductance is $0.25\ \text{mho/cm}^2$.
3. $2\ \Omega$.
4. In "mV" the "m" stands for *milli*, which means 10^{-3}. Did you slip a decimal point?
5. From Eq. 3-2 it is clear that specific resistance is $\rho L/1$. Hence

$$\rho = \frac{\text{specific resistance}}{L} = \frac{4 \times 10^3\ \Omega\cdot\text{cm}^2}{1 \times 10^{-6}\ \text{cm}} = 4 \times 10^9\ \Omega\cdot\text{cm}$$

7. Since $V_1 = IR_1$ and $I = E/(R_1 + R_2)$, Eq. 3-4 is obtained by direct substitution.

Now, when the switch in Fig. 3-5 is closed, charges (propelled by the electromotive force of the battery) will move and the top plate of the capacitor becomes positively charged while the bottom plate becomes negatively charged (Fig. 3-6). [1]

This process continues until the developing potential on the capacitor *V* equals the voltage of battery *E*, that is, until charge asymmetry on the capacitor becomes so great that additional charges moving toward the plate become repelled by their own kind. **When the number repelled equals the number attracted, then there will be no further change of the charge (potential) on the capacitor.** [2]

QUESTION: What is the essential difference between a capacitor and a battery? (Hint 9↓) [3]

Question: What number of charges is needed to charge a capacitor of 1 microfarad (1×10^{-6} farads), to 100 mV (100×10^{-3} V) if there are 6×10^{18} charges per coulomb? (Hint 8↓) [4]

Now refer to Fig. 3-6. Notice that **the current in this circuit is determined** not by E or V alone, but **by the difference between *E* and *V*,** that is, by $E - V$. Thus, for this circuit, Ohm's law can be written in the form

$$I = \frac{E - V}{R} = g(E - V) \qquad \text{Eq. 3-6}$$

Current will be zero when the capacitor is fully charged so that the charge on the capacitor has become equal to the battery voltage (i.e., when $E - V = 0$). **At all times, a current will flow that is directly proportional to the driving force** (i.e., the instantaneous value of $E - V$). [5]

QUESTION: What is the instantaneous driving force when the battery voltage is 10 V and the voltage on the capacitor at that instant is 9 V? (Hint 14↓) [6]

Many students complain that a current cannot flow "through" the capacitor of Fig. 3-6 since the dielectric separating the plates of the capacitor is an insulator through which electrons cannot pass. However, such students do not realize that **during any time the current is moving, as much charge is leaving the capacitor from one side as is coming into the other.** It does not matter whether a specific, "individually identifiable" charge has passed through the dielectric since one cannot tell one charge from another (an example of "Reagan's law"—Hint 13↓). What matters is that **the current has been the same in all parts of the complete circuit.** Thus, in modern electrophysiology and **in this book, we talk about a current "passing through" the** [7]

Fig. 3-6. Charge on capacitor when current *I* passes in direction shown.

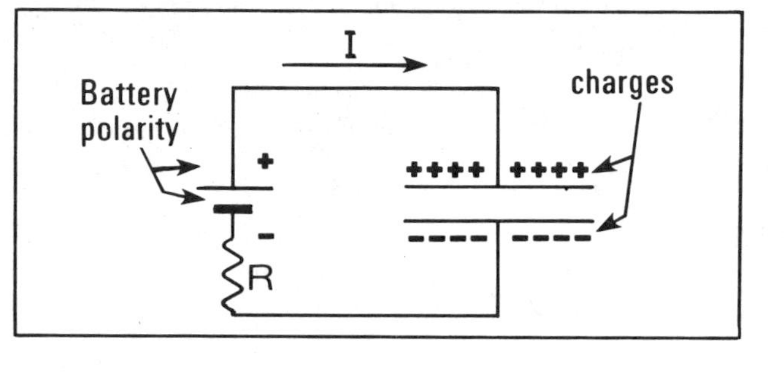

1 **membrane capacitance, without implying at all that any charges have physically penetrated the membrane itself.** Thus it is correct to say that current can pass through the membrane capacitance even though ions cannot!

2 Those who find it difficult to accept this terminology should take heart from the fact that such a "displacement" current through a capacitor was not accepted by physicists when it was first suggested in the 1860s (as a useful way to view how the capacitor functions) by a fellow named Maxwell (who later became famous for establishing the theoretical basis of electromagnetism)! Perhaps an analogy might help: "Laundering" money through a bank involves depositing "dirty" cash in one office and then withdrawing "clean" cash from another branch of the same bank. In this way, a given amount of "currency" can pass through the bank accounts, even though no specific greenbacks actually move from the deposit office to the withdrawal office. This process is analogous to the current that passes through the capacitance, even though a specific charge carrier does not.

3 QUESTION: In the bank analogy, what does the asymmetry between the tally sheets in the two banks (one showing a deposit, the other a withdrawal) represent in the electrical case? (Hint 12↓)

4 Just as resistance can be related to a specific resistivity, so capacitance can be related to a specific capacity (with units of farads per centimeter [F/cm]) determined by the nature of the plates and the dielectric constant of the intervening insulator. When the specific capacity is known, the actual capacitance can be calculated as follows:

$$C = \frac{\text{specific capacity} \times A}{D} \qquad \text{Eq. 3-7}$$

5 where A = area of one plate (cm^2)
D = separation of the plates (cm)
C = capacitance (F)

HINTS

6. Voltage V_1 drops to half the value of E, and the lights dim (if R_1 is a light).

18. Unlike the circuit described in Hint 19, **in this circuit the capacitor would not charge linearly,** since, just as with a constant-voltage source, current would be divided between C and R_1. Thus V will rise exponentially with a time constant determined by R_2C, but when $V = V_1$ and all the current is passing through R_1, V will stop increasing.

19. If a capacitor is exposed to a constant current, then the charge on the capacitor must increase linearly with time since $Q = CV$ and the conditions specify that Q is increasing linearly with time. (Remember that I is just Q/s.)

34

Notice that capacitance *increases* as the area of the plates increases, but *decreases* if the separation between them is increased. (You will find that this is significant when we discuss the conduction velocity of action potentials in Chap. 7.)

However, the concept most commonly used in membrane physiology is the **specific membrane capacitance** (measured in farads per square centimeter), defined as the capacitance of a 1-cm² area of membrane surface.

So far, we considered only **instantaneous** properties of circuits containing both resistors and capacitors. To understand cell membranes, it will be helpful to review briefly the **time-dependent** changes in current and voltage that occur in such circuits.

Let us consider what happens in Fig. 3-7 when the switch is closed. (Assume that there is no initial charge stored on the capacitor, i.e., that $V = 0$ before the switch is closed.) In the first instant after closing the switch, before any significant charge has flowed onto the plates of the capacitor, the driving force $(E - V)$ will be the full battery voltage E. At this time, current in the circuit is determined purely by the series resistor. By contrast, at some later time at which the capacitor has become fully charged, the driving force will have fallen to zero and current in the circuit will also be zero. At intervening times, **the rate at which the capacitor charges** (i.e., the rate of increase of V) **will be determined by the current flow in the circuit.** In turn, this is determined by the instantaneous value of the driving force $E - V$ and the circuit resistance. Fig. 3-8 shows the exponential form of the curve by which V approaches its final value in such a circuit.

Question: If we assume that V has reached its final value, what would happen to V in the circuit of Fig. 3-7 if E were increased from 10 to 20 V (by adding a second battery in series with the first)? (Hint 15↓)

An exponential curve such as that shown in Fig. 3-8 can be described by its characteristic **time constant,** which is the time that the curve takes to reach 63 percent $(1 - 1/e)$ of its final value. The same time constant would be obtained if the charged capacitor of Fig. 3-9 were allowed to discharge through a resistor equal to R in Fig. 3-7, and the point at which V falls to 37 percent $(1/e)$ of its initial value were used. Of course, V would be declining exponentially if the capacitor were discharged, so the curve of Fig. 3-8 would be inverted, starting at 100 percent charge and declining to zero.

Fig. 3-7. Circuit that charges a capacitor when the switch is closed; V measures charge buildup on capacitor.

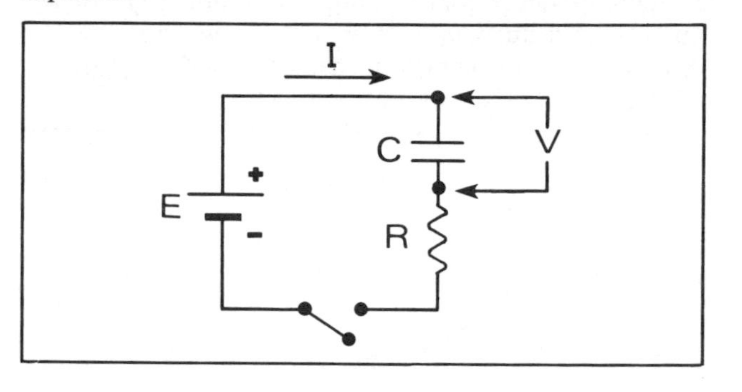

Fig. 3-8. Time course of voltage as a capacitor is charged, as in circuit of Fig. 3-7.

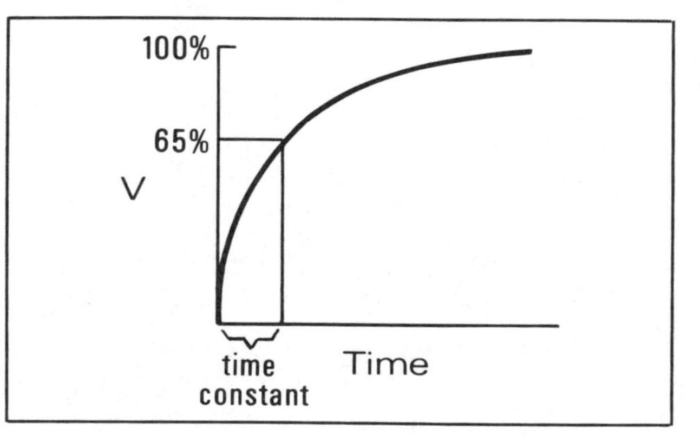

Fig. 3-9. Charged capacitor that will discharge through resistance R when the switch is closed.

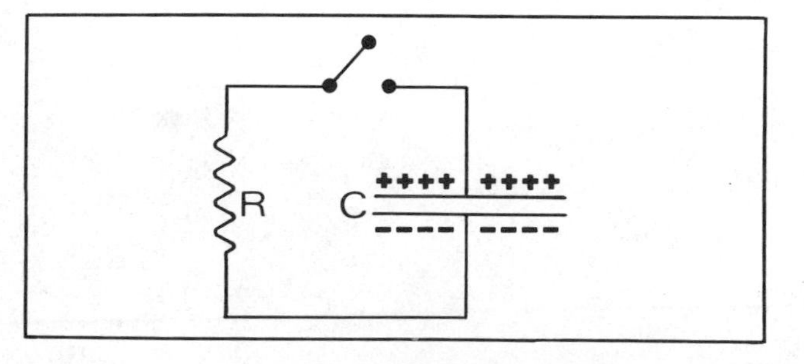

34

35

In such circuits, it can be shown that the time constant can be calculated as the product of R and C. The time constant has the unit of seconds if R is in ohms and C is in farads. In the calculation of membrane time constants, a more generally useful relationship is that RC has the unit of milliseconds if R is in kilohms times square centimeters ($k\Omega \cdot cm^2$) and C is in microfarads per square centimeter ($\mu F/cm^2$).

Fig. 3-10. Battery that will pass a current through R_1 and C when the switch is closed.

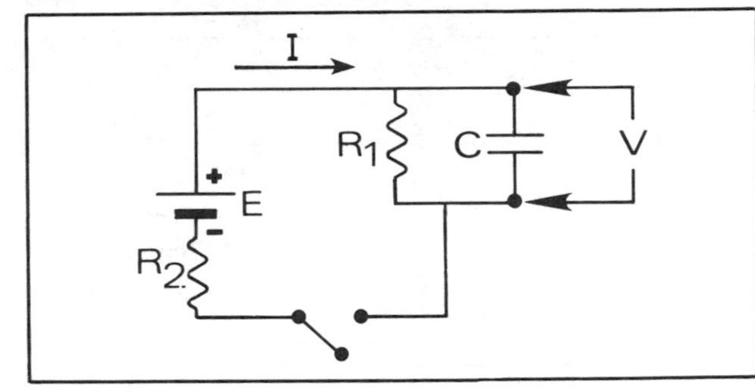

For electrophysiologists, a rather more realistic circuit is that shown in Fig. 3-10. Notice here that V will never reach the full value of E, even when the capacitor is fully charged, because in this circuit the maximal value of V occurs when the capacitor is charged to the same voltage that would appear across R_1 in a purely resistive series $R_1 + R_2$. Imagine that $E = 10$ V and $R_1 = R_2$. In the initial state, there will be no charge on the capacitor and the full 10 V will appear across R_2. In the final state, the voltage drop across R_1 (that is, V) must be 5 V. Between these two extremes, V will rise toward its final value and a steadily increasing current will be shunted through R_1 (rather than flowing onto the plates of the capacitor). The time constant for changing the capacitor can be shown to be $[R_1R_2/(R_1 + R_2)]C$.

However, when the switch is opened, the capacitor will not remain charged in this circuit but will discharge through R_1, with time constant R_1C.

Question: What is the time constant of a membrane whose specific resistance is $2\ k\Omega \cdot cm^2$ and whose specific capacitance is $1\ \mu F/cm^2$? (Hint 11 ↓)

Question: What is the consequence of making the current source hooked to the capacitor in the circuit of Fig. 3-7 a constant-current generator instead of a constant-voltage generator? (Hint 19↑)

Question: Would a similar effect be obtained if the current source in Fig. 3-10 were a constant-current generator? (Hint 18↑)

HINTS

8. 6×10^{11} charges. (See Hint 10↓ for a review of the calculation.)
9. The energy stored in a capacitor is in the form of a charge asymmetry. The energy stored in a battery is in the form of potential chemical energy.
12. The charge on the capacitance, with pluses on one side and minuses on the other!
13. "Once you've seen one positive charge, you've seen them all" (originally applied to redwood trees by a former governor of California).
14. The driving force $E - V$ is 1 V in this case. If the internal battery resistance is 1 Ω, what is the current flowing in this circuit? (Hint 16↓)

35

36

SOME BASIC ELECTRICAL PROPERTIES OF CELL MEMBRANES

Later, you will see the circuits that explain membrane functioning. At this point, we can indicate the **experimental basis for our knowledge that the membrane has both resistive and capacitative elements.**

1

Fig. 3-11 shows the typical result when an external stimulator is used to pass a *square-wave*, **constant-current pulse** across a cell membrane. Notice that **the cell's resting potential** (−90 mV in this example) **is affected by the current pulse, but this change of potential lags well behind the change of current,** finally reaching a new steady-state value if the current pulse is sufficiently prolonged. (Typically, the change of potential follows an exponential course with a time constant between 1 and 20 ms.)

2

These simple observations tell us a considerable amount about the properties of the membrane: **The membrane must have both resistance and capacitance.** Let's look at it this way: If the membrane were **purely resistive,** then the change of voltage would follow the same time course as the change of current; alternatively, if the membrane were **purely capacitative,** it would continue to charge with the continuing current and hence no new steady-state value would be reached. Since the membrane does neither of these things, it must be **both** resistive and capacitative. The most probable circuit would seem to be that of Fig. 3-10, in which the membrane resistance and capacitance are connected **in parallel.** Such a circuit is shown in Fig. 3-12. Notice that, as in Fig. 3-10, the membrane elements are **in series** with a second resistance (R_{ext}, the summed resistances of the electrodes, fluids, etc. in the external stimulating circuit). Initially, the current from the stimulator passes mainly through the capacitor; as the charge on the capacitor increases, more and more of the applied current passes through the resistor. The voltage across the membrane builds to a maximal value until all the applied current passes through the membrane resistance. In a circuit such as that shown in Fig. 3-12, if we know both I_{ext} (the applied current) and the V reached in steady state, then we can calculate R from Ohm's law. Then, once R is known, C can be obtained from the time constant.

3

Now look once more at Fig. 3-12. How can a steady **resting** potential be generated in such a circuit? Surely the *simplest* circuit that could account for all the membrane properties noted in Fig. 3-11 must be the circuit of Fig. 3-13. Here a battery located in the membrane charges the membrane capacitor to maintain the steady resting potential. The membrane resistance is shown as equivalent to the internal resistance of this *membrane battery.* (Look *now* at page 30.5 if you need to review the idea that a battery possesses an *internal resistance.*) The circuit in the dotted-line "box" is that which we present in Chap. 4 as the basis for understanding membrane potentials.

4

Fig. 3-11. Effect of applied current on membrane potential in hypothetical experiment. Note that both upper and lower graphs share the same time axis. The upper graph shows the applied current pulse (negative sign means that current is passed inward across cell membrane). The lower graph shows the change in membrane potential (from a resting potential of −90 mV) produced by the applied current. The time constant of potential change is the same when the current is turned off as when it is turned on.

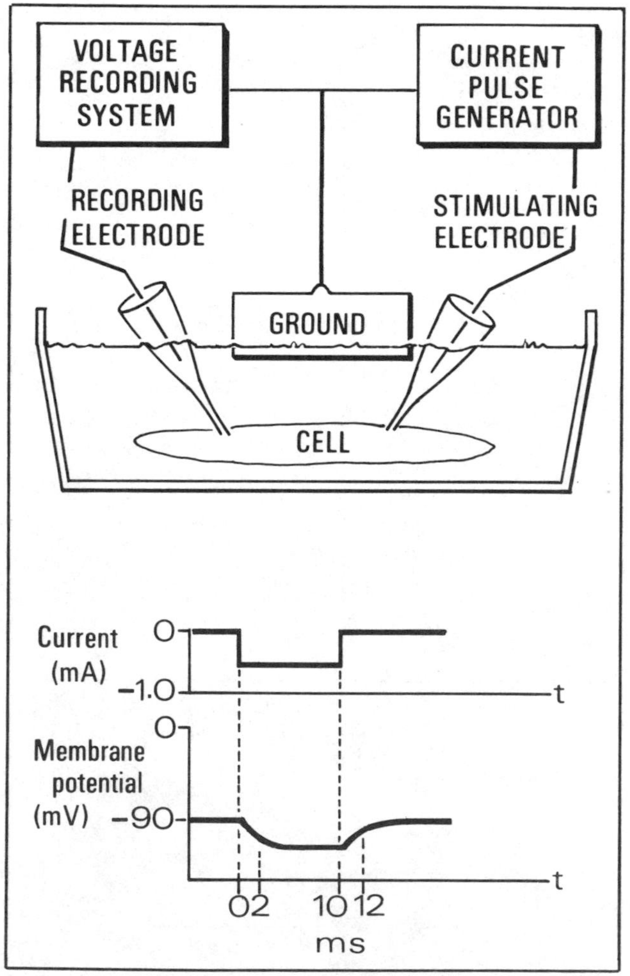

VOLTAGE RECORDING SYSTEM

CURRENT PULSE GENERATOR

RECORDING ELECTRODE

STIMULATING ELECTRODE

GROUND

CELL

Current (mA)

0

−1.0

t

Membrane potential (mV)

0

−90

t

0 2 10 12

ms

36

IONIC CURRENTS IN SOLUTIONS

1 **All movement of charge** (i.e., current) **in aqueous systems occurs by means of ions.**

The concept taught in your undergraduate physics course that current is carried **by electrons** is correct, but **only for wires,** not for aqueous solutions where charged ions (*both positive and negative*) can move and thus transfer charge (i.e., carry current). 2

The term **current is** used here (as in other biological texts) to mean **the direction of movement of a positive charge.** Thus, *current* is used in the same sense as in classical physics (e.g., current through a resistor moves from the positive pole of the battery to the negative pole). 3

Since in aqueous solutions there are both positive and negative ions, **a given transfer of charge** (current) **can occur by movement of positive ions in one direction, of negative ions in the other direction, or *both*** (as seen in Fig. 3-14). Thus, the phrase *an inward-directed current* can mean movement of positive charge into the cell, movement of negative charge out of the cell, or *both*. Fortunately, with respect to nerves, usually the positive ions are the most important physiologically, so that there is not much confusion or difficulty in conceiving of the current as being the movement of positive charge (i.e., positive ions). 4

Fig. 3-12. The elements enclosed by the broken lines are properties of the cell membrane. Here V is the potential seen by the recording system, and R_{ext} is the sum of the resistances in the external stimulating circuit.

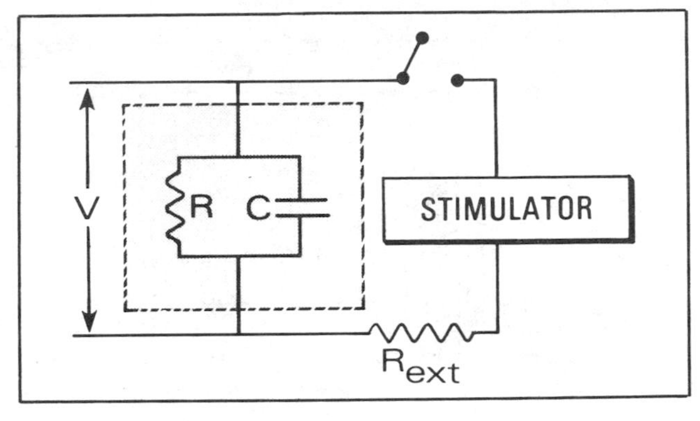

Fig. 3-13. Here E is the voltage of the "membrane battery." Other symbols are as in Fig. 3-12. The circuitry in the broken-line box represents the membrane.

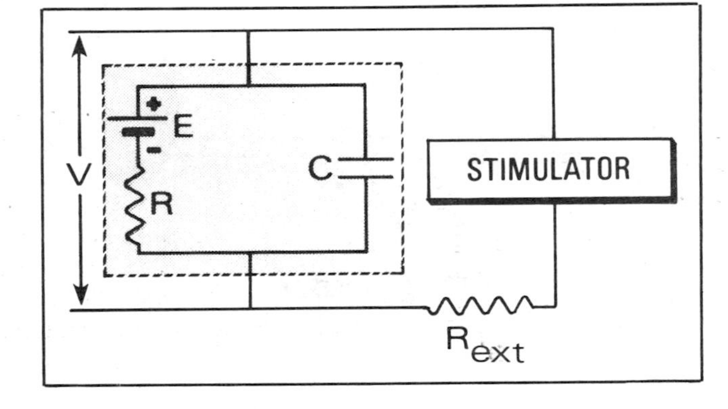

HINTS

10. $Q = CV$
$= (1 \times 10^{-6})(100 \times 10^{-3})$
$= 100 \times 10^{-9}$
$= 1 \times 10^{-7}$ coulombs
$(6 \times 10^{18}$ charges/coulomb$)(1 \times 10^{-7}$ coulombs$)$
$= 6 \times 10^{11}$ charges

11. 2 ms. Note that it is legitimate to use *specific* resistances and capacitances in this calculation since the centimeters cancel.

15. Voltage V would start at 10 V and increase from 10 to 20 V along a curve similar to that of Fig. 3-8.

16. 1 A, from $I = (E - V)/R$ (Eq. 3-6).

17. Initially, a current flows through the circuit consisting of the battery, the electrodes (anode and cathode), and the capacitance (an electrical insulator separating two conductors) of the plastic sheet and two adjacent solutions. This current results in a net buildup of Cl^- ions on the cathode side of the membrane (side A) and Na^+ ions on the anode side (side B). These "excess" ions (unmatched on their particular side) are attracted by their opposite numbers on the other side of the plastic sheet. Thus they spread out along either side of the sheet just as the charge asymmetry on the plates of a capacitor. This charge across the membrane ultimately becomes equal and opposite to the battery voltage, and then the flow of current ceases. Thus you can see that **electrical circuits can exist in aqueous solutions** as well as in metallic conductors. The circuit is equivalent to that of Fig. 3-7.

38

Free electrons cannot exist in an aqueous solution, where they are captured within nanoseconds by either water or the other constituents in the solution. Thus, **at an electrode-fluid interface** (diagrammed in Fig. 3-14) **there must be a chemical exchange from electrons** (in the wire) **to ions** (in the solution). Usually this chemical reaction is the rate-limiting step in such a situation, and special precautions are often necessary in constructing electrodes in order to prevent the *electrode polarization* that occurs when the amount of current being passed begins to exceed the rate of the reaction. Under these circumstances, the charges build up and like charges tend to electrostatically repel one another, so that the electrode-fluid interface begins to act as a capacitor, reducing the flow of current as the charges build up [27, pp. 17–19]. 1

A number of factors can affect the otherwise random movement of charged ions in solution. **We concentrate on the two factors of prime importance** in generation of the membrane potential: **voltage** and **concentration differences.** 2

Other factors such as ion-specific, energy-dependent transport processes, or osmotic gradients, which may be important under certain conditions, are considered in Chaps. 4 and 5. 3

Fig. 3-14. Movement of ions induced in solution by an external voltage source. Both negative and positive ions are carrying the current (i.e., positive charge from anode to cathode).

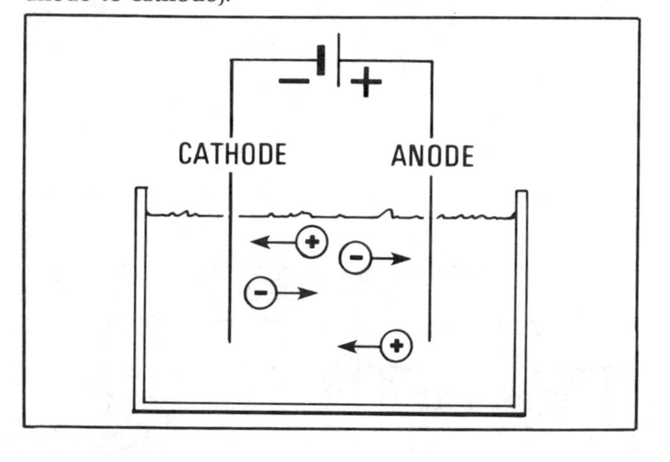

Voltage. When a voltage gradient is applied to a solution containing ions, the **positive ions move toward the cathode,** while the **negative ions are attracted to the anode** (Fig. 3-14). Obviously, the rate at which they move (the number of charges per unit time) is proportional to the applied voltage. The ionic current (number of charges per unit time) also is affected by the number of ions available (i.e., the ion **concentration**) and the **mobility** of the ions involved. 4

QUESTION: Relating the idea of ionic currents in solution to what you know about electrical circuits, what do you predict will happen when the switch in Fig. 3-15 is closed? The two solutions are separated by a very thin, impenetrable sheet of plastic. (Hint 17↑) 5

Fig. 3-15. Two solutions separated by a thin sheet of plastic.

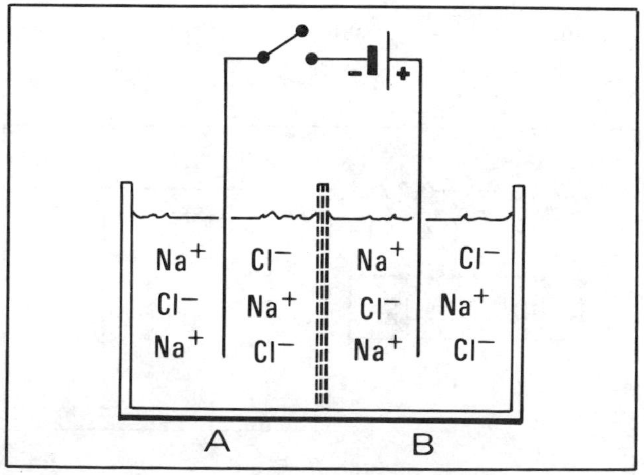

Concentration differences. Ionic movement due to concentration differences is also a simple phenomenon (although it may be more difficult to understand for the beginning student who has not had physical chemistry). Imagine a drop of ink gently placed in one corner of a bathtub of still water (Fig. 3-16). The ink diffuses slowly until it is uniformly distributed. Why should this be? After all, the only motion of the particles is Brownian motion (which is random), so why should the ink particles move from one end of the tub to the other? Why don't they just stay "randomly" in the corner? Well, consider an imaginary boundary (labeled *A* in Fig. 3-17) on one side of which are ink particles and on the other side is just clear water. Now, the random motion of the particles will cause *some* of the particles to cross the boundary. At the same time, there are *no* particles as yet on the other side of the boundary, so none are available to cross in the opposite direction. Thus, **there** 6

38

39

1 **must be a net flux** (i.e., net movement) **of ions across the boundary from the area of high concentration to the area of low** (zero) **ink concentration as a result of random motion alone!**

The same argument applies when there is *any* concentration difference across the boundary. Whenever there is a difference in the number of ions randomly moving in one direction compared to the other, the *net flux* (net movement) will be *from* the higher concentration to the lower concentration. Note that the position of the boundary A is entirely arbitrary; it could be imagined anywhere in the bathtub, wherever a concentration difference occurred. The *only* time that there is *no net flux* is when the *concentrations* of ink particles on both sides of the imaginary boundary are equal, which occurs only when the bath water *is* uniformly gray. 2

To restate the concept in more rigorous terms: The **net flux** J_{net} (in moles per second per unit area of the boundary) across this boundary will depend on the concentration difference $C_1 - C_2$ (in moles per cubic centimeter) and the ease with which the particles in question cross the solvent boundary measured by the **permeability** P (in centimeters per second). Thus

$$J_{net} = P(C_1 - C_2) \quad \text{Eq. 3-8}$$

In this equation, net flux can be seen to be the difference between the two unidirectional fluxes $J_1 = PC_1$ and $J_2 = PC_2$. Clearly, the simple relationship shown in Eq. 3-8 (which is equivalent to *Fick's first law of diffusion)* cannot apply to the movement of charged particles unless (1) the cations and anions concerned have exactly the same mobilities and (2) there is no externally applied electric field. 3

Note that this result is obtained only by considering **groups of particles;** i.e., contemplating the motion of any one particle leads to no useful result. From the motion of the one particle depicted in Fig. 3-18, you might conclude that particles will tend to stay in the same position! **The diffusional "force" and the resulting net ionic currents caused by concentration differences acting on charged particles are properties of masses of particles.** Motto: "Net flux is the crux of the current." 4

Those with a knowledge of thermodynamics will realize that the process of diffusion involves an increase in entropy. Of course, in keeping with the second law of thermodynamics, in order to decrease the entropy (in opposing the tendencies to move from high to low concentrations), an energy must be dissipated. 5

Fig. 3-16. Blob of ink particles in still water.

Fig. 3-17. Movement of various ink particles relative to imaginary boundary A.

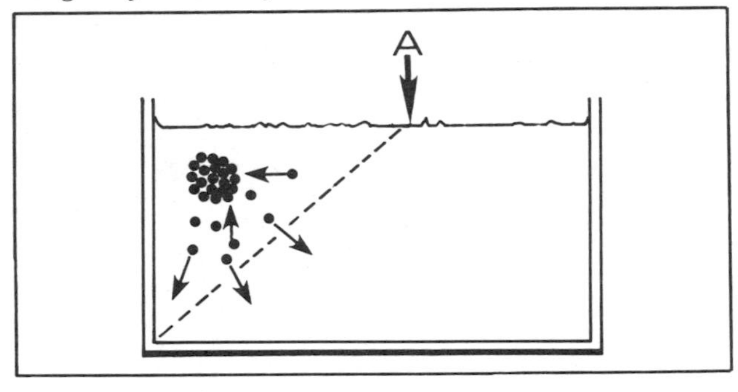

Fig. 3-18. Possible motion of a single ink particle due solely to random fluctuations.

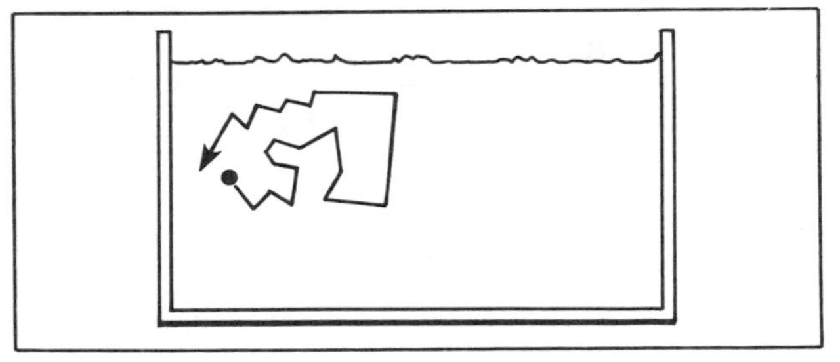

39

40

The pumping of ions against their electrochemical gradients requires chemical energy, as described in Chap. 5. 1

Intuitively, you can see that the diffusional *force* of concentration differences will be influenced by any other factors that also affect particle movement:

1. *Temperature*—increased temperature causes a greater rate of diffusion because of increased Brownian movement.
2. *Particle size*, causing differences in mobility—the larger particles will move slower than smaller particles because of their larger mass, greater frictional resistance, and smaller movements due to *averaging* the bombardment of the water molecules. 2

The preceding variables are found in the Goldman equation discussed in Chap. 5. 3

You may note that the *permeability*, as defined in Eq. 3-8 and Chap. 5, can be broken down into

$$P = \frac{MRT}{\Delta X} = \frac{D}{\Delta X} \qquad \text{Eq. 3-9}$$

where M = mobility of the particle [$cm^2/(S \cdot J)$]
R = gas constant (J/°K)
T = temperature (°K)
ΔX = thickness of boundary layer (Ω) 4

Thus D, which is often known as the *diffusion coefficient*, has units of square centimeters per second (rather than centimeters per second, as for P). 5

In Chap. 4 we show you how *ionic currents generate membrane potentials as ions are moved along their electrochemical gradients.* Ready? 6

40

41

Origins of the Resting Membrane Potential

CAUTION: If you have *not* read Chap. 3, you may find parts of this chapter difficult. Conversely, this 1chapter is not difficult if you have read Chap. 3. Be forewarned!

IMPORTANCE OF STUDY OF MEMBRANE POTENTIALS

The ideas presented in this chapter are the basis for understanding much of what goes on in all cells, not just cells of the nervous system. When you understand the principles involved in the 2 functioning of one cell type, you can apply them to many other cells as well.

We first study the axon and then extend the ideas involved to other examples of nervous systems, such as generator and receptor potentials (Chap. 9), neuromuscular transmission (Chap. 10), and synaptic transmission (Chap. 11) (see Fig. 4-1/2-14). All these neural mechanisms share this principle: **The membrane potentials are produced by movements of ions along electrochemical gradients. Changes in these membrane potentials are brought about by changes in the conductance of the membrane to one or more ions.** (Don't worry if you don't understand this generality right now. Obviously, if this idea were so simple as to be 3 immediately apparent, there would be no need for the remainder of this chapter.)

While all cells show membrane potentials, **only nerve and muscle cells show rapid changes in membrane potential directly related to their functions in the body.** These rapid changes in membrane potential allow us to understand what is happening in the cell much more easily than if we studied slower changes in cell function, e.g., secretory activity or response to circulating hormones. For this reason, at present neurophysiology provides a deeper understanding of cellular and subcellular biophysics than any other area of physiology. There is no reason to doubt that many of the general principles discovered in the study of nerve and muscle cells are also applicable to most of the cells of the body. For example, **the following** (which you will study with regard to nerve and muscle cells) **apply to all cells: diffusion, pumping against a concentration gradient, and ion selectivity by membranes.** Here are some examples: (1) The movement of nutrients to and from the cell can occur only by diffusion between the cell and the capillary. (2) There are many cells that expend energy in doing work against a concentration gradient, e.g., cells in the kid- 4 ney, many secretory cells, and cells that absorb nutrients from the intestines.

Thus, the moral is this: **A thorough understanding of the neuron is of considerable 5 benefit to understanding other cellular functions in the body.**

MEMBRANE STRUCTURE AND CIRCUITRY

The background is over, and now we begin to unfold the basic ideas of how membrane poten- 6 tials come about.

Membrane potentials, we now know, **are the obvious, natural consequence of the fluids on 7 either side of the membrane, which taken together act as *electrical circuits*.**

One possible biochemical structure of the membrane is indicated in Fig. 4-2. While the phosphate portion of the phospholipid is water-soluble (being charged), the central por- 8 tions are quite insoluble and thus must resist the tendency for ions to move between them.

Fig. 4-1/2-14. Chapters in this book and how they relate to functional parts of sensory and motor neurons.

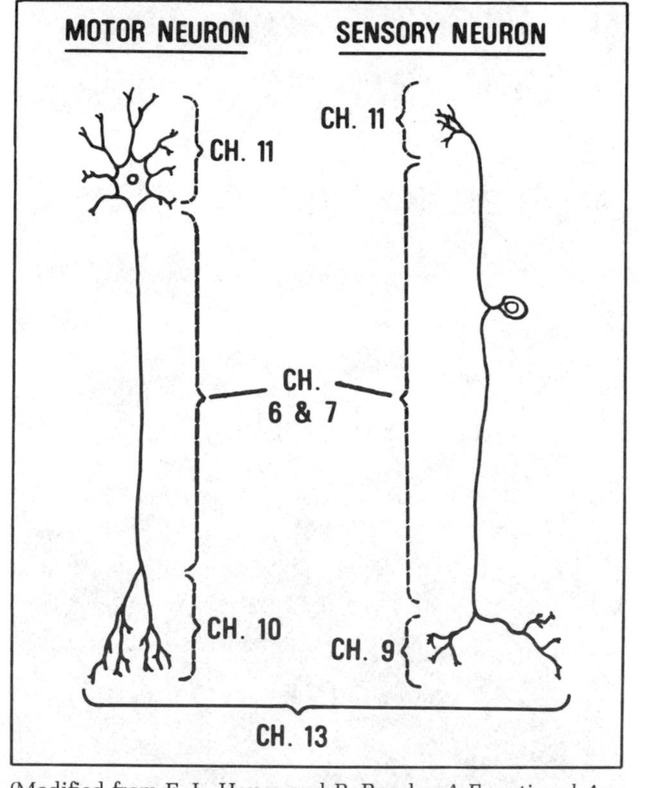

(Modified from E. L. House and B. Pansky, *A Functional Approach to Neuroanatomy* [2nd Ed.]. New York: McGraw-Hill, 1967.)

1 Question: When the cell membrane phospholipids are separated from the other constituents and spread over water as a **monomolecular** layer, the area of the layer is about twice that computed for the cells used. Why? (Hint 1↓)

2 The structure shown in Fig. 4-2 is not the only possible configuration of proteins and lipids consistent with known membrane properties. A number of other structures have been proposed based on different assumptions as to the tertiary structure of the proteins involved (see Fig. 4-3). Probably it is unwise to assume that there is only one membrane structure for all cells. Certainly, different membranes are known to differ markedly in both protein-lipid ratios and the nature of the lipids involved. Furthermore, physiological evidence (presented later in this book) suggests that substantial differences exist between different cell membranes and even between different membrane regions within the same cell with regard to what occurs in and across the membranes. (For a detailed account of the recent evidence concerning cell membrane structure, see Quinn [50].)

3 An obvious consequence of the lipid bilayer structure of **the membrane** is that it **must act as a capacitor.** Capacitance occurs whenever two conducting *plates* are separated by an insulator, the *dielectric*. Now **the cell membrane acts as an excellent dielectric** because of its high content of electrically insulating lipids, while the **charge-carrying fluids** on either side of the membrane (intracellular and extracellular) **act as the (conducting) plates of the capacitor.**

4 The dielectric properties of the membrane are best grasped by realizing that **the membrane has about the same electrical properties as glass of the same dimensions!**

5 Put in another way, the specific resistance of the membrane is about 100 million times greater than that of the fluids in contact with it [27, p. 48].

6 It should **not** be supposed, however, that the membrane is **completely** impermeable to ions; i.e., **some current can flow through the membrane.**

7 It is clear from experimental evidence (presented in the fourth level in the section "Some Basic Electrical Properties of Cell Membranes," and elsewhere in Chap. 3) that the dielectric of the membrane capacitance is "leaky"; i.e., currents can and do move through it. The simplest assumption is that the membrane is nonuniform, containing holes (better termed resistive *pores*) dispersed in a basically capacitative (dielectric) membrane. Such pores have not been definitely identified yet by electron microscopy, but calculations show that the pores would be very small (around the limits of resolution of current electron microscopy) and very far apart. It is likely that such pores make up less than 1 percent of the membrane surface [27, p. 90].

Fig. 4-2. Hypothetical model of a cell membrane. The membrane is shown as double row of phospholipid molecules sandwiched between two layers of protein. The outer layer is covered with a complex carbohydrate. The hydrophilic ends of the phospholipid molecules are in contact with the protein, and the hydrophobic ends abut in the center of the membrane.

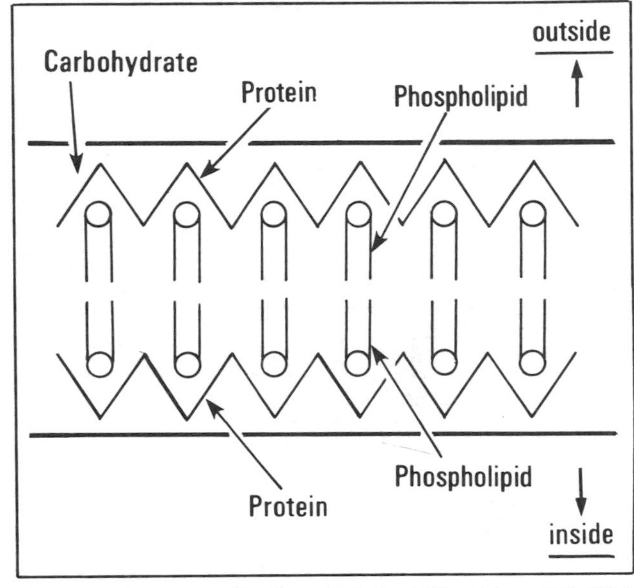

(From W. F. Ganong, *Review of Medical Physiology* [5th Ed.]. Los Altos, Calif.: Lange, 1971.)

44

The calculation is as follows: Assume, for example, a squid axon immersed in a seawater bathing medium. The specific resistivity of seawater is 15 $\Omega \cdot$cm; the specific resistance of the axon membrane is about $1 \times 10^3\ \Omega \cdot \text{cm}^2$. If we assume the membrane pores to be water-filled with the same resistivity as the extracellular medium, 3.5 Å in diameter and 100 Å long, then we can calculate the number of open pores in this membrane (L is length and A is area):

$$\text{Single pore area} = \pi \left(\frac{d}{2}\right)^2 = 1 \times 10^{-15}\ \text{cm}^2$$

$$\text{Single pore resistance} = \rho \frac{L}{A} = \frac{15\ \Omega \cdot \text{cm}\ (1 \times 10^{-6}\ \text{cm})}{1 \times 10^{15}\ \text{cm}^2} = 1.5 \times 10^{10}\ \Omega$$

$$\text{Single pore conductance} = 0.67 \times 10^{-10}\ \text{mho}$$

$$\text{Pore density} = \frac{\text{specific membrane conductance}}{\text{single pore conductance}} = \frac{1 \times 10^{-3}\ \text{mho} \cdot \text{cm}^2}{0.67 \times 10^{-10}\ \text{mho}}$$

$$= 1.5 \times 10^7\ \text{pores/cm}^2, \text{ or } \approx 1 \text{ open pore every } 7\ \mu\text{m}^2$$

$$\text{Total pore area as percentage of membrane area} = \frac{(\text{pore density})\ (\text{single pore area}) \times 100}{\text{unit membrane area}}$$

$$= \frac{(1 \times 10^{-15})\ (1.5 \times 10^7) \times 100}{1}$$

$$= 0.0000015\%$$

Even if the maximum membrane conductance were as much as 10,000 times greater than the conductance of the resting membrane, the part of the membrane occupied by open pores would still be substantially less than 1 percent.

You will realize that some of the assumptions made here are rather arbitrary. While we shall see (Chap. 5) that there is considerable evidence in favor of 3.5 Å as a reasonable estimate of pore diameter, pores may be as narrow as that for only a small part of their total length. Similarly, it may be unreasonable to assume that ions have the same mobility within such narrow pores as they would have in the bulk medium. Nevertheless, the calculation is instructive, and the answer may well be approximately correct although the single pore conductance calculated here is about ten times higher than recent, more direct measurements would indicate.

Fig. 4-3. One hypothetical scheme by which living membrane structure (upper diagram) might be converted by fixation (middle diagram) and subsequent metal staining to the typical "unit membrane" seen in electron micrographs (lower diagram).

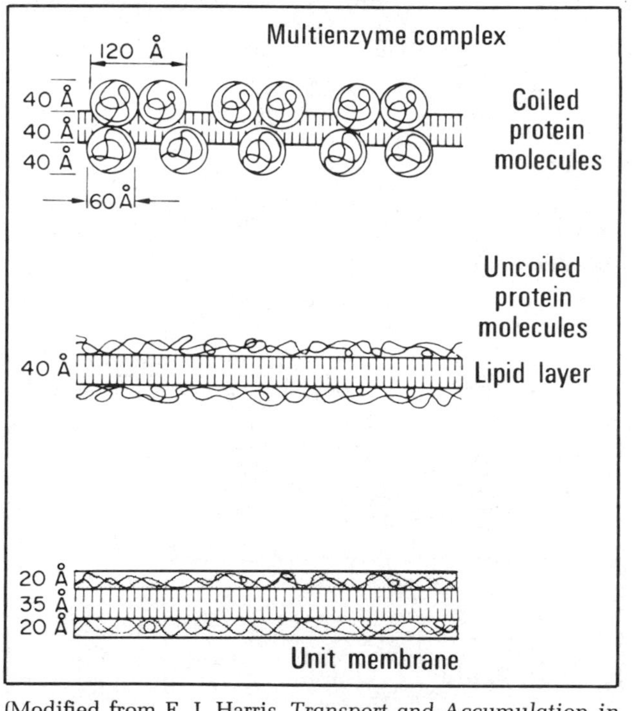

(Modified from E. J. Harris, *Transport and Accumulation in Biological Systems* [3rd Ed.]. Baltimore: University Park Press, and London: Butterworths, 1972.)

44

45

Movement of ions through the membrane must represent a resistance, electrically speaking. So the **membrane structure,** as it is imagined, **contains large dielectric areas** (more than 99 percent of the membrane) **and a few high-resistance "pores"** (less than 1 percent of the membrane).

1

This concept is shown in Fig. 4-4, where **the pore is a finite** (but present) **resistance and the remaining membrane acts as a capacitor.**

2

There is one other circuit element that occurs because of the structure of the membrane: a battery.

3

It will not be evident to you until later in this chapter why we can assert that the structure of the pore, in combination with the differences in concentration of ions on either side of the membrane, makes up a "battery" that helps to generate the membrane potential. For the moment, please take on faith that later we will be able to convince you how the battery works.

4

The placement of the battery as one of the elements allows us to diagram the membrane circuitry, as shown in Fig. 4-5. In Fig. 4-5 both the battery and the resistance are in the pore, and they are connected to the membrane capacitance as shown. The pore resistance must be the resistance to ion flow through the pore. But since the pore is also part of the battery, **the pore resistance must actually represent the *internal resistance* of the battery.** To make this clear, we have taken some liberties with conventional electrical symbols and have redrawn Fig. 4-5 as shown in Fig. 4-6, where the internal resistance of the battery is shown actually within the battery symbol. While this arrangement should help you visualize that the pore is both the battery and its internal resistance at the same time, we do not use this symbol again. Instead we return to Fig. 4-5 for the following, very important reason.

5

The **membrane potential** that is recorded by the experimenter (V in Fig. 4-5) **is the potential across the membrane capacitance, and *not* the potential of the battery.**

6

Such a relationship can be readily adduced from the circuit of Fig. 4-5 (with your prior knowledge of this circuit from Fig. 3-7). Unfortunately, some students might imagine from Fig. 4-6 that V measured the battery potential, but this would be an illusion caused by the misuse of the battery and resistor symbols in that case. In this way, a figure introduced to

7

Fig. 4-4. Section of membrane (cut through a "pore") showing the electrical equivalents of the structures.

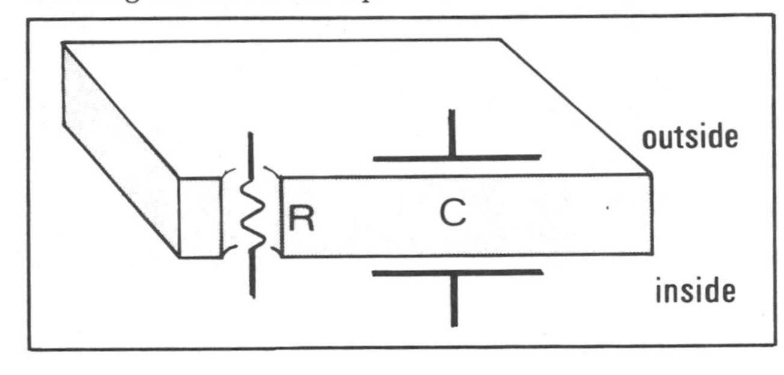

Fig. 4-5. Elements of battery, resistance, and capacitance in the membrane; V represents recordable membrane potential.

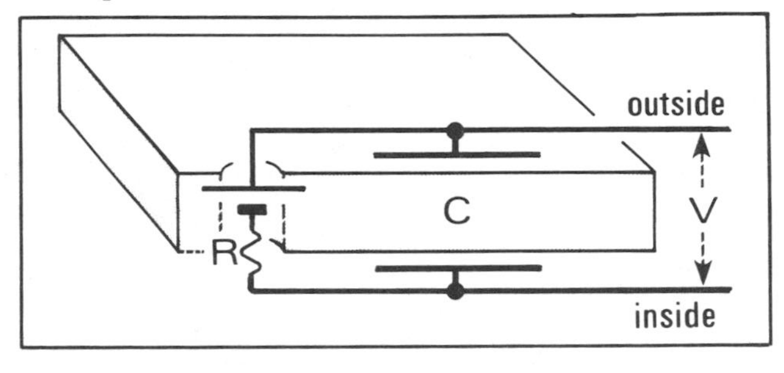

Fig. 4-6. "Internal resistance" of the battery relocated for visual effect.

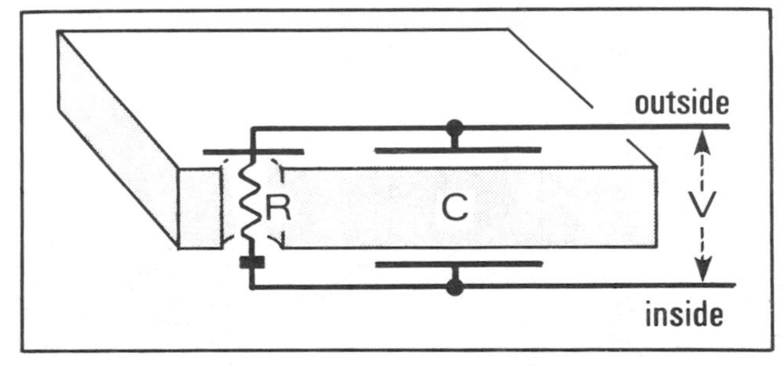

HINT

1. The phospholipids are in *duplicate*, with the hydrophobic ends toward the inside of the membrane. When the phospholipids are separated and spread on the surface of the water, they align themselves with the hydrophilic ends in the water and the hydrophobic ends sticking up; since they were in duplicate in the membrane and are spread on the water in a *monomolecular* layer, the area is doubled. Membrane structure is described further in various sources [28, pp. 272–278; 40, pp. 62–65].

45

46

clarify one point may confuse another. Thus, be sure you grasp this very important point: **The membrane potential is the potential on the membrane capacitance, and changes in membrane potential can occur only if there are changes in the charges on the membrane capacitance.** 1

It is to be regretted that most textbooks fail to make this important point, since a simple understanding of capacitors greatly facilitates understanding of membrane potentials. 2

At this point let us indicate what the membrane potential is **not!** It is not the potential on the battery, nor is it the voltage across the membrane resistance, as shown in Fig. 4-7. 3

MEMBRANE CURRENTS

You will soon see that membrane potentials are best understood by concentrating on the currents that flow through the membrane circuits. So it is time to build up a rigorous picture of the movements of ions (ionic currents) that occur across cell membranes (through the pores and capacitance). 4

As you will remember (see pages 38.6 to 39.2), ions may cross an imaginary boundary in two directions; **the net flux** across the boundary (i.e., ionic current) **represents the algebraic sum of these two unidirectional fluxes.** Similarly, the ionic current across a cell membrane represents the algebraic sum of two such unidirectional fluxes. Finally, it is possible to refer to the *net ionic current,* which is the algebraic sum of all the currents carried by the different ion species crossing the membrane in both directions. These terms are defined in Table 4-1. 5

Students interested in reading original literature in this field should note that just as membrane resistance and capacitance generally are referred to as the unit membrane area, membrane currents are also normally termed *current densities* (e.g., as microamperes per square centimeter). 6

Clearly, any description of ionic currents is complex since the charge of the ion must be specified as well as its direction of movement. **Carefully note the following conventions,** used throughout this book: 7

1. **The direction of a current is given as if the current were being produced by movement of positive particles.** Thus a movement of K^+ ions *into* a cell is described as an *inward current*. Similarly, a movement of Cl^- ions *out* of a cell is also described as an inward current since this is the direction in which positive particles would have to move to produce the same electrical effect. 8

Fig. 4-7. Some examples of what the membrane potential V is **not**.

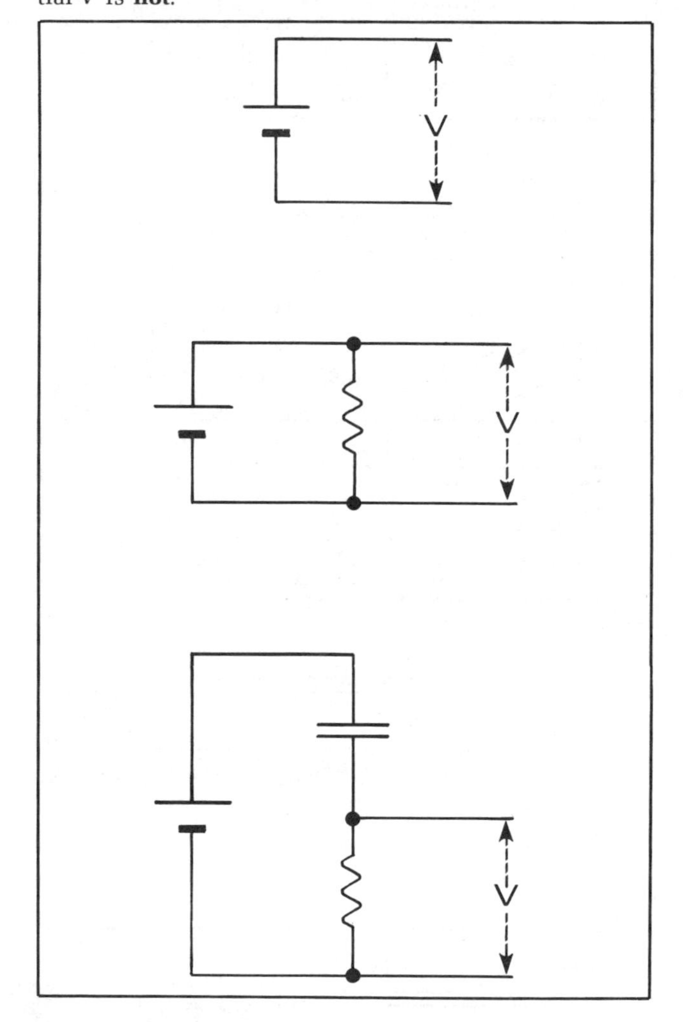

46

47

1 2. **"Sign" is used to indicate the direction of this conventionalized current: Outward currents are given positive sign; and inward currents have negative sign.** Thus the inward movement of K^+ ions is described as a *negative* potassium current; similarly, the outward movement of Cl^- ions is a *negative* chloride current.

2 Since we can assure you that a clear understanding of these conventions is going to be essential to your comprehension of this book, why not go ahead and get some practice?

3 QUESTION: Is inward movement of Ca^{2+} a positive or negative current? (Hint 5↓)

4 QUESTION: Is inward movement of Cl^- a positive or negative current? (Hint 3↓)

5 QUESTION: Given that I_{HCO_3} (the current carried by the negative bicarbonate ion) is *negative*, in which direction did bicarbonate ions move? (Hint 6↓)

6 **The same sign convention is used to describe the direction of induced capacitative currents across the cell membrane.**

7 QUESTION: In order for the inside of the cell (Fig. 4-8/4-5) to be negative, i.e., for the capacitor to have positive charges on the outside and negative charges on the inside, in which direction will a current have to have moved through the capacitor? (Hint 2↓)

8 QUESTION: In order for an inward-directed current to occur on the capacitor, in which direction must a current move through the pore-battery? (Hint 4↓)

9 QUESTION: What is the sign of I_{cap} (the current through the capacitor) necessary to bring about inside negativity as measured at V in Fig. 4-8/4-5? (Hint 7↓)

Since we know that currents always flow in complete circuits, whatever the direction of current flow through the circuit of Fig. 4-8/4-5, at all times the magnitudes of the currents through the pore battery must be same as those of the capacitor, *differing only in having opposite signs* (since one current will be "going out" when the other is "going in"). We can state this in symbolic terms as follows

10 $$I_{por} = -I_{cap} \quad \text{Eq. 4-1}$$

(See Table 4-2 for definitions of symbols in these equations.) Obviously, Eq. 4-1 can be rewritten as

$$I_{por} + I_{cap} = 0 \quad \text{Eq. 4-2}$$

which indicates that the net sum of all current in and out of the membrane is zero:

11 $$I_{mem} = I_{por} + I_{cap} = 0 \quad \text{Eq. 4-3}$$

Table 4-1. Basic Definitions of Current in an Aqueous Medium

1. The algebraic sum of two unidirectional ionic fluxes is a *net flux*.
2. A net ionic flux (per unit time) produces a specific *ionic current* (I_j), where the subscript indicates the specific ion (for example, I_K is the potassium ionic current).
3. The algebraic sum of all specific ionic currents through membrane pores is the *net pore current* I_{por}, where $I_{por} = \Sigma I_j$.

Fig. 4-8/4-5. Elements of battery, resistance, and capacitance in the membrane; V represents recordable membrane potential.

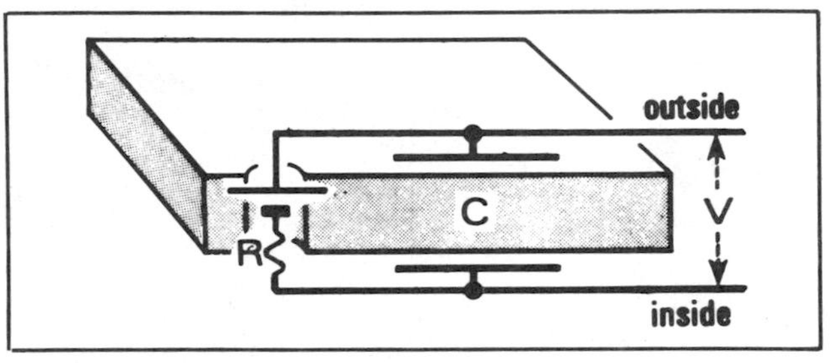

47

48

This is one of the central equations of neurophysiology. It is *true* for steady-state and transient situations, *with two limitations* (which need not concern us in this chapter):

1. **There must be no external circuit present to bypass the membrane circuit elements.**
2. **The currents must be summed over the whole membrane area** (unless it can be presumed that the whole membrane acts in a uniform manner).

1

Throughout this chapter and Chap. 6 we assume that both these limitations are met. Later we tackle the more difficult analysis required when different membrane regions are not in the same state at the same time (see Chaps. 7 and 8).

2

These equations are especially helpful when you realize that **if the membrane potential (V) is not changing, then I_{cap} is zero.** This follows directly from Chap. 3, since V will change if there is an I_{cap} and will not change when I_{cap} is zero.

3

The proof is as follows. You remember (from Eq. 3-5) that the voltage across a capacitor is directly proportional to the charge stored on its plates:

$$Q = CV$$

Then it follows that change of voltage must be proportional to the change of charge:

$$dQ/dt = C\, dV/dt \qquad \text{Eq. 4-4}$$

And since current across the capacitor I_{cap} must be the rate of change of charge dQ/dt at that time, this equation can also be written as

$$I_{cap} = C\, dV/dt \qquad \text{Eq. 4-5}$$

At the risk of restating the obvious, we can now point out that $I_{cap} = 0$ if $dV/dt = 0$. Alternatively, if V is *not* constant, then it follows that I_{cap} is *not* zero.

4

Returning to Eq. 4-2, now we see that **where V is constant (and hence $I_{cap} = 0$), I_{por} must also be zero.**

5

We have thus rigorously stated what you can probably intuit from Fig. 4-8/4-5: If V is not changing, then no current is flowing in this circuit ($I_{cap} = 0$ and $I_{por} = 0$).

6

Table 4-2. Symbols Used to Describe Currents in an Aqueous Medium

I_j = current carried by ions of jth ion species moving through membrane pores (for example, I_{Na} is current carried by Na^+ ions moving through membrane pores)

I_{por} = net sum of currents carried through membrane pores by ions; $I_{por} = \Sigma I_j$

I_{cap} = current across membrane capacitance

I_{mem} = total current moving across membrane circuit elements; $I_{mem} = \Sigma I_j + \Sigma I_{cap}$

Note: *Outward* current is *positive*; **this convention applies to all currents.**

48

PRINCIPLES OF MEMBRANE POTENTIALS

Now we are in a position to outline the principles that form the basis for a rather complete understanding of the mechanism of membrane potentials. Some of these principles you have already encountered, and others will be described. In the following list, can you see how far we have come?

1

1. **All currents are carried by charged ions whose net movements are the result of both electrical and diffusional (concentration difference) forces.**
2. **All currents always move in complete circuits, across both resistive and capacitative elements.**
3. **The membrane potential is the charge on the membrane capacitance.**
4. **Ion-selective pores in an otherwise impermeable membrane provide the equivalent of batteries with internal resistance.** The internal resistance is not necessarily constant. **Change in internal resistance will change the current flowing through the pore and hence the charge on the membrane capacitance** (since all currents travel in complete circuits).
5. **The intracellular and extracellular solutions may contain different concentrations of any given ion species, but at the same time these solutions have equal osmotic pressures and each is electroneutral** (contains the same numbers of anions and cations).
6. **Under steady-state conditions, the differences in ionic concentration on either side of the membrane are maintained by membrane "pumps" that use metabolic energy.**

2

Principles 1, 2, and 3 are described in the early parts of this chapter. Principles 5 and 6 are new ideas that are covered later in this chapter. That brings us to the point of describing principle 4 and how that mysterious pore-battery comes about!

3

IONIC DISTRIBUTIONS ACROSS CELL MEMBRANES

By now, you should have a general idea of the concepts presented, but you may be impatient for some facts on which you can exercise these principles. Now is the time! It is clear that details, such as which ions are present and which can pass through the membrane pores, are crucial. So let us consider some of the concentration gradients that exist across a typical cell membrane, as shown in Fig. 4-9.

4

Fig. 4-9. Typical ionic distributions. Numbers in boxes are concentrations in milliequivalents per liter; A^- is a collective term for all anions other than Cl^-. $[Y]_i$ = internal concentration of ion "Y," $[Y]_o$ = external concentration of ion "Y."

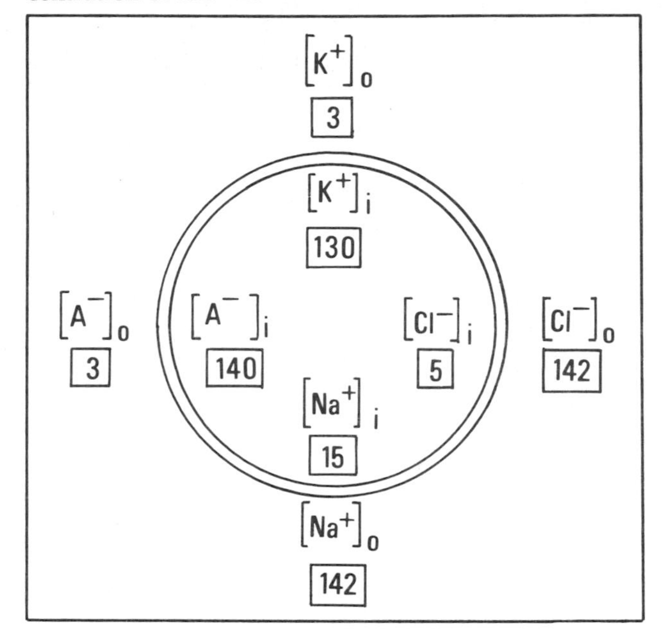

HINTS

2. An inward-directed current through the capacitor makes V negative inside with respect to the outside. If you missed this, look back at Fig. 3-6.
3. Inward movement of Cl^- is equivalent to outward movement of positive charge. An outward current has positive sign; hence this is a *positive* chloride current.
4. Outward-directed current in the pore will give a negative inside charge on the capacitor. If you missed this, go back to Fig. 3-6 and try again!
5. Negative.
6. A negative current means inward movement of positive ions; and therefore an *outward* movement of negative bicarbonate ions.
7. The current I_{cap} would have to be *negative*. Symbolically, this is written as $-I_{cap}$.

50

Unfortunately, because of uncertainties with respect to intracellular volume, the activities of intracellular ions, and binding and/or sequestration of ions in intracellular subcompartments (such as mitochondria), it is difficult to do complete studies on a single cell. Figure 4-9 is a composite diagram of a "typical" cell. 1

Actual examples of internal and external ion concentrations from four different sources are shown in Table 4-3. 2

Table 4-3. Extracellular and Intracellular Na^+ and K^+ Concentrations for Various Cells

	Concentration (mEq/L)	
	Na^+	K^+
Human muscle		
Intracellular	12	80
Extracellular	135	4.5
Rat muscle		
Intracellular	13	140
Extracellular	150	4
Cat nerve		
Intracellular	18	166
Extracellular	135	5
Squid axon		
Intracellular	50	400
Extracellular	440	22

Data from Kerkut, G. A., and York, B., *The Electrogenic Sodium Pump.* Bristol, U.K.: Scientechnica, 1971.

INTERACTION OF ELECTRICAL AND DIFFUSIONAL FORCES ACROSS CELL MEMBRANE (DIFFUSIONAL POTENTIAL)

We must now explain in some detail how **the pore and the ionic differences** on either side of the membrane can **act as a battery.** 3

Electrical forces act on ions because ions are charged particles. Diffusional forces will occur wherever there are concentration differences. In dilute solutions, one can regard ions as moving independently of one another down their concentration gradients, except insofar as this movement creates an electrical potential difference. 4

In the special case in which concentration gradients cause ions in dilute solution to cross a cell membrane, it has been possible to show that the different ion species move independently of one another *except insofar as their movements create electrical potentials across the membrane capacitance.* Where such potentials arise, *electrical coupling* then affects movement of other ion species. 5

Additional coupling between ion movements may occur through osmotic forces. This mechanism, and the limitations placed on ion movements by the electroneutrality requirement, are discussed in detail in Chap. 5. 6

To explain how concentration differences can create a "battery," that is, an electrical force that can charge the membrane capacitance, let's consider the simplest possible case. 7

MEMBRANE THROUGH WHICH ONLY K^+ CAN PASS

A membrane through which some but not all substances can pass is said to be **semipermeable.** As a first example, let's imagine the semipermeable membrane of our typical cell to be permeable to only K^+ ions (see Fig. 4-10). 8

Notice that the electroneutrality and osmolarity requirements are taken care of by the presence of the other ions (in Fig. 4-10), even though these ions cannot pass through the membrane pores. 9

To illustrate the development of a membrane potential, imagine first that our typical cell is bathed not in the extracellular solution shown in Fig. 4-10, but in some experimental isos- 10

50

51

molar medium containing 130 mEq/L of K^+ ions. Clearly, in such a solution there would be no concentration gradient for K^+. Since no source of energy exists to separate charge across the membrane under these conditions, no membrane potential would occur.

Now imagine that we take this typical cell and drop it into a beaker of extracellular solution. What will happen? We now have a steep concentration gradient for K^+ ions (see Fig. 4-10). Consequently, K^+ ions will tend to leave the cell down this concentration gradient (which is the same as saying that the statistical probability of a single K^+ ion leaving the cell must be greater than the probability of another K^+ ion entering from the extracellular solution). This net outward movement of K^+ ions is a net movement (net flux) of charged particles, and you remember that this constitutes an outward ionic current (if not, see p. 47.1). In the initial moments after the cell is added to the low potassium solution, I_K must be positive. (See Table 4-2 for the relevant definitions of the terminology used in describing currents and ion fluxes.)

What happens next can be described in two theoretically identical ways: by a physicochemical analysis and by an electrical analysis. By the end of this chapter we hope that you will be equally happy with either form. However, since most students have some initial difficulty thinking in terms of electrical circuits, we start with the physicochemical description.

The outward net K^+ flux tends to leave an excess of anions inside the cell and to create an excess of cations outside the cell. Thus *charge separation occurs.* Since the "unmatched" excess ions in the two solutions (inside and outside) are mutually attracted, they collect opposite one another on either side of the membrane. This charge separation establishes a membrane potential (i.e., an electrical gradient across the membrane), in which the inside is negative relative to the outside. As the electrical gradient builds up, positive ions are increasingly attracted to the inside of the cell, so the efflux of K^+ diminishes while the K^+ influx increases. As long as the *net* K^+ flux is outward, the membrane potential continues to become more negative since there is charge separation involving more and more ions. **Finally, a point is reached at which the electrical gradient (the transmembrane potential) is sufficient to bring efflux and influx into balance. That is, the outward flux of K^+ through the pores due to the outward diffusional force is balanced by the additional inward flux of K^+ (through the pores) caused by the electrical force.** Since the electrical effect of even a small ion movement is large (we prove this shortly), equilibrium between efflux and influx is reached **without any experimentally detectable change in the internal K^+ concentration,** although obviously some (very small) net loss of K^+ must have occurred.

The magnitude of the membrane potential is a function of the magnitude of the concentration difference (diffusional force) **across the membrane.**

Thus, a greater concentration difference will create a larger, longer net flux, resulting in a larger membrane potential. This analysis is expressed mathematically in the *Nernst equation* (Eq. 4-6), which gives the membrane voltage E_K at which the net flux becomes zero for

Fig. 4-10. The "broken" region of cell membrane indicates a specific permeability of K^+ ions. Numbers in boxes are concentrations in milliequivalents per liter.

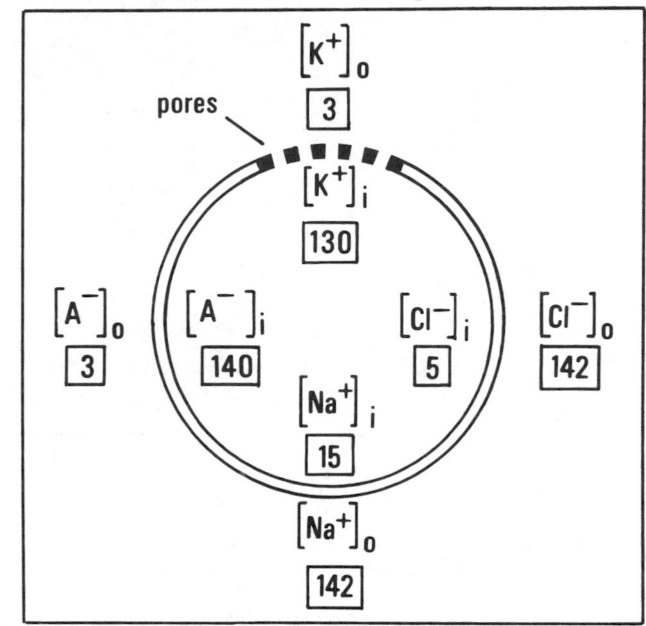

51

any given concentration gradient. Said in another way, **the Nernst equation gives the electrical force necessary to counteract the diffusional force produced by the concentration gradient across the membrane:**

$$E_K = \frac{RT}{FZ} \ln \frac{[K^+]_o}{[K^+]_i} \qquad \text{Eq. 4-6}$$

1 (See Table 4-4 for an explanation of these symbols.)

Where E_K is required in *millivolts*, assuming the temperature is 37°C and converting to logarithms to the base 10, we can obtain a more convenient form of the Nernst equation by combining all the constants into a single value:

$$E_K = \frac{61}{Z} \log \frac{[K^+]_o}{[K^+]_i} \qquad \text{Eq. 4-7}$$

From Eq. 4-7 we can calculate E_K in Fig. 4-10; the steps are as follows:

$$\frac{[K^+]_o}{[K^+]_i} = \frac{3}{130} = 0.023$$

$$\log 0.023 = -2 + 0.36 = -1.64$$

$$E_K = \frac{61}{+1}(-1.64) = -100 \text{ mV}$$

Note that the Nernst equation applies to an equilibrium condition, where free energy is at a minimum. 2

A purist might complain that E_K in these equations is the *dependent variable,* which therefore should have been noted as V_K, following the convention introduced in Chap. 3. However, later we use E_K as an *independent variable equivalent to the concentration ratios,* and it seems less confusing to maintain a single terminology at the expense (in this case) of mathematical rigor. (This is the exception to the V and E convention mentioned in Chap. 3.) 3

An exactly similar Nernst equation may be written for any other permeant ion species. In the case of the Na^+ ion, this would be 4

Table 4-4. Symbols in Eqs. 4-6 and 4-7

E_K = internal potential relative to an external ground (volts in Eq. 4-6, millivolts in Eq. 4-7). Students concerned with dimensional analysis of this equation should remember that voltage can be expressed as joules per coulomb (see page 28.1).

R = universal gas constant (joules per °K)

T = absolute temperature (°K)

F = Faraday constant (coulombs per mole of univalent ions)

Z = valence of ion in question (e.g., for Cl^-, $Z = -1$)

ln = natural logarithm (to base e)

log = logarithm to base 10

Note: Since the *ratio* of internal to external concentrations is used in these equations, the units of concentration need not be specified. See Aidley [1, p. 17] and Woodbury [62, p. 12] for derivations of the Nernst equation.

53

1 $$E_{Na} = \frac{61}{Z} \log \frac{[Na^+]_o}{[Na^+]_i}$$ Eq. 4-8

QUESTION: If you use the logarithms in Table 4-5, what would be the Nernst equilibrium potential for Na if $[Na^+]_o$ = 140 mEq/L and $[Na^+]_i$ = 14 mEq/L? (Hint 8↓) 2

3 QUESTION: Under what conditions will the Nernst potential be zero? (Hint 9↓)

The Nernst equation for anions is similar to that for cations:

$$E_{Cl} = \frac{61}{Z} \log \frac{[Cl^-]_o}{[Cl^-]_i}$$ Eq. 4-9

Note that **since the valence is −1, the sign of E_{Cl} is typically negative even though the denominator is smaller than the numerator.** 4

To avoid having to include the valence in the simplified Nernst equation, many books show the Nernst equation for chloride ions as

$$E_{Cl} = \log \frac{[Cl^-]_i}{[Cl^-]_o}$$ Eq. 4-10

Note that inverting the numerator and denominator changes the sign of E_{Cl} and thus compensates for the omission of the negative valence. We prefer, however, to leave the equations in their more general forms (in case you should ever want to calculate the equilibrium potentials for Ca^{2+} or SO_4^{2-}, for example). 5

The general version of the Nernst equation that is true for both cations and anions of any valence is thus (at 37°C)

$$E_j = \frac{61}{Z} \log \frac{[j]_o}{[j]_i}$$ Eq. 4-11

6 **You can solve any Nernst equation problem if you memorize the equation in this form.**

QUESTION: Calculate the Nernst potential for Cl when $[Cl^-]_o$ = 130 mEq/L and $[Cl^-]_i$ = 13 mEq/L. (Hint 10↓) 7

Question: Calculate the value of E_{Ca}, where $[Ca^{2+}]_o$ is 0.2 mEq/L and the free intracellular calcium ion concentration is 0.02 mEq/L. (Hint 12↓) 8

Table 4-5. Easy-To-Remember Logarithms

log 100 = +2.0
log 10 = +1.0
log 1 = 0
log 0.1 = −1.0
log 0.01 = −2.0

53

54

In the preceding examples, the concentrations were made multiples of 10 for easier calculation. **The Nernst potentials that would result from the concentrations of Fig. 4-11/4-9 if only one ion at a time were permeable are shown in Table 4-6.** Note that you still can predict the polarity and approximate magnitude of the potential on the basis of your work with the preceding examples.

To summarize this section: If a membrane has pores that allow only a single ion through the membrane and there is a concentration gradient for that ion across the membrane, then the resulting ionic fluxes can be readily imagined and the membrane potential that will develop can be quantitatively predicted by the Nernst equation for that ion. Memorize: **The Nernst equation gives the "battery potential" for any given ion.**

EQUIVALENT PHYSICOCHEMICAL AND ELECTRICAL CIRCUIT DESCRIPTIONS OF IONIC MOVEMENTS

Although the description just given provides an adequate picture of the physicochemical approach to equilibrium, some useful additional insights are obtained when we describe the same process in electrical terms. **The electrical circuit analogue that follows is fully as valid as the previous physicochemical description.** Then let us return to the case in which only K^+ can cross the membrane (Fig. 4-10).

We have shown (see page 51.4) that a *net flux* occurs where the influx of an ion is different from the efflux, with this net flux being equivalent to an electrical current (since it is a net movement of charged particles). When our typical cell is placed in a low-potassium medium, an outward (positive) I_K occurs because of the differences in concentration of K^+ on either side of the membrane. But how can a current flow unless an electromotive force is applied? Where is the elusive "battery" that can create an outward potassium current through resistive membrane pores?

Think back to the physicochemical description. The larger the concentration gradient across the cell membrane, the larger the initial net flux will be and hence the larger this "hidden electromotive force" must be. Can you intuit that **this battery potential is none other than the Nernst potential calculated for the concentration gradient in question?**

Fig. 4-12 shows the appropriate electrical circuits drawn onto Fig. 4-10; it represents the electrical analog of the typical cell permeable only to K^+. It may be more familiar if we "unbend" it, since it is electrically identical to Fig. 4-13/4-5. Let's see how much you remember about this simple circuit from your reading of Chap. 3:

1. Notice that as a result of the battery E_K, current I flows in a complete circuit through the battery, the resistive pore R, and the membrane capacitance C. Thus I_{por} and I_{cap} must be of the same absolute magnitude but opposite sign. (Reread pages 47.9 to 48.1 if you found that last sentence difficult to understand.)

Fig. 4-11/4-9. Typical ionic distributions. Concentrations are in milliequivalents per liter; A^- is collective term for all anions other than Cl^-. $[Y]_i$ = internal concentration of ion "Y," $[Y]_o$ = external concentration of ion "Y."

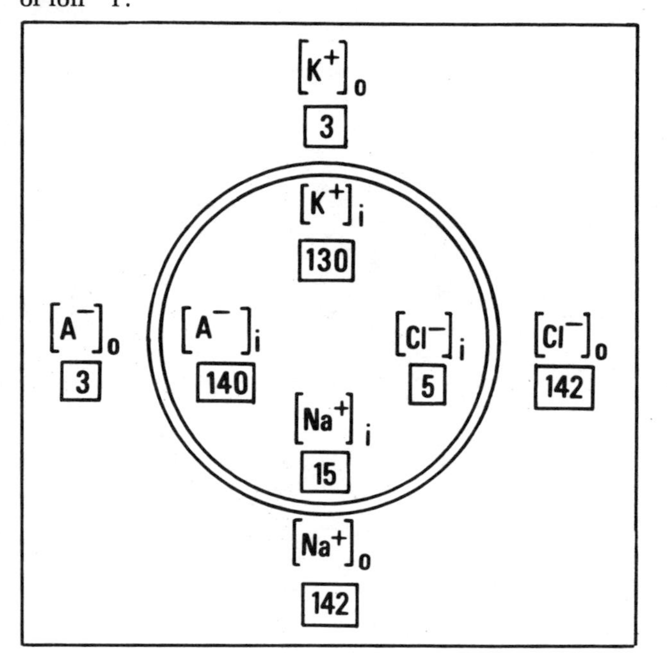

Table 4-6. Nernst Potentials for the "Cell" of Fig. 4-11/4-9

$$E_K = \frac{61}{+1} \log \frac{3}{130} = -100$$

$$E_{Na} = \frac{61}{+1} \log \frac{142}{15} = +60$$

$$E_{Cl} = \frac{61}{-1} \log \frac{142}{5} = -89$$

54

2. Notice also that current flow in this circuit will be determined by the *driving force* and the *pore resistance*. Remember that the driving force is simply the *difference* between the battery voltage E_K and the charge on the membrane capacitor V. Current flow can be obtained from a simple equation that turns out to be of major importance in this and later chapters. We call it the **ionic Ohm's law:**

$$I_K = g_K(V - E_K) \qquad \text{Eq. 4-12}$$

whose symbols are defined in Table 4-7. (This is derived directly from

$$I_K = \frac{V - E_K}{R} \qquad \text{Eq. 4-13}$$

by exchanging conductance g_K for resistance R, since $g_K = 1/R$.)

Clearly, when the cell is first placed in the low-potassium extracellular solution, the capacitance will be unchanged and V will be zero. However, as K^+ ions flow *outward* across the membrane ($+I_{por}$), capacitative current will flow *inward* across the membrane ($-I_{cap}$), tending to charge the capacitor toward E_K. As V approaches E_K, $V - E_K$ will go to zero and hence I_K will fall to zero. Finally, equilibrium will be established when V becomes equal to E_K. Thus in both the electrical and physicochemical explanations, net current (net flux) tends to zero as equilibrium is approached.

QUESTION: According to Eq. 4-12, what is I_K at equilibrium, when the flux due to diffusional force is equal and opposite to the flux due to the electrical force? (Hint 13↓)

Fig. 4-12. Electrical current pathway in hypothetical cell permeable only to K^+.

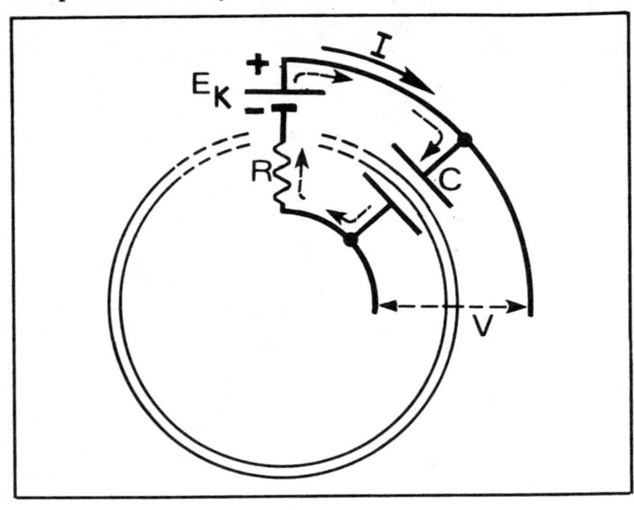

Fig. 4-13/4-5. Elements of battery, resistance, and capacitance in the membrane; V represents recordable membrane potential.

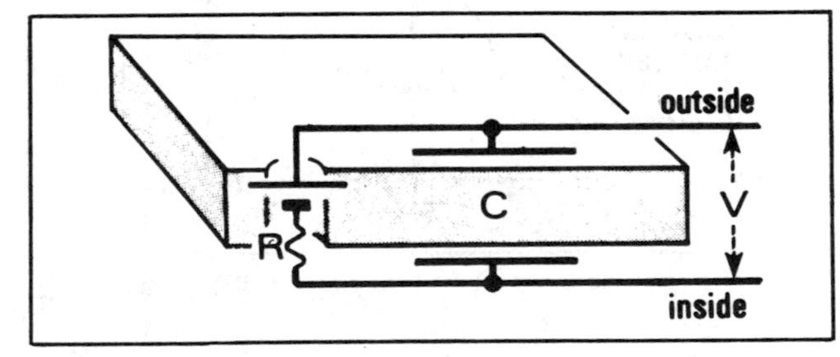

HINTS

8. $E_{Na} = 61 \log 10 = 61(+1) = +61$ mV. Thus, you can readily see that **if** in the Nernst equation for a given cation X^+ **the numerator is larger** than the denominator (as in the case of Na^+ just given), **then the potential E_X will be positive,** whereas **if the denominator is the larger, the Nernst potential will be negative** (as in the case of K^+).
9. When the numerator and denominator are equal, since $\log 1 = 0$ (Table 4-5). Does this agree with what you have already learned from the physicochemical description? (Hint 11 ↓)
10. $E_{Cl} = (61/-1) \log 10 = -61$ mV. If you missed this, go back to Table 4-5 and Eq. 4-9 and check your arithmetic.
12. $E_{Ca} = (61/+2) \log 10 = 30.5$ mV. Do you find it strange that a divalent ion gives a smaller Nernst potential for a given concentration gradient? If so, see Hint 14. ↓

56

Note that **Eq. 4-12 holds at all times, both before and at equilibrium.** Thus you can see that you have already verified that Eq. 4-12 holds at the instant when the cell is immersed (when $V = 0$) and when $V = E_K$ (at equilibrium). Needless to say, Eq. 4-12 also *holds for all the times in between.*

1

Note that **since both conductance and resistance are always positive, the sign of I_K must be the sign of the difference between V and E_K.** Thus $V - E_K$ defines the direction of the pore current I_K. Note further that according to this result, I_K is always positive in our typical cell as it goes from $V = 0$ to $V = E_K$, since under these circumstances V is always less negative than E_K. This means that for our typical cell, the pore current is always outward (positive), which in turn will drive an inward current through the capacitance [since $I_{por} = -I_{cap}$ (Eq. 4-1)]. The inward capacitative current makes V more negative, so V approaches E_K.

2

Table 4-7. Definitions of Terms of Eq. 4-12

I_K = pore current for K^+
g_K = conductance for K^+ (inverse of resistance of K^+ pore)
V = membrane potential
E_K = Nernst potential for K^+

At first glance some students may find it strange that a negative battery E_K gives a positive pore current I_K. A few moments' reflection reveals that in an electrical circuit, **current** (positive charges) **moves from the positive pole of the battery to the negative pole in the circuit *external* to the battery.** However, **the current *within* the battery moves in the opposite direction:** from the negative to the positive pole. Thus, our diagrams showing E_K as negative (negative pole inside the cell) are correct since I_K is the pore current, i.e., the current within the battery.

3

Since V is the charge on the membrane capacitance (for which $Q = CV$ holds) and since I is just dQ/dt, you can see that the magnitude of I is proportional to dV/dt. Thus, when the membrane potential is far from E_K, it moves more rapidly toward E_K than when it is closer: it follows the typical exponential curve of an RC circuit, as in Fig. 3-8.

4

Fig. 4-14. Extra battery added to K^+ circuit.

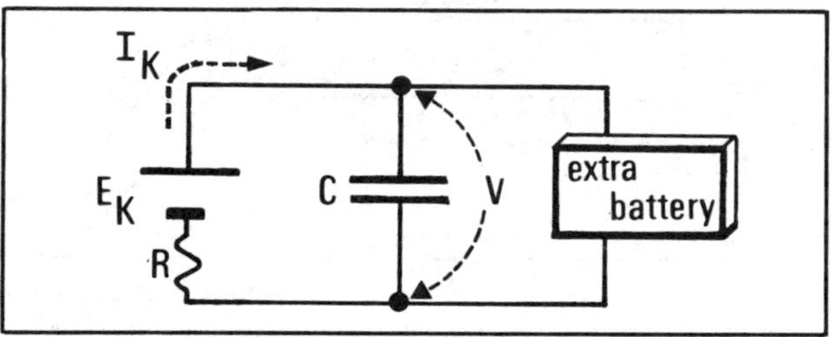

Equation 4-12 also holds when the membrane potential is being maintained at some value other than E_K by another battery, as in Fig. 4-14. If the "extra battery" maintains the membrane potential V at E_K, then I_K will be zero, as predicted by Eq. 4-12. But the farther away from E_K that V is maintained by the extra battery, the greater I_K will be.

5

The details of how such an extra battery operates are given later in this chapter.

6

QUESTION: If V is maintained at a value *less* negative than E_K, what will be the direction of I_K? (Hint 16↓)

7

QUESTION: If V is maintained at a value *more* negative than E_K, what is the direction of current through the pore? (Hint 15↓)

8

If you have been following closely, you have noticed that driving force was defined as $E - V$ in Eq. 3-6 and as $V - E$ in Eqs. 4-12 and 4-13. Both versions are equally legitimate

9

56

57

since the driving force is the *difference* between E and V. **The formula used here is followed throughout the remainder of this book since the sign of the current obtained from Eq. 4-12 then conforms to our standard convention whereby outward is positive and inward is negative.**

1

For every ion there can be an ionic Ohm's law. The general expression for the ion j is

$$I_j = g_j(V - E_j) \qquad \text{Generalized ionic Ohm's law} \quad \text{Eq. 4-14}$$

You will find this generalized ionic Ohm's law very useful because it clearly indicates the following, which *we recommend that you memorize:*

1. *The sign* of a pore current is determined by the *sign* of the driving force.
2. The *magnitude* of a pore current is determined by both the *conductance* and the *driving force*.

Doubters are invited to try for themselves different combinations of V and E_j (with both anions and cations) to discover that this equation yields the correct polarity of current for all cases. If the membrane is not permeable for an ion, that just means that the conductance is zero; hence the current must be zero, whatever the driving force. Alternatively, the current will also be zero when the ion is not present, since its driving force is then zero, no matter what the value of the conductance is.

2

IONS THAT CARRY I_{cap}

As you will now see, whereas I_{por} is always a specific ion, **I_{cap} is very nonspecific, being carried by any available charge carriers near the membrane.**

3

Although many students understand that the electrical and physicochemical descriptions are formally identical, nevertheless they have difficulty in visualizing I_{cap} in terms of ion movements. Let's go through a physicochemical description of I_{cap} in the situation just considered. If K^+ ions have left the cell, the intracellular and extracellular solutions must contain excess anions and cations, respectively. These excess ions initially confer an equal, but opposite, net charge on the two solutions. Cations from the extracellular solution and anions from the intracellular solution thus are at-

4

Fig. 4-15. Possible ionic composition of I_{cap} in Fig. 4-12. Width of each arrow indicates an estimate of contribution of that ion to I_{cap}. Note that the relative importance of a given ion is different on the two sides of the membrane. This reinforces the idea that the ions which carry I_{cap} do not necessarily cross the membrane.

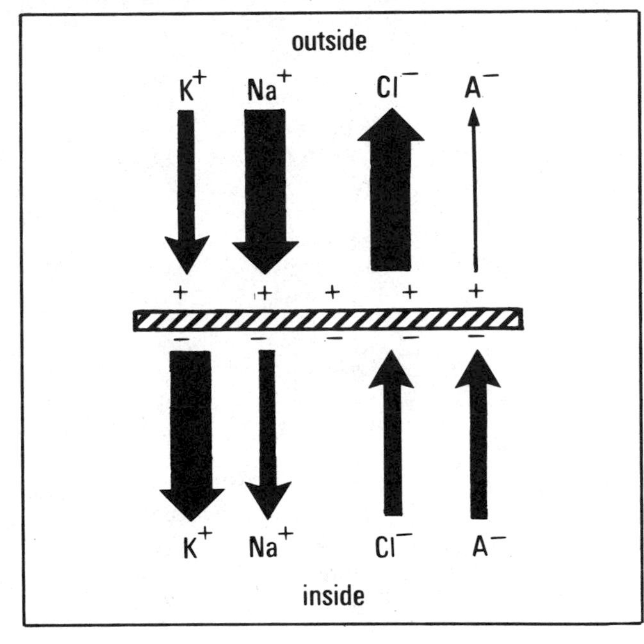

HINTS

11. You bet! The only time that the inward and outward fluxes can be equal across a membrane permeable only to a single ion *without* a membrane potential is when the concentrations inside and outside are equal.

13. $I_K = 0$, since at equilibrium $V = E_K$, and hence $V - E_K = 0$.
14. Try thinking of it this way: The divalent ion is more strongly attracted by a given electrical field, hence the fluxes become equal at a lower electrical gradient.

57

58

tracted to one another across the membrane and line up on opposite sides of the capacitative membrane until the bulk of each solution no longer carries a net charge (i.e., is electroneutral once more). 1

But when cations approach the membrane from the outside and cations leave the region immediately inside the membrane (producing an anion excess in that region), clearly **as long as this process continues, it is as if positive charge actually crossed the membrane.** Thus, **even though no ions can pass through the membrane capacitance, current *can* pass through!** Notice that it makes no difference which specific ions contribute to this I_{cap}. For example, Na^+ may approach the membrane from the outside while K^+ may leave the area inside the membrane; identical results are obtained by anions moving in the opposite directions. The actual ion movements that might be expected to contribute to I_{cap} in Fig. 4-12 are shown in Fig. 4-15. 2

Question: What do you guess was the basis for the width of the arrows in Fig. 4-15? (Hint 18↓) 3

SUMMARY OF EQUIVALENT DESCRIPTIONS OF THE MEMBRANE POTENTIAL

You have now seen the same phenomenon—the development of the membrane potential from a basic difference in ionic concentrations across a membrane with pores—described in two different ways: the physicochemical description and the equivalent electrical circuit description. 4

A comparison of the main points of the two descriptions is summarized in Table 4-8, together with the symbolic language that we require for quantitative analysis of the electrical analog (fourth level). While the two descriptions are completely equivalent, as the story develops and additional details are introduced, the electrical description becomes increasingly easier to deal with. This is the reason for placing such emphasis on electrical circuits in the preceding material. 5

NUMBER OF IONS NEEDED TO ESTABLISH MEMBRANE POTENTIAL

It is important to note that **the concentration gradient** (and hence the Nernst potential) **is not noticeably affected by the number of ions that must cross the membrane to produce the electrical gradient required to bring net flux to zero.** 6

It turns out that **the membrane capacitance is so small that the number of ions necessary to change the charge on the capacitance by about 0.1 V** (100 mV) **is very, very small compared with the number of ions available in the intracellular and extracellular solutions.** Thus there is no discernible concentration change when our typical cell is suddenly immersed in a low-potassium solution. 7

Fortunately, you do not have to accept this assumption simply as an article of faith. In Chap. 3 we showed that it takes 6×10^{11} charges (ions) per square centimeter to charge a 1-μF capacitor (membrane capacitance is typically 1 μF/cm^2) 8

Table 4-8. Three Equivalent Descriptions of the Movement of Ions across Membranes

Physicochemical Description	Electrical Circuit Description	Symbolic Description
1. Mobile ions in the extracellular and intracellular fluids	1. Conductive paths between the battery (with internal resistance) and the membrane capacitance	1. Line connecting the circuit elements (assumed to have negligible resistance unless otherwise stated)
2. Diffusional force (resulting from concentration differences)	2. Battery (resulting from concentration differences)	2. E_j
3. Pore restrictions on ion movement	3. Resistance to ionic flow	3. $R_j, \frac{1}{g_j}$
4. Charge separation (excess ions collecting on either side of membrane) gives membrane potential.	4. Membrane potential V: the potential due to the charge stored on the membrane capacitance	4. V
5. Net flux occurs for a given ion species whenever electrical and diffusional forces on that ion species are not equally balanced.	5. Ionic current flows whenever the battery potential for a given ion species is not the same as the back electromotive force on	5. $I_j = \frac{V - E_j}{R_j}$ $I_j = g_j(V - E_j)$

to a potential of 100 mV. Thus it would take a total flux of 6×10^{11} K^+ ions per square centimeter of membrane surface to charge one typical cell to E_K from $V = 0$ mV. Let's assume this cell is a sphere of diameter 10 μm. Then its surface area is πd^2, or 3.14×10^{-6} cm^2, and the total K^+ efflux will be 1.9×10^6 K^+ ions. However, cell volume is $\pi d^3/6$, or 5.2×10^{-10} cm^3; hence the total number of K^+ ions contained in this cell will be $(130 \times 10^{-6}$ mol/cm$^3) \times (5.2 \times 10^{-10}$ cm$^3) \times (6 \times 10^{23}$ ions/mol$) = 4 \times 10^{10}$ ions. *Thus the loss of K^+ ions is about $[(2 \times 10^6)/(4 \times 10^{10})]$ 100 = 0.005 percent of the potassium content of the cell—clearly well below the limits of experimental detection by chemical analysis.*

Thus, in calculating the final membrane potential from the Nernst equation, you need to know only the concentrations inside and out; **these concentrations will not change significantly when there is a temporary net flux of ions as the membrane potential develops.**

MEMBRANE THROUGH WHICH BOTH Na^+ AND K^+ CAN PASS

The previous section was something of a warm-up, since it has been known since the late 1940s that **excitable cells are rather freely permeable to both Na^+ and K^+.** (However, Cl^- conductance varies quite markedly among different types of cells, being very low in some cells and higher than the resting potassium conductance in others.) These observations might lead you to suppose (quite correctly) that **the key to the resting membrane potential lies primarily in the behavior of Na^+ and K^+.**

Let us start by considering the cell shown in Fig. 4-16, which has two sets of pores, one set permeable to K^+ and another set permeable to Na^+. Now imagine that such a cell is immersed in a solution containing 130 mEq/L of K^+ and 15 mEq/L of Na^+ instead of the extracellular solution shown in that figure. **Such a cell would have no membrane potential since there would be no net flux of either Na^+ or K^+.** Now we take this cell and drop it into a beaker of "extracellular solution" with the concentrations shown in Fig. 4-16. What will happen?

Clearly, Na^+ will enter the cell and K^+ will leave the cell, but will any potential develop across the cell membrane? Can you see that this depends on the relative rates of Na^+ entry and K^+ loss? If Na^+ entry exactly balances K^+ loss *at all times*, then there can be no charge separation, no matter how large the fluxes nor for how long they occur!

	the membrane capacitance (membrane potential).	
6. Conversely, no net ion flux occurs if the electrical and diffusional forces are equal and opposite.	6. No ionic current flows when the battery potential is equal in magnitude to the potential across the capacitance.	6. If $\lvert V \rvert = \lvert E_j \rvert$, then $I_j = 0$.
7. All net flux through the pores contributes to charge separation.	7. All battery currents travel in complete circuits through the membrane capacitance.	7. $I_{por} + I_{cap} = 0$, also implying $I_{por} = -I_{cap}$ and $\lvert I_{por} \rvert = \lvert I_{cap} \rvert$.

Note: The third column (Symbolic Description) is developed more fully later in this chapter.

HINTS

15. The current I_K is negative, which means that there is net influx through the membrane resistance. This can be verified by assuming that $V = -120$ mV while $E_K = -100$ mV. If you can't finish this idea, then go on to Hint 17.↓

16. Positive $(+I_K)$; that is, there will be a net current out of the cell. For example, if $V = -40$ mV and $E_K = -100$ mV, then I_K is positive since $V - E_K = (-40) - (-100) = +60$.

However, if initially Na^+ entry were slightly slower than the K^+ loss, then the cell would become negative inside as a result of the net flux of cations in an outward direction. The resultant internal negativity would tend to speed Na^+ entry and retard K^+ loss. **A balance under these conditions would finally be achieved when the membrane potential increased sufficiently to equalize the Na^+ and K^+ fluxes—then no further charge separation would occur.**

QUESTION: How can K^+ and Na^+ give opposite effects, one making the inside of the cell negative, the other positive, when they are both positive ions? (Hint 19↓)

At this point you can pretty well predict that since we have described a membrane permeable to two ions from the physicochemical standpoint, now we will launch into the electrical description! If you predicted it, you were right!

The electrical description facilitates *quantitative* understanding of the interaction between the two ionic fluxes and markedly simplifies description of the *dynamic* aspects of the membrane potential (which we take up in later chapters).

Figure 4-17 shows the electrical analog of Fig. 4-16, where the membrane has two kinds of pores: one permeable to K^+ and the other permeable to Na^+. If we "unwind" Fig. 4-17, we get Fig. 4-18, in which we also replaced the symbol *R* with its equivalent, 1/*g*. Since the ionic gradients for Na^+ and K^+ shown in Fig. 4-16 are in opposite directions across the cell membrane, it is clear that the Nernst potentials for these ions will be of opposite sign. Thus in Figs. 4-17 and 4-18, **the "sodium battery" is inwardly directed while the "potassium battery" faces outward across the cell membrane.** If these batteries face in opposite directions, then the ionic currents that they generate will tend to charge the membrane capacitor in opposite directions.

Now it can be revealed: The "extra battery" described earlier was none other than the Na^+ battery in disguise!

If we define the current generated by the Na^+ battery as I_{Na} and the current generated by the K^+ pores as I_K, then **the net current across the membrane capacitor (I_{cap}) is determined by the net pore current I_{por})**, that is, by the *algebraic sum* of I_{Na} and I_K. (Review definitions provided in Table 4-2, and remember that all currents flow in complete circuits.)

Thus it remains true that

$$I_{por} + I_{cap} = I_{mem} = 0 \qquad \text{Eq. 4-15/4-3}$$

But since $I_{por} = I_{Na} + I_K$, we can now write

Fig. 4-16. "Broken" regions of the cell membrane indicate specific permeabilities to K^+ and Na^+. Concentrations are in millequivalents per liter.

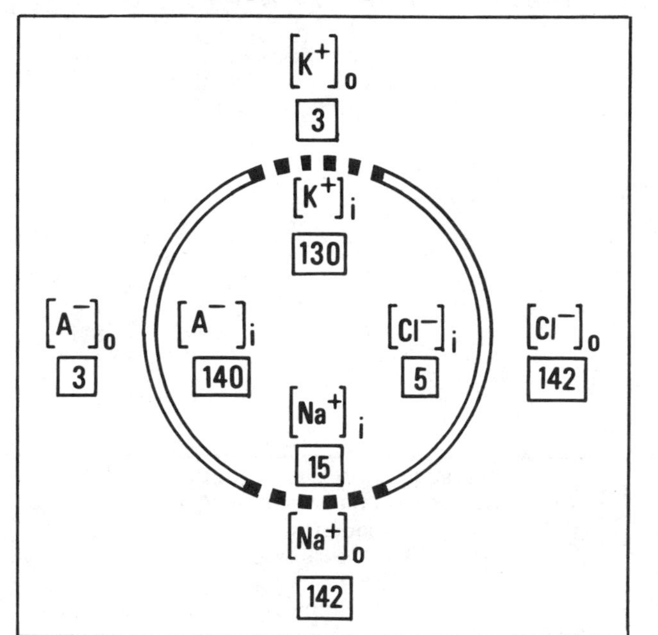

Fig. 4-17. The electrical analog of the "typical cell" permeable to both Na^+ and K^+. Symbols are as in Fig. 4-12. Note that E_{Na} and E_K face in opposite directions relative to cell membrane.

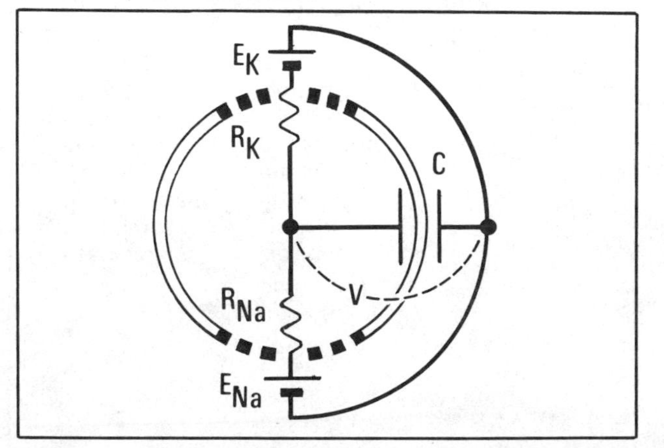

$I_{Na} + I_K + I_{cap} = 0$ Eq. 4-16

QUESTION: If $I_{Na} = -0.3$ mA and $I_K = 1$ mA, what is the magnitude of the capacitative current across the cell membrane? Is this an inward or an outward current? (See Hint 21.↓)

In our physiocochemical description (page 60.1), we concluded that a stable, resting membrane potential would be reached when Na^+ entry became equal to K^+ loss. Similarly, **Eq. 4-16 points out that when $I_K + I_{Na} = 0$, I_{cap} must be zero. Where there is no current across the capacitor, the membrane potential will remain constant, since the charge on the capacitor cannot change** (recall this previously, page 48.3).

For the special condition that the membrane potential is not changing (i.e., that the steady-state resting membrane potential has been reached), Eq. 4-16 reduces to

$I_{Na} + I_K = 0$ Eq. 4-17

But each of these ion-specific currents can be defined in terms of a specific conductance and driving force (see page 55.1). Thus, from the equations of the ionic Ohm's law

$I_{Na} = g_{Na}(V - E_{Na})$

$I_K = g_K(V - E_K)$ Eq. 4-18

(If the derivation of these is unclear, see Eq. 4-14.)

Equations 4-18 are especially valuable since they **hold at all times**—whether the membrane potential is changing or not; whether V is more than, less than, or equal to any given E_j; and even when V is determined by some other "battery" (or even by a stimulating electrode).

Question: In Fig. 4-19/4-14, could the extra battery shown be E_{Na}? (Hint 23↓)

Even though the ionic Ohm's law holds at all times, you must remember that we are using it here for a very special condition, the situation of Eq. 4-17, in which $I_{cap} = 0$. To remind you of this, we add a subscript to indicate that V is here V_s, the **steady-state resting potential.**

Fig. 4-18. Electrical circuit drawing of Fig. 4-17.

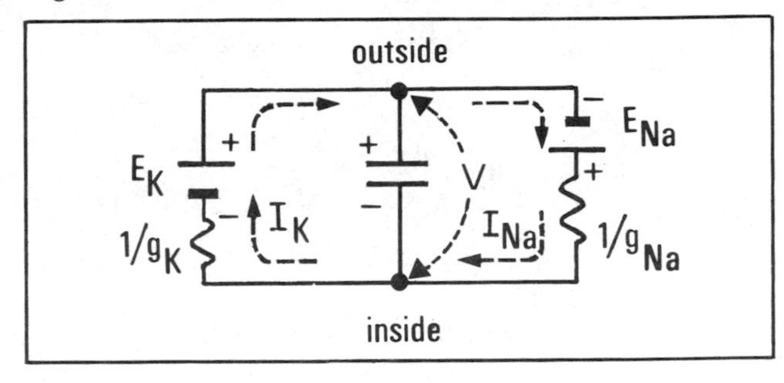

Fig. 4-19/4-14. Extra battery added to K^+ circuit.

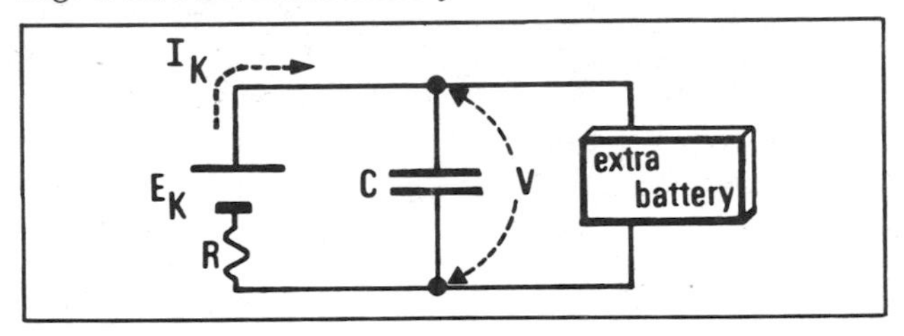

HINTS

17. Now $V - E_K = -120 - (-100) = -20$; hence the sign of I_K must be negative. This is the equivalent of the "charging" of E_K by the "extra battery."
18. Both the availability of the ion (concentration) and its ability to move readily through the solution (mobility).

When Eqs. 4-18 are substituted into 4-17, we get

$$g_{Na}(V_s - E_{Na}) + g_K(V_s - E_K) = 0 \qquad \text{Eq. 4-19}$$

which can readily be solved algebraically for V_s, the resting membrane potential:

$$V_s = \frac{g_{Na} E_{Na} + g_K E_K}{g_{Na} + g_K} \qquad \text{Steady-state equation} \quad \text{Eq. 4-20}$$

Equation 4-20 is so useful for the estimation of resting potentials in steady-state conditions that we can dignify it with the name **steady-state equation.** (When you use this equation, V_s is obtained in millivolts if Nernst potentials are given in millivolts. The units used for the conductances must be consistent throughout the equation, but it does not matter what unit is chosen. A correct result can be obtained simply from the *ratio* of g_{Na} to g_K!) 1

QUESTION: If $g_{Na} = 0.02$ mmho/cm², $g_K = 0.26$ mmho/cm², $E_{Na} = +60$ mV, and $E_K = -100$ mV, what is the value of the resting membrane potential in this cell? (Hint 24↓) 2

QUESTION: What would V_s have been if g_K had been the same as g_{Na} (other values as in the previous question)? (Hint 26↓) 3

At this point, you might very well ask (incorrectly, on the basis of intuition alone) why V_s is not zero when the conductances are equal. The answer is that the magnitude of I_j is determined not only by g_j, but also by $V - E_j$. Since E_{Na} is not equal to E_K, this means that the midpoint between E_{Na} and E_K is not equal to zero. (Notice, once again, that both physicochemical and electrical descriptions are equivalent and lead to the same conclusion.) 4

Question: At what value of g_{Na}/g_K does $V_s = 0$, assuming $E_K = -100$ and $E_{Na} = +60$? (Hint 27 ↓) 5

It is not the absolute value but **the ratio of conductances** that **determines the resting potential,** as you can see from the previous examples. 6

This fact is easily shown algebraically by dividing both the numerator and denominator of the right-hand side of Eq. 4-20 by g_K:

$$V_s = \frac{(g_{Na}/g_K)E_{Na} + E_K}{g_{Na}/g_K + 1} \qquad \text{Eq. 4-21}$$

Since E_{Na} and E_K are constants, only g_{Na}/g_K can affect V_s. 7

63

The steady-state equation not only provides a simple, *quantitative* approach to the steady-state resting potential, but also **provides a method to determine the effects on the resting potential of conductance changes or changes in ion concentration.** 1

You can readily verify the following generalizations, which are implicit in the steady-state equation:

1. **If the conductance for an ion is increased, the effect of that ion on resting membrane potential is increased.** As g_{Na} becomes extremely large relative to g_K, V_s approaches E_{Na}. Conversely, as g_K becomes large relative to g_{Na}, V_s approaches E_K. (Don't accept this on faith—assume $E_{Na} = +60$ and $E_K = -100$, substitute conductances into the steady-state equation, and see what happens.) 2

QUESTION: If $E_{Na} = +60$ and $E_K = -100$, what is the approximate value of V_s if g_{Na} is 100 times g_K? (Hint 25↓) 3

2. **If the *gradient* for an ion is increased, the effect of that ion on the resting membrane potential is increased.** If the concentration gradient across the membrane increases, then the absolute value of the Nernst potential must increase (i.e., become either *more* positive or *more* negative and hence the driving force increases (to be more positive or more negative). For example, if $[Na^+]_o$ goes up, then E_{Na} will become more positive than 4

HINTS

19. Charge separation across the membrane occurs when ion fluxes produce any departure from electroneutrality in the intracellular and extracellular solutions. Thus **net cation loss** (K^+ loss larger than Na^+ entry) **leads to internal negativity, whereas net cation gain** (K^+ loss less than Na^+ entry) **leads to internal positivity.** Remember, however, that (1) the excess cations or anions in the extracellular and intracellular solutions are not freely distributed, but line up on both sides of the membrane, and (2) the total number of such excess ions is extremely small, for example, only 6×10^{11} ions/cm² if the membrane capacitance is 1 μF/cm².

    But how can we be so sure that the excess ions are spread out along the membrane rather than being distributed through the intracellular and extracellular solutions? If you cannot give yourself a satisfactory answer to this question, see Hint 22. ↓

21. From Eq. 4-16,

$$I_{cap} = -(I_{Na} + I_K)$$
$$= 0.3 \text{ mA} - 1 \text{ mA}$$
$$= -0.7 \text{ mA}$$

    Hence I_{cap} must be an *inward* current. In which direction would this current tend to charge the cell membrane capacitance? (Hint 20↓)

23. Why not? As we stated several times, the ionic Ohm's law equations hold under all circumstances.

28. $$V_s = \frac{0.02(+60) + 0.26(-100)}{0.02 + 0.26}$$
$$= \frac{1.2 - 26.0}{0.28} = -\frac{24.8}{0.28}$$
$$= -88.6 \text{ mV}$$

29. Note that the calculation is the same no matter what the absolute values of g_{Na} and g_K are:

$$\frac{100(+60) + 1(-100)}{100 + 1} = \frac{6000 - 100}{101}$$
$$= \frac{5900}{101} = 58.4 \text{ mV}$$

63

64

+60 mV and $V_s - E_{Na}$ will have a larger absolute value (in this case, a larger negative number). The effect will be to increase I_{Na}. If g_{Na} and g_K remain unchanged, balance can be achieved only when the potassium driving force has increased to match the increase in I_{Na}, so V_s will move toward E_{Na}. You needn't accept this on faith. Hold g_{Na} and g_K constant and substitute some values for E_{Na} and E_K into the steady-state equation to see what happens.

1

These two rules provide a sufficient understanding to carry you a long way through both clinical and experimental problems. Now we offer you some practice in their use.

2

Here are the concepts you will need to be familiar with in order to answer easily the questions to follow:

1. The direction of the concentration gradients of Na^+ and K^+.
2. How to calculate the "battery potential" E_j for a given concentration gradient (from the Nernst equation). In many instances, you need only estimate the direction in which the battery potential changes as a result of a given change in the ionic gradient.
3. What determines I_j (that is, g_j and $V_s - E_j$).
4. How the direction of current flow through the membrane capacitance is related to direction of flow through a pore (that is, $I_{por} = I_{cap}$).
5. How a given current flow through the capacitance affects V.
6. How all the actions of g_j and E_j can be summarized in the steady-state equation.

If you find any of the above puzzling, we recommend a brief, mind-refreshing review of the previous parts of the chapter!

3

Because we are going to be dealing with the physiology of membrane potentials, we might as well phrase the questions in the standard physiological terminology shown in Table 4-9.

4

In the following questions, assume we are dealing with a membrane that is permeable to two ions, K^+ and Na^+, and that the concentration gradients are as diagrammed in Fig. 4-20/4-16. While these questions are meant to illustrate how easy it is to use the two rules described above, remember that **if you get stuck, you can always use the steady-state equation to find the answer.**

5

QUESTION: If $[K^+]_o$ is increased, the membrane will become hyperpolarized—*true or false?* (Hint 30↓)

6

QUESTION: If g_{Na} is increased, the membrane will become depolarized—*true or false?* (Hint 34↓)

7

QUESTION: If $[Na^+]_i$ is increased, the membrane becomes depolarized—*true or false?* (Hint 32↓)

8

Table 4-9. Definitions of Terms Used to Describe Membrane Potentials

Membrane potential	= *potential difference* across the membrane. Thus the term *increase in membrane potential* properly describes either a change of $V = -70$ to $V = -90$ or a change of $V = +20$ to $V = +40$ mV.
$\lvert V \rvert$	= *absolute* value of membrane potential, regardless of sign
Polarized (unless otherwise qualified)	= *normal direction of membrane polarization* (i.e., inside is negative)
Depolarized	= *less polarized*. Thus a cell whose normal resting potential is −80 mV would be described as depolarized if $V = -70, -10, +20$, or even +100 mV.
Hyperpolarized	= *more polarized*. Thus a cell whose normal resting potential is −80 mV would be described as hyperpolarized if $V = -85, -100$ mV, etc.

Note: The terms *depolarization* and *hyperpolarization* describe *unambiguously* the direction of change of membrane potential. Apparently equivalent words such as *decrease* or *increase* and *fall* or *rise* are *ambiguous* and do not always make clear the direction of change that has occurred. So we confine ourselves to the terms defined above. (But you should be aware that the real world, including many examination questions, may prove more confusing!)

64

Now don't give up! If you can work these problems with confidence, you have understood the essence of this chapter: what factors interact to determine the membrane potential. If you have mastered this terminology and these ideas, you have learned what is considered by many to be the hardest part of neurophysiology!

QUESTION: If $[Na^+]_o$ decreases, the membrane will become depolarized—*true or false?* (Hint 31↓)

QUESTION: If g_K decreases, the membrane depolarizes—*true or false?* (Hint 33↓)

At this point, you can see that the two batteries E_K and E_{Na} oppose each other in their actions. *Electrically* speaking, what determines which one predominates? Given a circuit such as that of Fig. 4-21/4-18, it is easy to predict that the one with the lowest internal resistance (highest conductance) will be the major factor, although both must always play a role. The battery with a high resistance (low conductance) can deliver only a small amount of current (for a fixed, small electromotive force), compared to the amount of current delivered by a battery (with a similar electromotive force) that

Fig. 4-20/4-16. "Broken" regions of cell membrane indicate specific permeabilities to K^+ and Na^+. Concentrations are in milliequivalents per liter.

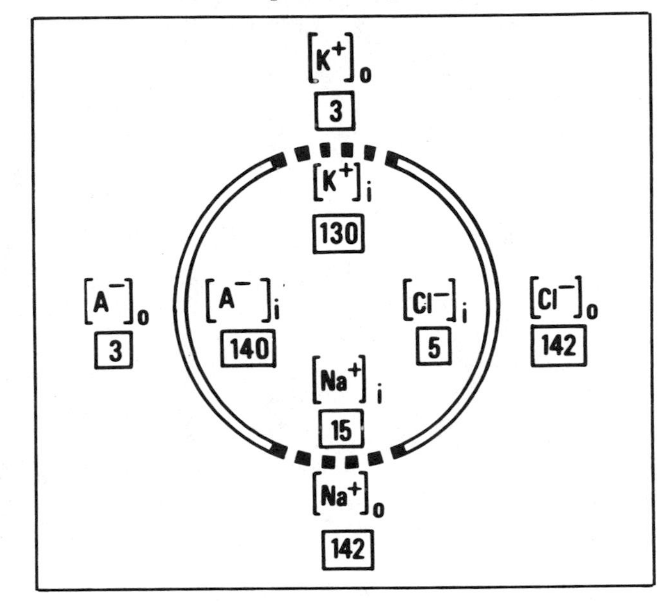

Fig. 4-21/4-18. Electrical circuit applicable to Fig. 4-20/4-16.

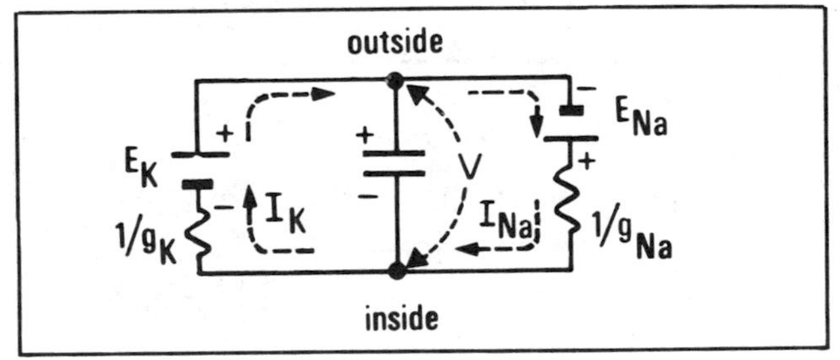

HINTS

20. The inside is negative relative to the outside.
22. Electroneutrality is not absolute in the sense that anions and cations are always exactly equally spaced. On the contrary, it is a statistical concept in the sense that any given area must average out to be electroneutral. Any area with too few anions or cations will produce an electrical field tending to attract those ions required to restore the balance. Thus if some cations leave the cell and enter the extracellular solution, the only final *stable* situation will occur when the bulk of the extracellular solution is once more statistically electroneutral with the excess cations *matched up across the membrane* against the excess anions left behind in the intracellular solution. The cations that line up along the membrane are, of course, not likely to be the same ions that left the cell; any ion will do as long as it carries positive charge (see page 58.2). If you are not clear on this, now would be a good time to reread the section describing the ionic basis of I_{cap} (pages 57.3 to 58.3).
24. −88.6 mV. To check your calculation, see Hint 28.↑
25. +58.4 mV (which *approaches* $E_{Na} = +60$). If you still doubt that V can come even closer to E_{Na}, try it again with g_{Na} being 1000 times g_K. Calculations for the "100 times" example are shown in Hint 29.↑
26. Whether you assume g_K and g_{Na} to be 0.02 or 0.26 or any other value, V_s will turn out to be −20 mV. Work it through yourself if you find this difficult to believe!
27. The answer can be obtained by substituting in Eq. 4-20. However, it is easier to start with Eq. 4-19 as follows:

$$g_{Na}(V_s - E_{Na}) + g_K(V_s - E_K) = 0$$

$$\frac{g_{Na}}{g_K} = -\frac{V_s - E_K}{V_s - E_{Na}} = -\frac{+100}{-60} = 1.67$$

66

is a *constant-voltage* source, i.e., that has a low resistance. Since the charge on the membrane capacitance is determined by the amount of charge over time (total current) that has passed through it, **the battery that can deliver the greater current will have the greater influence on the membrane potential.** 1

BALANCE-BEAM MODEL

Now that you have learned the rules, we can offer you a visual model that may be helpful. **Remember, it is only a model;** if it seems difficult to understand, forget it. We include it here simply as a service to those who find mechanical models easier to comprehend than the behavior of electrical circuits: 2

Recall that in the steady state $|I_K| = |I_{Na}|$ (from Eq. 4-17). Thus, this *balance* between the two currents can be rearranged from Equations 4-18 to give

$$|g_{Na}(V_s - E_{Na})| = |g_K(V_s - E_K)| \qquad \text{Eq. 4-22}$$

Thus, under steady-state conditions there is a balance between two products, each being a conductance times a driving force. Thus, the balance between these products can be presented visually as a "balance beam," where the balance beam is not in equilibrium unless the products of the weight times the distance on both sides are equal. Thus, in Fig. 4-22 we show the distance from the fulcrum as the driving force $V_s - E_j$ and the weight as the magnitude of the conductance g_j. 3

Just as in a real balance beam, it is not the absolute value of the weights, but their ratio that must balance the ratio of the distances from the fulcrum:

$$\left|\frac{g_{Na}}{g_K}\right| = \left|\frac{V_s - E_K}{V_s - E_{Na}}\right| \qquad \text{Eq. 4-23}$$ 4

From Fig. 4-22 you may be able to visualize that when in the resting state g_K is much larger than g_{Na}, V_s approaches E_K; that is, $V_s - E_K$ is small, the fulcrum having moved to the left in Fig. 4-22. However, if g_K and g_{Na} were equal, you would also be able to see that the fulcrum would have to shift so that V_s was halfway between the ends. Hence under these special conditions,

$$|V_s - E_{Na}| = |V_s - E_K| \qquad \text{Eq. 4-24}$$

You can also deduce this from Eq. 4-23, when $g_{Na}/g_K = 1$. 5

Fig. 4-22. Representation of the "balance" of the resting membrane potential with a large g_K (and small driving force) equal (and opposite) to a small g_{Na} (and large driving force).

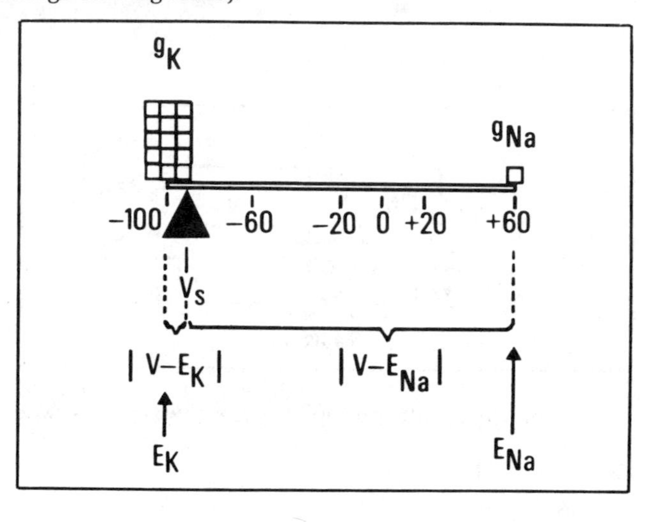

66

67

Finally, if g_{Na} increases markedly, you can see that the fulcrum V_s will have to move to the right for the beam to be once more in balance.

Since the mathematics of the balance beam and of the membrane permeable to two ions is the same in the steady state (for both systems), clearly the intuitions you can gain from the diagram are the same as those from working with the steady-state equation (which we also recommend). To be sure that the balance-beam analogy is complete, let us define the points clearly by allowing the beam to be some measuring device (such as a ruler). Now, in terms of the calculations, it does not matter where the zero point on the scale is; points can be measured on either side by using plus and minus values. Thus, in Fig. 4-22 we show the balance beam with a scale. At either end of the scale are E_K and E_{Na}.

The sign (direction) of current flow that occurs if the equilibrium is disturbed can be determined by the direction in which the beam tips: A counterclockwise movement occurring when a variable (or variables) is changed indicates a positive I_{por} accompanied by $-I_{cap}$ and polarization (or hyperpolarization) of the membrane.

If you imagine an "automatic" beam balance that has an automatic means of moving the fulcrum *toward* an end that has *tipped down*, then such a balance will reach equilibrium whether the changes are in the weights (conductances) or in the distances along the beam at which the weights are placed (diffusional forces). Such an automatic balance would act as an analog of the membrane as we described it.

Remember, ignore this balance-beam analogy if you find it confusing. (The same understanding can be obtained directly from the steady-state equation or, alternatively, from a thorough understanding of the physicochemical principles from which the equation was derived.)

QUESTION: If V is halfway between E_{Na} and E_K, what is the magnitude of g_{Na} if g_K is known to be 1 mmho/cm^2? (Hint 35↓)

HINTS

30. False. If $[K^+]_o$ is increased, the K^+ gradient will be *reduced*, the effect of K^+ will be reduced, and V will move toward E_{NA}; that is, the membrane will be *depolarized*. You can get the same result from the steady-state equation.
31. False. The gradient *decreases*, right?
32. False. Increasing $[Na^+]_i$ reduces the Na^+ gradient and so moves the membrane toward E_K; that is, the membrane *hyperpolarizes*. Remember this one!
33. True. If g_K decreases, this is equivalent to an increase in g_{Na} in terms of changing the value of g_K/g_{Na}. In either case, V_s moves toward E_{Na}, and the membrane *depolarizes*.
34. True. Increasing g_{Na} moves V toward E_{Na}; that is, the membrane becomes depolarized.

67

68

If we could promise you that no more than two ions (for example, Na^+ and K^+) were important in the production of cell membrane potentials, then by now you would possess a very complete understanding of how such potentials arise. Unfortunately, at times even more ions are involved. So now we must consider the contribution of other ions that may be able to pass through the membranes of living cells.

1

CONTRIBUTION OF OTHER IONS TO RESTING MEMBRANE POTENTIAL

Many cell membranes have been shown (by radioisotope flux studies) to be permeable to not only Na^+ and K^+ ions but also to Ca^{2+}, Mg^{2+}, H^+, HCO_3^-, and Cl^-, in addition to other less commonly available ion species. **How, then, can it be true, for many cells, that the resting potential can be calculated from Na^+ and K^+ movements without taking these other ions into account?**

2

Since capacitative current I_{cap} must be zero when the membrane potential is in steady state, it follows from the argument on page 61.3 that the sum of the ionic currents also must be zero. The approximation used in the previous section (namely, $I_{por} = I_{Na} + I_K = 0$) thus requires that **all other ionic currents be so small as to be of negligible importance.**

3

This approximation is valid for many types of cell. However, it is not true for smooth muscle cells or spinal motor neurons, to name two, and hence the general equations should be introduced at this point.

4

If I_{por} is the sum of all the specific ion currents, then clearly

$$I_{por} = \Sigma I_j = I_{Na} + I_K + I_{Cl} + I_{Ca} + I_{Mg} + \cdots \qquad \text{Eq. 4-25}$$

Now for every specific ion current I_j, an ionic Ohm's law can be written as

$$I_j = g_j(V - E_j) \qquad \text{Generalized ionic Ohm's law} \quad \text{Eq. 4-26/4-14}$$

(Compare Eqs. 4-26 and 4-12.) For the special case of $V = V_S$, we can substitute the individual ionic Ohm's law equations into Eq. 4-25 and solve for V_S:

$$V_S = \frac{g_{Na}E_{Na} + g_KE_K + g_{Cl}E_{Cl} + g_{Ca}E_{Ca} + g_{Mg}E_{Mg} + \cdots}{g_{Na} + g_K + g_{Cl} + g_{Ca} + g_{Mg} + \cdots} \qquad \text{Eq. 4-27}$$

Fortunately, this equation can be contracted to the more easily remembered general form:

$$V_S = \frac{\Sigma g_jE_j}{\Sigma g_j} \qquad \text{Generalized steady-state equation} \quad \text{Eq. 4-28}$$

5

68

69

IONIC CURRENTS AND IONIC OHM'S LAW

By now you will probably agree that the ionic Ohm's law (Eq. 4-26) is of major importance in the understanding of membrane potentials. Here are a few more consequences of this equation that should be introduced at this stage:

1

Notice that the Ohm's law equation states that **I_j will be negligible if either conductance for that ion species is negligible or the driving force on the ion ($V - E_j$) happens to be close to zero.** We shall see that we can find examples of negligible ionic currents due to one or the other of these reasons. For example, the lack of any contribution of I_{Ca} to the resting potential in most cells is due to low g_{Ca}. However, while g_{Cl} is high in frog muscle, I_{Cl} is zero, since the driving force $V_s - E_{Cl}$ is zero at the resting potential.

2

Similarly, there are two ways in which the *conductance* for an ion can be low: Its concentration is very low, or its *permeability* (i.e., mobility within the membrane substance) is low. Notice that conductance, which is a proportionality factor between current and driving force, depends on two factors—the permeability of the membrane for the ion concerned *and the availability of ions to carry the current* (i.e., the ion concentration). However, *permeability*, which is defined as a proportionality factor between concentration gradient and flux, is independent of concentration. Thus, these two terms (which are often used completely interchangeably) are, in fact, quite different.

3

QUESTION: Table 4-10 shows values of the battery potential and the conductances for the major ions contributing to the resting potential in a typical skeletal muscle cell (Part A) and a typical smooth muscle cell (Part B). Using Eq. 4-28, calculate the steady-state resting potentials for these two cells. What do you notice about the relationship between V_s and E_{Cl} is these cells? (Hint 39↓)

4

Notice that **in the skeletal muscle** cell **$V_s = E_{Cl}$, hence the driving force on chloride ions must be zero in steady state.** On the other hand, **in the smooth muscle** cell V is considerably more negative than E_{Cl}, and it follows that **a finite chloride current will flow when the cell is in the steady state.** The condition for balance here must be that $|I_K| = |I_{Na} + I_{Cl}|$.

5

QUESTION: In the smooth muscle cell just described, in the steady state are Cl^- ions moving into or out of the cell? (Hint 40↓)

6

Table 4-10. Values of "Battery Potential" and Conductances for Major Ions Contributing to Resting Potential

A. Typical values for skeletal muscle cell

Potential (mV)	Conductance (mmho/cm²)
$E_{Na} = +60$	$g_{Na} = 0.1$
$E_K = -100$	$g_K = 0.15$
$E_{Cl} = -90$	$g_{Cl} = 0.30$

B. Approximate values for typical smooth muscle cell

Potential (mV)	Conductance (mmho/cm²)
$E_{Na} = +50$	$g_{Na} = 0.01$
$E_K = -90$	$g_K = 0.20$
$E_{Cl} = -30$	$g_{Cl} = 0.20$

HINT

35. You could work out an answer by substitution in the steady-state equation, but isn't it easier to say that if $|V - E_K|$ is the same as $|V - E_{Na}|$, then $g_{Na} = g_K$ and therefore $g_{Na} = 1$ mmho/cm²?

69

70

Returning now to consider one typical cell (see Fig. 4-23), we can ask some questions that should, at this stage, seem quite straightforward.

QUESTION: Turn to Fig. 4-18. Now draw the comparable equivalent circuit for the cell shown in Fig. 4-23. (Hint 36↓)

QUESTION: Given that $V_s = E_{Cl}$ initially, if $[Cl^-]_o$ is suddenly increased by 10 mEq/L, (1) in which direction will Cl^- move across the cell membrane and (2) will this hyperpolarize or depolarize the cell? (Hint 38↓)

QUESTION: Does it worry you that addition of an extra 10 mEq/L of extracellular anions might *reduce* the membrane potential? Why doesn't that happen? If you cannot think of an answer, see Hint 42.↓

REVIEW OF MECHANISM OF RESTING MEMBRANE POTENTIAL

We stated (page 42.3) that "membrane potentials are produced by movements of ions along electrochemical gradients." *By now, you should thoroughly understand this statement.* At this point, you may find it interesting to go back to page 49.2 to reread the six principles of membrane potentials to measure what you learned (since we are sure you understand them much better now than you did then!).

Given sufficient information, you should be able to calculate the resting membrane potential, the magnitude of the electrical driving force on any given ion, the magnitude of the resulting ionic currents, etc. Perhaps more importantly, now you should have a sound theoretical basis for your intuitions as to the effects on the resting membrane potential of changing ion concentrations and conductances. If you have any lingering doubts about your competence in this area, reread the review on page 64 and test yourself again on the questions on pages 63.1 to 64.3 and 64.5 to 65.3.

If in doubt, always remember that by using the steady-state equation (Eq. 4-20 or 4-27) you can easily determine the effects of any conductance or concentration changes from resting conditions!

IONIC PUMPS

The mechanism of the membrane potential has been explained in the preceding pages by **assuming that constant concentration differences for K^+ and Na^+ are maintained across the membrane.** The question naturally arises, How are these concentration differences produced? The simple answer is that they are produced by *membrane ionic pumping.*

In the uniform resting membrane, $I_{mem} = 0$, but this does not necessarily mean that I_K and I_{Na} are zero, as you know. In fact, $|I_K| = |I_{Na}|$ in the resting membrane. So if Na^+ entry substitutes for K^+ loss in the maintenance of intracellular cation concentrations, there seems no logical reason why a cell should not "run down" until the concentration gradients for K^+ and Na^+ disappear.

Fig. 4-23. Hypothetical cell permeable to K^+, Na^+, and Cl^-, but not to A^-.

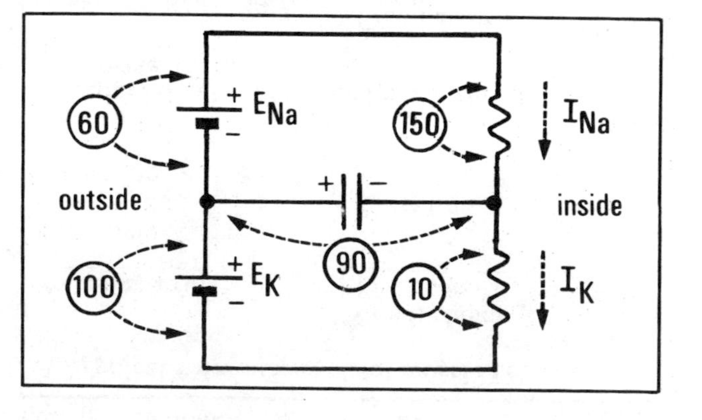

Fig. 4-24. Voltage changes across different circuit elements, indicated by numbers in circles. Note that here $E_K = -100$, $E_{Na} = +60$, $V = -90$.

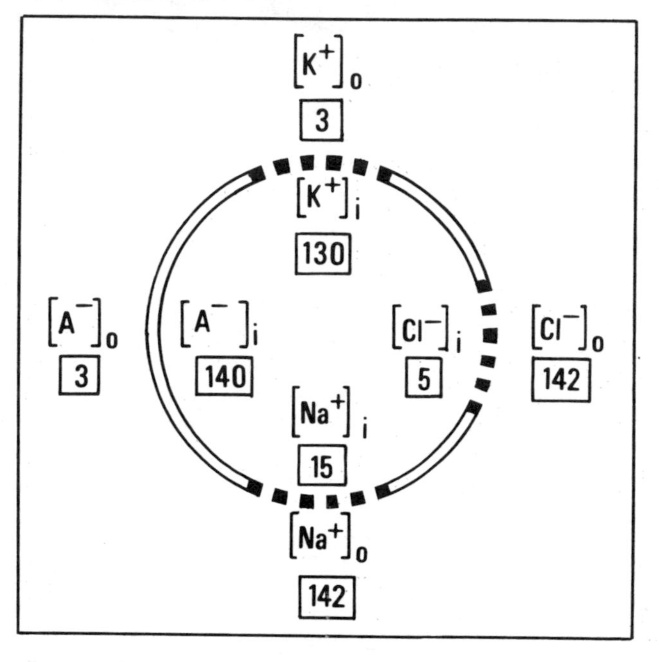

70

71

Putting the problem in yet another way, we can point out that the circuit in Fig. 4-18 can be redrawn (see Fig. 4-24) to make clear that the Na^+ and K^+ batteries act in series to pass a constant steady-state current across the voltage-divider network represented by $1/g_{Na} + 1/g_K$ acting in series. **In such a circuit, the batteries would have to run down!** Why doesn't this happen in life? It seems necessary to assume that the batteries are connected to some kind of "trickle charger."

1

For *steady state* to be maintained, some mechanism must exist that extrudes Na^+ ions and recaptures K^+ ions **at the same rate as these ions pass through the membrane as a result of electrodiffusional forces.** This situation is diagrammed in Fig. 4-25.

2

To maintain a steady state one simply requires a *metabolic pump* that will pump out Na^+ ions and pump in K^+ ions at a rate equal to the *steady-state* ion fluxes. If, as we assume here, $|I_{Na}|$ and $|I_K|$ are equal under steady-state conditions, then the pump should extrude one Na^+ ion for every K^+ ion carried into the cell. **Such a 1:1 pump would separate no charge and therefore could make no direct contribution to the membrane potential.** However, if the pump were poisoned, the membrane potential would slowly decay as the Na^+ and K^+ concentration gradients slowly disappeared.

3

A vast body of evidence now exists to suggest that some form of coupled sodium-potassium pump does indeed exist in cell membranes:

1. As predicted previously, metabolic inhibitors produce slow changes in the resting steady-state membrane potential. The observed rate of change correlates well with calculated changes in E_{Na} and E_K.
2. Radioisotope experiments show that when the membrane potential is constant, the sodium influx is equal to the sodium efflux while the potassium influx is also equal to the potassium efflux (within the limits of accuracy of such experiments).
3. Metabolic inhibitors, which cause slow decay of the membrane potential, cause reduction in the sodium efflux and potassium influx [1, pp. 23–28; 11, pp. 1123–1124; 27, pp. 59–68; 62, pp. 14–18]. It can also be shown that (1) pump activity is increased after repetitive firing of an axon, (2) pump activity is increased if $[Na^+]_i$ increases, and (3) pump activity is markedly depressed by removing K^+ from the extracellular solution. **Such observations confirm the concept of a *coupled* pump moving K^+ in and Na^+ out and suggest that the activity of the pump is in some way controlled by $[Na^+]_i$.**

4

Fig. 4-25. Mechanical model of membrane in steady state. Passive electrodiffusional fluxes occur through nonspecific pores (top and bottom). Pumped fluxes are dependent on metabolic energy derived from ATP.

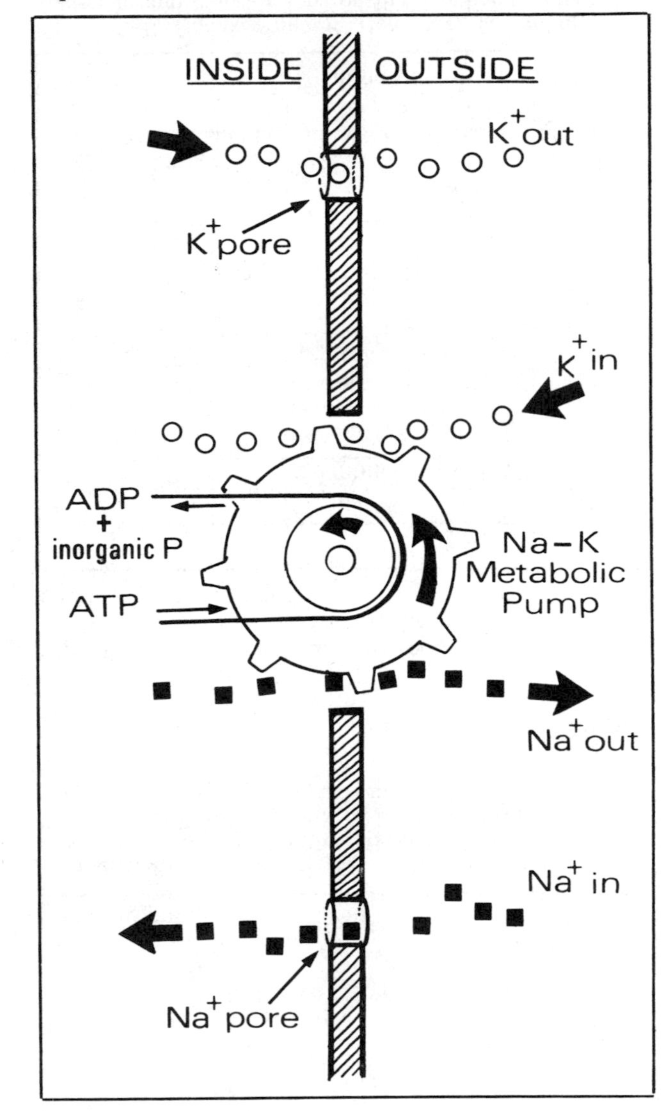

HINTS

39. There is no simple way to intuit an answer to this question unless V_s is already known. So go ahead and calculate the answers. Then see Hint 37.↓

40. I_{Cl} is in the same direction as I_{Na}, that is, *negative.* A negative current implies a movement of positive particles *into* the cell and hence a movement of negative ions *out* of the cell.

71

Although we assume a 1:1 pump ratio throughout this chapter, advanced students may well be aware that the available evidence overwhelmingly favors a ratio of $3Na^+$ to $2K^+$ for almost all tissues studied. The rather minor modifications required to take this observation into account are discussed in Chap. 5.
1

The term *pump* is particularly apt since, as diagrammed in Fig. 4-25, **the process of moving ions to develop a concentration difference requires energy.**
2

Indeed, the concentration difference can be viewed as a *storage* of diffusional energy, which is then used by the cell to produce the membrane potential. This justifies the K^+ and Na^+ batteries we diagrammed in the electrical circuits.
3

We noted the important principle that the cell membrane contains a metabolic pump whose function is to regulate the internal concentrations of Na^+ and K^+. We also noted that the ionic gradients for Na^+ and K^+ would decay in the absence of such a pump. Therefore, it seems reasonable to ask: **What happens to the concentrations of a permeant ion that is not being metabolically pumped?**
4

If metabolic energy is not used to distribute an ion, that ion must be at equilibrium in the steady state (i.e., the net flux of the ion must be zero). **The only circumstance under which this can occur is if the driving force on the ion is zero.** Hence it must be that $E_j = V_s$, where E_j is the Nernst *battery potential* for the ion concerned and V_s is the resting potential. We can state the principle as follows: **If a freely permeable ion is passively distributed, its concentrations must be such that it is in equilibrium at the resting membrane potential.**
5

QUESTION: Which of the ions listed in Table 4-11 are passively distributed? (Hint 41↓)
6

Our analysis permits another curiously powerful generalization concerning the effects of changing external ion concentrations on the cell's resting potential. **Unless a cell actively regulates the internal concentration of an ion, changing the extracellular concentration of that ion can have only transitory effects on membrane potential.**
7

Figure 4-26 shows the change of membrane potential after a sudden change in $[Cl^-]_o$ in a frog muscle fiber (which is not known to have an active "chloride pump"). Notice that the initial rapid change of potential associated with the change of $[Cl^-]_o$ is followed by a gradual return to the original resting potential. The ion is simply redistributed until E_{Cl} is once more equal to V_s, where V_s is the diffusion potential set by the steady-state values of I_{Na} and I_K.
8

Let's trace this out in more detail (unfortunately, the full physicochemical explanation must be delayed until the next chapter; only the electrical analog is given here). Reduction in $[Cl^-]_o$ causes a dramatic positive shift in E_{Cl}; I_{Cl} be-
9

Table 4-11. Typical Equilibrium Potentials for a Mammalian Muscle Cell

Ion	Potential
Na^+	+66 mV
K^+	−97
H^+	−32
Cl^-	−90
HCO_3^-	−32
Resting potential	−90 mV

Data from J. W. Woodbury, The Cell Membrane: Ionic and Potential Gradients and Active Transport. In T. C. Ruch, H. D. Patton, J. W. Woodbury, and A. L. Towe (Eds.), *Neurophysiology*, 2nd Ed. Philadelphia: Saunders, 1976.

Fig. 4-26. Effect of sudden reduction in external chloride concentration on membrane potential of isolated frog muscle fiber.

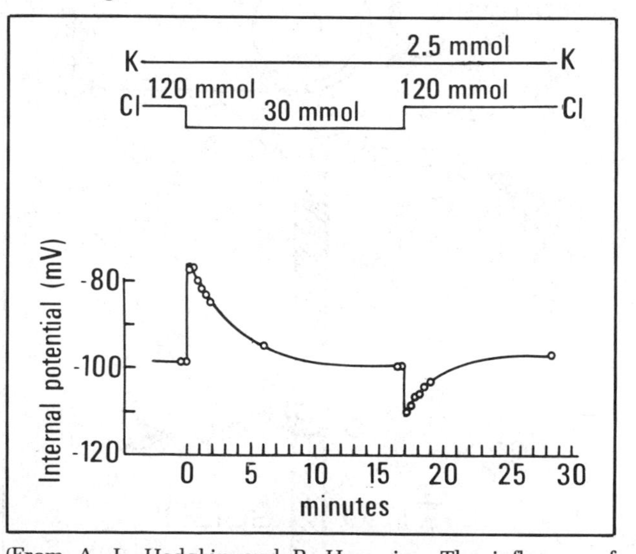

(From A. L. Hodgkin and P. Horowicz, The influence of potassium and chloride ions on the membrane potential of single muscle fibres, *J. Physiol.* [*Lond.*] 148:127, 1959.)

73

comes *negative*, and the cell depolarizes. However, a negative I_{Cl} is equivalent to an outward movement of Cl^-, and such a net loss of Cl^- (in the absence of any pump to regulate $[Cl^-]_i$) must slowly reduce the intracellular chloride concentration. So $[Cl^-]_i$ falls until it, like $[Cl^-]_o$, is one fourth its initial level and E_{Cl} is once more equal to V. These changes are reversed when $[Cl^-]_o$ is returned to its initial value.
1

Thus the logical sequence in the development of the resting membrane potential is as follows: (1) A metabolic pump creates concentration differences across the semipermeable cell membrane. (2) Moving down their electrochemical gradients and through complete circuits, K^+ and Na^+ determine the charge separation across the membrane capacitance that gives rise to the membrane potential. (3) The metabolic pump remains continually active to prevent the constant "running down" of the concentration differences.
2

CLINICAL ESTIMATION OF CHANGE IN RESTING POTENTIAL

As we see in Chap. 6, the level of the resting potential has major effects on the excitability of nerve cells, of muscle cells, and, of course, of the heart. **In clinical medicine, ion concentrations**
3

HINTS

36. In which direction should the chloride battery face? Think about this before completing your drawing. If necessary, reread page 60.5, and then turn to Hint 43.↓
37. **For the skeletal muscle cell,**

$$V_s = \frac{0.01(60) + 0.15(-100) + 0.3(-90)}{0.46} = -90 \text{ mV}$$

For the smooth muscle cell,

$$V_s = \frac{0.01\ (50) + 0.2(-90) + 0.2(-30)}{0.41} = -57 \text{ mV}$$

Did you get these right? Now return to the text for the significance of this difference.

38. (1) Chloride ions will move *inward*, which is equivalent to an outward current of positive particles. (2) An outward ionic current will induce an inward capacitative current and hence *hyperpolarize* the cell. If you find this difficult, look at the circuit contained in Hint 43.↓ **Alternatively, the physicochemical argument can be employed:** If Cl^- tends to enter the cell, then the inside becomes more negative and the outside more positive (i.e., the cell hyperpolarizes). This increase in potential will reduce Cl^- entry and favor Cl^- efflux, so that a new balance will tend to be achieved.
42. Of course, we cannot add negative ions without also adding an equal number of cations. What such a statement means is that **Cl^- was added as the salt of an impermeant cation.** In this case, we might have used Tris Cl. The Tris ion, being impermeant, could then be ignored in subsequent calculations.

73

74

in the extracellular fluid often are markedly altered by disease or doctor, sometimes to the detriment of the patient. Since it should now be very clear to you that changes in ion concentrations may affect the resting potential, it is a good time for you to review the most common of these effects. 1

First, we have the rules:

1. Increasing the gradient for an ion increases that ion's effect on the resting potential.
2. Increasing the conductance for an ion increases that ion's effect on the resting potential.
3. Ions with the greatest relative conductances have relatively greater effects on the membrane potential.
4. Effects are transitory unless the internal ion concentration is physiologically regulated (as by an ion "pump").

Next, remember the following:

1. Resting g_K is high in all excitable cells.
2. Resting g_{Cl} is low in axons but high in skeletal and smooth muscles and in cell bodies of most neurons.
3. Intracellular chloride concentration is regulated in only smooth muscle and some neuron cell bodies.
4. Resting g_{Na} is low in all excitable cells.

So we conclude thus:

1. Physiological changes in serum Na^+ levels produce only minor changes in the resting potential.
2. Marked changes in the resting potential may occur as a result of changes in the serum K^+ level.
3. Changes in serum Cl^- levels may affect the resting potentials in some tissues. 2

We hope that the following questions will seem very straightforward. If you have problems with them, return to the relevant section in the text. 3

QUESTION: What happens to the resting membrane potential if $[K^+]_o$ is increased, say, by an intravenous (IV) injection of K_2SO_4? (Note: SO_4^{2-} is essentially impermeant.) (Hint 45↓) 4

QUESTION: Patients using certain diuretics may lose large amounts of K^+ in their urine; thus their serum $[K^+]$ is lowered if K^+ supplementation has not been instituted (or if the patient fails to take the K^+ supplement as instructed). What would happen to resting potentials under these circumstances? (Hint 49↓) 5

74

75

QUESTION: It is very common for patients to receive intravenous injections of NaCl. If both $[Na^+]_o$ and $[Cl^-]_o$ are increased by the same amount, what happens to the membrane potentials of the skeletal muscle? (Hint 46 ↓)

SOME COMMON POINTS OF CONFUSION

After many years of teaching this material, we found that certain pernicious logical problems can present themselves, like tormenting demons, to the more conscientious students. These unresolved problems then fester in such students' minds, preventing further study (even of unrelated disciplines!). In the following section, we set out to exorcise some of these demons. However, we warn you not to underestimate their power. **Do not read this section unless you have a problem.** If you do have a problem, look for it in the following list and read the "Answers" paragraph bearing the corresponding number. (If your problem is not listed, write to us; we promise you a sympathetic hearing.)

1. How can one talk about a resting potential "increasing" when it is becoming more negative?
2. When $[K^+]_o$ increases, more positive charge is added to the outside of the membrane, so how can this depolarize the cell?
3. Aren't there times when currents don't move in complete circuits?
4. Why do you always include a capacitor when drawing the equivalent circuit for the diffusion potential (e.g., Fig. 4-24)?
5. How can you be so sure than an indiffusible anion (A^-) does not contribute to the resting membrane potential? Why do you ignore the older "double Donnan" explanation of the membrane potential?
6. How can you regard the "induced currents" as being purely capacitative? Surely they should have both a resistive and a capacitative component.
7. Why is it always *capacitative* current that changes the membrane potential?
8. What happens to I_K if $[K^+]_o$ is zero?

HINTS

41. Of the ions listed, only chloride appears not to be actively pumped, since this is the only ion whose Nernst potential equals its resting potential. Are you surprised to see that H^+ and HCO_3^- are not passively distributed? (See Hint 44.↓)

43.

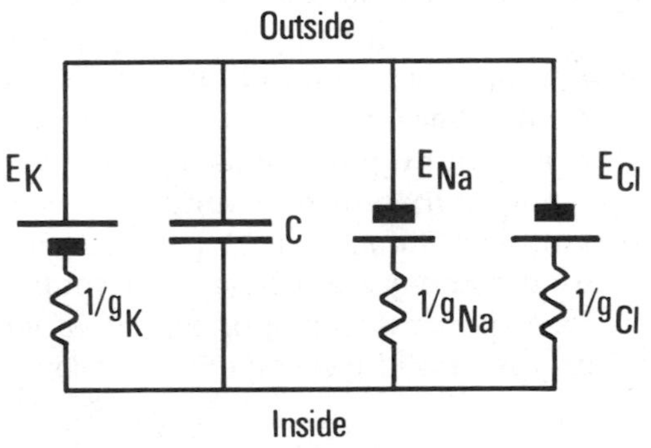

Outside
E_K
E_{Na}
E_{Cl}
C
$1/g_K$
$1/g_{Na}$
$1/g_{Cl}$
Inside

75

76

ANSWERS

1. That the resting membrane potential is given as a negative number is a purely arbitrary convention deriving from the fact that the intracellular microelectrode necessarily has a high resistance. Thus the external medium is grounded and considered the reference point, compared to which the inside of the cell is indeed negative. But what is being measured is the *potential difference across the membrane*. This has *magnitude and direction*, but until the reference point is established, it has no "sign." Think of someone standing in a valley looking at an eagle perched on a cliff. The person thinks the cliff is *high*; to the eagle, the valley is *deep*. However, we tried to keep such confusions to a minimum by referring to increase in $|V|$ and by using the relatively unambiguous terms defined in Table 4-9.

2. Now, $[K^+]_o$ cannot increase all by itself. There is no way in which we can add simply the positive ions. When $[K^+]_o$ increases, it is because we have added an electroneutral quantity of cations and anions. To produce the effect of increased $[K^+]_o$, you would add the potassium salt of an impermeant anion (for example, K_2SO_4). Thus it is incorrect to think that the positive charge has been added to the outside of the membrane; actually an electroneutral salt has been added. The effects on membrane potential are produced by the change in the K^+ gradient, which in turn causes a change in the K^+ movement across the membrane. It is this movement that drives the current through the capacitor and changes the membrane potential. We wish it were simpler, but it isn't!

3. This question usually stems from an incomplete understanding of the physics of capacitance. You have to convince yourself that current flows across a capacitor in a circuit and that all current flows in complete circuits. Read a good physics book (which uses calculus). Read Maxwell himself. Talk to a physicist. And/or ponder the following example.

   An ebonite rod is rubbed on a suitable surface. The rod becomes charged. Where is the capacitor? Between the charged rod and ground is air, a good insulator. There is a charge separation built up between the rod and the ground; i.e., the rod becomes one plate of a charged capacitor, as can be shown by discharging (shorting) the charge by touching the rod to the ground. Even in classic electrostatics, there are capacitors and complete circuits.

   Become a believer. **Current always flows in complete circuits.**

4. Clearly, it makes no difference whether the capacitor is removed from the circuit shown in Fig. 4-24. The "membrane potential" would still be generated across the voltage-divider circuit represented by the two resistors. But does this modified circuit bear any resemblance to the real membrane?

   Remember the following: (*a*) Both the batteries and the resistors are located in the membrane pores; (*b*) these pores are extremely sparsely distributed over the membrane surface; (*c*) the potential is due to a separation of charge across the membrane; (*d*) this charge is represented by excess cations spread out over one surface of the membrane and excess anions spread out over the other surface; and (*e*) these charges remain separated owing to the capacitance of the nonpore regions of the membrane. Thus removing the capacitor from Fig.

76

4-24 is equivalent to stating that the entire membrane area is composed of either sodium or potassium pores. Since it is known that this is not the case, the capacitor is better retained in Fig. 4-24. (In Chap. 5 we see that the size of the membrane capacitance determines the rate of change of voltage produced by a given current flowing across the membrane; so membrane capacitance is an important determinant of the velocity of propagation of the action potential.)

5. The presence of intracellular *indiffusible anions* certainly contributes to the membrane potential in that it allows $[Cl^-]_i$ to be small without compromising internal electroneutrality. However, these anions do *not* contribute to an equilibrium membrane potential as was theorized in the early (now discredited) "double Donnan" model of the membrane potential. That model hypothesized the existence of an equilibrium state in which K^+ was retained intracellularly and Cl^- was excluded by the presence of the indiffusible anion. Let us briefly review the evidence against this concept. (*a*) It requires that the membrane be impermeable to Na^+—and that the resting potential be a true equilibrium state. First, evidence against this hypothesis was provided by isotope experiments showing relatively large Na^+ fluxes; second, the decay of V_s in the presence of metabolic inhibitors demonstrates that V_s is not an equilibrium condition. (*b*) It has been found possible to squeeze the axoplasm from a squid giant axon. Then the axon may be internally perfused with any solution the experimenter chooses. It has been shown that near-normal resting potentials are produced when the axon is perfused with KCl solution (g_{Cl} is very small in squid axons) [18, pp. 36–42; 40, p. 59]. **The potential is not caused by the presence of indiffusible anions;** the perfused axon experiment

HINTS

44. If you were surprised, then it probably didn't occur to you that cells might regulate their intracellular pH. Conductances for these ions are very low, however, and they do not contribute significantly to the cell's resting potential.
45. *Always* think of the gradient. The K^+ gradient is *reduced*, so all excitable cells will be *depolarized*.
46. Now $g_{Cl} > g_{Na}$, so the chloride effect predominates: Increasing the chloride gradient will hyperpolarize the cell. But **the chloride effect is transient in this tissue.** When it has disappeared, the cell will have *depolarized very slightly* because of the increased E_{Na}.
48. The heart—quite an important muscle.
49. The increased K^+ gradient would be expected to hyperpolarize cells. However, if the loss of K^+ has been very gradual—say, over many months—*intracellular* K^+ may have fallen, maintaining a near-normal K^+ gradient across the cell membranes. What would happen if you suddenly restored normal serum electrolyte levels by rapid IV injection of KCl? (Hint 47 ↓)
50. Don't be confused by this question. Part (*a*) is *correct* since if $V_s = E_{Cl}$, the driving force must be zero. If the driving force is zero, then $I_{Cl} = 0$ regardless of the magnitude of g_{Cl}. Part (*c*) is *incorrect*; decreasing $[Cl^-]_i$ increases the chloride gradient. Thus both decreasing $[Cl^-]_i$ and increasing $[Cl^-]_o$ should have the same effect.
51. (*b*). Increased $[Na^+]_i$ reduces the Na^+ gradient and hence reduces I_{Na}. The cell *hyperpolarizes* until the balance between I_{Na} and I_K is restored at the new value of V.
52. (*d*). Cells should depolarize as $[K^+]_i$ falls and $[K^+]_o$ rises.
54. Compare your answer with Table 4-8.

shows that normal potentials can be produced even when the only internal anion is actually permeant. **The potential is caused by differences in the rates of diffusion of K^+ and Na^+ ions** (just as we described on pages 60.7 to 62.6).

6. All the current that flows across the membrane must flow in complete circuits. So (provided no external circuits are introduced by experimental equipment, etc.) it follows that **when current is summed over the whole membrane area, net current flow is zero.** In other words, what goes in must come out. Since there can be only two kinds of current, ionic and capacitative, it must be true also that the sum of these currents is zero (see Eq. 4-15/4-3). Hence **net ionic current *must* be balanced by an equal and opposite capacitative current.** There is just one catch! In Chap. 7 we end up considering real, dynamic situations in which the membrane potential is not necessarily the same in all parts of the cell membrane. We show that under such circumstances, *local* induced currents do have a resistive component.
7. This question grows out of the "voltage-divider" view of the membrane potential discussed in problem 4. In the "voltage-divider" circuit of Fig. 4-24, it is clear that changing the conductances of the Na^+ or K^+ pores would affect the membrane potential. So why do we complicate matters by introducing the capacitative current?

   If an external current were applied to a purely resistive membrane, then the change of voltage would be instantaneous. However, it has long been known that cell membranes behave as resistor-capacitor networks [1, pp. 42–44; 9, p. 1070; 27, pp. 34–36] under these conditions. Therefore, the capacitor must be included in the circuit, and it immediately follows that the membrane potential must be proportional to the charge on the membrane capacitance, as shown in Eq. 3-5. But in such a circuit it is clear that the rate of charge of membrane potential is determined by I_{cap}, the rate of current flow onto or off the membrane capacitor [27, p. 69; 40, p. 75; 63, p. 36]. For further help on this problem, reread the discussion of the basic electrical properties of cell membranes in the section of that name in Chap. 3.
8. Touché! If $[K^+]_o$ were zero, then from the Nernst equation E_K should become infinite and I_K would be given by $I_K = g_K (V - \infty)$. Obviously this is not what actually happens; there are two reasons, one theoretical and the other purely practical. First, we know that the Nernst equation describes the equilibrium where the K^+ influx equals the K^+ efflux. So how can an equilibrium be reached if $[K^+]_o$ is zero and therefore the K influx is zero? Thus the Nernst equation is not applicable where $[K^+]_o$ is zero. Nevertheless, $+I_K$ can be obtained by calculating the unidirectional K^+ efflux (in which case it turns out to be a quite reasonable value). Second, it has long been suggested that a diffusion barrier exists (as a result of the Schwann cell and/or "basement membranes"?) outside the surface membrane of the axon or muscle cell. The existence of such a diffusion barrier was proposed to take into account certain minor anomalies in quantitative studies of transient phenomena. This region is often called the *Hodgkin-Frankenhaeuser space*. Thus even where $[K^+]_o$ is zero in the external medium, some K^+ ions leaving the cell will be restrained in the Hodgkin-Frankenhaeuser space, and the value of $[K^+]_o$ *seen by the membrane* will not become zero until $[K^+]_i$ is zero.

The problem presented in the question in terms of K^+ must apply equally to any other ion species. However, in all cases so far the experimental results can be satisfactorily described by one or the other (or both) of the explanations given.

Finally, there is the additional, subtle point that pores may not be totally selective for their chosen ion (see Chap. 5). Thus, if $[K^+]_o$ is zero, some external Na^+ ions may be able to wriggle in through the potassium pores when the driving force is high enough, reducing the expected outward I_K.

1

EXAM QUESTIONS

To help you evaluate your understanding of this chapter, we collected a number of questions taken from midterm examinations given in a first-year medical school neurophysiology course; it is hoped that you will find these questions quite easy.

2

1. A certain drug is known to produce a generalized inhibition of the sodium-potassium pumps located in cell membranes. Which of the following would *not* occur after administration of the drug?
   a. Increase in serum K^+ concentration
   b. Decrease in serum Na^+ concentration
   c. Increase in intracellular Na^+ concentration
   d. Hyperpolarization of cell resting potentials (Hint 52↑)

3

2. Assuming in each case that only the one change in permeant ion concentrations occurs, which of the following statements is *incorrect?*
   a. Increased $[K^+]_i$ would hyperpolarize a cell.
   b. Increased $[Na^+]_i$ would depolarize a cell.
   c. Increased $[K^+]_o$ would depolarize a cell.
   d. A typical skeletal muscle cell would be only transiently affected by increased $[Cl^-]_o$. (Hint 51↑)

4

3. Given that V_s (the resting membrane potential) is initially equal to E_{Cl}, which of the following statements is *incorrect?*
   a. V_s is not affected by the change in g_{Cl}.
   b. The membrane becomes transiently hyperpolarized when $[Cl^-]_o$ is increased.
   c. The membrane becomes transiently depolarized when $[Cl^-]_i$ is decreased.
   d. No chloride pump is active in this membrane. (Hint 50↑)

5

4. Compare and contrast the physicochemical and electrical circuit descriptions of the resting membrane potential in a membrane permeable to K^+ and Na^+. (Hint 54↑)

6

HINTS

47. Right! *Depolarization.* We pointed out that sometimes doing the right thing can have disastrous consequences! Now name an important muscle that would probably be among the first affected by your goof. (Hint 48 ↑)

53. (*a*) +; (*b*) +; (*c*) −; (*d*) +; (*e*) 0.

5. Assuming a membrane permeable to only K^+ and Na^+, with the concentration gradients of normal cells and with the usual resting membrane conductances of g_K and g_{Na}, classify the following *changes* with respect to whether the membrane potential will become depolarized (+), unchanged (0), or hyperpolarized (−):
   a. $[K^+]_o$ increased.
   b. $[K^+]_i$ decreased.
   c. g_K increased.
   d. g_{Na} increased.
   e. Both g_K and g_{Na} increased proportionally. (Hint 53↑)

1

81

Advanced Topics: Steady-State Potentials

82

The fundamental concepts required to explain the origin and maintenance of the resting membrane potential are presented in Chap. 4. Here we take up a number of additional topics which, while not essential to any of the core ideas developed in later chapters, nevertheless may be of considerable interest to those seeking a more thorough mastery of the basic principles of membrane physiology.

1

SEPARATE PORES FOR DIFFERENT ION SPECIES?

The concept of noninteracting ion fluxes, the so-called **independence principle**, is of very considerable theoretical interest.

2

In Chap. 4 the efflux of K^+ down its concentration gradient was presumed to occur without these K^+ ions jostling any Na^+ ions that might be attempting to enter through the same pore. Since such interactions usually were not apparent, it became acceptable to use the terms *sodium pore* and *potassium pore* as if separate ion-specific pores actually existed in the cell membrane.

3

After all, one of the basic underlying *assumptions* of the steady-state equation (Eq. 4-20) is that **ions move separately down their ionic gradients, interacting only insofar as these movements create an electrical potential across the membrane.** Similarly, both the Goldman equation (page 90) and the analysis of the action potential (Chap. 6) are heavily dependent on this concept. In Chaps. 6 and 8 we present evidence that the selective voltage and time-dependent changes in g_{Na} and g_K, which give rise to the action potential, result from changes in the relative proportions of "open" or "closed" pores within separate and distinct populations of sodium-selective or potassium-selective pores.

4

Important evidence as to the correctness of this general assumption has been provided over the last 10 or 15 years through the discovery of an increasing number of pharmacologically active substances that exert their effects by opening (or closing, as the case may be) one or another pore type.

5

Thus tetrodotoxin blocks sodium pores, whereas batrachotoxin opens them. Neither agent has any effect on g_K. By contrast, tetraethylammonium (TEA) ions block potassium efflux without affecting g_{Na}.

6

These observations provide very strong support for the concept of separate sodium-specific and potassium-specific pores.

7

More detailed analysis suggests that these two types of pores are *not absolutely specific* for sodium and potassium, respectively. If the conductance for different ions is measured under conditions where only one pore type is functional, the following series are found:

8

82

83

1 Na$^+$ pore: $g_{Li} > g_{Na} \gg g_K > g_{Rb} > g_{Cs}$

K$^+$ pore: $g_K > g_{Na} > g_{Li}$; blocked by Cs$^+$, Rb$^+$, and TEA$^+$

2 **The possibility of a separate CA^{2+}-specific pore must also be raised** since (although some Ca^{2+} ions might enter through the Na$^+$ pores) good evidence exists in heart muscle and smooth muscle that g_{Ca} is controlled independently of g_{Na}. The lanthanum ion appears to be a specific blocking agent for this membrane channel; La^{3+} does not affect g_{Na} or g_K.

3 The order of conductances at the Ca^{2+} pore appears to be the following:

Ca^{2+} pore: $g_{Na} > g_{Ca} > g_{Mg}$; blocked by La^{3+} and Mn^{2+}

4 Finally, it should be noted that **there may be at least two types of anion-selective pores** in addition to the three major cation-selective pores just described. Cell membranes have been found to differ quite markedly in their relative permeabilities to Cl$^-$ and HCO_3^- ions, which suggests that these ions may move through separate and distinct ion-specific pores.

5 Nitrate ions are known to reduce g_{Cl} selectively without affecting g_{HCO_3}.

6 It seems a fair conclusion that separate **ion-specific pores** exist at least for the cations and anions of major biological significance.

7 Until quite recently there seemed to be no simple physicochemical model that could account for the properties of the ion-specific pore, although the cation selectivity of certain resins had been observed to follow a series similar to the relative conductances of the Na$^+$ and K$^+$ channels. In the last five years, however, cyclic antibiotics (e.g., valinomycin) have been found to produce selective conductance channels in artificial lipid bilayer membranes. Further study of such compounds suggests that eventually it may be possible to obtain chemical structures for the "imaginary" pores described here.

8 Recent studies of the nature of pore selectivity suggest that two distinct "filters" may be involved:

9
1. **Selectivity filter.** A major, charged site is presumed to exist in the walls of the pore. This site selectivity accelerates or retards different ions according to sequences determined theoretically as consequences of the "strength" of the site and verified experimentally in ion-exchange resins. Since the permeability sequences for Na$^+$, K$^+$, and Ca$^+$ pores are in exact correspondence with

83

predicted sequences in these series, it seems highly probable that the major component of pore selectivity is determined by the charge strength of an included site, rather than by such factors as hydrated ion radius. The present assumption is that hydrated inorganic ions lose their hydration shells as the ions penetrate the pore. This loss would not require a great energy expenditure if the pores were oxygen-lined; the effect would be like sliding an iron bar from one Teflon-coated magnet onto another (only frictional forces are involved) as compared with pulling the bar away perpendicular to the magnet (against the full magnetic force).

1

2. **Sieving filter.** Information as to the true limiting internal diameter of pores may be obtained from studies of the organic cations that these pores may admit. Such evidence has suggested a typical size range of 3 to 5 Å in diameter. That this sieve size is *not* a major factor in the determination of cation selectivity sequences follows from such observations as that the sodium pore of a squid axon is markedly less permeable to choline than is the sodium pore of frog muscle, whereas the sodium pore of crustacean muscle is quite freely permeable to choline ions.

2

Thus while the selectivity filter appears similar in each of these pores, apparently the sieve size is variable. The sodium pore of the crustacean muscle must have a larger sieve diameter than that of the squid axon. If the diameter of a pore may vary, so, too, may its effective length, that is, the length of that portion containing the sieve and selectivity filters. Hille and his coworkers have suggested that the major restriction of Na^+ ion movement may be of quite limited extent. It is even possible that the sieve filter of the sodium pore may be little longer than the minimum pore diameter. By contrast, the potassium pore may contain as many as three sequential ion binding sites within a channel whose narrowest region is several times longer than its minimum diameter. K^+ ions hop from site to site as they travel through the pore (possibly with more than one ion in the pore at once).

3

The parameters of a typical pore are shown in Fig. 5-1. Remember that this is purely a diagrammatic presentation of those aspects of the pore that might, eventually, prove measurable. It should not be interpreted as suggesting the actual shape or dimensions of any presently known pore type.

4

Fig. 5-1. Cross-section of generalized pore structure.

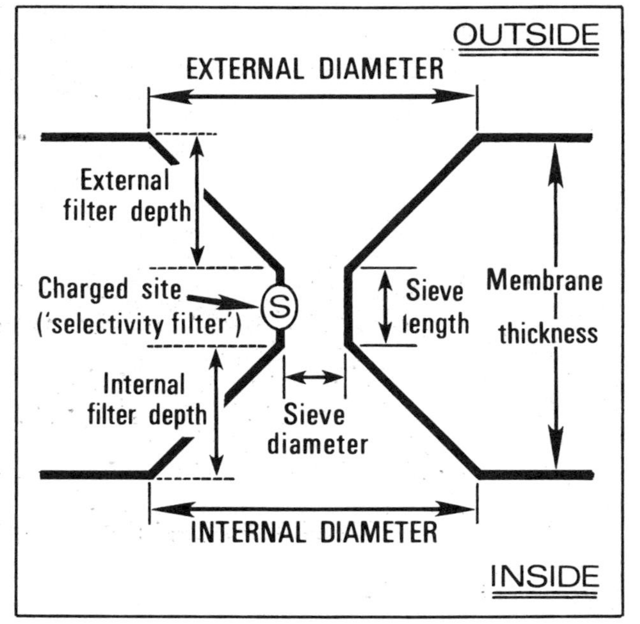

While it seems that the major ion species move independently of one another through separate ion-specific pores, we should now remind you that there is complete interaction of ionic movement with regard to the membrane capacitance. That is, **all ions can contribute to and be influenced by the charge on the membrane capacitance.**

5

In this sense, the membrane potential not only is the result of currents, but also determines them.

6

1 This fact implies a type of feedback loop, which is described in Chaps. 6 and 12.

PUMP POTENTIALS

2 In Chap. 4 we assumed that the metabolic Na-K pump is **electroneutral** (i.e., does not separate charge since it moves one ion in for every ion moved out), as diagrammed in Fig. 5-2.

3 It is possible, at least theoretically, that a metabolic pump might be **electrogenic** (i.e., contributes to separation of charge) and that **such electrogenicity would be expected for anything other than a strict 1:1 pumping ratio.**

4 Under certain conditions (e.g., when the Na-K pump is particularly active), **membrane potentials have been found that could be shown to be more negative than E_K!** Such a situation could not arise from a diffusion potential and points to **electrogenic** action of the Na-K pump (as if the pump moved **unequal** numbers of Na^+ and K^+ ions). Similarly, work on red blood cell membranes supports the idea that each "turn of the pump" (Fig. 5-3) involves 3 Na^+ ions for every 2 K^+ ions entering the cell. Recent studies have tended to confirm the 3:2 rather than 1:1 nature of membrane pumping. How can such an unequal pump ratio lead to steady state? To see how this works out, compare Figs. 5-2 and 5-3. Note that in both cases influx and efflux are equal for each ion. In Fig. 5-3, the electrogenic action of the pump hyperpolarizes the membrane toward E_K, thus reducing I_K and increasing I_{Na}: **steady state is reached when the ratio of the diffusional currents is the same as the ratio of the pumped fluxes.**

5 Since it still must be true that $I_{mem} = 0$, it follows that Eq. 4-3 should be modified to include I_{pump}:

$$I_{mem} = 0 = I_{por} + I_{cap} + I_{pump} \quad \text{Eq. 5-1}$$

But in steady state $I_{cap} = 0$, so

$$I_{por} + I_{pump} = 0 \quad \text{Eq. 5-2}$$

Expanding this equation in the general form (cf. Eq. 4-26), we find

$$I_{pump} = -\Sigma(g_j V_s - g_j E_j) \quad \text{Eq. 5-3}$$

Fig. 5-2. Steady state with 1:1 pump ratio. Numbers indicate ions moving per "turn of the pump." Simplified from Fig. 4-25.

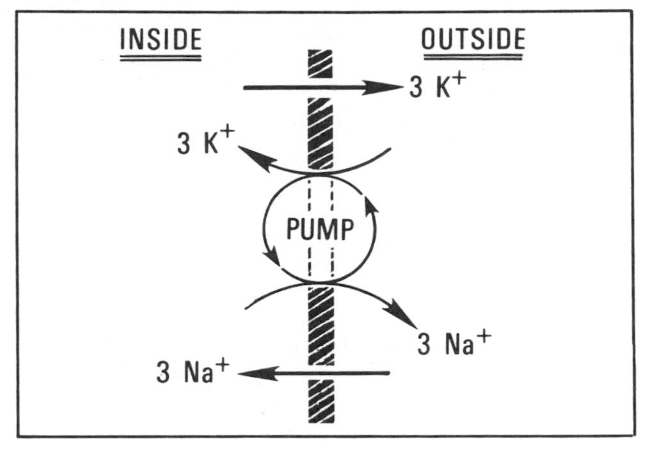

Fig. 5-3. Steady state for a 3:2 pump ratio. Symbols are as in Fig. 5-2.

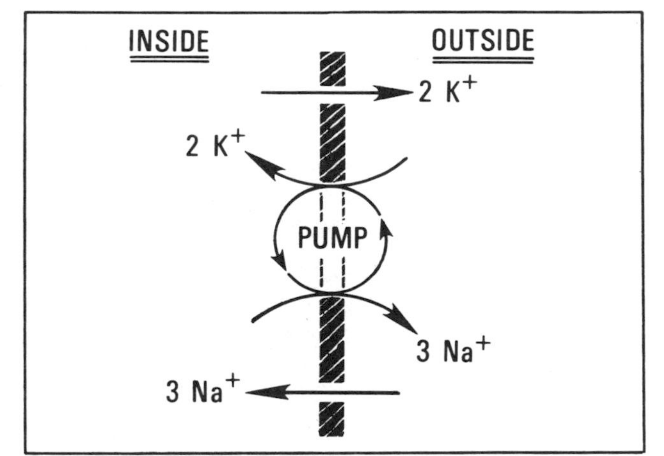

86

which yields

1 $$V_s = \frac{\Sigma g_j E_j - I_{pump}}{\Sigma g_j}$$ Eq. 5-4

But we also know that the ratio of the electrodiffusional currents must be the same as the ratio of the pumped fluxes (see Fig. 5-3); hence in steady state I_{Na} must be larger than I_K in the ratio 3:2. If we assume, for example, that $I_{Na} = -3$ mA and $I_K = 2$mA, then it is easy to see that $|2I_{Na}|$ must be equal to $|3I_K|$. And V_s must be that potential at which the electrodiffusional fluxes are unbalanced to exactly this extent. Thus in steady state

2 $$2I_{Na} + 3I_K = 0$$ Eq. 5-5

and we can expand this to get the steady-state equation that would occur if a 3:2 pump were active:

3 $$V_s = \frac{2g_{Na}E_{Na} + 3g_K E_K}{2g_{Na} + 3g_K}$$ Eq. 5-6

QUESTION: Where $E_{Na} = +60$ mV, $E_K = -100$ mV, $g_{Na} = 0.02$ mmho/cm^2, and $g_K = 0.28$ mmho/cm^2, calculate the resting membrane potential, first for a 1:1 pump ratio and then for a 3:2 pump ratio. (Hint 2↓) 4

Finally, note that even if the Na-K pump normally should be electrogenic, it is always possible that this electrogenicity might be counteracted by the activity of other possibly electrogenic ion pumps regulating other ions, for example, $[Ca^{2+}]_i$, $[Mg^{2+}]_i$, $[H^+]_i$, $[HCO_3^-]_i$, and even $[Cl^-]_i$ in a number of cells. **It seems probable that charge separation due to metabolic pumping makes only a very minor contribution to the steady-state resting potential under normal circumstances in most cells.** 5

QUESTION: A 3:2 (Na^+-K^+) pump is clearly *electrogenic*; however, if, in addition, 1 chloride ion were pumped out for every 3 sodium ions, then *such a 3:2:1 Na^+-K^+-Cl^- pump would be electroneutral*. How would you set about calculating V under these circumstances? (Hint 1↓) 6

86

87

CELL VOLUME AND INTRACELLULAR ANIONS

In our "typical cell" (Figs. 5-4/4-23), the major ions responsible for the resting membrane potential are listed as Na^+, K^+, Cl^- and A^-, with A^- being a collective term for all intracellular anions other than chloride. Most of these anions have extremely low conductances (many are completely impermeant); therefore, they can be ignored even in expanded versions of the steady-state equation. The electrophysiological contribution of **these anions** is simply that they **balance the intracellular cations to produce an isosmotic, electroneutral intracellular medium.** 1

However, the presence of impermeant anions has interesting consequences related to the maintenance and regulation of cell volume. Some of these consequences are listed here: 2

1. **Cell volume may be affected by changes in external ion concentrations.** Let us look back to the section "Ionic Pumps," in Chap. 4, where the effect of sudden changes in extracellular chloride concentration is discussed. Note that where $[Cl^-]_o$ falls, $[Cl^-]_i$ must also fall because of the resultant net efflux of Cl^- ions. This fall of $[Cl^-]_i$ results in a return of E_{Cl} to its initial value and hence a return of membrane potential to the value predicted by consideration of Na^+ and K^+ distributions alone. This explanation (although valid) appears to have a serious logical flaw when it is translated to physicochemical terms. Surely the large shifts in intracellular chloride concentration (required by the electrical explanation) could not occur without seriously disturbing intracellular electroneutrality and thus producing enormous transmembrane potentials! Note the extremely small magnitude of the ion fluxes required to produce potentials on the order of 100 mV (see page 58), whereas here we consider changes in $[Cl^-]_i$ that could be as large as tens of milliequivalents. There is no doubt that they occur without producing huge potentials; how can that be? *The answer is that cell volume changes.* To show how this works, let's return to the example of a sudden reduction in $[Cl^-]_o$: Falling $[Cl^-]_o$ reduces the influx of Cl^- ions, thus creating a net Cl^- efflux. However, as Cl^- leaves the cell, the cell becomes depolarized and the driving force on K^+ ions is increased while the driving force on Na^+ ions is reduced. The net result is an increase in both I_K and I_{Cl}—hence a net loss of KCl from the intracellular medium. But the cell membrane is freely permeable to water (actually more permeable than to ions!); therefore, water follows the movement of KCl, and cell volume falls. Such volume changes act on a fixed amount of A^-, and the intracellular concentration of A^- rises as the cell volume decreases. Since movements of water are very fast compared with the rather slow concentration changes that would be produced by the resting ion fluxes, volume changes keep pace with the KCl shift, thus preserving both osmotic equilibrium and gross electroneutrality. 3

Fig. 5-4/4-23. Hypothetical cell permeable to K^+, Na^+, and Cl^-, but not to A^-.

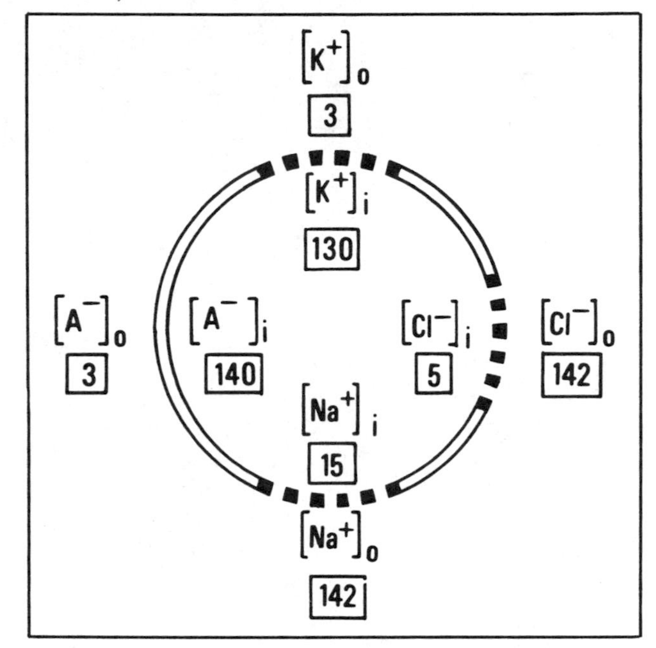

87

88

QUESTION: If the typical cell of Fig. 5-4/4-23 is placed in an extracellular solution containing 71 mEq/L of indiffusible SO_4^{2-} ions and only 71 mEq/L of Cl^- ions, can you think of any way in which you could calculate the resulting percentage reduction in cell volume? (Hint 4↓)

1

2. Cell volume is affected by changes in the steady-state resting potential. For the typical cell of Fig. 5-4/4-23, we may imagine that $V_s = E_{Cl} = -88.6$ mV. But what would happen if a drug were introduced that increased g_{Na} such that V_s tended to a new steady-state value of -61 mV? Surely $[Cl]_i$ would rise until E_{Cl} became equal to this new value of V_s. Once again, a large apparent anion shift is "buffered" by water movement: what happens is that cell volume rises as water enters along with the net entry of KCl and NaCl. This entry of water dilutes the finite quantity of intracellular A^-, thus allowing an increase in $[Cl^-]_i$ without change of total intracellular anion concentration [62, pp. 19–23].

2

QUESTION: Where $E_{Cl} = -61$ mV, $[Cl^-]_i$ must be 14.2 mEq/L in the typical cell. How large a percentage increase in cell volume is required to accommodate this degree of depolarization? (Hint 5↓)

3

The advanced student should be aware that the rather simple method of calculating cell volume change suggested here predicts enormous volume increase (several thousand percent!) if the cell is depolarized to near-resting potential. In fact, such changes do *not* occur. The reason is that we have assumed that E_{Na} and E_K remain constant despite the changing membrane potential. This assumption is perhaps valid for hyperpolarizations or small depolarizations. The assumption is certainly not valid, however, following large depolarizations (which typically are found to be accompanied by a rise of $[Na^+]_i$ and a fall in $[K^+]_i$).

4

3. The rate of passive chloride redistribution may be regulated by g_K, not g_{Cl}! In the previous examples, we noted that chloride movements were accompanied by K^+ ions and water. Now let us suppose that $g_K < g_{Cl}$. The rate of Cl^- movement would be limited by g_K, since any movement of Cl^- alone would transiently change the membrane potential to the point where the K^+ and Cl^- ions moved at the same rate. In frog muscles $g_K < g_{Cl}$, and such an effect has been noted in the transient changes of membrane potential shown following change of $[Cl]_o$ in this tissue.

5

In each of the given examples of volume change, the volume change has been precipitated by the requirement for gross changes in the internal concentration of the passively distributed chloride ion. Two questions come to mind at this point:

6

88

89

a. Why always chloride? The answer is that **chloride is about the only permeant ion of major physiological importance whose concentration may *not* be regulated by some kind of metabolic pump.**

1

b. Would such volume changes occur in cells known to contain a metabolic chloride pump? Probably only small changes would occur, since it is reasonable to suppose that such a pump would tend to change its rate so as to maintain a fairly constant intracellular chloride concentration. It might be (teleologically speaking) that pumping chloride is the price that a cell must pay to avoid large volume changes associated with prolonged alteration of membrane potential away from E_{Cl}. Indeed, it is noticeable that **cell membranes that normally face periods of prolonged depolarization** (e.g., the cell bodies and dendrites of central neurons, smooth muscle cells, and some sensory endings) **typically pump chloride ions,** whereas **membranes that face only very transient depolarizations** (axons and skeletal muscle membranes) **typically do not possess active chloride pumps.**

2

HINTS

1. If the total pump is electroneutral, then in steady state

$$I_{pump} = 0 = I_{por}$$

and it follows that V could be obtained directly from Eq. 4-27. But we know more than this about the system. If the cell is in steady state, then the ratio of the pumped fluxes must be the same as the ratio of the passive fluxes. Hence if the "pump ratio" is 3:2:1, then where $I_{Na} = -3$ mA, I_K would be +2 mA and I_{Cl} would be +1 mA. Thus it must also be true that V_s could be calculated from Eq. 5-6 or from an equivalent equation based on the following:

$$I_{Na} + 2I_K + 3I_{Cl} = 0$$

If you don't believe that all three equations predict the same value of V_s, go ahead and substitute some appropriate values! (The answers come out right, *but only if you have chosen values such that the passive flux ratios are 3:2:1.)*

2. For the 1:1 pump,

$$V_s = \frac{0.02(60) + 0.28(-100)}{0.30}$$

$$= \frac{-26.8}{0.3} = -89.3 \text{ mV}$$

For the 3:2 pump,

$$V_s = \frac{2(0.02)(+60) + 3(0.28)(-100)}{2(0.02) + 3(0.28)}$$

$$= -\frac{81.6}{0.88} = -92.7 \text{ mV}$$

Thus the difference in the calculated value of V_s is only 3.4 mV. Do you think that this difference would be detectable experimentally? (Hint 3↓)

89

90

GOLDMAN EQUATION

The **Goldman equation** has been so important in the historical development of the concept of the **diffusion potential** that it would be irresponsible to omit it from any discussion of basic neurophysiology. However, **the later sections of this book do not require that you understand the Goldman equation,** and you need not read this section unless you are especially interested in this concept.

1

The Goldman equation was developed to describe the steady-state "nonequilibrium" potentials that might develop across cell membranes as a result of diffusion of ions down their concentration gradients. In the situation shown in Fig. 5-5, if K^+ and Na^+ ions had the same ability to cross the membrane, diffusion would occur until the concentration gradients ran down; no potential would be produced. If, however, the mobility of K^+ ions within the membrane M_K were greater than the mobility of Na^+ ions in the membrane, an initial charge separation would occur, which would tend to slow the movement of K^+ ions and speed the movement of Na^+ ions. A "steady-state potential" would be achieved when the fluxes became equal; clearly the magnitude of this potential would depend on the difference in the mobilities of the ions and the size of their concentration gradients.

2

Presuming that no hydrostatic or osmotic gradients exist across the membrane and that the electric field is linear within the membrane, Goldman was able to derive the following general equation to describe the potential produced by diffusion of univalent ions when net flux across the membrane is zero (see Fig. 5-6):

$$V_s = \frac{RT}{F} \ln \frac{\Sigma M_C C_2 + \Sigma M_A A_1}{\Sigma M_C C_1 + \Sigma M_A A_2} \quad \text{Eq. 5-7}$$

where V_s = potential in solution 1 with respect to an electrode placed in solution 2

M_C, M_A = membrane cation and anion mobilities, in micrometers per second divided by volts per centimeter.

C_1, C_2 = cation concentrations on sides 1 and 2 of membrane, respectively

A_1, A_2 = anion concentrations on sides 1 and 2 of membrane, respectively

3

The advanced student should note that Goldman's equation for the diffusion potential across an ion-selective membrane is *not* identical to the often-quoted equation for the diffusion potential at a liquid-liquid interface [10, p. 1084; 27, p. 49].

4

Goldman's equation was subsequently adapted by Hodgkin and Katz, who utilized the subtle concept of *permeability* (with units of centimeters per second), such that permeability P is given by

5

Fig. 5-5. Diffusion across ion-selective membrane. Assume the membrane to be impermeable to SO_4^{2-} (Concentrations in mEq/L.)

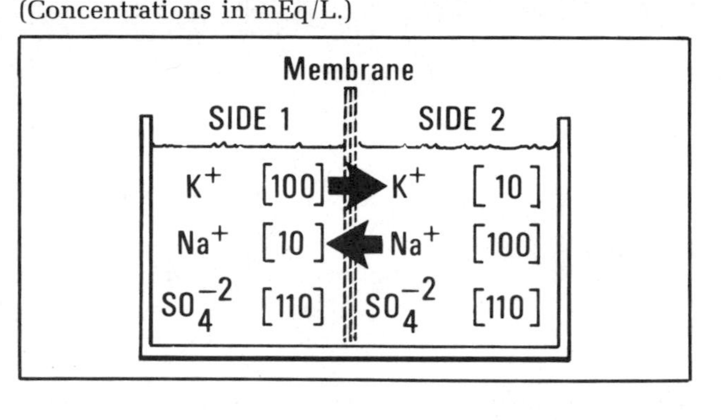

Fig. 5-6. Potential produced by diffusion across ion-selective membrane separating different concentrations of univalent ions. In a cell of this type, osmotic gradient could be balanced (temporarily) by inclusion of uncharged substance (e.g., sucrose) on the side with lower concentration.

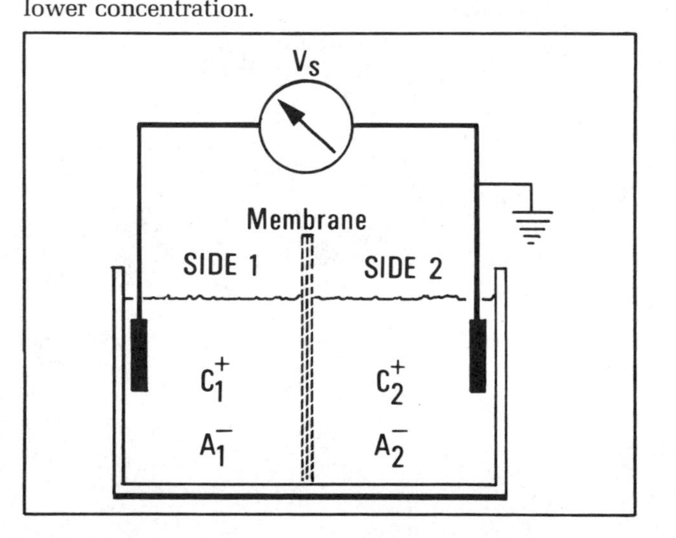

90

91

$$P = \frac{M\beta RT}{\Delta x\, F}$$

where M = mobility as previously defined
β = proportionality factor relating ion concentration at surface of membrane to concentration in bathing solution
Δx = thickness of membrane.

1

Assuming the major monovalent ions involved in the diffusion potential to be Na^+, K^+, and Cl^-, Hodgkin and Katz derived this familiar form of the Goldman equation:

$$V_s = \frac{RT}{F} \ln \frac{P_K[K^+]_o + P_{Na}[Na^+]_o + P_{Cl}[Cl^-]_i}{P_K[K^+]_i + P_{Na}[Na^+]_i + P_{Cl}[Cl^-]_o} \qquad \text{Eq. 5-8}$$

2

Remember that the main assumptions underlying this equation are as follows:

1. Net current across the membrane is zero.
2. No hydrostatic or osmotic pressure gradients exist across the membrane.
3. Ions cross the membranes independently.

3

(Do these assumptions sound familiar?)

4

QUESTION: Assuming $P_{Na}/P_K = 0.01$, what is the value of the potential expected between sides 1 and 2 in Fig. 5-5 at 37°C? (Hint 6↓)

5

HINTS

3. Probably not. Internal ion concentration is not easy to measure accurately; hence estimates of E_{Na} and E_K may be off by several millivolts. Similarly, there may be some doubt as to the accuracy of measurements of resting ion conductances, and even the measurement of the cell membrane potential is hardly accurate to more than ±1 mV. As a result, although the possibility of a 3:2 pump has been suggested many times over the last 25 years, it is only quite recently that neurophysiologists have seriously considered that the Na-K pump might be normally electrogenic (even in the squid axon where so much research has been done!).

4. Steady state will be reached when $[Cl]_i$ = 2.5 mEq/L. To accept this loss of anion, $[A^-]_i$ must rise to 142.5 mEq/L; that is, cell volume must be reduced to (140/142.5) 100 percent of its original size. The reduction in cell volume is about 1.8 percent.

5. If $[Cl]_i$ increases from 5 to 14.2 mEq/L, then A^- must fall from 140 to 130.8 mEq. The percentage volume change is thus (140/130.8)100 − 100 = 7 percent.

7. Assume $P_{Na} = 1$. Then if $P_{Na}/P_K = 0.01$, $P_K = 100$. Now substitute into Eq. 5-8. Alternatively, you can divide both numerator and denominator by P_K (as was done for Eq. 4-21) to get

$$V_s = 61 \log \frac{[K^+]_2 + (P_{Na}/P_K)\,[Na^+]_2}{[K^+]_1 + (P_{Na}/P_K)\,[Na^+]_1}$$

Now substitute directly into this equation. See Hint 8.↓

9. Well, that's one intelligent possibility. Why not go check it out, either in the literature or on your favorite membrane?

91

The major advantage of this equation over the steady-state equation is that it involves the *membrane property of permeability P* rather than the *system property of conductance C*. In other words, since conductance is merely a proportionality factor between current and driving force, it confuses low concentration with low permeability. The current would be small both for a plentifully available but barely permeant ion and for a scarce but highly permeant ion: both situations would indicate a low conductance if the driving force were large.

1

Unfortunately, although the Goldman *permeability equation* gives V_s directly from membrane permeabilities and ionic concentrations, it becomes extremely cumbersome in the description of transient events (covered in Chaps. 6, 7, and 8). Compare, for example, the simplicity of the Ohm's law equations for ionic currents, $I_K = g_K(V - E_K)$, with the equation for I_K derived from the Goldman equation:

2

$$I_K = P_K \frac{F^2V}{RT} \frac{[K^+]_i e^{FV/(RT)} - [K^+]_o}{e^{FV/(RT)}} \qquad \text{Eq. 5-9}$$

Complexity may not be the only problem with the Goldman equation. The potassium conductance that would be calculated from the Goldman equation (called G_K here, to differentiate it from the experimentally determined g_K) can be simplified to

$$G_K = \frac{P_K F^3 [K^+]_o V}{R^2T^2(1 - e^{-VP/(RT)})} \qquad \text{Eq. 5-10}$$

for small potassium currents [27, p. 72]. But when $[K^+]_o$ and P_K do not change, this equation reduces to

3

$$G_K = \text{const}\left(\frac{V}{1 - e^{VF/(RT)}}\right) \qquad \text{Eq. 5-11}$$

It is now patently clear that **G_K in the resting membrane is predicted to be a nonlinear function of voltage,** rather than a constant value independent of voltage as we previously assumed (and as has been implicit in our utilization of the ionic Ohm's law). Thus the Goldman equation predicts that $|I_K|$ would be different if a driving force of the same absolute magnitude were applied to produce first an outward, and then an inward, potassium current! To put this yet another way: If the driving force were the same in each direction and if P_K were similarly

4

93

symmetrical, why would the nearly 50-fold K^+ concentration difference across the membrane not result in a larger probability of K^+ ions leaving the cell than of their entering the cell? Obviously, this is a serious discrepancy between two approaches to the analysis of membrane function. The pleasant simplicity of the ionic Ohm's law should not be enough to justify its use if it is not an adequate representation of membrane properties. 1

Fortunately, an impressive array of experimental studies has confirmed that *in most* instances an entirely adequate analysis of membrane function may be obtained on the basis of the Ohm's law approach. Conductance measurements on excitable membranes under voltage-clamp conditions (described in detail in Chap. 8) confirm the wide applicability of this analysis. Where necessary, the Ohm's law approach can be modified by including a more complex Goldman conductance term; however, often this has not provided any major improvement in the analysis of membrane function. 2

QUESTION: Does this imply that permeabilities may not be symmetrical and that permeability in one direction might be different from the permeability which would be measured for the same ion moving in the opposite direction? (Hint 9↑) 3

HINTS

6. Just as we simplified the Nernst equation, the Goldman equation can be simplified by the same "fudge factor" at 37°C to give V_s in millivolts:

$$V_s = 61 \log \frac{P_K[K^+]_2 + P_{Na}[Na^+]_2}{P_K[K^+]_1 + P_{Na}[Na^+]_1}$$

So calculate the potential! If you are still having trouble, see Hint 7.↑

8. $$V_s = 61 \log \frac{10 + 0.01(100)}{100 + 0.01(10)}$$

$$= 61 \log 0.11 = 61(-0.96) = -58.6 \text{ mV}$$

10. Surface area of sphere $= \pi d^2 = \pi(10 \times 10^{-4} \text{ cm})^2$

Applied current density $= \dfrac{7.85 \times 10^{-12} \text{ A}}{3.14 \times 10^6 \text{ cm}^2}$

$= 2.5 \times 10^{-6} \text{ A/cm}^2$

$$R_m = \frac{10 \times 10^{-3} \text{ V}}{2.5 \times 10^{-6} \text{ A/cm}^2}$$

$= 4 \times 10^3 \ \Omega \cdot \text{cm}^2$

$$C_m = \frac{\tau_m}{R_m} = \frac{4 \text{ ms}}{4 \text{ K}\Omega \cdot \text{cm}^2}$$

$= 1 \ \mu\text{F/cm}^2$

11. First, calculate applied current density from applied current and cell surface area. Next, calculate specific membrane resistance directly from Ohm's law, using the maximum ΔV_s as the driving force. Now, obtain the time constant of the rise and/or fall of ΔV_s when the current pulse is turned on or off. Finally, obtain C_m from τ_m/R_m.

93

94

EXPERIMENTAL TECHNIQUES

For those students especially interested in experimental methods we include some details here and others in Chaps. 7 and 8.

How does one set about actually *measuring* specific membrane resistance or conductance? How can one *measure* membrane capacitance? These and a whole number of other questions may have occurred to you, and there is no good reason why we should not satisfy your curiosity. However, when such measures are attempted in real cells, a variety of problems arise that cannot be discussed in any depth at this stage of the book, since such discussion would have to draw on material presented in Chaps. 7 and 8.

One might have supposed from an experiment such as that shown in Fig. 5-7/3-11 that it would be a relatively simple matter to obtain R_m from the applied current and the observed maximal voltage deflection seen in this figure. Let's see how we could do that. If we knew the cell surface area, we could calculate the applied current density (current per unit membrane area) *but only if we assume that current is uniformly distributed over the cell surface.* Here is the catch. We have, as yet, no way to deal with *nonuniform* situations where currents are *not* the same in all parts of the cell. We cannot yet even define the circumstances under which nonuniform currents would be expected. We take up this problem in Chap. 7.

QUESTION: Presuming we *could* assume a uniform current density, can you figure out how to obtain R_m (in ohm-cm^2) and C_m (in μF-cm^{-2})? (Hint 11↑)

QUESTION: In a spherical cell 10 μm in diameter in which a 10-mV hyperpolarization occurs with a time constant of 4 ms following a current pulse of 7.85×10^{-6} μA, what is the value of the membrane specific capacitance in microfarads per square centimeter? (See Hint 10.↑)

We therefore offer you a raincheck; read as far as Chap. 8 and then you will be in a position to tolerate a thorough exposition of the experimental techniques required to measure basic cell membrane properties.

After the curious and interesting (to us anyway) digressions of this chapter, now we return to the mainstream of the logical development of your understanding of membrane potentials. In Chap. 6 we consider the mechanisms of cell electrical excitability by which an action potential can be induced after triggered depolarization of the cell membrane.

Fig. 5-7/3-11. Both upper and lower graphs share the same time axis. The upper graph shows the applied-current pulse (negative sign means that current is passed inward across the cell membrane). The lower graph shows the change in membrane potential (from a resting potential of −90 mV) produced by the applied current. The time constant of potential change is the same when the current is turned on as when it is turned off.

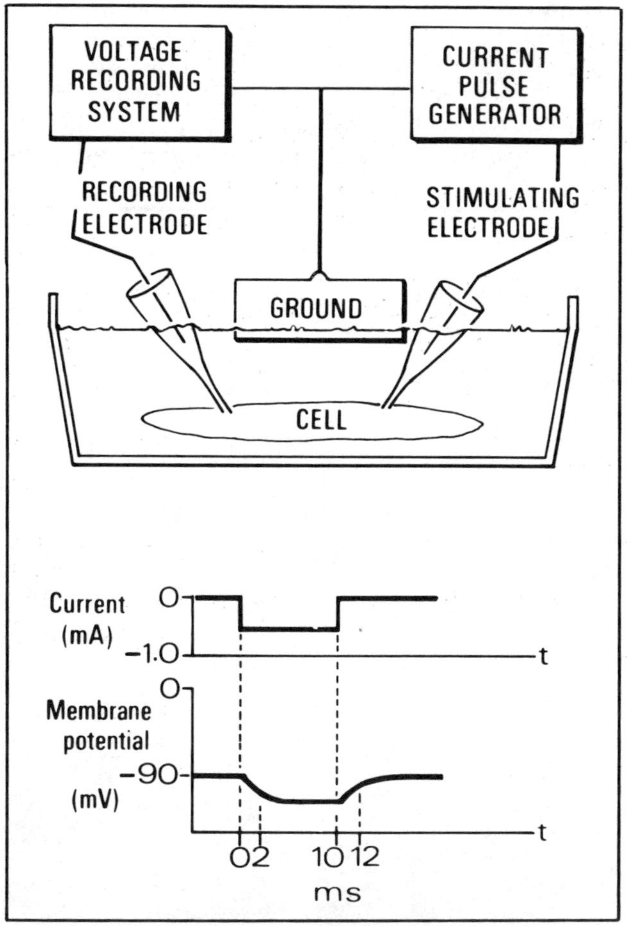

94

95

Introduction to the Action Potential

96

Up to this point, we dealt with **steady membrane potentials,** that is, those that do not change rapidly. We now describe a class of naturally occurring, rapid changes in membrane potential, the *action potential,* that can propagate along the membrane surface. Such action potentials serve the function of transmitting information from one location to another, over distances ranging from millimeters to more than 30 m (in the great whales). 1

The mechanism by which an action potential can travel down an axon is described in detail in Chap. 7. Here we are concerned with the membrane mechanisms that produce the action potential itself, that is, by what means the membrane potential can change so rapidly, from negative inside to positive inside and back again, as shown in Fig. 6-1. The details of this figure will become clearer as we progress through the chapter. Note that in this chapter we describe only the mechanisms involved in the generation of the action potential **at one region** of the axon surface. The mechanisms of interaction between neighboring regions of cell membrane are described in Chap. 7. 2

To further simplify our initial presentation of the action potential, we presume that the whole membrane region under study experiences the same potential changes at exactly the same time and that this membrane region is in some way isolated from the remainder of the cell. 3

This assumption can be duplicated experimentally by a technique known as **space clamping.** In its simplest form, this technique involves inserting a low-resistance (*axial*) wire electrode down the middle of the axon. The interior of the axon thus remains isoelectric; regional differences in transmembrane potential cannot occur where the extracellular medium is also essentially isoelectric. The action potential that is seen under these experimental conditions is referred to in the literature as a **membrane action potential.** Such membrane action potentials show only minor differences in shape and time course from the more commonly recorded **propagated action potential.** The basic underlying mechanisms are exactly the same in both instances, but rigorous mathematical analysis is enormously simplified by the space clamp conditions. 4

Finally, **a word of warning:** We are convinced that even simple concepts cannot be fully understood until utilized in a number of different ways. Therefore, we deliberately made the early sections of this chapter repeat concepts from previous chapters. These basic concepts form the only foundation on which a thorough understanding of the action potential can be based. 5

Most students find the review of concepts that follows is beneficial in reaffirming the importance of these ideas in the logical progression that we present. We assure you that if the next few pages are not familiar (even slightly boring), then you may well find the understanding of the action potential unnecessarily difficult. 6

Fig. 6-1. Transmembrane potential in squid giant axon during action potential. The shaded regions in this diagram have the names *overshoot* and *after-hyperpolarization.* Note that each has both magnitude (mV) and duration (ms).

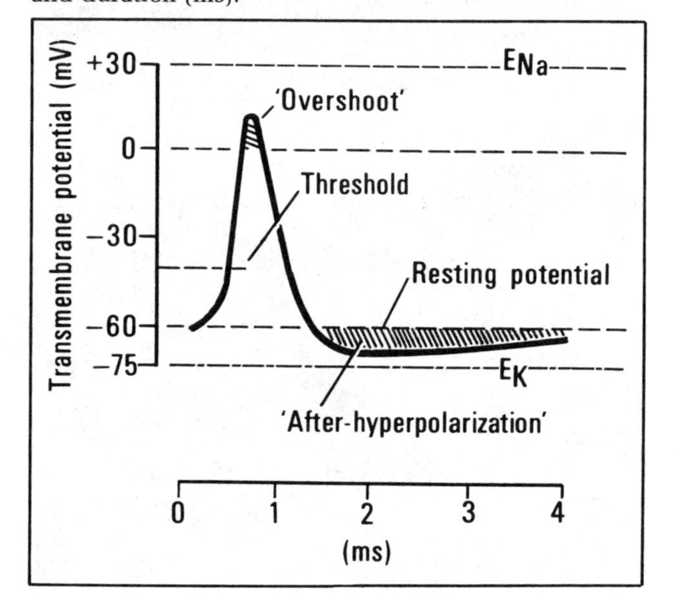

96

97

REVIEW OF RESTING POTENTIALS

1 As you will soon see, the ionic mechanisms of the action potential are, for the most part, merely extensions of the concepts you learned in Chap. 4, so why not a quick review as a warmup?

2 (*Note:* It will be easiest for you to understand the material of this and subsequent chapters in terms of electrical circuit models of the membrane. So if you never made the transition from the physicochemical membrane model to the ideas described in this section, you may have to review Chap. 4.)

3 **The membrane potential is the potential that exists across the membrane capacitance.** The membrane capacitance is charged by current flow through membrane "pores" as ions move along their electrochemical gradients. A given ionic current is determined by the product of that ion's conductance, g_j, and the driving force, $V - E_j$. Thus, $I_j = g_j(V - E_j)$.

4 **The pores together with the existing concentration differences for permeable ions** (assumed to be K^+ and Na^+ for the time being) **form "batteries" whose currents, flowing in complete circuits** (as shown in Fig. 6-2/4-18), **charge the membrane capacitance.** Since the conductance of the K^+ battery is higher than that of the Na^+ battery, the membrane potential is closer to E_K.

The influence of membrane conductance on membrane potential *V* is most easily remembered by the steady-state equation:

5
$$V_s = \frac{g_{Na}E_{Na} + g_K E_K}{g_{Na} + g_K} \qquad \text{Eq. 6-1/4-20}$$

Since E_{Na} and E_K are constants (let's choose +60 and −100 mV, respectively), Eq. 6-1/4-20 can be restated (for our typical axon) as

6
$$V_s = \frac{+60g_{Na} - 100g_K}{g_{Na} + g_K} \qquad \text{Eq. 6-2}$$

7 From this equation you can readily see that ***V* must depend on only g_{Na} and g_K.** In fact, **the g_{Na}/g_K ratio determines *V*.** (If you do not believe this last statement, see Hint 1.↓)

Fig. 6-2/4-18. Current flows from Na^+ and K^+ batteries.

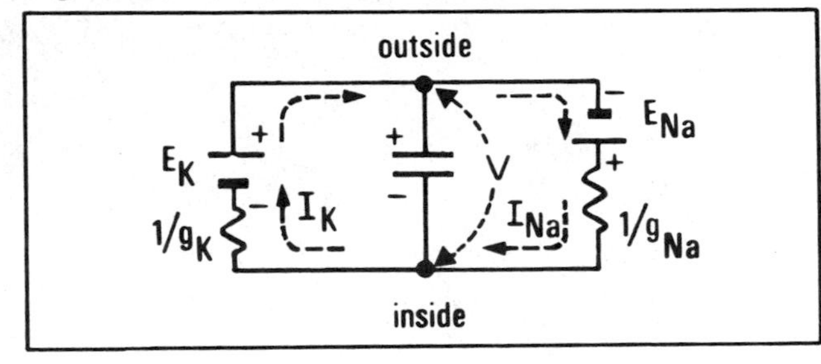

97

The importance of the g_{Na}/g_K ratio can be seen easily by dividing both numerator and denominator of Eq. 6-2 by g_K and simplifying to obtain:

$$V_s = \frac{60(g_{Na}/g_K) - 100}{g_{Na}/g_K + 1} \qquad \text{Eq. 6-3}$$

Again, it should be apparent that the g_{Na}/g_K ratio determines V. (Those who are still in doubt should see Hint 2.↓)

1

Of course, V_s is also dependent on the ionic concentrations that determine E_{Na} and E_K, but a change in equilibrium potentials could hardly be the mechanism responsible for the rapid, millisecond-duration events of the action potential (see Fig. 6-1). It has not been possible to demonstrate *any* mechanism by which such relatively enormous ionic movements could be accomplished within the limited time available, without total disruption of the cell membrane.

2

LIMITS OF MEMBRANE POTENTIAL THAT CAN RESULT FROM CONDUCTANCE CHANGES

From the preceding section, you undoubtedly see the general line of the argument: **Rapid changes in membrane conductance ratios change the membrane potential rapidly.** What is the range within which V can vary as a result of changes in g_{Na} and g_K?

3

Since only the g_{Na}/g_K ratio matters, all we need to do is figure out the value of V at the minimum and maximum of the ratio, zero and infinity, respectively.

4

QUESTION: O.K., what are the limits as revealed by either Eq. 6-2 or 6-3 (or both)? (Hint 3↓)

5

Thus, you can see **by varying the E_{Na}/E_K ratio, the membrane potential can vary between the limits of −100 and +60 mV.** That is, a **membrane potential determined by g_K and g_{Na} can never exceed the limits of E_K and E_{Na}.**

6

While *rapid* changes in membrane potential are produced by ionic conductance changes, *slower* changes in membrane potential may involve changes in either equilibrium potentials or the rate of electrogenic pump activity of the type discussed in the section "Pump Potentials" in Chap. 5.

7

ACTION POTENTIAL: A FIRST APPROACH

An action potential, in this case from the giant axon of a squid, is shown in Fig. 6-1. The following points should be noted:

8

99

1. The resting potential in this axon is only −60 mV, and both E_{Na} and E_K are different from the values assumed for our typical axon.
2. The action potential will occur if the membrane potential is depolarized (reduced) past "threshold" by some "external energy source" (to be explained later). The mechanism of the initial upsweep (before threshold is reached) is described in Chap. 7.
3. At the peak of the action potential, the inside of the cell is **positive** with respect to the outside. This portion is called the **overshoot.**
4. The falling phase of the action potential typically hyperpolarizes beyond normal resting potential, that is, it is more negative than the resting potential. This hyperpolarized period is called the **after-hyperpolarization.**
5. Note the rapidity of the potential changes. The upsweep takes about 0.5 ms, and the whole depolarization phase is complete within about 1 ms.
6. All the changes in membrane potential occur within the limits set by E_{Na} and E_K.

1

QUESTION: Carefully examine Fig. 6-1. (*a*) What is the approximate magnitude of the overshoot in this action potential? (*b*) What is the duration of the after-hyperpolarization? (Hint 6↓)

2

QUESTION: Remembering that the steady-state equation (Eq. 6-1/4-20) can be applied only when the membrane potential is not changing (even for just an instant), can you determine the three points in Fig. 6-1 where this equation can be used? (Hint 4↓)

3

Since the steady-state equation can be applied at three points during the action potential, it is instructive to examine these points before progressing to any more complex analysis. Table 6-1 shows how the g_{Na}/g_K ratio (right-hand column) varies at these three points. The values of V_s (left side) and of driving forces and currents are also given, along with the magnitude in change in conductance.

4

At Resting Potential. Here g_K must predominate while g_{Na} is small. Table 6-1 shows g_K to be about 15 times greater than g_{Na} at rest ($g_{Na}/g_K = 0.067$).

5

HINTS

1. If you substitute a few numbers, you will be readily convinced! Try any values you like; solve for V; multiply g_{Na} and g_K each by a factor of 10, and solve again. Convinced?

5. Yes, the third point is the point of maximum after-hyperpolarization.

99

100

Table 6-1. Appropriate Values for Typical Axon at Those Points in Action Potential to Which Steady-State Equation Can Be Applied

		$E_{Na} = +60$ mV				$E_K = -100$ mV				
	V_s	$V - E_{Na}$	g_{Na}	Δg_{Na}	I_{Na}	$V - E_K$	g_K	Δg_K	I_K	g_{Na}/g_K
Resting potential	−90	−150	0.01	1	−1.5	+10	0.15	1	+1.5	0.067
Peak overshoot	+40	−20	10.0	1000	−200	+140	1.43	9.5	+200	7.0
Maximum after-hyperpolarization	−95	−155	0.04	4	−6.2	+5	1.24	8.3	+6.2	0.032
Units	mV	mV	mmho/cm²	ratio to resting g_{Na}	μA/cm²	mV	mmho/cm²	ratio to resting g_{Na}	μA/cm²	

At Peak Overshoot. Clearly the situation must be reversed, and g_{Na} must be much greater than g_K. Table 6-1 shows that g_{Na} is now 7 times greater than g_K. But notice that the ratio has increased (g_{Na} now predominating), even while g_K was also increasing! Thus g_{Na} has had to increase by as much as 1000 times to achieve the sevenfold ratio.

At Maximum Hyperpolarization. Now g_K must predominate again and to an even greater extent than at resting potential (since V_s is now so close to E_K). In Table 6-1, g_K is, in fact, 30 times greater than g_{Na} even though g_{Na} has still not returned to its resting level.

You will notice (probably with surprise) that g_K is almost the same at maximum hyperpolarization as it was at the peak overshoot. The reason is that g_K reaches its maximum value during the falling phase of the action potential, between the peak overshoot and maximum hyperpolarization. At peak overshoot it is still increasing, but here at maximum hyperpolarization it is slowly falling back to its resting level. But we are getting ahead of ourselves; all this is explained later in this chapter.

At some risk of redundancy, let us describe the mechanism by which the sodium and potassium currents are balanced in each of these instances. Remember that **where the membrane potential is not changing there can be no current passing through the membrane capacitance, and it necessarily follows that $|I_{Na}| = |I_K|$** (Fig. 6-2/4-18).

At Resting Potential. Here g_{Na} is smaller than g_K, but the Na and K currents are balanced since the sodium driving force $V - E_{Na}$ is substantially greater than $V - E_K$. Look now at Table 6-1 to see the driving forces, conductances, and currents for each ion at the resting potential. Remembering (from Chap. 4) that current is the product of conductance and driving force, you can easily verify the values for current, and then note that **at all three steady-state points the sum of the currents is zero.** (This is a good time to check whether you can calculate driving forces correctly. If you have trouble, reread pages 55 to 57.)

Fig. 6-3/4-22. Beam-balance model for resting membrane.

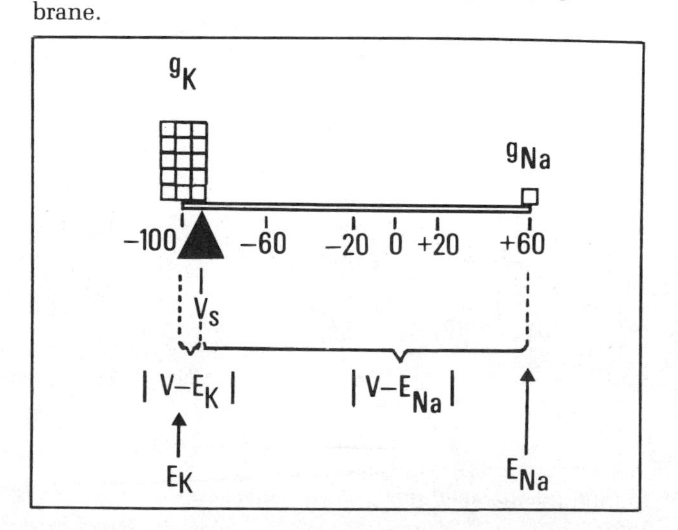

100

In Fig. 6-3 this relationship is diagrammed according to the balance-beam analogy introduced in Chap. 4.

At Peak Overshoot. Here g_{Na} is much larger than g_K. Fig. 6-4 shows the equivalent balance beam. Despite the marked changes in g_{Na} and g_K, the sodium and potassium currents are once again exactly balanced by the reduction in $V - E_{Na}$ (since V is near E_{Na}) and compensatory increase in $V - E_K$ (since V is far from E_K). Again, see Table 6-1 to note that both the sodium and potassium currents are markedly increased compared with their values at rest.

At Maximum After-hyperpolarization. Now g_K is greater than g_{Na}, but the disparity is more marked than at resting potential. The balance beam for this condition is shown in Fig. 6-5. Nevertheless, sodium and potassium currents are still balanced since $V - E_{Na}$ is larger and $V - E_K$ is even smaller than at resting potential (see Table 6-1).

To summarize, if we follow the course of the action potential starting from resting potential, it is clear that

1. **g_{Na} must increase relative to g_K during the depolarizing (rising) phase** of the action potential in order to reach the peak overshoot condition.
2. **g_{Na} must fall relative to g_K during the falling phase** to reach the maximum after-hyperpolarization.
3. **g_K must fall relative to g_{Na} in recovery from after-hyperpolarization** so as to return to the normal resting condition.

CURRENT FLOW DURING ACTION POTENTIAL

Up to this point, our analysis has been limited to the three points of the action potential at which the steady-state equation holds. Now, in order to understand better what occurs during an action potential, let us describe the portions *between* the steady-state conditions.

HINTS

2. If you substitute a few numbers, you will be convinced! Try any values you like. Solve for V; then multiply g_{Na} and g_K each by a factor of 10 and solve again. Or just realize that multiplying by 10/10 gives no change, so the absolute magnitude isn't crucial. **Now** are you convinced?
3. When $g_K \ggg g_{Na}$, then V approaches -100 mV (that is, E_K). When $g_K \lll g_{Na}$, then V approaches $+60$ mV (that is, E_{Na}).
4. The resting potential is obviously the first point. The peak of the overshoot is a second point. If you know the third point, you won't have to look at Hint 5.↑
6. (*a*) About 10 to 15 mV. (*b*) About 2.5 ms in this figure.

101

Fig. 6-4. Beam-balance model for peak overshoot of action potential. Quite different from Fig. 6-3, isn't it?

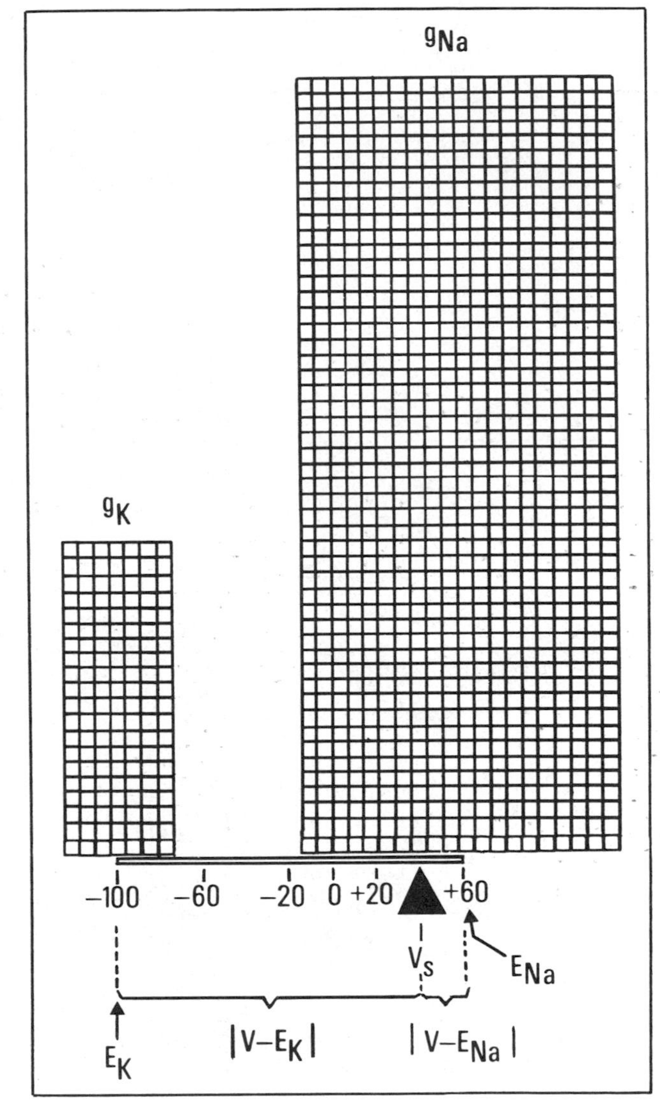

101

102

Since any flow of current into or out of the membrane capacitance affects the membrane potential, you can readily guess that we are about to describe currents flowing across the membrane capacitance. 1

QUESTION: If the membrane potential is depolarized (goes from negative to zero), what was the direction of current flow across the membrane capacitance? (Hint 7↓) 2

Where the steady-state equation holds, you remember that I_{cap} **= 0 and** hence $|I_{Na}| = |I_K|$. We now consider those regions of the action potential where I_{cap} does not equal 0 and hence where $|I_{Na}|$ does not equal $|I_K|$. 3

During the upsweep of the action potential the membrane depolarizes, which is equivalent to saying that **there is an outward current across the membrane capacitance.** Where does this outward I_{cap} come from? Isn't this just what you would expect if $|I_{Na}| > |I_K|$? There will be a net *inward* I_{por} and hence an *outward* I_{cap} to depolarize the membrane capacitance. Compare the situation in the depolarizing membrane (Fig. 6-7) with the resting condition (Fig. 6-6). 4

At the peak of the action potential, there is an instant when the membrane potential is not changing, and, again, $I_{cap} = 0$ and $|I_{Na}| = |I_K|$. Of course, **both currents are greater than in the resting state,** as you saw in Table 6-1. The current flow at the peak of the action potential is shown in Fig. 6-8. Note that the membrane capacitance is now positive inside. 5

As the membrane starts to repolarize after the peak of the action potential, there must be an inward I_{cap}, driven now by an outward current through the K^+ "battery," as shown in Fig. 6-9. Such a current through the membrane capacitance implies that $|I_K| > |I_{Na}|$, and this must remain the case throughout the falling phase of the action potential. 6

At the instant of maximum after-hyperpolarization, the membrane potential is not changing, so again $I_{cap} = 0$ and $|I_K| = |I_{Na}|$. While g_K is higher than at rest, g_{Na} is near the resting level (see Table 6-1). The current flows at this point are somewhat higher than at rest (Fig. 6-10). 7

Finally, as the membrane *depolarizes* back to the original resting level, there is again an outward capacitative current, and it must be that $|I_{Na}| > |I_K|$. The current flows are shown in Fig. 6-11. The charge on the membrane capacitance starts out as hyperpolarized (many negative charges inside) and then is reduced somewhat during return to the resting level. 8

By now you might appreciate (and even enjoy!) a summary diagram that correlates the preceding sections: Fig. 6-12. The preceding descriptions and diagrams describe the different regions of the action potential, as shown in Fig. 6-12. So you can reread any parts that are not clear at this point. 9

It is of utmost importance (as you will realize later) that you note that **all currents traveled in complete circuits,** i.e., that **the total current flow across the entire membrane, I_{mem}, was zero at all times!** 10

Fig. 6-5. Beam-balance analog of maximum after-hyperpolarization. Contrast this situation with Figs. 6-3 and 6-4.

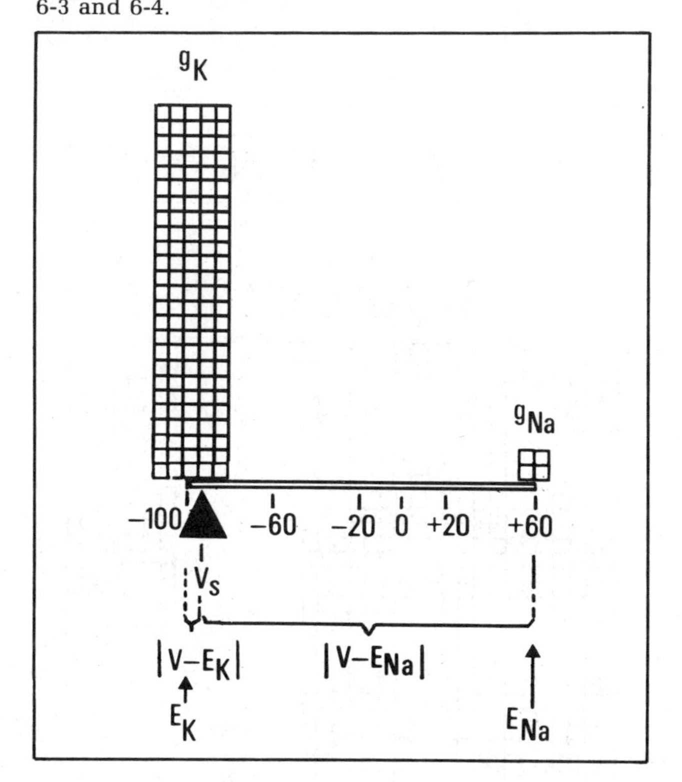

102

103

Fig. 6-6. Current flow during resting state, where $|I_K| = |I_{Na}|$, and $I_{cap} = 0$.

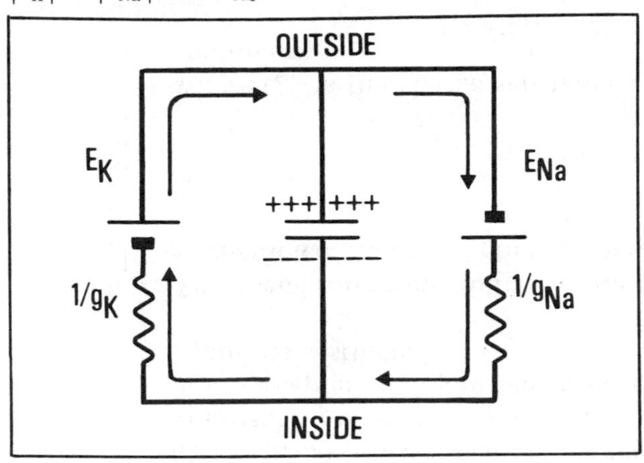

Fig. 6-7. Outward capacitative current due to increased g_{Na} (i.e., decreased $1/g_{Na}$) during depolarization.

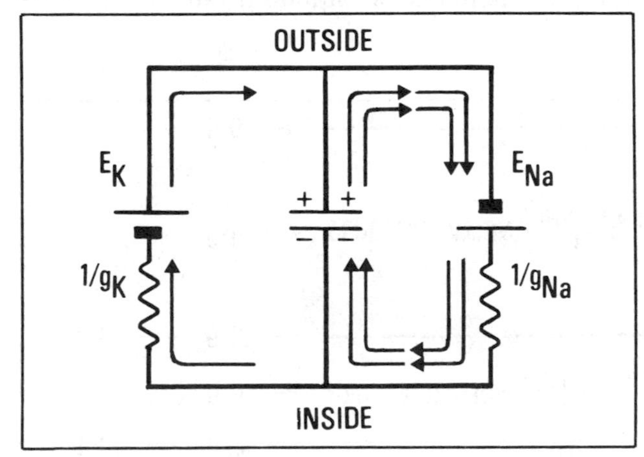

Fig. 6-8. Current flow at peak of action potential.

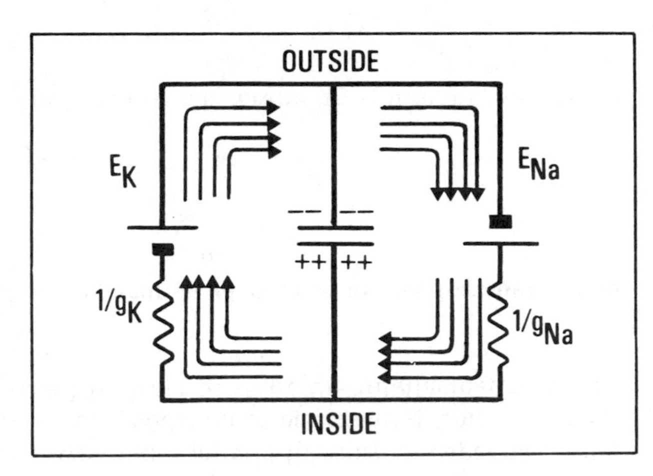

Fig. 6-9. Membrane currents during repolarization of membrane after depolarization of action potential.

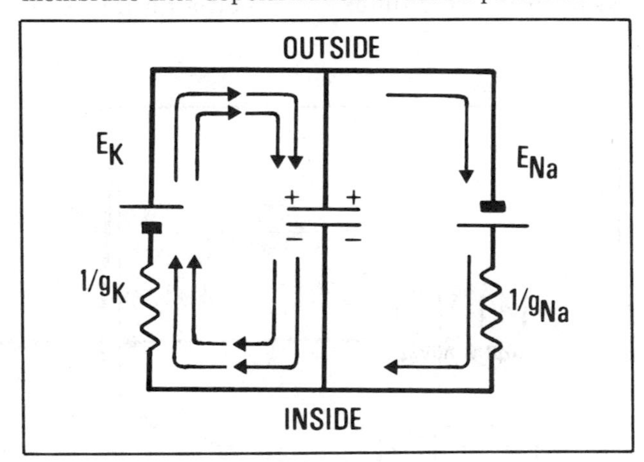

Fig. 6-10. Current flows at maximum of after-hyperpolarization.

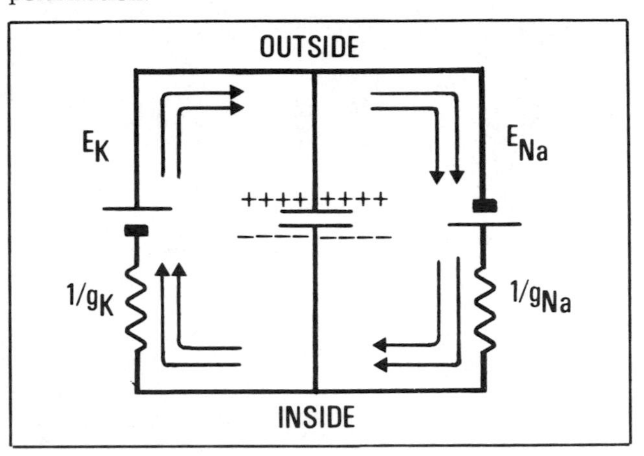

Fig. 6-11. Membrane currents during depolarization to the resting level following maximum after-hyperpolarization, as g_K decreases to resting level.

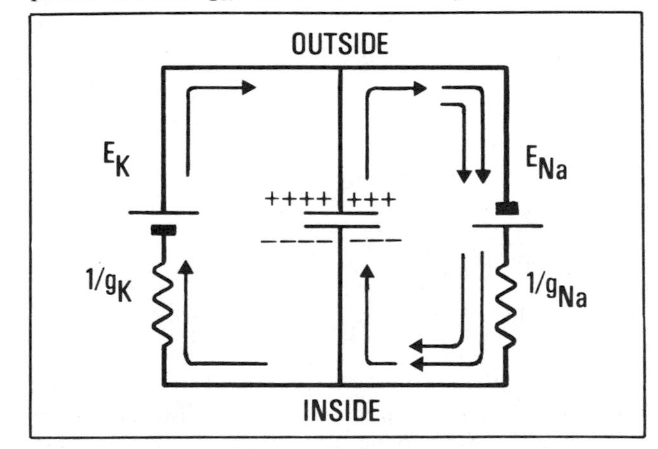

103

104

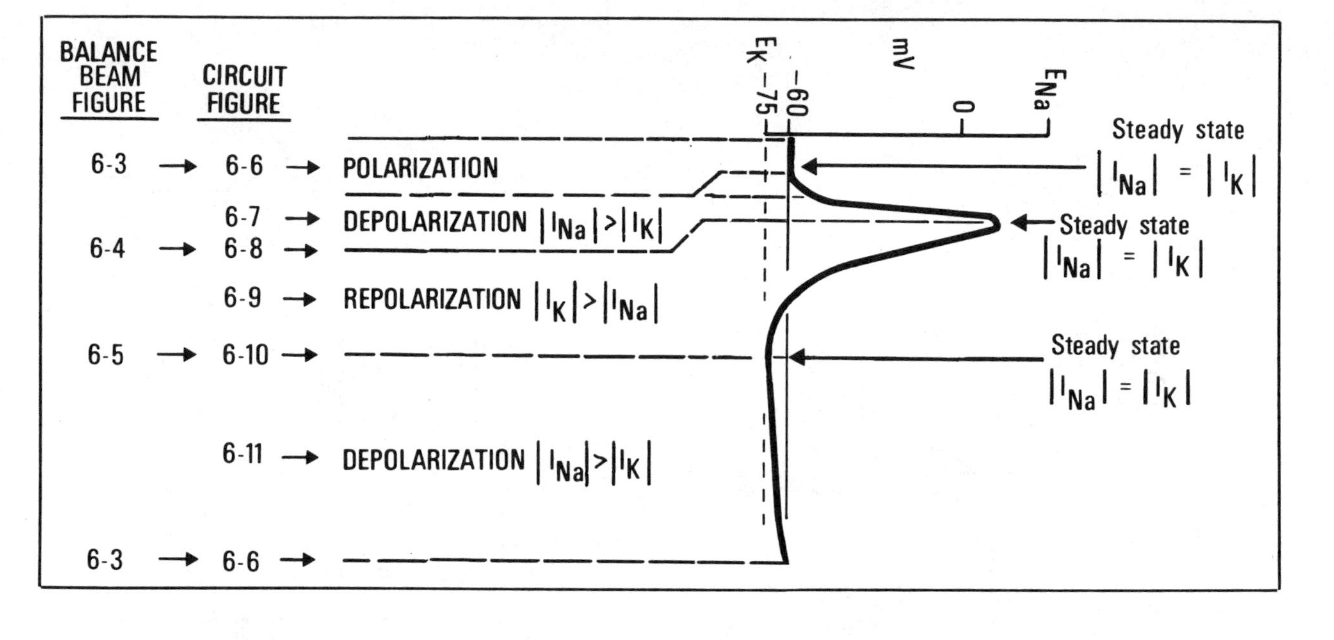

Fig. 6-12. Temporal course of depolarization and repolarization during action potential, showing relationship between the membrane potential at specific times and the underlying changes in both sodium and potassium currents. Corresponding circuit figures and beam-balance figures are indicated. Note that there is a region between initial membrane polarization and first depolarization that is not adequately described by these figures; this region is described later in this chapter.

To put it another way, any current going in (or out of) a pore (for example, I_{Na} or I_K) was matched by an equal current moving out (or in) through either the capacitance or other pores. (The reason for emphasizing this point is brought out in the next section, "Some Problems in Semantics.")
1

At this point we reveal the immense power and beauty of the current-capacitance approach to membrane potentials. Here is that modicum of background from the previous chapters that you will need to recall in order to enjoy the intellectual sequence and its aesthetics.
2

1. **The membrane potential is produced by separation of charge across the membrane capacitance.** As in any other capacitor,

$$Q_m = C_m V \qquad \text{Eq. 6-4}$$

where Q_m is the charge separated per unit area of membrane, in coulombs per square centimeter.
3

104

105

2. It follows from differentiating Eq. 6-4 that

$$\frac{dQ_m}{dt} = C_m \frac{dV}{dt} = I_{cap} \qquad \text{Eq. 6-5}$$

(Compare this equation with Eq. 4-5.) Thus we see that **dV/dt must be zero when I_{cap} is zero and** note that **dV/dt is directly proportional to I_{cap}.**

1

3. **Since all currents flow in complete circuits, the total current summed across the membrane surface must be zero at all times.** This total membrane current I_{mem} is made up of capacitive and pore components such that

$$I_{mem} = 0 = I_{por} + I_{cap} \qquad \text{Eq. 6-6/4-3}$$

We see that when $dV/dt = 0$, then $I_{cap} = 0$ and $I_{por} = 0$, which is the condition under which the steady-state equation (Eq. 6-1/4-20) holds.

2

Equation 6-6 is valid throughout the membrane action potential of the space-clamp condition we are describing, since **all regions of the membrane are at the same potential at the same time.** (The more complex case, when this restriction does *not* hold, is considered in Chap. 7.)

3

Now let us use the preceding sequence of ideas to gain further insight into how the currents must flow in order to bring about the changes in V during the action potential.

4

Rearranging Eq. 6-5 to solve for dV/dt, we obtain

$$\frac{dV}{dt} = \frac{I_{cap}}{C_m} \qquad \text{Eq. 6-7}$$

where C_m is known (we can assume a value of 1 $\mu F/cm^2$ here). **This allows us to calculate I_{cap} directly from measurement of dV/dt at any point throughout the action potential!**

5

HINT

7. An *outward* current through the capacitance depolarizes. If you have any qualms about accepting this statement, you should immediately return to Chap. 4 for more intellectual and psychological reinforcement (and do not collect $200!).

105

106

You can do this not only for any point in Fig. 6-1, but also for any point in any published record of any action potential!

We do not have to stop here. Since from Eq. 6-6/4-3 we know that

$$I_{por} + I_{cap} = 0$$

by rearranging and substituting from Eq. 6-5 we get

$$I_{por} = -I_{cap} = -C_m \frac{dV}{dt} \qquad \text{Eq. 6-8}$$

Notice that **I_{cap} has the same sign as dV/dt, but the opposite sign to I_{por}.** Thus, an inward (negative) net pore current produces an outward (positive) capacitative current, which must tend to drive the membrane potential to the positive-inside, negative-outside condition. In other words, when V becomes more positive (less negative), dV/dt is positive and the membrane depolarizes.

In the case where only I_{Na} and I_K are important,

$$I_{por} = I_{Na} + I_K \qquad \text{Eq. 6-9}$$

and hence

$$I_{Na} + I_K = -C_m \frac{dV}{dt} \qquad \text{Eq. 6-10}$$

It follows that

1. Wherever $|I_{Na}| > |I_K|$, I_{por} is negative, and the membrane will depolarize since dV/dt is positive.
2. Wherever $|I_K| > |I_{Na}|$, I_{por} is positive, and the membrane will repolarize or hyperpolarize since dV/dt is negative.

106

107

Reversing the preceding logic, we can approach the action potential knowing that **wherever depolarization is seen, $|I_{Na}| > |I_K|$**. Similarly, **during hyperpolarization or repolarization $|I_K| > |I_{Na}|$**. This brings us back to Fig. 6-12. But we can go even further!

Figure 6-13 differs from Fig. 6-12 by the inclusion of I_{cap}.

QUESTION: What is the relationship of the lower curve to the upper one in Fig. 6-13? Use either inspection or mathematical derivation. (Hint 8↓)

Now look back to Eq. 6-8; clearly, dV/dt is dependent on the degree of disparity between the inwardly and outwardly directed ionic currents, that is, I_{por}.

Inspection of Figs. 6-13 and 6-1 suggests that the maximum rate of depolarization is about 60 mV in 0.2 ms, or about 300 mV/ms or 300 V/s. When $C_m = 1\ \mu F/cm^2$,

$$I_{cap} = (1 \times 10^{-6}\ F/cm^2)(3 \times 10^2\ V/s)$$

$$= 3 \times 10^{-4}\ A/cm^2$$

Hence the disparity between I_{Na} and I_K must reach at least 300 $\mu A/cm^2$ in that action potential.

Let's sum up the ideas to this point. **During an action potential there are changes in both g_{Na} and g_K, which produce rapid changes in the g_{Na}/g_K ratio. When the resulting I_{Na} and I_K currents are unequal, the difference between the two pore currents flows across the membrane capacitance, changing the membrane potential.** Thus the action potential is generated.

(It really is relatively clear-cut when you understand enough of the basics to appreciate the preceding sentences, isn't it?)

Before we discuss how the changes in g_{Na} and g_K come about (called *membrane excitability*), there are a few, largely semantic problems which should be cleared up at this stage.

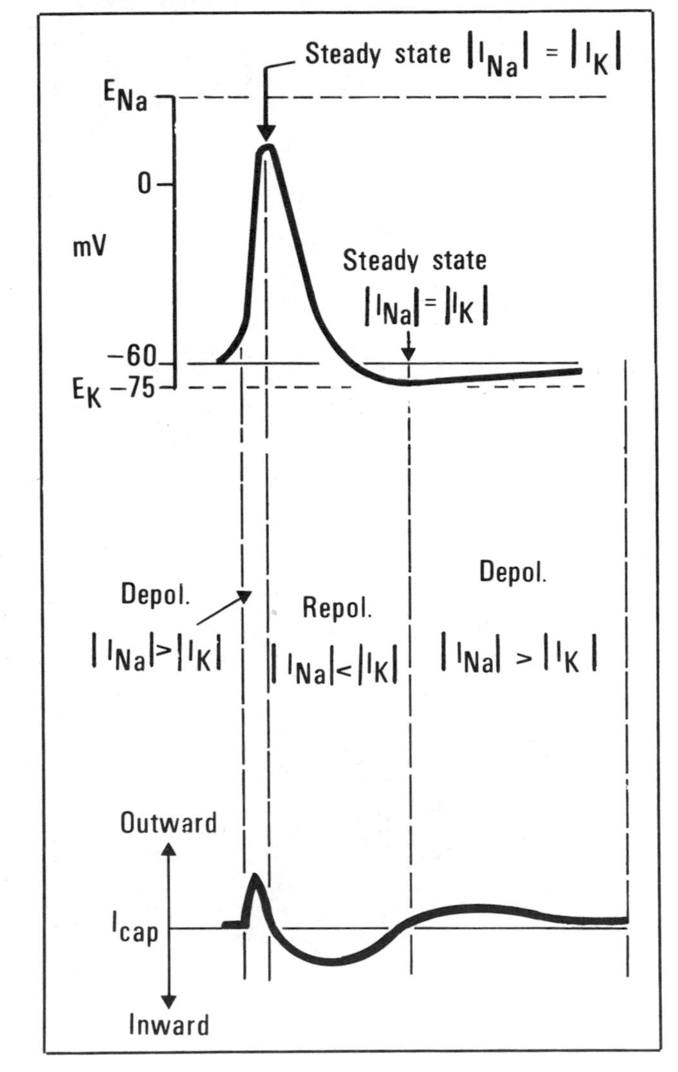

Fig. 6-13. Capacitative current analysis of Fig. 6-1. Ordinate is potential in upper curve and current density in lower curve.

107

SOME PROBLEMS IN SEMANTICS

Nothing in this section contradicts what you have just learned. However, you should be aware of a problem in the semantics of descriptive detail that you *may* encounter sooner or later, in either a lecture or a textbook.
1

Textbook descriptions of the mechanism of the action potential generally say something like this: "The depolarization of the membrane during the action potential is due to an inward sodium current." Now this statement, though correct, is sufficiently *lacking in detail* that a careful student (like you) could get confused. You see, since membrane potential is the potential across the membrane capacitance, **only an outward current across the capacitor can depolarize the membrane!**
2

The complete statement should be: **On the upstroke of the action potential, the net inward ionic current** (resulting from increased $|I_{Na}|$) **induces an outward capacitative current I_{cap}, which depolarizes the membrane.**
3

Remember, **although the current can be considered to cross the capacitor, the ions do not.** The intracellular capacitative current would be composed mainly of K^+ ions moving *toward* the membrane and Cl^- ions moving *away* from it. The extracellular capacitative current would be composed of Na^+ ions moving *away* from the membrane and Cl^- ions moving *toward* it.
4

The origin of the confusion is that where $I_{cap} = -I_{por}$, clearly it is mathematically legitimate to substitute one term for the other (as we have done in a number of places, for example, in Eq. 6-8). Hence, a positive dV/dt appears to result *directly* from an increase in I_{Na}.
5

There is no harm in using the shorthand form, "depolarization is due to the inward sodium current," *provided you remember the logical steps used to obtain this form!* Remember, **this shorthand statement in no way violates the rule that the membrane can be depolarized only by outward capacitative current.**
6

Unfortunately, many teachers (including ourselves in former years!) have carried around the contradictory idea that a membrane can be depolarized in two ways, by *either* inward *or* outward currents! In our experience, this argument rapidly becomes very fuzzy, for few people distinguish clearly between inward ionic currents through pores and outward capacitative currents across the membrane regions between the pores. **The key point to remember is that all currents must flow in complete circuits.** From this all else follows (see Chap. 8 to note the importance of this concept in the derivation of the successful Hodgkin-Huxley equations).
7

109

The frequent statement (in other texts) that "sodium rushes in" in the early stages of the action potential is *additionally misleading* in that it implies a relatively large-scale movement of sodium ions. In fact, however, **the whole marvel of the action potential occurs without any *detectable* change in the internal or external concentrations of any ion! The number of ions that must move across the membrane is only that number needed to change the charge on the membrane capacitance.**

1

By means of radioactive tracers, it has been possible to show that about 3 pmol (picomoles) of ions cross per square centimeter of membrane per action potential (1 pmol = 10^{-12} mol). This is about twice what would be needed on the basis of calculations of the membrane capacitance. Now what proportion of the cell ion contents is this? For the squid giant axon, it is about 1 part in 300,000; for a 50-μm axon, it would be about 1 part in 30,000 [63, p. 57]. Thus, you should not be surprised if a myelinated nerve fiber can transmit many thousands of impulses even after it is removed from the body!

2

While a *single* action potential in a large fiber does not change ionic concentrations significantly, you will find (in Chap. 9) that very rapid, *repetitive action potentials* can sometimes give modest changes in ionic concentrations, even in large axons, before the "pumps" catch up.

3

The smallest unmyelinated fibers (C fibers) are sufficiently small that a *single* action potential may change the internal concentrations by as much as 10 percent [63, p. 57]. Thus, there is a need for rapid "pumping" of the ions to return the concentrations to normal. The prolonged after-potential observed in C fibers may be an indication of increased activity of an electrogenic Na-K pump.

4

Finally, you should understand that many textbooks talk about membrane *permeability* rather than membrane *conductance*. **Permeability and conductance are qualitatively similar, but they differ quantitatively** (as we explained in Chaps. 4 and 5).

5

HINT

8. The bottom curve is the first derivative of the upper one. This follows directly from Eq. 6-8. Note that the zero crossings of the lower curve correspond directly with the locations in the upper curve where the steady-state equation holds. To us there is simplicity and beauty here, as the story unfolds.

109

110

VOLTAGE-DEPENDENT CONDUCTANCES

Up to this point, we mentioned changes in sodium and potassium conductances without discussing how such changes might occur. Although the true physicochemical nature of the control of membrane conductance is still unknown (as is the molecular structure of the pores themselves), it is clear that **the "cause" of the changes in g_{Na} and g_K is an electrical event, the change in the membrane potential itself!** 1

Axon membranes thus fall into a rather *rare* class of cell membranes which can be called **electrically excitable,** which means that conductance changes within these membranes are brought about in response to changes in membrane potential. By contrast, **the ionic conductances of the vast majority of cell membranes are almost entirely unaffected by a change in membrane potential.**

1. Some membranes are excited by *specific physical forces* (light, heat, mechanical distortion).
2. Other membranes are excited by *specific chemicals,* from either outside the body (odors, tastes) or inside the body (chemical transmitters, hormones).
3. The rare membranes are those (including the axon) excited by a *change in membrane potential* itself. 2

The job of the student is to remember, for a given membrane, what stimulus brings about the conductance changes for which ions. Sounds simple, doesn't it? In this chapter we deal with the third class; you learn in detail about the first two types from Chap. 9 on. 3

What, then, is the "electrical excitability" of the axonal membrane? It is best understood by first considering the **upsweep** of the action potential: 4

When the membrane is depolarized past threshold (see Fig. 6-1), this changes the conductance ratios, which alters the I_{Na}/I_K ratio, so that an outward I_{cap} is created and thus **the membrane is depolarized further.** You shouldn't be surprised that this additional depolarization increases g_{Na} still further, and then isn't it obvious that this increase in g_{Na} must depolarize the membrane further still? 5

This is one situation in which circular reasoning is legitimate! Here we go: An initial depolarization of the membrane causes increased g_{Na}, which causes $|I_{Na}|$ to increase, which further increases outward I_{cap}, which causes further membrane depolarization, which in turn causes a further increase in g_{Na}, and so on. This interaction is probably best summarized by a diagram (Fig. 6-14). 6

It is not hard to see that if the feedback cycle of Fig. 6-14 were the only mechanism operating, a cell would become depolarized and stay depolarized (which does *not* happen). In fact, after about 1 ms, g_{Na} shuts off and returns to its resting level, even if an experimenter deliberately maintains the membrane potential in a depolarized condition. **So, the circular interaction of Fig. 6-14 can occur only during the early phase of the action potential.** 7

Fig. 6-14. Feedback cycle of rising phase of action potential.

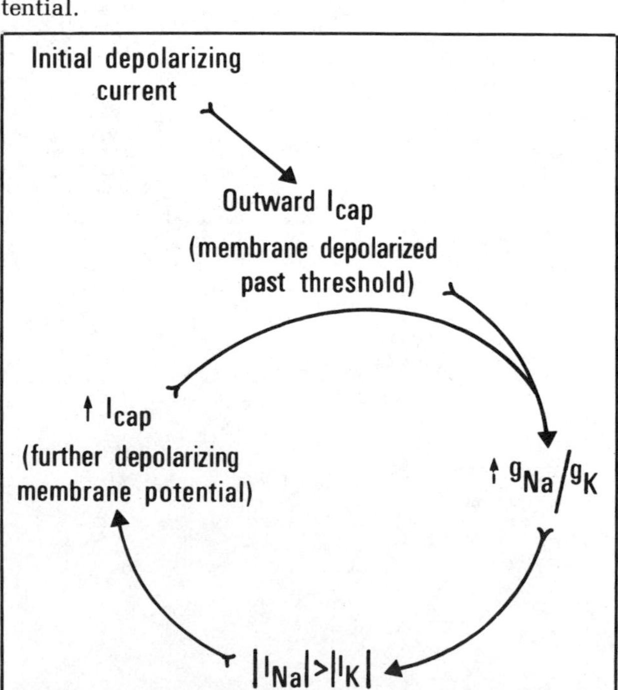

110

Technically, this is a positive feedback loop and is one of the instances in which positive feedback occurs in a normal (i.e., nonpathological) biological process. 1

Experimentally, the presence of a loop such as in Fig. 6-14 can best be studied by "opening the loop," as in the voltage-clamp experiments described later in this section. See Chaps. 8 and 12 for further discussion. 2

In the later phases of the action potential, **the membrane potential becomes repolarized** not only **because g_{Na} has returned to resting levels,** but **also because I_K has increased** as a result of two factors [recall that $I_K = g_K(V - E_K)$]: first, $|V - E_K|$ **has increased** since V is less negative than in the resting state, and second, g_K **has increased.** Such changes in g_K also result from the electrical excitability of the axonal membrane, but they occur more slowly than the comparable changes in g_{Na}. 3

This "feedback cycle" system sounds so powerful that one might wonder why the membrane potential does not "take off" during the slightest depolarization from resting potential! 4

There are two reasons why the membrane potential is not unstable at the slightest random depolarization:

1. In the region close to resting potential, the increase in g_{Na} for a given ΔV is quite small; hence $|I_{Na}|$ increases by only a small amount. This increase is less than the increase in $|I_K|$ that is brought about by the same ΔV, because of the increase in K^+ driving force.
2. In the region close to resting potential, the change in the potassium driving force is large for a given ΔV. Thus $|I_K|$ increases even without a change in g_K.

Thus for *below-threshold* depolarizations $|I_K| > |I_{Na}|$, and the membrane is repolarized by inward I_{cap} (this is an example of **negative feedback**). Only **beyond threshold** does g_{Na} increase sufficiently that $|I_{Na}| > |I_K|$, which allows the positive feedback cycle to occur. (This concept is discussed in greater detail in a later section of this chapter, so do not worry if you do not fully understand it at this time.) 5

The nature of the axon membrane's response to a change in membrane potential could *not* be predicted by reference to any of the principles set forward in Chap. 3 or 4. These properties, therefore, are best described in relation to those experimental observations that were so crucial in the development of modern neurophysiology (in that they provided the first clear demonstration of voltage-dependent conductance change), the **voltage-clamp** experiments of Hodgkin and Huxley. 6

These experiments are discussed in detail in Chap. 8. We present here only a simplified version of the major conclusions reached in Hodgkin and Huxley's Nobel Prize–winning work. Students interested in the experimental techniques and the mathematical analysis by Hodgkin and Huxley of the control of 7

112

voltage-dependent conductances are encouraged to read Chap. 8, but be advised that a thorough understanding of this chapter is almost a prerequisite to reading Chap. 8. [1]

An obvious (though technically challenging) method to study voltage-dependent conductance changes would be to make rapid stepwise changes in membrane potential and measure the resulting changes in membrane conductance. This has been achieved by a technique called **voltage clamping,** in which the membrane potential is held constant (i.e., "clamped") before and after a rapid, stepwise change in membrane potential. The conductance changes are measured indirectly by measuring the membrane current of each ion separately. [2]

If V is held constant, then g_{Na} must be proportional to the measured I_{Na} since $I_{Na} = g_{Na}(V - E_K)$. Further details of the method can be found in Chap. 7. [3]

Results of such an experiment, in which g_{Na} and g_K have been separately identified, are shown in Fig. 6-15 for a depolarizing voltage step from −65 to −9 mV. Notice that

1. Sodium conductance g_{Na} rises rapidly to reach a maximum value at about 0.8 ms after the voltage step. **Thereafter g_{Na} declines even though the membrane remains depolarized.**
2. Potassium conductance g_K rises much more slowly and shows no subsequent decline within the time course of this experiment.

The initial voltage-dependent conductance increases are called **activation.** Both g_{Na} and g_K show voltage-dependent activation. The delayed voltage-dependent decrease in Na conductance is called **inactivation.** The time-dependent changes in g_{Na} were thus analyzed as resulting from the interaction of two processes, rapid activation and slower inactivation. [4]

The essential correctness of this analysis was confirmed by later experiments in which internal perfusion of squid giant axons with solutions containing proteolytic enzyme resulted in selective removal of the inactivation mechanism without change in the time course of the activation response. [5]

By varying the size of the depolarizing voltage step in voltage-clamp experiments, it is possible to study the extent to which the activation and inactivation processes are brought into play by any given membrane potential. **Small voltage steps produce smaller effects than larger voltage steps;** both processes become saturated at potentials more positive than about +40 mV. [6]

Perhaps more surprising was the discovery that some 40 percent of the maximum sodium conductance is already *inactivated* at normal resting poten- [7]

Fig. 6-15. Time course of g_{Na} and g_K after a 56-mV depolarizing step.

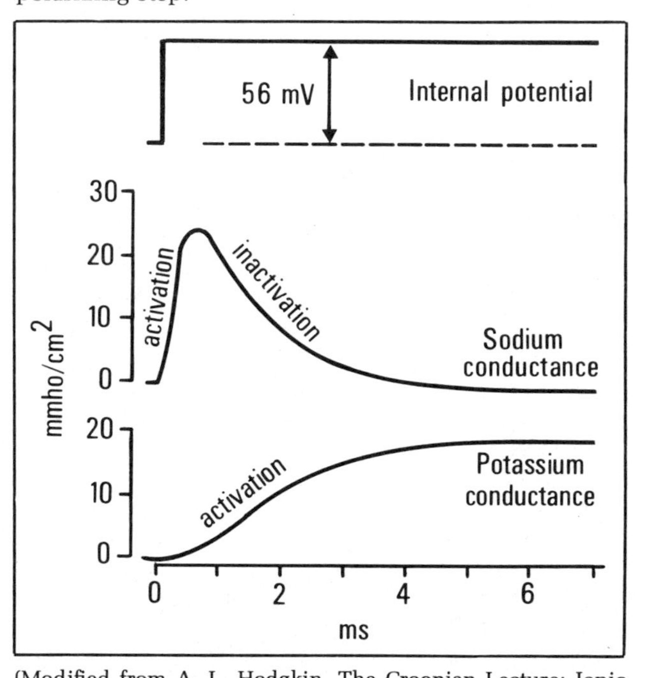

(Modified from A. L. Hodgkin, The Croonian Lecture: Ionic movements and electrical activity in giant nerve fibres, *Proc. R. Soc. Lond.* [*Biol.*] 148:1, 1958.)

112

113

tial. The interesting implications of this observation are discussed in the section "Factors Affecting Threshold" later in this chapter.

QUESTION: How is a sodium pore like the mechanical leaf shutter of an old-style, nonautomatic camera? (Hint 9↓)

Question: Do you think the potassium pores ever inactivate? (Hint 11↓)

If conductance *increases* after membrane depolarization, it seems reasonable to suppose that it would *decrease* again if the membrane potential were returned to resting level. Figure 6-16 shows the results of an experiment similar to that of Fig. 6-15 except that the membrane potential was rapidly returned to the resting potential after depolarization. Notice in the upper half of Fig. 6-16 that g_{Na} falls rapidly—more rapidly than would have occurred as a result of the inactivation process (compare with Fig. 6-15). In the lower half of Fig. 6-16, g_K also falls when the normal resting potential is experimentally restored, but more slowly than g_{Na}. We conclude that **the activation process is readily and rapidly reversible.**

QUESTION: What is the difference between Na inactivation and reversal of Na activation? (Hint 12↓)

Is the inactivation process also reversible? Let's rephrase that question: If a large depolarizing voltage step were used to first activate and then inactivate the sodium pores, *under what conditions could that inactivation be removed so that the original excitability could be recovered?* By using the camera-shutter analogy, this same question becomes, How do you reset the shutter?

Hodgkin and Huxley posed and answered this question in a particularly satisfying series of experiments (see Chap. 8): **Inactivation is removed when the membrane is repolarized to near resting potential.**

As we already noted, some 40 percent of inactivation remains when the membrane is at a resting potential of −60 mV, and at this potential the time constant for removal of inactivation is longer than 10 ms. However, complete removal of inactivation is achieved relatively rapidly if the membrane is hyperpolarized to −100 mV.

The important message here is twofold: First, **reactivation of sodium conductance** (the resetting of the Na^+ shutters) **can be achieved only when the membrane is repolarized to near-resting potential.** Second, **the more hyperpolarized the membrane becomes, the more rapidly this recovery process takes place.**

Fig. 6-16. Reversal of sodium activation (upper curves) and potassium activation (lower curves) after sudden repolarization of nerve membrane.

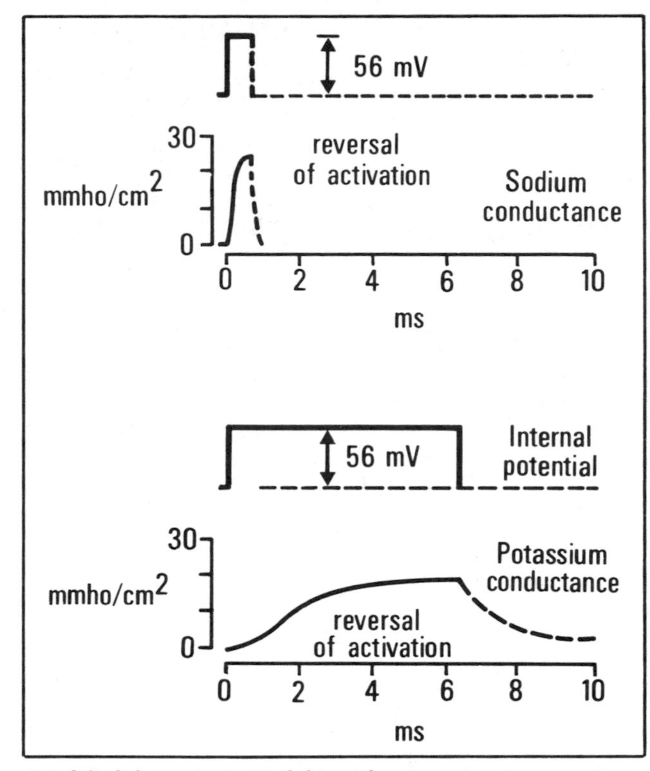

(Modified from A. L. Hodgkin, The Croonian Lecture: Ionic movements and electrical activity in giant nerve fibres, *Proc. R. Soc. Lond.* [*Biol.*] 148:1, 1958.)

113

Now we can combine the information presented in this section with the principles developed in earlier sections of this chapter. Our earlier analysis, based on dV/dt, could give us the absolute magnitude of I_{cap}, and hence the absolute magnitude of I_{por}, but only the *difference* between $|I_{Na}|$ and $|I_K|$. Except at the points at which the steady-state equation could be applied, it was not possible to solve for g_{Na}/g_K. Now, however, that we know so much more about the voltage dependence of the sodium and potassium conductances, we can begin to fill in the picture. First, we consider **only those parts of the action potential that occur when I_{cap} is prominent,** i.e., to the right of the first vertical broken line in Fig. 6-17/6-13. (In the next section we deal with the more complicated problem of understanding the initial portion of the upsweep, before threshold is reached for the all-or-nothing processes that are triggered.)

1. **Upsweep of the action potential.** Let us start with the point just above threshold on the upsweep of the action potential. Since the membrane is already depolarized, this depolarization will cause voltage-dependent changes in the conductances: g_{Na} will be rising steadily, whereas g_K, with its slower rate of change, will have increased only slightly. Therefore, I_{Na} will increase to a greater extent than I_K, and an increased outward I_{cap} will further depolarize the membrane. Now, this further depolarization must cause a further, and more rapid, increase in g_{Na}. Obviously, this change is in a direction to produce further depolarization and hence a still further increase in g_{Na}. This cycle, by which depolarization increases g_{Na} and further depolarizes the membrane, is the primary mechanism of the upsweep of the action potential. This is shown diagrammatically in Fig. 6-14.
2. **Peak of the action potential.** Such a cycle would be expected to drive the membrane potential up to E_{Na} until the increased driving force $V - E_K$ for I_K and the reduced driving force of I_{Na} brought the currents into balance at a maximum value for g_{Na}. But we can already see why this balance point, the peak of the action potential, must be extremely precarious:
   a. **Sodium inactivation,** which occurs most rapidly at positive values of the membrane potential, will start to reduce g_{Na} and hence **will reduce I_{Na}.**
   b. **Potassium activation,** although slower than sodium activation, becomes fast enough to cause a significant increase in g_K and hence a further **increase** in I_K.
3. **Falling phase of the action potential.** Fall in $|I_{Na}|$ coupled with rise in $|I_K|$ must initiate the repolarizing phase of the action potential by reversal of I_{cap}, since $|I_K| > |I_{Na}|$. As the membrane potential returns to rest, g_{Na} falls as a result of not only the inactivation process but also the even more rapid reversal of activation (see Fig. 6-16). The low g_{Na} in the falling phase ensures that $|I_{Na}|$ remains less than $|I_K|$ despite an increase in the driving force $(V - E_{Na})$ and a fall in g_K (by reversal of activation).
4. **After-hyperpolarization.** Obviously, during the after-hyperpolarization, $|I_{Na}|$ must be very close to its resting level, while $|I_K|$ must be abnormally large. By this

Fig. 6-17/6-13. Capacitative current analysis of Fig. 6-1. Ordinate is potential in upper curve and current density in lower curve.

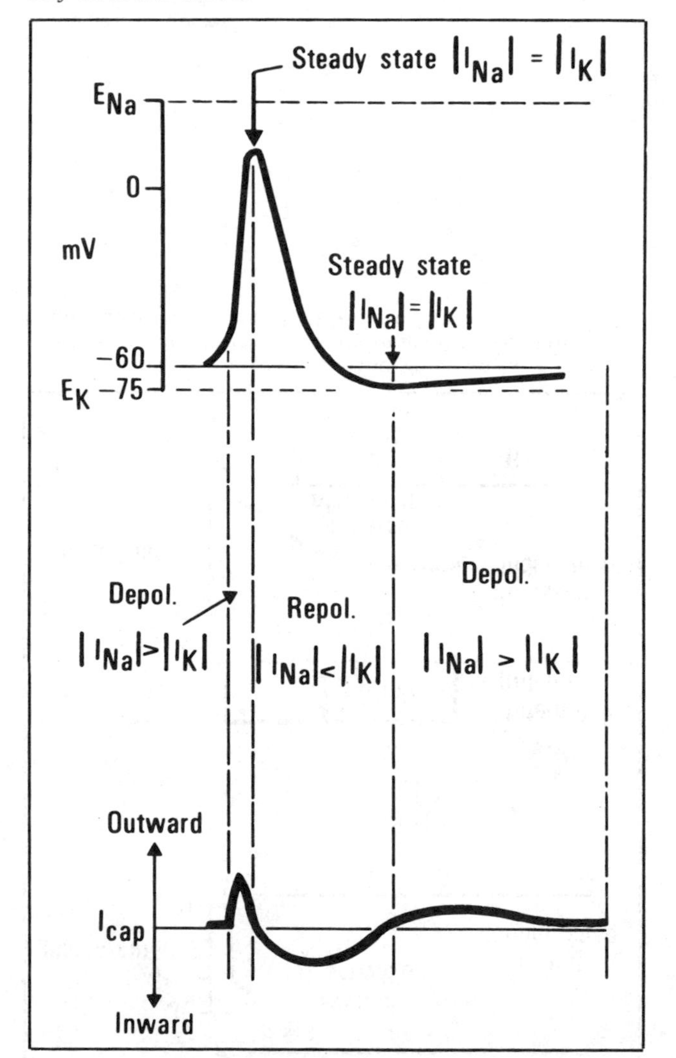

115

stage, $|I_{Na}|$ will be small, in part as a result of Na inactivation and in part as a result of reversal of activation as the membrane repolarizes during the falling phase of the action potential. In the case of the potassium current, while reversal of activation lowers g_K, this process is sufficiently slow that g_K remains higher than normal throughout the period of after-hyperpolarization.

As g_K slowly returns to its original level, $|I_K|$ falls relative to $|I_{Na}|$ and the membrane potential returns to its resting value. (A rather elegant mechanism, don't you think?)

The changes in g_{Na} and g_K, and consequent changes in I_{Na} and I_K, that occur during the action potential have been presented here in a qualitative rather than quantitative manner. A more quantitative approach is taken in Chap. 8, and you should be aware that rigorous quantitative description of these voltage-dependent conductance changes is readily available. Computer simulation and correct prediction of the form of the action potential to be expected under a wide variety of experimental conditions provide further evidence of the essential validity of this analysis.

The changes in g_{Na} and g_K occurring during an action potential in a squid axon are shown in Fig. 6-18. We suggest that you examine this figure very carefully, reviewing the preceding explanation of these conductance changes, before attempting the questions that conclude this section.

QUESTION: Does g_{Na} rise faster than g_K in Fig. 6-18 solely because of the feedback cycle of Fig. 6-14? (Hint 14↓)

QUESTION: In Fig. 6-18 the peak of the action potential occurs slightly *before* the peak of the sodium conductance curve. Can you give any explanation for this curious observation? (Hint 13↓)

Fig. 6-18. Conductance changes during membrane action potential.

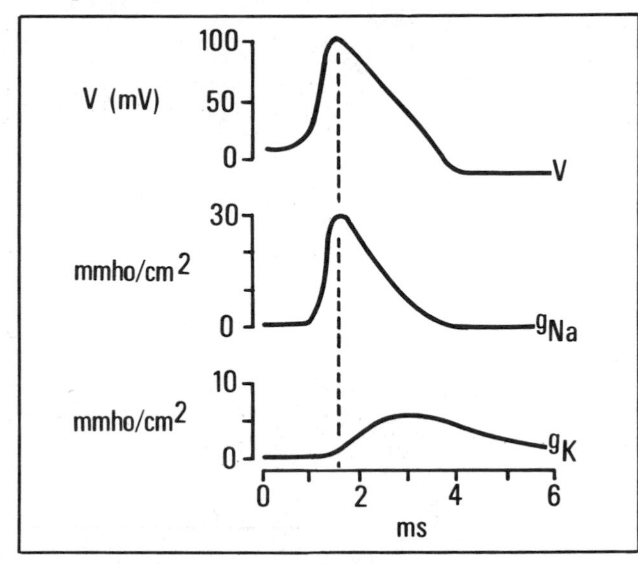

(Modified from J. Zachar, *Electrogenesis and Contractility in Skeletal Muscle Cells.* Baltimore: University Park Press, 1971.)

HINTS

9. When the pore is triggered by a sufficient depolarization, it opens (activation). However, the same stimulus that initiates the opening of the pore also initiates a slower process, leading to closing (inactivation) of the pore. Thus, like a camera shutter, an appropriate triggering stimulus first opens and then closes each membrane sodium pore. Can you guess what the "exposure time" would be for these pores? (See Hint 10.↓)

11. Potassium pores do start to inactivate if depolarization is maintained for as long as 100 ms (not shown in Fig. 6-15). However, potassium inactivation is so slow that it cannot possibly affect an event as rapid as the action potential and can, therefore, be ignored at this stage of our discussion.

12. Na inactivation occurs when the membrane is depolarized; reversal of Na activation occurs when the membrane is repolarized. *But note:* When the membrane repolarizes past the resting level (hyperpolarizes), the reversal of Na inactivation also occurs! (See the next two paragraphs in the text.)

115

1 Question: Is the delay in the peak of g_K relative to the peak of the action potential due to a sluggish response of g_K to V? (Hint 15↓)

2 Question: Where does the peak I_{Na} occur? (Hint 17↓)

ELECTRICAL STIMULATION AND THRESHOLD

3 Up to this point, for reasons that gradually will become clear, we have not described the events in the earliest part of the action potential, the part that immediately preceded the "depolarized" portion of the recording in Fig. 6-19. We merely assumed that the membranes somehow were magically depolarized. Actually, there is no magic about it. **The depolarization needed to bring about the action potential is derived from an external source.** So, first we must describe what happens when an external stimulator applies a current to the membrane.

4 It is easiest to consider, first, what happens when a *hyperpolarizing* current passes through the membrane (Fig. 6-20). It should be clear from Fig. 6-20 that **the membrane capacitance will be hyperpolarized when the current passes inward across the membrane circuit elements.**

Notice that the **external current $-I_{ext}$,** which is generated by the stimulator, **flows across both the resistive and the capacitative circuit elements.** All currents in solutions are carried by moving ions; thus this external current must be carried either by ions moving up to and away from the membrane ($-I_{cap}$) or by ions moving through membrane pores ($-I_{por}$). **When a current from an external source flows across the membrane, the sum of the ionic and capacitative currents is no longer zero, but must equal the external current.** Thus,

5 $$I_{mem} = I_{por} + I_{cap} = I_{ext} \qquad \text{Eq. 6-11}$$

Initially, the external current is carried mainly as I_{cap}. But as the charge on the membrane capacitor changes, the ionic currents depart from their steady-state values [following their individual Ohm's law equations, that is, $I_j = g_j(V - E_j)$]. For the situation shown in Fig. 6-20, as the membrane potential becomes more negative (hyperpolarizes), a point is reached at which $+I_K$ has *decreased* (since $|V - E_K|$ is less) and $-I_{Na}$ has *increased* (since $|V - E_{Na}|$ is greater) to such an extent that the net inward $-I_{por}$ is now exactly equal to $-I_{ext}$. When this occurs, the entire external current is carried by ions flowing *through* membrane pores. So I_{cap} must be zero, and **there can be no further change in the membrane potential.** At this point,

6

7 $$I_K + (-I_{Na}) = I_{ext} \qquad \text{Eq. 6-12}$$

Fig. 6-19. Current analysis of Fig. 6-1.

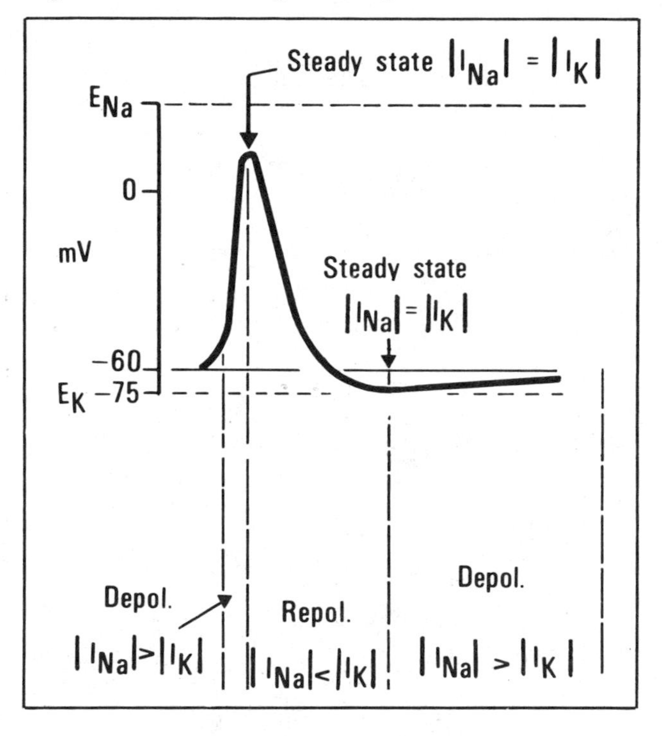

This new "steady state" in the presence of a constant I_{ext} can be handled by a variant of the steady-state equation. When $I_{cap} = 0$,

$$I_{por} - I_{ext} = 0 \quad \text{Eq. 6-13}$$

Expanding this equation and solving for V_s give the general form,

1 $$V_s = \frac{\Sigma g_i E_i + I_{ext}}{\Sigma g_i} \quad \text{Eq. 6-14}$$

QUESTION: When E_{Na} and E_K are +60 and −100 mV, respectively, $g_{Na} = 0.03$ mmho/cm², and $g_K = 0.3$ mmho/cm², and assuming the conductances for all other ions are negligible, calculate the steady-state hyperpolarization that would be produced by $-I_{ext} = 2\ \mu A/cm^2$. (Hint 18↓)
2

Question: How long will it take this potential to go from starting voltage to $1 - 1/e$ of its final value? (Hint 19↓)
3

Fig. 6-20. Stimulator passes hyperpolarizing current across the cell membrane. Notice that the sign of the external stimulating current is determined by the direction in which it crosses membrane circuit elements, and not by its direction within microelectrode.

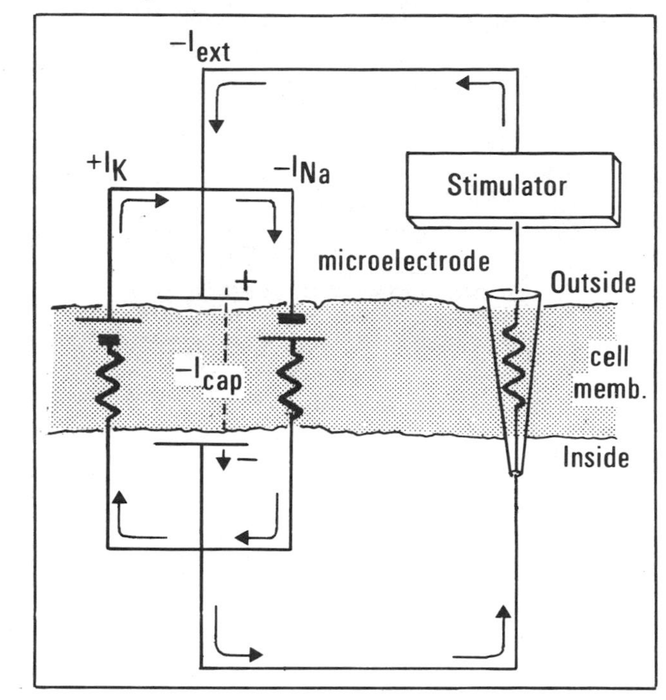

HINTS

10. *The average "exposure time" varies with the size of the depolarizing step.* For the squid axon data shown by Hodgkin and Huxley, it would be about 1/1000 s for smaller depolarizing steps, 1/5000 s at the largest step sizes, and about 1/2500 s in Fig. 6-15.
13. Look at it this way: At the peak of the action potential, a balance is reached between g_{Na} and g_K. In the next fraction of a millisecond, the further small increase in g_{Na} is balanced by a compensating increase in g_K. Notice that the membrane starts to repolarize (as a result of increasing g_K) slightly *before* the peak of g_{Na} is reached. Thereafter, falling sodium conductance assists g_K in initiating repolarization (and you can see that $-dV/dt$ increases just where g_{Na} starts to fall).
14. No. The voltage-clamp data of Fig. 6-15 show that g_{Na} responds much more rapidly to depolarization than does g_K, even when the membrane voltage is constant. But why does that happen? Now *that* is a really good question! At present, there is no answer to this question at the molecular level. Perhaps one of you will answer it someday.

118

Now that we have some idea of what happens when an external current is switched *on*, we must try to figure out what to expect when the current is switched *off*. Recall that with current flowing, as shown in Fig. 6-20, in steady state

$$I_{mem} = I_{por} + I_{cap} = I_{ext} \qquad \text{Eq. 6-15/6-11}$$

As soon as I_{ext} becomes zero, I_{mem} becomes zero (Eq. 6-15/6-11), and it follows that a capacitative current will flow equal to $|I_{por}|$; that is, I_{cap} will be determined by the difference between $|I_K|$ and $|I_{Na}|$. In other words, in the first instant of time after I_{ext} ceases, I_{cap} will be the same size as I_{ext} was, **but of opposite sign.** Hence the membrane will tend to repolarize back to its original steady-state resting potential, following a simple exponential decay curve. 1

QUESTION: What will be the sign of I_{cap} immediately after a hyperpolarizing external current is turned off? (Hint 22↓) 2

We assumed here that the external current produces no change in E_j or g_j. Usually this is a legitimate assumption in the case of hyperpolarizing currents. However, in the case of *prolonged* depolarizing currents, E_K may change surprisingly fast, *even if* g_K *remains constant.* Since resting $g_K > g_{Na}$, the resulting net K^+ efflux may produce substantial changes in $[K]_i$ (within 1 to 10 min, depending on the size of the cell). No significant Na^+ influx occurs after a hyperpolarization of similar magnitude and duration because g_{Na} is relatively small. 3

Question: Why doesn't $[K^+]_o$ change significantly with a prolonged depolarizing current? (Hint 23↓) 4

Question: Why doesn't $[K^+]_i$ change significantly with prolonged hyperpolarizing currents? (Hint 24↓) 5

Thus, **when no change occurs** in either the conductances or the equilibrium potentials of the ions involved, **the resting cell membrane behaves as a stable system** when its potential is hyperpolarized by an external current. As soon as the external current is turned off, the potential returns to its steady-state value. The membrane responses to brief hyperpolarizing currents are shown in the V_H portion of Fig. 6-21. Note that as soon as I_{ext} is turned off (at about 0.2 ms), the membrane potential follows an approximate exponential decay to the resting level. 6

Now, **from a strictly electrical standpoint, all the foregoing descriptions can be reversed for** the case of **depolarizing currents.** That is, if g_{Na} and g_K remain constant, then during the $+I_{ext}$ step, $|I_{Na}|$ would be *reduced* and $|I_K|$ would be *increased* as the membrane capacitance charges to its new level. Under such circumstances, *if* g_{Na} and g_K remain unchanged, then the membrane 7

Fig. 6-21. Above: membrane responses (V_D or V_H) to brief (0.2-ms), small electrical currents, either depolarizing or hyperpolarizing. Below: the differences between the two responses.

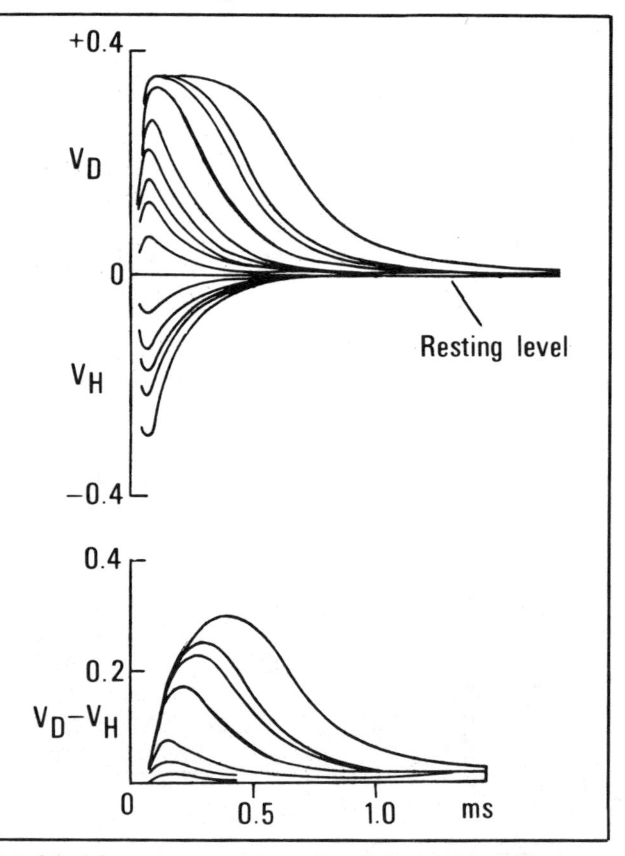

(Modified from A. L. Hodgkin, The subthreshold potentials in a crustacean nerve fibre, *Proc. R. Soc. Lond.* [*Biol.*] 126:87, 1938.)

118

responses to depolarizing and hyperpolarizing currents should be mirror images of each other, but this is **not** the case!

A good way to demonstrate that the responses are not symmetrically opposite is by comparing the responses of the membrane to *small* depolarizations and hyperpolarizations (V_D and V_H in the top part of Fig. 6-21). The responses look quite similar, but on close inspection you can see that the **depolarizing responses are always larger.** This is shown more readily by subtracting the corresponding depolarized and hyperpolarized responses (bottom of Fig. 6-21 with expanded voltage axis), which shows that **when the membrane is depolarized, there is an extra change in the membrane over and above what is caused by**

HINTS

15. A simplistic answer to this question would be, yes, that g_K reacts slowly to changes in membrane potential and merely "fails to get the message" that the peak of the action potential has passed. Although we must confess to having supplied this answer to students in our earlier years, *it is simply not true!* Check Fig. 6-18, where you will note that both g_{Na} and g_K react instantaneously to repolarization; note the reduction in the rate of rise of g_K as the membrane starts to repolarize. The true answer is considerably more beautiful and vastly more instructive. See if you can figure it out; if not, see Hint 16.↓
17. The peak I_{Na} occurs well before the peak overshoot (and therefore substantially earlier than the peak g_{Na}). Think of it this way: I_{Na} must neutralize I_K before it can contribute to I_{cap}. Thus the "cost" of high dV/dt is considerable—it can be achieved only by high I_{Na}. The peak I_{Na} is approximately coincident with the peak I_{cap}, which necessarily coincides with maximum dV/dt (whereas I_{cap} must be zero at the peak of the action potential).
18. Normal resting potential

$$= \frac{0.03(60) + 0.3(-100)}{0.03 + 0.3}$$

$$= -85.5 \text{ mV}$$

Notice that $\Sigma g_i E_i$ also has units of microamperes per square centimeter. Hence the hyperpolarized steady-state potential

$$= \frac{0.03(60) + 0.3(-100) + (-2)}{0.03 + 0.3}$$

$$= -91.5 \text{ mV}$$

Thus the hyperpolarization produced by 2 μA/cm^2 is $-91.5 - (-85.5) = -6$ mV. *But there is another way* to reach the same result. Ohm's law tells us that when the current is 2 μA/cm^2 and the resistance is $1/(g_{Na} + g_K)$ ohms centimeter square, the voltage change can be obtained from

$$\Delta V = IR$$
$$= (-2 \times 10^{-6} \text{ A/cm}^2)(3 \times 10^3\ \Omega \cdot \text{cm}^2)$$
$$= -6 \text{ mV}$$

19. 3 ms. For the explanation, see Hint 20.↓
21. Since

$$R_m = \frac{1}{\Sigma g} = \frac{1}{0.33 \text{ mmho/cm}^2}$$

$$= 3.03 \text{ k}\Omega/\text{cm}^2$$

we know that

$$\tau_m = (1 \times 10^{-6})(3.03 \times 10^3 \text{ s})$$

$$= 3 \text{ ms}$$

120

the electrical stimulus itself. This difference is **due to the "active" response of the membrane**—the change in g_{Na} and g_K due to the depolarization of the membrane by the electrical stimulus. This active response of the membrane is called the **local response.**

1

Let's state the same idea in a different way. When a depolarizing current is applied to a membrane, the change in membrane potential is due to the applied current—as in the hyperpolarizing response V_H of Fig. 6-21. This change in membrane potential is called the **electrotonic potential,** which *is* a mirror image for depolarizing and hyperpolarizing currents.

2

Now **if** the **current depolarizes** an electrically excitable membrane, **in addition to the electrotonic potential there will be a local response** (bottom of Fig. 6-21) as a result of an increase in g_{Na}. Thus V_D **in Fig. 6-21 is made up of both the electrotonic potential** (from the stimulus) **and the local response** (from the change in membrane conductance).

3

This idea is expressed in Fig. 6-22, where you can see that **the membrane potential can be depolarized by two factors: an electrotonic depolarization and a "local response" depolarization,** both of which can cause a further increase in g_{Na} **since g_{Na} is affected by the membrane potential, however it is changed.**

4

Note that the evidence of Fig. 6-21 rests on the hypothesis (proved in other experiments) that there is no active response of the membrane to the hyperpolarizing current.

5

Note that the changes in membrane potential shown in Fig. 6-21 are small—generally less than 10 mV. Such **small stimuli do not cause full-blown action potentials,** i.e., ones in which the membrane potential goes above zero.

6

Figure 6-23 shows responses to both weak and strong stimuli. In Fig. 6-23, the stimulus pulse is longer (2ms) than in Fig. 6-21, so you can see what happens to the membrane potential during the stimulus. If the stimulus is below threshold, the potential (which combines the electrotonic potential and the local response) returns to the baseline (*C*, and curves below *C*, in Fig. 6-23). However, if the stimulus is just strong enough to depolarize the membrane potential past threshold (−40 mV in this cell), then the "overshooting" action potential develops (*A* in Fig. 6-23). At stimulus strength well above threshold action potential, *B* in Fig. 6-23 is generated.

7

Now you know from the feedback cycle of Fig. 6-22 that depolarization leads to an increase in g_{Na}, which depolarizes the membrane still further, and so on. Why doesn't this cause a full action potential for all stimuli, no matter how weak? As stated on page 111, there are two reasons why, **at the end of a "subthreshold" stimulus, the membrane potential returns directly to the resting level: g_{Na} changes very little if the change in V is small, and near the resting membrane**

8

Fig. 6-22. Feedback cycle of below-threshold stimulus. Compare with Fig. 6-14, from which it differs only in that depolarizations due to I_{cap} are labeled differently. Not shown is fact that the changes are much less when below threshold than when above threshold.

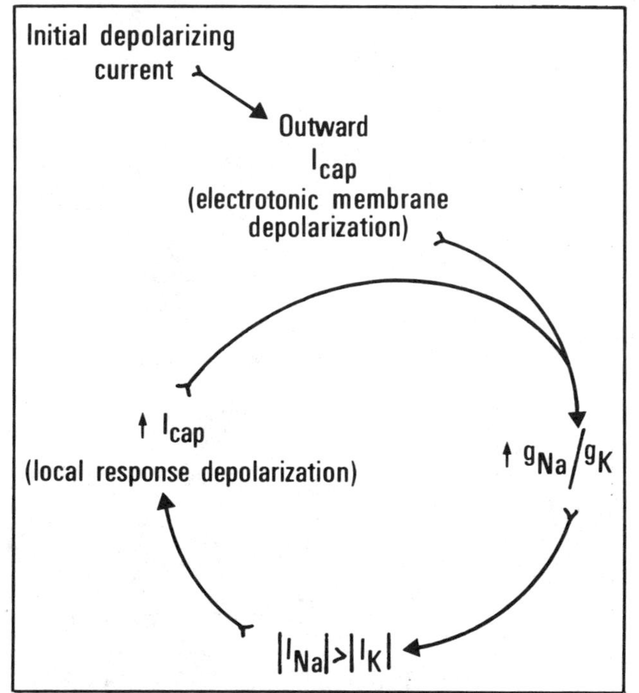

120

121

potential a small depolarization causes a relatively large increase in I_K [for example, if V goes from -60 to -45 when $E_K = -75$, this doubles I_K (since $V - E_K$ rises from 15 to 30), even if there is no change in g_K].

If g_{Na} changes by only a very small amount, then I_{Na} may actually **decrease** since $|V - E_{Na}|$ necessarily decreases with depolarization. At the same time, I_K has increased. You already know from the early part of this chapter that, when $|I_K| > |I_{Na}|$, there will be a net $-I_{cap}$ (which repolarizes the membrane back to the resting level).

Moreover, if a stimulus (together with any local response that has had time to develop) is large enough to depolarize the membrane past threshold, then at this point the depolarization is so large that it brings about a *large* change in g_{Na}, sufficient to overcome any decrease in I_{Na} that results from the decreased driving force $|V - E_{Na}|$.

Thus, if the g_{Na} change is large enough that $|I_{Na}| > |I_K|$, then the **positive feedback cycle** of Fig. 6-22 obviously will take hold and an action potential must occur. But what, then, is this elusive **threshold potential?** It must be that point between the *sub*threshold potential (where $|I_K| > |I_{Na}|$) and the *supra*threshold potential (where $|I_{Na}| > |I_K|$), which can mean only that **the threshold potential is where $|I_{Na}| = |I_K|$! Memorize this definition.** It is the starting point for many discussions in later sections of this chapter. This is a good moment to reread the last few paragraphs if you are not sure you followed the logic here.

QUESTION: If $|I_{Na}| = |I_K|$ at threshold, then dV/dt must be zero at that time. Does this mean that the threshold is a *stable* membrane potential? (See Hint 26.↓)

Fig. 6-23. Membrane responses to stimuli of different strengths, both above and below threshold.

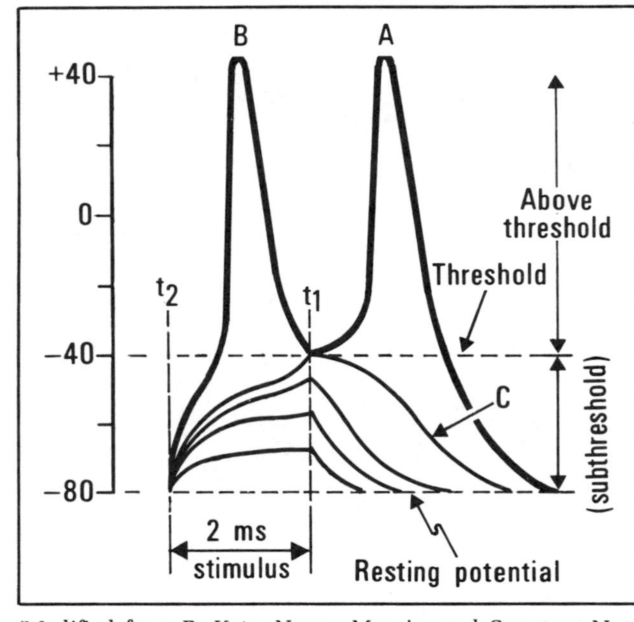

(Modified from B. Katz, *Nerve, Muscle, and Synapse.* New York: McGraw-Hill, 1966.)

HINTS

16. The peak g_K is not that value appropriate to the *peak* of the action potential, but rather the value appropriate to that point on the falling phase with which the peak coincides! Now, notice how flattened the rising part of g_K becomes following the peak of the action potential. Here the rate of rise of g_K is continually being readjusted by the falling membrane potential. Finally, the rise of g_K is reversed when the membrane potential falls sufficiently to block any further potassium activation. Although the subsequent fall of g_K is slow, it occurs at a rate which is instantaneously controlled by the membrane potential, just as was shown in Fig. 6-16.

20. In Chap. 3, the time constant of a resistive-capacitative network τ is defined as the time taken for V to rise to $1 - 1/e$ of its final value. It is given that $\tau_m = R_m C_m$. So what is τ_m for this membrane, assuming $C_m = 1 \times 10^{-6}$ F? See Hint 21.↑

22. A hyperpolarizing I_{ext} has a negative sign; therefore, I_{cap} will be positive until V is returned to the steady-state membrane potential.

23. $[K^+]_o$ is the *extracellular* fluid whose concentrations are determined by the capillary diffusion from the (relatively) enormous circulating blood volume.

24. Hyperpolarization *decreases* I_K. We just wanted to see if you were getting groggy.

121

122

QUESTION: If threshold is defined as a point at which $dV/dt = 0$, why is no such point visible in the action potential *B* of Fig. 6-23? (Hint 27↓)

1

Do you understand why we did not explain the earliest portion of the action potentials shown in Figs. 6-13 and 6-19 at an earlier point in this chapter? It *is* rather complicated!

2

Now, you can understand that the early portion of the action potential (Fig. 6-13 or 6-19 before the rising phase) is that part of the depolarization that occurs **below threshold.** It thus **represents a depolarization that could not occur in the absence of external current.** Exactly how that external current is generated in a normal, propagated action potential is explained in Chap. 7. For now, you can presume that an external stimulating current was applied to depolarize the membrane to threshold.

3

Two Types of Response to Electrical Stimulus

If you think about it, you will realize that you already know that **the axon gives one of two responses** to an electrical stimulus: **If** the stimulus is **below threshold,** the cell gives **a graded, local response. If** the stimulus is **greater than threshold,** the cell gives **a maximal, all-or-nothing response.**

4

5

Both types of response are shown in Fig. 6-23.

1. *Local response.* As the strength of a subthreshold stimulus is increased, the active response of the membrane (the local response) becomes greater. While this response is present in Fig. 6-23, it is best seen at the bottom of Fig. 6-21, where the electrotonic effects of the stimulus on the membrane have been subtracted. The same sort of active response is occurring in Fig. 6-23 as a result of *subthreshold* stimuli, but the potentials shown are combinations of the effect on the membrane of the stimulus and the response of the membrane itself. The important point to note is that **the local response is a graded response;** i.e., it is a function of stimulus strength: **the greater the stimulus, the greater the response.**

6

2. *Action potential.* Compare the local response and the action potential. **With the action potential, no matter what the strength of the stimulus** (as long as it is **above threshold**), **the amount of depolarization of the cell is the same.** This is clearly seen by comparing *A* and *B* in Fig. 6-23, where *B* is the membrane response to a much stronger stimulus than that which initiated *A*. In spite of differences in stimulus strength, the two responses are very similar (differing only in latency—explained in the next subsection). This is a clear example of the all-or-nothing law of axonal conduction: If an axon is stimulated **beyond threshold, the response of the membrane is maximal, given the existing conditions.** Essentially, **the magnitude of the response** of the membrane **is independent of stimulus strength,** once the stimulus is **above threshold.**

7

122

123

The reason for the qualifier *given the existing conditions* in the "law" just stated is that **under some conditions the height of the action potential can vary from one time to another.** For example, if there has been extremely rapid firing of the cell, or the pump has been poisoned, or drugs have been applied, or the axon is diseased or injured, then you might expect the ionic concentration differences to be abnormal or even the normal ability of the membrane to change its conductances to be impaired.

The converse corollary of the law is as follows: **If an axon is stimulated at an intensity below its threshold, there will be a response, but it will not generate an action potential.**

While we have described two responses of the axon as if they were two "types"—local response and action potential—it is very important to realize that **both the local response and the action potential are generated by the same mechanism: an increased g_{Na} (and g_K) in response to an electrical depolarization of the membrane.**

Changes in Latency Due to Stimulus Strength

There is a noticeable difference in latency between the two action potentials *A* and *B*, shown in Fig. 6-24. This latency is caused by two factors: when the membrane is barely above threshold, the membrane response is slower; a stronger stimulus brings the membrane to threshold sooner.

In Fig. 6-24, the first factor is the difference between the intervals *h* and g, while the second factor is the difference between the intervals *f* and *e*. Considering these two factors in greater detail, you should note the following:

1. The slowness of the membrane response around threshold can be seen by noting the difference in the shape of the action potentials *A* and *B*, shown in Fig. 6-24, just past threshold. The response of the membrane when barely stimulated (*A*) shows a slow upsweep. In fact, had the stimulus strength been very slightly less, the response might have been subthreshold, with a rather prolonged return to baseline (*C* in Fig. 6-24 also shown by *X* in Fig. 6-25). The length of time before it is clear whether the membrane will generate an action potential or just return to the resting level can be quite prolonged, as seen by *Y* in Fig. 6-25.

Fig. 6-24. Further analysis of responses obtained with stimuli of different strengths. See Fig. 6-23 for additional information as to duration of stimulus, etc. Symbols are described in the text.

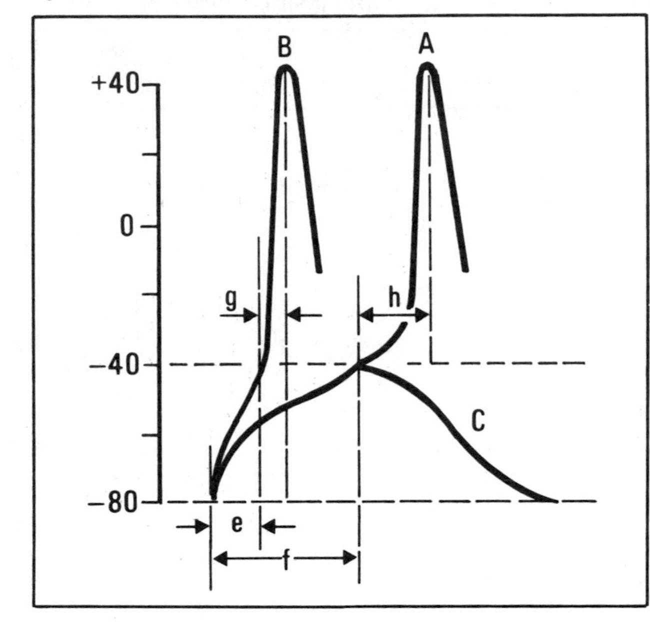

Fig. 6-25. Changes in membrane response to increasing stimulus strength.

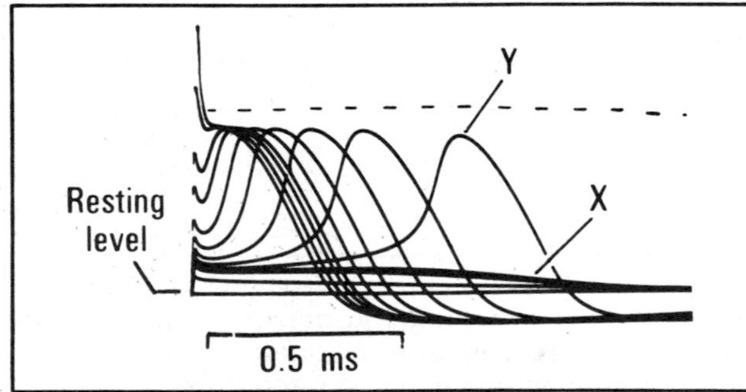

(From A. L. Hodgkin, A. F. Huxley, and B. Katz, Measurement of current-voltage reactions in the membrane of the giant axon of *Loligo*, *J. Physiol.* [*Lond.*] 116:424, 1952.)

HINT

26. No, of course not. Even if $|I_{Na}| = |I_K|$ at the instant the threshold is reached, both g_{Na} and g_K are changing with time. Thus it is highly unlikely that the currents will *remain* equal for more than a fraction of a millisecond.

123

2. A stronger stimulus will bring the membrane to threshold sooner. Why? This is easily understood if you realize that (as emphasized in Chap. 5) **a change in membrane potential is a change in the charge on the membrane capacitance. The stronger stimulus can change the charge on the capacitance sooner.** All other things being equal, a stronger voltage will move charge faster onto a given capacitor than a weaker voltage.

It can be shown that for a threshold stimulus, the product of the stimulus duration and the stimulus strength is relatively constant over a considerable range (see the next subsection, "Strength-Duration Curve"). This suggests that the preceding analysis is reasonable, that a given amount of charge must be "moved" on the membrane capacitance before the membrane will reach threshold.

Strength-Duration Curve

We now take up the characteristics of the stimulus necessary to "stimulate." **There is an inverse relationship between the strength of a stimulus and the duration of that stimulus needed to bring a cell to threshold** (also known as the *firing level*).

The strength-duration curve for a nerve is shown in Fig. 6-26. The longer the stimulus, the less strength that is needed—until a minimum electric potential is reached. This minimum is called the **rheobase** (strength *A* in Fig. 6-26). As the duration of the stimulus is decreased, stronger and stronger stimuli are needed—until a minimum duration is reached (above the maximum strength shown in Fig. 6-26). The explanation for the shape of the curve has been described in the preceding paragraphs.

A convenient way to compare strength-duration curves from different experiments or tissues is by means of the **chronaxie,** which is defined as the *duration* of the stimulus whose strength is *twice* the rheobase (duration C in Fig. 6-26). The chronaxie had clinical use, for nerves and muscles have different chronaxies. Therefore, when a nerve had been injured, it was possible to determine the strength-duration curve (and the chronaxie) by means of electrical stimulation through the skin (or by needles), as an aid in understanding the extent of the injury and possible recovery from it.

This is beautifully shown in Fig. 6-27, where you see the strength-duration curves obtained from a patient's injured arm. Compare the curves from the normal arm with those from the injured arm before operation. As the nerve recovered after the operation, the intermediate curves were obtained. It was not until 24 weeks after injury that any voluntary movement returned. However, the beginning phases of recovery were evident more than a month before that by means of the electrical measurement (!), as you can see from the graph. Note that the chronaxie also could be used as a measure of recovery; however, in modern medicine, the electromyogram (recording of muscle potentials) is used to determine the extent of denervation and reinnervation.

Fig. 6-26. Strength-duration curve relating strength of stimulus to time for which it must be applied to an excitable tissue to produce a response. (*A*) Minimum strength of stimulus needed to reach threshold, the rheobase. (*B*) Minimum stimulus strength at the rheobase. (*C*) Duration of stimulus at twice rheobase strength, the chronaxie.

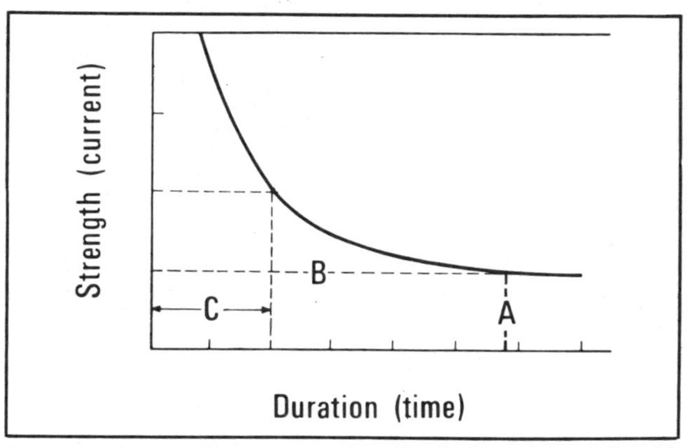

(From W. F. Ganong, *Review of Medical Physiology* [9th ed.]. Los Altos, Calif.: Lange, 1979.)

Fig. 6-27. Strength-duration curves from injured and uninjured human arms.

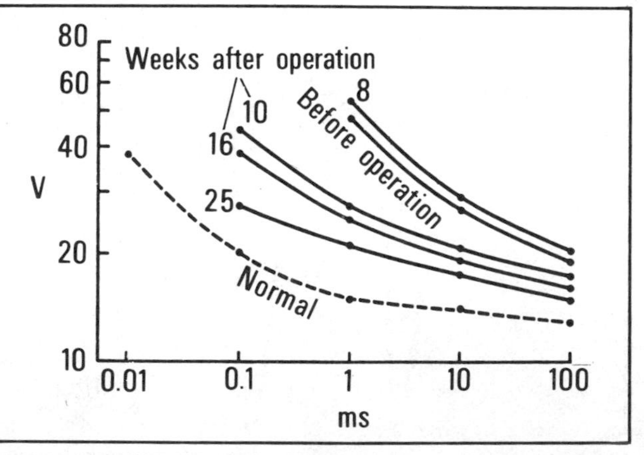

(After A. E. Ritchie, The electrical diagnosis of peripheral nerve injury. *Brain* 67:322, 1944; from C. Keele and E. Neil [Eds.], *Samson Wright's Applied Physiology* [12th Ed.]. London: Oxford University Press, 1971. Copyright © Oxford University Press 1961, 1965, 1971.)

125

From the shape of the strength-duration curve (Fig. 6-26), it is not hard to see why the product of strength and duration, SD, is relatively constant. The curve approximates an inverse relationship: $D = 1/S$, from which it immediately follows that $SD = 1$ (or some other constant).

1

Question: If you did not have an oscilloscope to record action potentials, is there any other way that you could study the strength-duration curve? (Hint 28↓)

2

If the process of excitation involved only the change in charge on a capacitor, then one would expect the strength-duration curve not to flatten out at the rheobase, but, instead, to continue to decrease to zero at infinite duration. The reason that there is a **minimum stimulus strength** (no matter what the duration) **is due to the membrane resistance** which, though large compared with water, can still carry enough current to affect these experiments. The membrane resistance is in parallel with the membrane capacitance and can be considered to represent a "leaky" capacitor in a physical sense. Biologically, the membrane resistance is mainly due to $1/g_K$ (and, in some tissues, $1/g_{Cl}$). At minimal stimulus strengths, the stimulus current is passing through the membrane resistance as fast as it is supplied by the stimulating electrodes, and there is no continuing change in charge on the membrane capacitance, so the cell cannot be brought to threshold.

3

Temporal Summation

Returning now to Fig. 6-23, we see another feature of considerable importance. Note that **the membrane response to a subthreshold stimulus outlasts the stimulus.** This provides a basis for temporal summation and subliminal excitation.

4

Recall that the membrane potentials **following subthreshold stimuli** in Fig. 6-23 are composed of **both** an **electrotonic potential and** a **local response.** Both these potentials **outlast the stimulus.** That is, the membrane potential returns to the resting level well after the end of the stimulus. This has interesting consequences.

5

HINT

27. Threshold is *always* defined on the basis of the following question: What would happen if I_{ext} were switched *off* at this potential? If I_{ext} is switched off *above* threshold, depolarization continues; if *exactly at* threshold, then, as predicted, $dV/dt = 0$; if *below* threshold, repolarization occurs. So, what happens when, as in action potential *B* of Fig. 6-23, threshold is passed while I_{ext} is still flowing? (Hint 25↓)

125

126

Fig. 6-28. Two subthreshold responses. Note that the duration of even a subthreshold response considerably outlasts the stimulus duration.

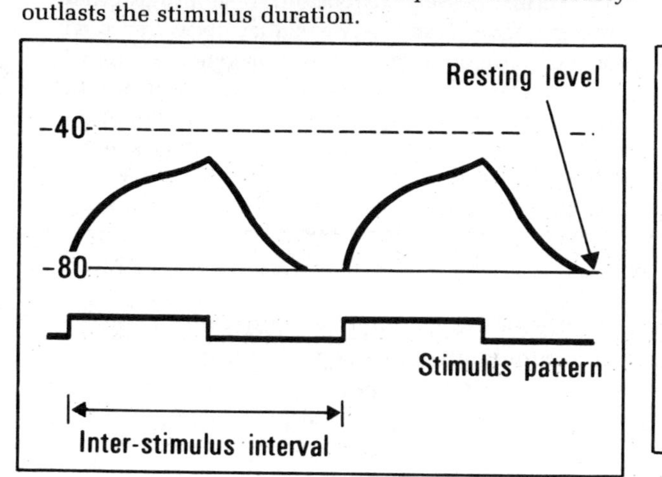

Fig. 6-29. Two subthreshold stimuli at reduced interstimulus interval. Note additive effects between first and second responses.

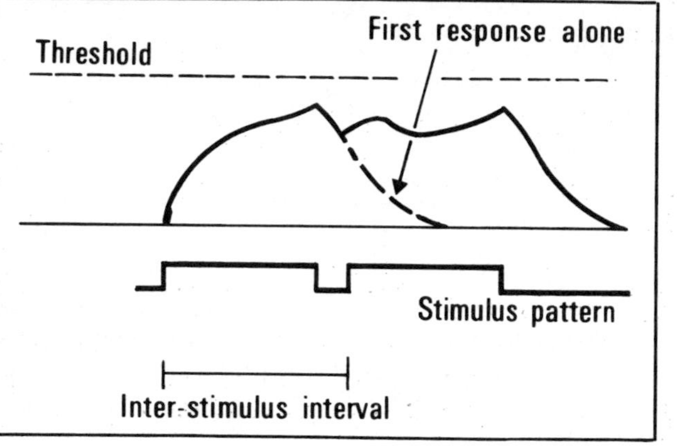

Fig. 6-30. An all-or-nothing action potential appears when the stimulus interval is further reduced, since the combined response now exceeds the threshold level.

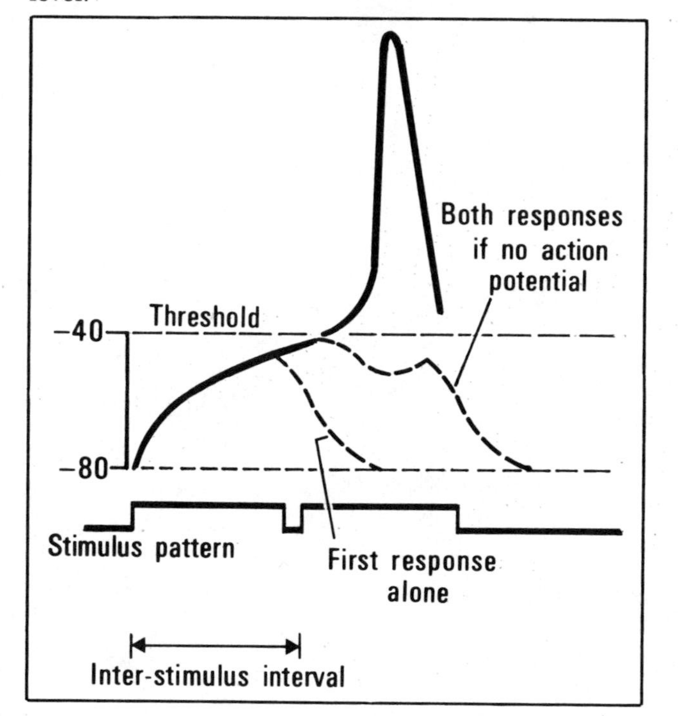

126

127

First, let's discuss temporal summation. What would happen if we selected *one* subthreshold stimulus *strength* shown in Fig. 6-23 and gave *two stimuli*, separated by a short interval? A possible result is shown in Fig. 6-28. There is nothing very exciting here (pardon the pun). *But*, what would happen if the two stimuli were brought closer together in time (i.e., the interstimulus interval were shortened)? **The two subthreshold membrane responses can add,** so we may get the results shown in Fig. 6-29. (The funny shape of the second half of the wave results from the rising portion of the second wave adding to the falling phase of the first.) Finally, **if the two stimuli are close enough together, the responses may add to reach threshold and give an action potential** (Fig. 6-30). The effects of two stimuli, each inadequate by itself to cause an action potential, are summed in time by the membrane, so that threshold is reached. Hence this phenomenon is called **temporal summation.** You will find that temporal summation and the ideas underlying it are very important in understanding mechanisms of sensory coding and how synapses operate (see Chaps. 9 and 10). 1

(You may notice that in Fig. 6-29 the start of the second stimulus is *very* close to the end of the first stimulus. This is *not* a requirement for temporal summation. Shorter stimuli could be used. But, of course, then a stronger stimulus must be used to get the same response.) 2

Question: Assume that the interval between a pair of subthreshold stimuli (shorter and stronger than those shown in Fig. 6-28) is slowly reduced until an action potential is obtained. The latency of the action potential is measured from the start of the first stimulus. Will the latency change if the interval between the two stimuli is further reduced? (Hint 29↓) 3

Note that we assumed no alteration of threshold itself throughout this discussion. The additional complications produced by change in threshold can be dealt with after you read the next section. 4

FACTORS AFFECTING THRESHOLD

Up to this point, the *threshold* has been treated as if there were, for each membrane, some characteristic, fixed potential at which the all-or-nothing action potential would be triggered. This is 5

HINTS

25. At the threshold $|I_{Na}| = |I_K|$; therefore, $I_{por} = 0$. Clearly, then, $I_{cap} = I_{ext}$, and dV/dt at threshold is entirely dependent on the external current. Beyond threshold, I_{por} becomes negative and assists in generating I_{cap}. Not surprisingly, therefore, no clear inflection occurs on the rising phase of action potential B as it passes threshold, driven both by I_{ext} and I_{por}.

28. The study of the process of excitation antedates the oscilloscope by decades! A common method was to record the twitch response of the muscle (attached to the nerve) on smoked paper. In this way, the muscle acted as a form of "biological amplifier."

127

128

absolutely **not** the case. **In all membranes, the threshold is highly variable, can be established only by test stimuli, and is intimately dependent on both the history of the axon and the exact parameters of the stimulating current pulse.** 1

The lability of the threshold is not difficult to understand if you recall the definition of threshold arrived at in the previous section. **Threshold is the usually brief quasistable state at which $|I_{Na}|$ comes to equal $|I_K|$** (as a result of the reactivity of the voltage-dependent conductances). Remember that the increase in $|I_{Na}|$ with depolarization could not occur unless g_{Na} increased more than enough to compensate for the reduction in driving force $(V - E_{Na})$. Similarly, if g_K reacted more to depolarization than is usually the case, a larger $|I_{Na}|$ would be required to reach the threshold condition of equality between sodium and potassium currents. 2

Thus, **threshold is affected by any factor that alters the reactivity of either the sodium or the potassium conductance to change in membrane potential.** In the remainder of this section, we examine a number of commonly occurring situations that are manifested as changes in threshold. But, first, let's ponder a few questions about the underlying principle just stated. 3

QUESTION: If g_{Na} becomes *more* reactive to ΔV, will the threshold increase or decrease? (Hint 32↓) 4

QUESTION: If g_K becomes more reactive to ΔV, will the threshold increase or decrease? (Hint 34↓) 5

Question: If g_K becomes *less* reactive to ΔV, will the threshold increase or decrease? (Hint 33↓) 6

Finally, then, we can restate the rules:

1. **The threshold increases whenever $|I_K|$ tends to become larger relative to $|I_{Na}|$.** Thus, if g_{Na} became *less* responsive to V or if g_K became *more* responsive to ΔV, the threshold would *increase.*
2. **The threshold decreases if $|I_{Na}|$ becomes larger relative to $|I_K|$.** Thus, if g_{Na} becomes more responsive to ΔV, the threshold decreases. 7

Since the threshold potential depends on a balance being reached between $\Delta|I_{Na}|$ and $\Delta|I_K|$, the interplay between steady-state membrane potential and threshold potential can be expected to be quite subtle. For example, will prolonged depolarization (1) *reduce* the threshold by making g_{Na} more responsive to change in membrane potential, (2) *reduce* the threshold by bringing the steady-state potential closer to the threshold potential, or (3) *increase* the threshold by bringing about *inactivation* of g_{Na} 8

Table 6-2. Some Terms Used in Description of Threshold Phenomena in Membranes

Threshold:	A membrane potential V at which the membrane elicits an all-or-nothing action potential on only 50 percent of the stimulations (also called *threshold potential*)
Threshold stimulus:	A stimulus which elicits all-or-nothing action potentials 50 percent of the time
Threshold depolarization:	The ΔV required to reach threshold
Increase in threshold:	Normally used to mean an increase in the required ΔV (for example, when the threshold is less negative and hence farther from the resting potential)
Decrease in threshold	Opposite of increase in threshold (i.e., when the threshold is more negative and closer to the resting potential, there is a decrease in ΔV required for stimulation)

128

129

and hence a reduction in change in g_{Na} for given change in voltage? In fact, any of these situations can occur in any given cell, depending on the magnitude and duration of the depolarization! As a general rule, short depolarization *reduces* the threshold by the first method; prolonged depolarizations of less than, say, 10 mV may *reduce* the threshold by the second mechanism; and larger prolonged depolarizations *increase* the threshold by inactivating sodium conductance. Similarly, hyperpolarization may *increase* the threshold by moving the resting potential further from the threshold, but may *reduce* the threshold by moving the threshold potential to a more negative value (by removing Na^+ inactivation). 1

In practice, this terminology usually is not as confusing as it may seem at this moment! The important point is to be somewhat on your guard whenever statements are made as to effects on threshold, unless it is perfectly clear whether the statement refers to threshold stimulus, threshold potential, or threshold depolarization, as delimited in Table 6-2. 2

In spite of these semantic problems, threshold remains an extremely important concept since so many of the classic membrane properties described here can be represented as the result of changes in threshold. For example, see the next subsections! 3

Fig. 6-31. Changes in excitability following a *subthreshold* stimulus. This curve does *not* refer to changes in excitability following an action potential.

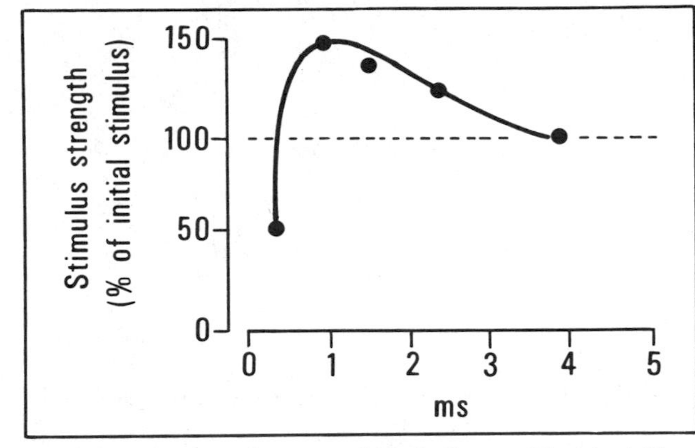

(Modified from J. Erlanger and H. S. Gasser, *Electrical Signs of Nervous Activity.* Philadelphia: University of Pennsylvania Press, 1937.)

Subliminal Excitability

The use of two sequential stimuli to study membrane excitability is known as the *double-shock technique.* We just described in the preceding section the use of this technique to study temporal summation. A further refinement, in which the experimenter varies the strength of the second stimulus until threshold is reached, **makes it possible to study the variation of the threshold with time after the first stimulus.** 4

In the case to be discussed now, changes in threshold are measured immediately after an initial **subthreshold** stimulus (hence the term *subliminal excitability*). Figure 6-31 shows two phases: an initial period of increased excitability and a later period in which a larger-than-normal stimulus is required. 5

Since only stimulus strength is measured here, two types of mechanism are encountered: initial effects of temporal summation (see pages 125 to 127) and both coincident and delayed changes in the reactivity of the voltage-sensitive conductances. 6

HINTS

29. Yes, it will be reduced, and for two reasons. Do you know the reasons? See Hint 30.↓

31. The "stimulus" is made up of the applied stimulus and the residual membrane depolarization. The sum will be greater so the membrane potential will "take off" faster, as in Fig. 6-23. So, the closer the second stimulus occurs to the end of the first stimulus, the greater the combined effect.

35. The height of the action potential would increase, the falling phase of the action potential would be slower, and the duration of the action potential would increase.

129

1 QUESTION: What part of the subliminal excitatory curve corresponds to the period during the action potential at which g_{Na} is increased? (Hint 36↓)

2 Question: What do you consider the most probable cause of the *decrease* in excitability that occurs between 0.5 and 4 ms in Fig. 6-31? (Hint 37↓)

Absolute and Relative Refractory Periods

3 The double-shock technique just used to describe the changes in excitability after a subthreshold stimulus also can be employed to observe the excitability changes after a stimulus that generates an action potential.

4 The first stimulus of each stimulus pair must be *above* the threshold of the resting membrane. Then both the latency and the size of the second stimulus can be varied in order to investigate the changes in excitability following the all-or-nothing action potential generated by the first stimulus. One finds that the period during and immediately after an action potential is characterized by a *decreased* excitability (i.e., *increased* threshold), which can be separated into two parts: the *absolute refractory period* and the *relative refractory period.*

5 Figure 6-32 diagrams the strength of second stimulus needed to reach threshold for that stimulus. At the start of the action potential, there may be a very *short* period in which temporal summation may occur; that is, the membrane is more excitable than normal. However, **during most of the action potential, the membrane is refractory;** that is, excitability is low. It is very low (zero) during the **absolute refractory period,** a time when the axon cannot be stimulated a second time, no matter how strong the second stimulus!

6 We noted that g_{Na}, which increases rapidly following a strong depolarizing stimulus, will return to resting levels (as a result of Na inactivation) even if the membrane is held in a depolarized state (see Fig. 6-15). Thereafter, a second depolarizing stimulus elicits only a negligible response unless the sodium pores are first "reactivated" by allowing the membrane to repolarize to the resting potential. The absolute refractory period lasts until a sufficient number of pores have been reactivated for g_{Na} to increase enough to reach threshold depolarization for a second action potential. Otherwise, $|I_{Na}|$ cannot increase to counter the elevated $|I_K|$ from the depolarization *and* increased g_K.

7 Note that the absolute refractory period is not as immutable as its name might suggest. The refractory period can be doubled by applying a burst of stimuli in rapid succession [40, p. 43].

8 After the absolute refractory period there is a **relative refractory period** during which the membrane can be stimulated a second time. But this **requires a stimulus larger than usual.**

Fig. 6-32. (*A*) Strength of second stimulus required to initiate an action potential at different times after initial stimulus. Assume both stimuli have the same duration. (*B*) Waveform of action potential initiated at time zero.

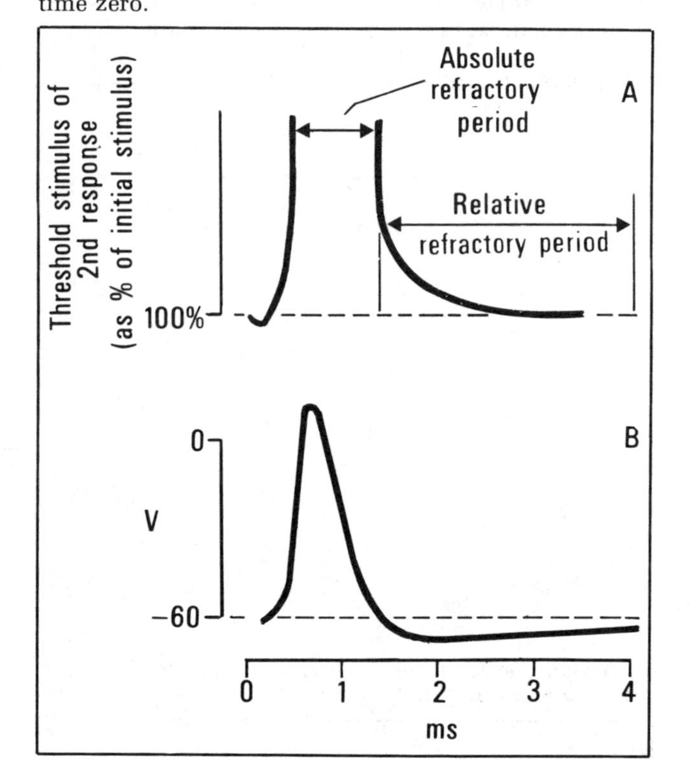

131

1 While the **relative refractory period corresponds to the period of increased g_K** (see Fig. 6-32), there need not be hyperpolarization during the relative refractory period.

2 Question: Why does an increase in g_K make it harder to reach threshold? (Hint 38↓)

An increase in g_K is not the only mechanism involved in the relative refractory period. To some extent, lingering **Na inactivation** will reduce the ability of g_{Na} to increase when the membrane is depolarized. Such lingering Na inactivation is important, particularly in the early part of the relative refractory period. 3

4 QUESTION: Does a subthreshold response have an absolute refractory period? (Hint 39↓)

QUESTION: Would you like an easy way to remember the difference between absolute and relative refractory periods? (Hint 41↓) 5

After the relative refractory period, smaller and longer-lasting changes in nerve excitability are known to occur. Such changes are referred to as periods of **supernormality** and **subnormality.** Apparently, the membrane excitability undergoes underdamped oscillations. Such oscillations are often seen in feedback loops (see Chap. 12). 6

Accommodation and Break Excitation

A classic definition of **accommodation** is that property of the membrane by which its excitability *decreases* with prolonged depolarizing stimuli. However, an exactly analogous *increase* in excitability also can be demonstrated with prolonged hyperpolarizing stimuli. Both the increase and the decrease in excitability are clearly related to changes in threshold, and we discuss these in this section. So we had best define **accommodation** as **a change in threshold that is due to a prolonged stimulus.** 7

HINTS

30. Obviously, the second stimulus is closer to the first stimulus than before, so that will shorten the latency. If you know the second reason, you don't have to look at Hint 31.↑

32. If g_{Na} becomes more reactive, $|I_{Na}|$ will be larger for any given ΔV, so the threshold must *decrease.*

33. You were going to answer *decrease,* right? Well, it's not quite so simple. Under normal circumstances, the increase in $|I_K|$ that occurs during the stimulus pulse is due primarily to an increase in the driving force $V - E_K$, rather than to the rather low, voltage-sensitive increase in g_K. Thus a *decrease* in the reactivity of g_K is likely to have little or no effect on the threshold. What *would* be affected by the reduced reactivity of g_K? See Hint 35.↑

34. If g_K becomes more reactive, $|I_K|$ will be larger for any given ΔV. Thus a larger ΔV will be required to allow $|I_{Na}|$ to increase sufficiently to match $|I_K|$, and the threshold will *increase.*

131

It should come as no particular surprise that accommodation is frequently investigated by a variant of the double-shock technique—a short, variable test pulse is superimposed on a long subthreshold pulse. Three representative positions of the test pulse are shown superimposed in Fig. 6-33, in *A*, on a prolonged subthreshold *depolarization* and, in *B*, on a similar but *hyperpolarizing* background stimulus. Clearly, the size of the test pulse can be adjusted to measure changes in threshold at different times in relation to the background depolarization or hyperpolarization. The results of such an experiment, in which changes of threshold are shown in relation to prolonged background stimuli, are seen in Fig. 6-34. [1]

You should particularly note, in Fig. 6-34, that the changes in threshold (Th) both are *slower than* and *outlast* the changes in membrane potential *V* produced by the prolonged background pulses. The rather slow onset of accommodation is the reason why this phenomenon can be ignored when short stimulating pulses of, say, less than 2-ms duration (in typical membranes) are used. [2]

Also observe that during a strong hyperpolarizing stimulus, threshold actually may fall below the normal resting potential. Then, since the change in threshold outlasts the period of hyperpolarization of the membrane potential, **break excitation** may occur at the end of such a stimulus. That is, the return of membrane potential to normal, after the hyperpolarizing pulse is turned off, may trigger an action potential since **the threshold is exceeded at the resting potential!** [3]

The term *break excitation* stems from the use by early investigators of telegraph key "switches" to control the stimuli in their experiments. When the switch was closed on the "make," the stimulus began. The stimulus ended when the switch opened on the "break." [4]

Question: What would you guess to be the mechanisms involved in the changes in threshold occurring in Fig. 6-34? (Hint 42↓) [5]

Not all prolonged depolarizations result in reduced excitability. The background depolarization of Fig. 6-34 is about 20 mV; depolarizations smaller than about 10 mV actually may *increase* excitability and *reduce* the threshold. [6]

A very slight depolarization simply **moves the resting potential closer to the threshold without producing** a marked **alteration in** the absolute value of **the threshold potential.** It's all a question of "balance"! Small depolarizations affect the resting potential more than the absolute value of the threshold potential. Thus the change in *V* needed to reach threshold is reduced, and excitability increases. Larger depolarizations affect the threshold more than the resting potential. The change in *V* needed to reach threshold increases, and the membrane becomes less excitable. [7]

Fig. 6-33. Series of test pulses superimposed on longer background stimuli (depolarizing in *A*, hyperpolarizing in *B*) to study "accommodation" in a nerve membrane. Sizes of test pulses indicate the strength of stimulus needed to reach threshold as diagrammed in Fig. 6-34.

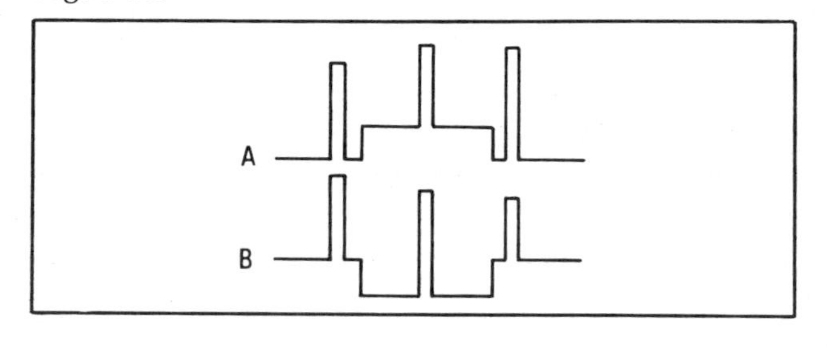

Fig. 6-34. Accommodation (changes in threshold potential "Th") after (*A*) depolarization of the membrane potential *V*, (*B*) hyperpolarization, (*C*) hyperpolarization sufficient to bring about "break excitation" at the arrow, which will generate an action potential since the resting membrane potential is then above threshold.

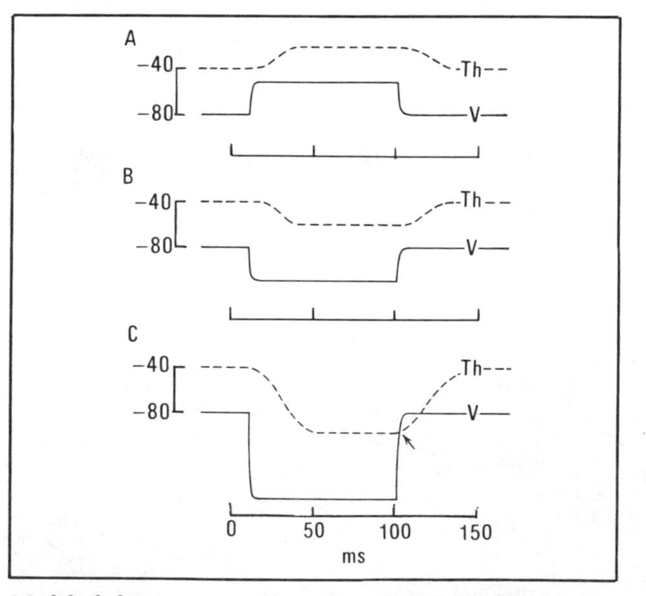

(Modified from D. J. Aidley, *The Physiology of Excitable Cells*. Cambridge: Cambridge University Press, 1971.)

133

Obviously, what is small and what is large in the way of depolarization will vary considerably from cell to cell. The suggestions above are intended to be qualitative rather than reliable, quantitative information.

Prolonged depolarization (and hence accommodation) is not just an experimental curiosity. As we point out later, changes in extracellular K^+ concentration produce changes in membrane potential (and hence changes in threshold) exactly comparable to those discussed here.

Question: In C of Fig. 6-34, if we had drawn the threshold curve turning sharply upward after the arrow (when break excitation triggers an action potential), can you guess the mechanism of such a threshold change? (See Hint 45.↓)

Spontaneous and Repetitive Activity

In some cases, nerve membranes show **spontaneous activity.** That is, **nerve membranes generate action potentials at some rather regular characteristic frequency without any obvious external stimulus.** Under the appropriate circumstances, most electrically excitable membranes are capable of responding "repetitively" (i.e., with a train of action potentials after a single stimulus). These types of activity do not require any new or mysterious mechanisms to explain them.

For example, since the after-hyperpolarization phase of a squid action potential somewhat resembles a long hyperpolarizing pulse of the kind described in the previous section, you

HINTS

36. The first 0.2 to 0.5 ms during which the excitability is increased, i.e., when the membrane is already partially excited, or when the g_{Na} is already a *little* increased—not enough to cause an action potential, but enough to increase the size of the *local-response* component of the second subthreshold depolarization.
37. The predominant effect here is increased g_K triggered by the depolarizing subthreshold response to the first stimulus (although some inactivation of g_{Na} also may be evident). Thus $|I_K|$ will be larger than normal, while $|I_{Na}|$ may be smaller. Obviously, the threshold would be increased; here, however, all that is measured by this method is a reduced reactivity of the membrane (reduced *local response*) to a depolarizing pulse.
38. First, during the relative refractory period, a given applied current produces a smaller-than-expected response. The mechanism is as follows: Increased g_K reduces R_m and hence increases the external current required to produce a given change in membrane potential. Second, the addition of an outward ionic current *generated by the membrane* must cause an *inward* I_{cap}, which will reduce the *net outward* capacitative current across the membrane. And *then* there's the mechanism you thought of. (See Hint 40.↓)
39. No. If it did, you wouldn't get much temporal summation, would you?
41. The *absolute refractory period* is when you try again too soon and no matter how hard you try, you *can't* do it. The *relative refractory period* is when you *can* do it again, but you've got to *try harder* to make it.

133

134

can probably imagine a situation in which that after-hyperpolarization might be large enough to trigger a second action potential by the break-excitation mechanism.
1

This is almost certainly the mechanism by which a damaged axon fires repetitively for a short time, as when penetrated by a blunt microelectrode.
2

While spontaneous activity involves predictable responses of the system, which we have described, we do not consider this to be "core" material. Thus further explanation is offered at the third and fourth levels.
3

Probably the easiest way to explain the mechanism of **spontaneous activity** is to describe the properties of a quiescent cell that continues to fire spontaneously for an indeterminate time following a single input stimulus. Consider the case of the cell that

1. Has about 40 percent or more of its sodium conductance inactivated at resting potential
2. Has a rather high resting sodium conductance relative to its resting potassium conductance, such that the steady-state resting potential will be more than, say, 10 mV from E_K
3. Has almost complete reactivation of g_{Na} if the membrane potential hyperpolarizes close to E_K

This is just the sort of cell which might be expected to show easy stimulation by **break excitation** (see the previous subsection, "Accommodation and Break Excitation"). Hyperpolarization would increase the sensitivity of the cell to depolarization to such an extent that the threshold potential would become more negative than the resting potential. To produce continued spontaneous activity, the hyperpolarization phase that follows each action potential must be able to produce a similarly effective reactivation of g_{Na} and hence a lowering of the threshold. Where this occurs, each action potential must initiate the next action potential; the tail end of one response stimulates the next response (see Fig. 6-35).
4

This mechanism is sufficient to explain rather rapid rates of spontaneous activity (about 100/s in Fig. 6-35) and the bursts of repetitive firing after a single input stimulus, such as are commonly seen in cells exposed to low-$[Ca^{2+}]_o$ media. In the latter case, the burst is limited in duration by the failure of the after-hyperpolarization to fully remove Na inactivation. Inactivation builds up with each succeeding action potential until the membrane no longer responds.
5

Such bursts have a very characteristic appearance (see Fig. 6-36): Both spike height and after-hyperpolarization *decrease* throughout the burst. Spike height falls because the maximum g_{Na} is falling; hence I_{Na} cannot carry V so close to
6

Fig. 6-35. (*A*) Membrane potential and (*B*) conductance changes in spontaneously firing nerve cell.

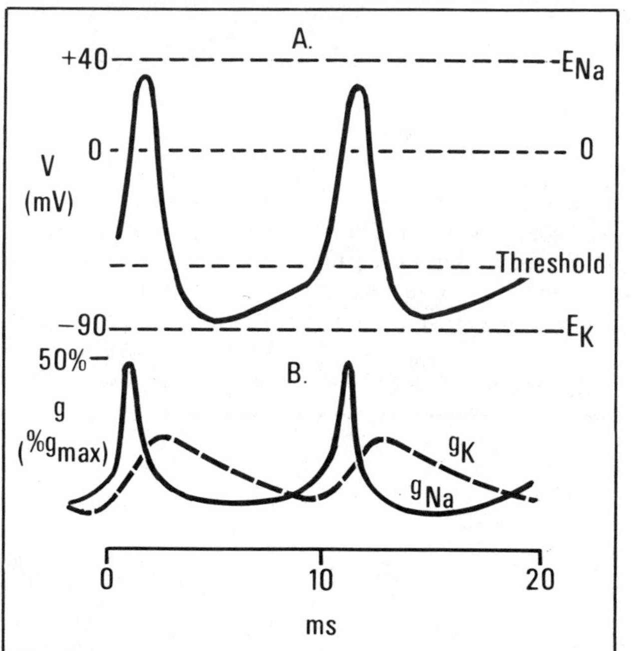

Fig. 6-36. Transmembrane potential during repetitive response to single stimulus applied at time indicated by arrow. Note reduction in overshoot and after-hyperpolarization and increase in interspike interval during the burst.

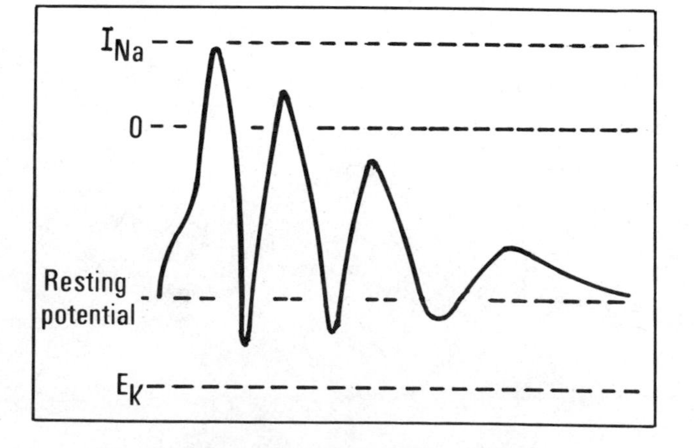

134

135

E_{Na}. The size of the after-potential also decreases since a smaller spike initiates a smaller increase in g_K.

1

Spontaneous activity also can be observed in some cells at frequencies so low that the mechanism just described could not possibly be operative. In axons firing at two impulses per second, it would be necessary to suppose an after-hyperpolarization lasting for close to 500 ms! Presumably, some other mechanism must be involved.

2

In most instances where low-frequency spontaneous activity occurs, it seems probable that one area of cell membrane is being continually excited by a neighboring, relatively depolarized region of the cell membrane. (Such interactions are considered in detail in Chap. 9, where we discuss the mechanism by which depolarization of receptor regions causes changes in firing rate of the sensory axon, and are clearly outside the scope of this chapter.) An alternative mechanism would be that slow potential changes resulting from the activity of electrogenic ion pumps (see Chap. 5) are the trigger mechanisms that control the spontaneous rhythm.

3

Question: Can you think of two other types of spontaneously active, "pacemaker" cells that are extremely important to the physiologic function of an organ? (Hint 46↓)

4

Calcium Ions and Membrane Excitability

The final example (but by no means the least important) of the lability of the threshold in electrically excitable membranes is the effect of calcium ion concentration. Although Ca^{2+} is not important as a carrier of charge in mammalian nerves and skeletal muscle fibers, **the extracellular calcium concentration is of major importance in the control of excitability in these membranes.** Both membrane potential and threshold are dependent on $[Ca^{2+}]_o$.

5

HINTS

40. When the threshold is elevated by a larger-than-normal $|I_K|$, more depolarization is required to achieve a sufficient activation of g_{Na} to generate an equivalent $|I_{Na}|$.
42. Notice the slow time course of these changes, typical of the time course of both sodium inactivation and removal of inactivation, and of changes in g_K at these membrane potentials. Does this give you any ideas? See Hint 43.↓
44. Remember, we pointed out that the squid axon g_{Na} is 40 percent inactivated at resting potential (see page 112 if you forgot that). So, when the membrane is hyperpolarized, you can get, with time, an extra 40 percent, more than half as much again, in reactivity of the sodium system. Now does the phenomenon of break excitation seem so surprising?
45. Did you forget about the *absolute refractory period*?

135

Calcium has been described as a *membrane-stabilizing agent.* Thus membranes become virtually inexcitable in high-calcium solutions and may be spontaneously active in low-calcium solutions. However, *some* external calcium seems necessary for maintenance of excitability, and **membranes become inexcitable in media that do not contain at least some calcium ions.** (The critical concentration is typically around 10^{-4} *M* Ca^{2+}.)

The mechanism by which Ca^{2+} ions affect membrane excitability has intrigued neurophysiologists for the last few decades. It has become clear, however, that the **effects of changing $[Ca]_o$ can be accurately expressed as equivalent to the effects of prolonged hyperpolarization or depolarization.**

Such effects are manifested as a change in threshold:

1. **An increase in $[Ca^{2+}]_o$ mimics membrane hyperpolarization.** The **threshold** is **elevated**, and a larger depolarization is required to achieve a given increase in g_{Na}.
2. **A decrease in $[Ca^{2+}]_o$ mimics membrane depolarization.** For changes down to about one-fourth of the normal calcium concentration, the **threshold** is **reduced,** just as might be expected from small depolarizations (see pages 132 to 133). Some cells may even become spontaneously excitable under these conditions. Since serum Ca^{2+} concentrations lower than this are scarcely compatible with life, **the normal clinical experience is hyperexcitability in hypocalcemia.** More drastic reductions in $[Ca^{2+}]_o$ produce an elevated threshold and reduced excitability, just as we noted in Fig. 6-34 in response to the prolonged depolarizing background pulse.

Until very recently, calcium concentrations for clinical tests were measured as *total serum* Ca^{2+}, rather than *free ionized* Ca^{2+}. However, excitability is related *not* to total serum Ca^{2+}, but to the free ionized Ca^{2+} concentration. At normal pH, free ionized Ca^{2+} is about 60 percent of total serum Ca^{2+}, and **physiologically significant changes in the free ionized calcium concentration occur if there are changes in pH within the physiological range.**

Alkalosis increases the proportion of bound and complexed Ca^{2+} and so **reduces the available ionized Ca^{2+} concentration.** Thus, **hyperventilation, which produces alkalosis, can produce symptoms of hypocalcemia in normal subjects.** Clearly, alkalosis must exaggerate the effects of low serum Ca^{2+}. (Thus a useful clinical emergency treatment of hypocalcemia may be to induce acidemia.)

The relationship between serum pH and serum ionized Ca^{2+} concentration is an often-overlooked point of considerable clinical importance. Frequently it may be advisable to supply Ca^{2+} ions when severe alkalosis is present. Similarly, since the toxicity of cardiac glycosides, such as the drug digitalis, may be increased markedly by elevation of ionized serum Ca^{2+}, one should be concerned, when

137

initiating digitalis therapy, about not only *total* serum calcium, but also the patient's acid-base status.

QUESTION: Would the effects of hypocalcemia be exaggerated or reduced by hyperkalemia (elevated serum potassium levels)? See Hint 48.↓

QUESTION: Would the effects of hypercalcemia be reduced by alkalosis? See Hint 49.↓

Although this section has been written with regard to the effects of Ca^{2+} ions, **Mg^{2+} ions may substitute for calcium to stabilize cell membrane excitability.**

The relative effectiveness of Mg^{2+} and Ca^{2+} ions in the control of membrane excitability varies from tissue to tissue and from species to species, although calcium is either as effective as or more effective than magnesium.

Detailed physicochemical models are available for the mechanisms by which divalent cations affect membrane excitability. Unfortunately, these models must lie beyond the scope of this book. We can simply offer the hint that changing the concentrations of these ions alters the voltage gradient across the inner, voltage-sensitive membrane regions (by changing the degree to which surface charges, intrinsic to the membrane structure, are neutralized by the extracellular medium).

AFTER-POTENTIALS, POSITIVE AND NEGATIVE

One of the more confusing pieces of terminology commonly accepted among neurophysiologists is the labeling of after-potentials as "positive" and "negative" (see Fig. 6-37).

Fig. 6-37. Illustration of the membrane potential shapes associated with after-potential. Solid line represents an intracellular recording of an action potential in which no after-potential is apparent.

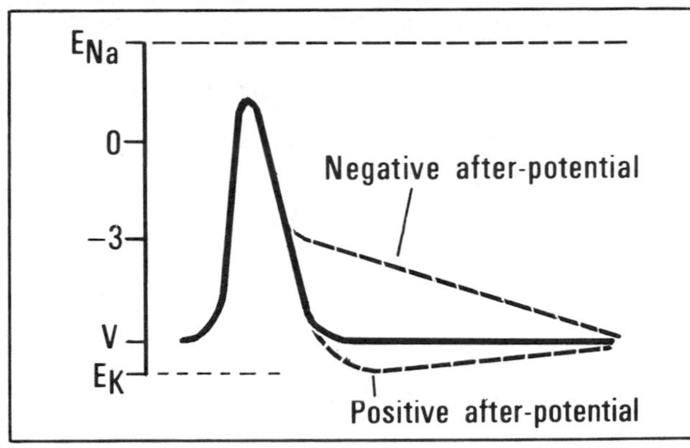

HINTS

43. Prolonged depolarization must alter threshold by two mechanisms: Inactivation of g_{Na} will reduce the reactivity of the sodium system to subsequent depolarization, and increased g_K will oppose both the external stimulus and the ability of the sodium system to reach a current equal to $|I_K|$. The response to hyperpolarization contains a curious wrinkle that only the most *wide-awake* students will anticipate. So think for a moment before you read Hint 44.↑

46. First, the easy one: the heart. The mechanism of pacemaker activity in the cardiac action potential is considered in the next chapter. Can you think of any other important pacemakers? See Hint 47.↓

137

In Fig. 6-37, the solid line represents an action potential in which *no after-potential* was seen. The upper broken line represents an alternative form of action potential in which the membrane does not repolarize as rapidly as expected. For historical reasons, this after-depolarization is called a **negative after-potential,** although the membrane potential clearly is more positive than its steady-state value. Conversely, the more typical nerve action potential (shown by the lower broken line in Fig. 6-37), in which a period of after-hyperpolarization is seen, is described as having a **positive after-potential.** By now, you are probably quite familiar with the mechanism of the positive after-potential. The mechanism of the negative after-potential (which typically is seen only in skeletal muscle cells) is discussed in a later section of this chapter.

1

The reason for this confusing terminology is that the after-potentials were first described in *extracellular* records, where the polarities are reversed by comparison with intracellular recording.

2

In this chapter, we consistently refer to the **positive after-potential** as **after-hyperpolarization.** While this term is exact and avoids the confusion just described, we cannot pretend that it is widely used. Therefore, be warned that you may encounter other terminology elsewhere.

3

A VARIETY OF ACTION POTENTIALS

The action potential as described for the squid giant axon remains a reference point against which all other forms of action potential are compared. In general, the special variants found in other tissues have been found to involve basic mechanisms similar to those discussed so far. The particular properties of some of these cells are, however, sufficiently interesting to warrant further discussion.

4

In the rest of this chapter, we present action potentials that are *not* "space-clamped," but this will not affect the points we are making. However, the advanced student should note that space clamping cannot be applied to all membranes because of technical difficulties, and this limits the conclusions at times, as will be clearer after you have read Chap. 7.

5

Node of Ranvier

Action potentials at a node of Ranvier (in Fig. 6-38) have a shape somewhat different from that of the squid giant axon (e.g., Fig. 6-39). This difference does not seem to be due to any fundamental difference in ionic mechanism.

6

Fig. 6-38. Action currents (equivalent to potentials) at a single node of Ranvier in the frog. Stimulus parameters are as follows:

	Strength (mV)	Duration (ms)
A	200	0.05
B	46	1.6
C	44	6.4

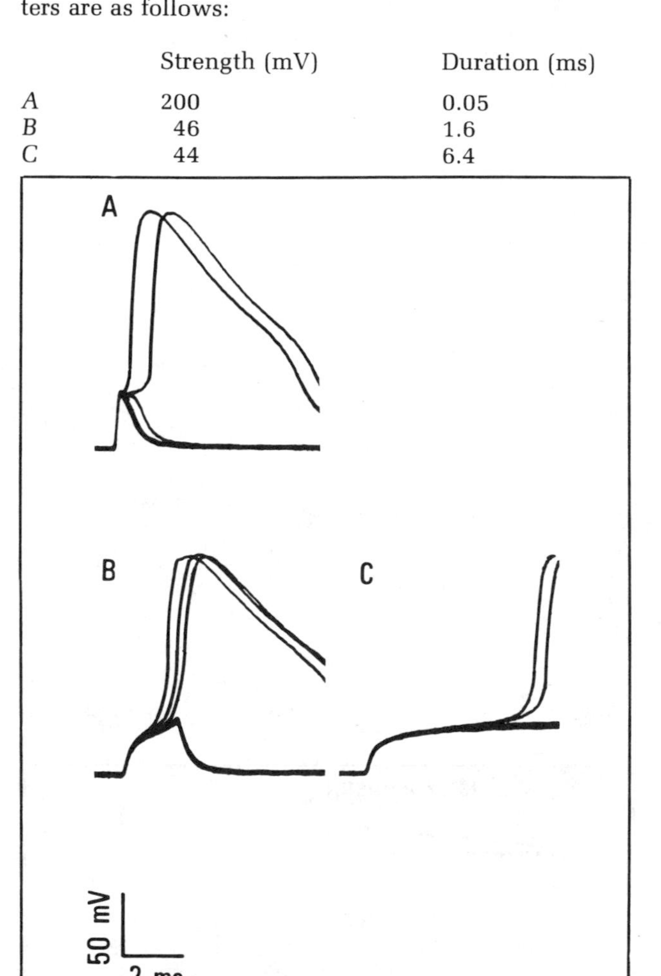

(Modified from I. Tasaki, Initiation and abolition of the action potential of a single node of Ranvier, *J. Gen. Physiol.* 39:377, 1956. By copyright permission of The Rockefeller University Press.)

Part *A* of Fig. 6-38 shows several subthreshold responses (no action potential) and the two threshold responses (action potentials) to the same-strength stimulus in a **single node of Ranvier.**

Parts *B* and *C* of Fig. 6-38 show the constancy of the threshold of firing when the strength and duration of the stimulus are varied (only the first parts of the action potentials are shown). Note that the latency of firing is affected markedly if the pulse is long and barely at threshold (part *C*). (Familiar?)

It has been shown that this action potential also results from changes in g_K and g_{Na}, with its **different shape** being **due to slight differences in the time course of the conductance changes.** Subsequent work has resulted in the properties of the frog node membrane now being as well understood as those of the squid axon.

Skeletal Muscle

The action potential in skeletal muscles is similar to that in nerves except that the falling phase of the action potential is relatively prolonged and there is no after-hyperpolarization during the relative refractory period (see Fig. 6-40).

Instead, there is an after-depolarization! The negative (i.e., depolarizing) after-potential seen in muscle membranes seems best explained by assuming that those potassium pores activated by depolarization are *less selective* than the potassium pores of the resting membrane; they allow some sodium ions to enter the cell through "potassium" channels! Thus while the g_{Na} of the sodium pores may have returned to the resting level, the membrane remains unduly permeable to sodium until g_K has returned to *its* resting value.

Fig. 6-39. Action potential in squid axon. Compare difference in waveshape with that of Figs. 6-38 and 6-40.

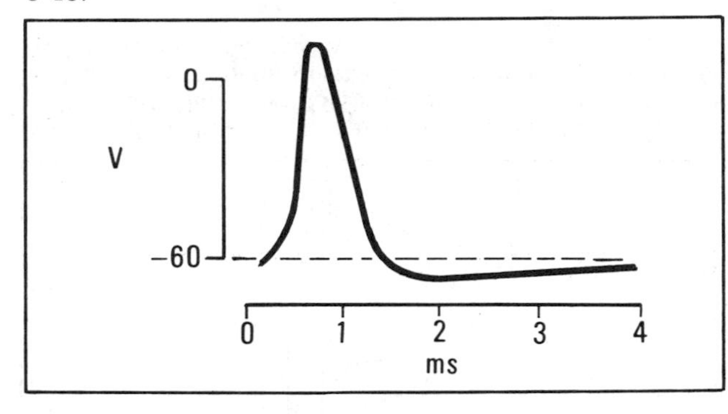

Fig. 6-40. *A* and *B*: Two examples of different types of negative after-potential. *C*: Four action potentials evoked in quick succession. Note the absence of summation of after-potentials.

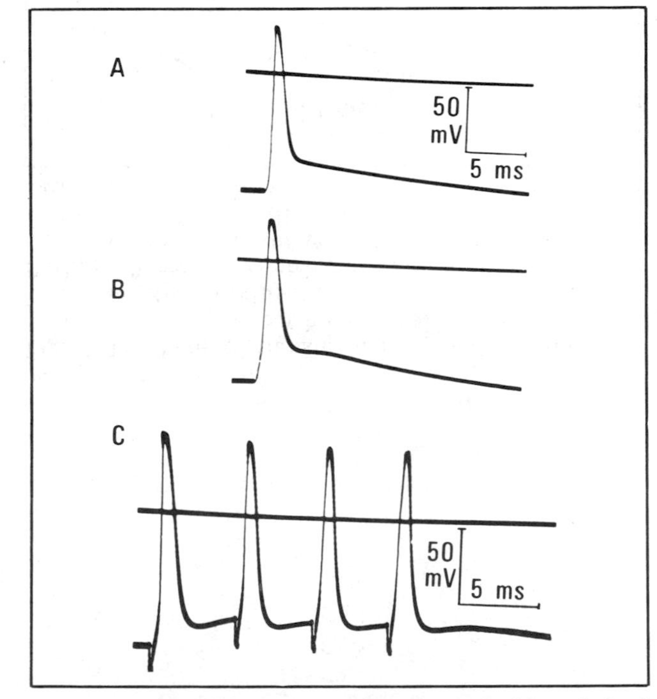

(From A. Persson, The negative after-potential of frog skeletal muscle fibres, *Acta Physiol. Scand.* 58[Suppl. 205]: 1958.)

HINTS

47. We were thinking of the smooth muscle of the gut. The mechanisms of pacemaker activity in this tissue also are discussed in Chap. 7. Other pacemaker cells may well be involved in the regulatory control centers of the CNS. It would be difficult to believe that the rhythm-generator function of the pacemaker cell, so important in peripheral excitable cells, is not also utilized within the CNS (since "spontaneous" activity occurs in many regions—not only in the cortex, but also in such brainstem locations as the respiratory center).

48. Since hyperkalemia must tend to depolarize cells, the effects of hypocalcemia on membrane excitability would be increased.

49. If we presume an *increase* in serum pH, more binding sites for Ca^{2+} ions would become available as a result of the increased ionization of serum proteins. Thus the ionized calcium concentration would fall, even though the total serum calcium concentration would not change. Cell excitability would be expected to return to normal.

A further property of cell membranes that was first noted in skeletal muscle was given the forbidding name **anomalous rectification.** This describes the observation that *steady-state potassium conductance decreases when the membrane is depolarized and increases with hyperpolarization.* The "anomalous" response is due, in part, to the existence of very slow *potassium inactivation* (similar to sodium inactivation, but with a vastly longer time course) and, in part, presumably, to the existence of a separate population of K^+ pores that *close* when the cell is depolarized and *open* with hyperpolarization. The function of anomalous rectification in skeletal muscle is obscure. 1

We noted earlier that chloride conductance tends to be high in skeletal muscle. High chloride conductance must tend to return the membrane potential to E_{Cl}, which is, in this case, the resting membrane potential. The high chloride conductance of muscle membranes must therefore increase the threshold (since $|I_{Na}|$ must equal $|I_K + I_{Cl}|$) and reduce the rate of rise of the action potential. Why might muscle be set up like this? One speculation might be that the system is designed to make the muscle membrane as stable as possible, so as to ensure that this important effector organ does not escape from the control of the CNS. Or can you think of a better hypothesis? 2

This hypothesis, originally written as a piece of "armchair" physiology, now seems almost legitimate! Recent experimental work has demonstrated a hereditary loss of chloride permeability in the muscles of a particular strain of goats. These goats, known as "myotonic goats" because of a tendency to go into muscle spasms, show exactly those symptoms that we would have predicted to result from inadequate "damping" of the excitability of muscle fiber membranes. 3

Cardiac Muscle

Unfortunately, most textbooks of physiology are noticeably reticent on the subject of the ionic mechanisms of the cardiac action potential (an exception is Marshall [30, pp. 40–44]), even though this is probably the most important electrophysiological phenomenon in the future lives of medical students (and their patients), since disorders of electrolyte balance (and toxic drugs) may cause death by their effects on the cardiac action potential. 4

One problem has been that there is no equivalent of the squid giant axon among vertebrate heart muscles! Consequently, experimental approaches to the measurement of conductance changes have been necessarily complex—and hence suspect even to other workers in this field. The following is an attempt to synthesize what is known in a useful framework. 5

Fig. 6-41. Typical transmembrane action potentials from (top to bottom) SA node, atrial muscle, AV node, bundle of His, Purkinje fiber of false tendon, terminal Purkinje fiber, and ventricular muscle fiber (all drawn on same time axis, but with different zero points on vertical scale). Note differences in configuration and sequence of activation.

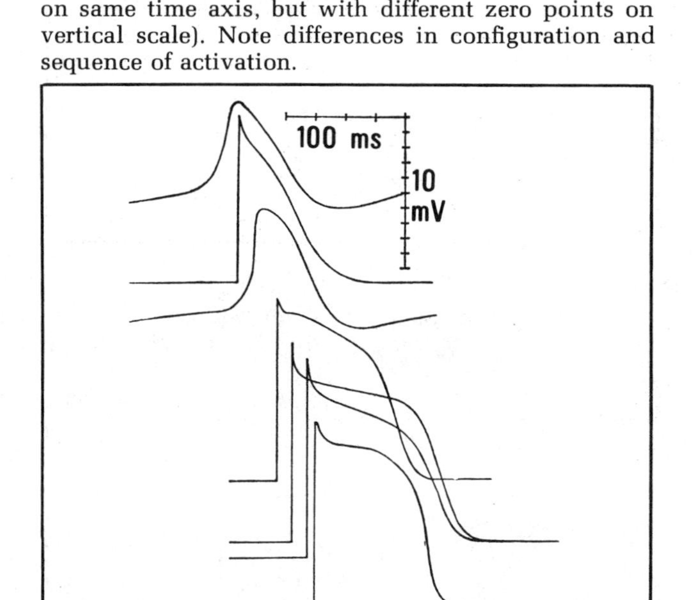

(From B. F. Hoffmann and P. F. Cranefield, *Electrophysiology of the Heart.* New York: McGraw-Hill, 1960. Copyright 1960 by McGraw-Hill Book Company. Used with permission of McGraw-Hill Book Company.)

141

While the cardiac action potential varies considerably in shape in different regions of the heart (see Fig. 6-41), all regions show four separable phases:

1. **Rapid initial depolarization** at the start of the action potential
2. **Plateau phase** of variable length
3. **Rapid depolarization,** which terminates the action potential
4. **Recovery phase** between action potentials.

All regions of the heart *may* show inherent rhythmicity. But under normal conditions, the heart rate is controlled by specialized "pacemaker" regions. In these regions, a gradual depolarization is seen during the recovery phase. This depolarization continues until the threshold is reached from initiation of the next action potential.

The initial rapid depolarization is due (primarily, if not entirely) to Na^+ ions moving *inward* through membrane pores and thus inducing an *outward,* nonspecific depolarizing current across the membrane capacitance. It seems that the initial inward pore current involves two components that have different time courses, that is, a fast component and a slow component. The plateau phase results from the prolonged "tail" of inward current, which is determined by the slowly changing component (See *B* in Fig. 6-42).

The evidence is not clear-cut as to the extent to which this second, slow inward current may be carried by Ca^{2+} ions. Free ionized Ca^{2+} is present in only minute concentrations within relaxed muscles, and (as we see in Chap. 9) contraction is initiated by a sharp rise (approximately 100-fold) in $[Ca^{2+}]_i$. Even the higher concentration in active muscle is, however, 10- to 100-fold lower than $[Ca^{2+}]_o$. Hence, the appropriate gradient exists to drive an inward I_{Ca}. While the sensitivity of the plateau potential to changes in $[Ca^{2+}]_o$ varies markedly from species to species, removal of external calcium ions typically prevents contraction of cardiac muscle, and it is generally accepted (from tracer studies) that *some* entry of Ca^{2+} always occurs during the action potential. However, even if all the slowly inactivated current were carried by Ca^{2+}, this would not be enough to bring about the observed increases in $[Ca^{2+}]_i$. It seems probable that while the importance of the *charge* carried by entering Ca^{2+} ions varies widely from species to species, this Ca^{2+} entry is more importantly a requirement for *release* of Ca^{2+} ions from the main intracellular Ca^{2+} storage sites, for the purpose of initiating mechanical contraction of the muscle fiber.

Now, it is generally agreed that in the typical mammalian heart, the **inward ionic currents** can be divided into two components:

1. *An early inward current,* carried primarily by Na^+ ions, which is relatively rapidly and completely inactivated during the action potential
2. *A late inward current,* carried primarily by Ca^{2+} ions, which is only slowly inactivated during the plateau of the action potential. The interaction of these two currents is shown in Fig. 6-43.

Fig. 6-42. (*A*) Cardiac action potential computed by assuming both that the inward Na^+ current consists of two components with different time course and that the outward K^+ current is rapidly shut off by depolarization. (*B*) Changes in potassium and sodium conductance (g_K and g_{Na}) during the computed action potential. Compare this predicted mechanism with the true ionic mechanism established in much later work (Fig. 6-43).

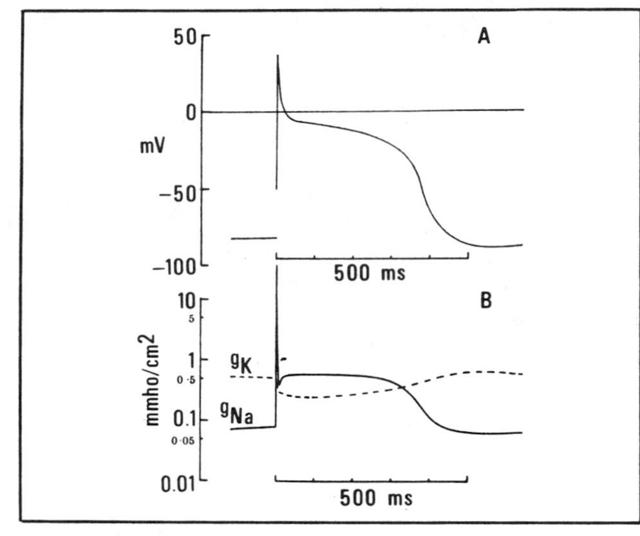

(From D. Noble, Cardiac action and pacemaker potentials based on the Hodgkin-Huxley equations, *Nature* 188:495, 1960.)

Fig. 6-43. Changes in conductances of major ions involved in cardiac action potential.

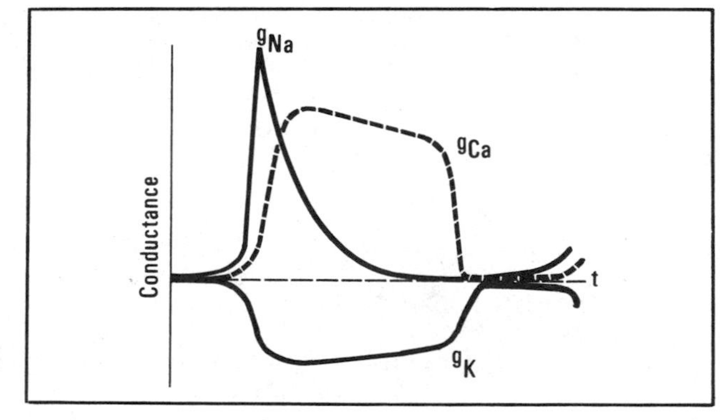

141

Although the major voltage-dependent *outward* current is carried by K^+ ions, just as in nerve and skeletal muscle, the voltage dependence of that current is exactly reversed. Potassium conductance *decreases* during depolarization and *increases* during hyperpolarization (see *B* in Fig. 6-42 and Fig. 6-43). 1

The apparent teleological advantage of the low g_K throughout the prolonged plateau phase is that the same g_{Na}/g_K ratio (and hence the same membrane potential) can be achieved with less Na^+ entry and less K^+ loss—hence with less recovery work for the Na-K pump to do. Neat? 2

The mechanism of the cardiac action potential involves rapid transitions of membrane potential between two separate **semistable states** (dashed lines *A* and *B* in Fig. 6-44). Once the thresholds have been reached at which the membrane "flips" from the one state to the other, rapid transitions are achieved by positive feedback mechanisms. 3

Electrical engineers will note that this system closely resembles a classic oscillator circuit driven into saturation. 4

1. During the *creep* phase, in which membrane potential approaches the depolarizing threshold, inward currents are almost exactly balanced against outward currents. However, recovery from sodium inactivation leads to a slow but steady increase in inward current. Increased reactivity of g_{Na} inexorably drives membrane potential to a threshold point since any depolarization achieved must *reduce*, rather than increase, g_K. 5
2. At the *depolarizing threshold*, g_{Na} rapidly increases and g_K falls. Necessarily $|I_{Na}|$ becomes substantially greater than $|I_K|$, and the membrane depolarizes to the peak of the action potential. 6
3. Although g_{Na} falls by sodium inactivation after the peak of the action potential, sufficient inward current is carried by the slower calcium system to maintain the plateau of the action potential. However, this state is not fully stable either. Inactivation of both sodium and calcium currents continues. Such inactivation reduces the net inward currents, increasing the relative size of the potassium current, and so the membrane slowly repolarizes. But **the further this "creeping repolarization" progresses, the larger g_K becomes,** while at the same time repolarization must reduce g_{Ca} by reversal of activation as well as by inactivation. 7
4. At the **repolarizing threshold,** retreat turns into rout! As g_K rapidly increases, g_{Ca} just as rapidly falls; the situation is the exact reverse of that occurring during the rapid depolarization phase. But as the membrane potential returns to the lower semistable level, inactivation of g_{Na} starts to be removed. And so the process must repeat itself for as long as the ionic gradients and the voltage sensitivity of the conductances can be maintained. Now, isn't that beautiful? 8

Fig. 6-44. Bistable mechanism of cardiac action potential. See text for full explanation of symbols and mechanism. (*A*) Upper semistable state; (*B*) lower semistable state.

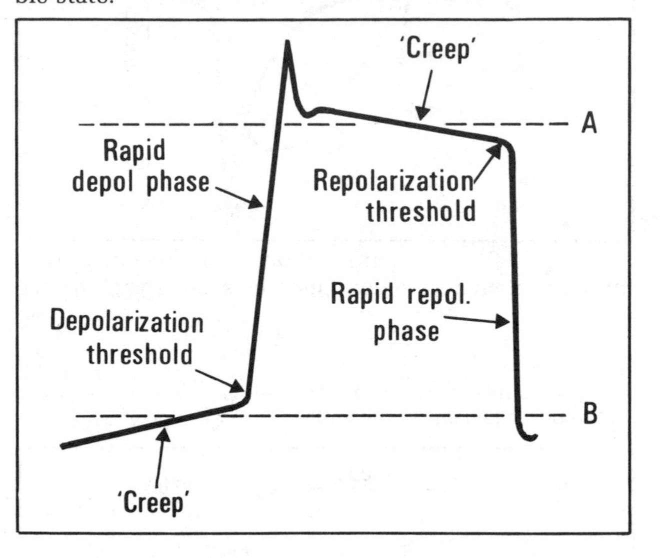

You will notice that the instability of the upper and lower semistable states is quite complex. The major determinant is the gain in the loops by which V affects g_{Na}, g_K, and g_{Ca}. The greater the gain, the more rapid the repetition rate. Relative gain changes in different components of these loops govern the relative time spent in either semistable state.

QUESTION: Why aren't all regions of the heart pacemakers? (Hint 50↓)

We are now ready to consider the mechanisms of some substances that affect the cardiac action potential: stimulation of the vagus nerve causes release of the chemical transmitter **acetylcholine,** which slows or even momentarily stops the heart. It is known that acetylcholine **stabilizes pacemaker membranes by increasing g_K**. (In contrast, **epinephrine** and **norepinephrine,** which accelerate the heart rate, may well increase the response of g_{Na} to depolarization.) An increase in serum K^+ levels (hyperkalemia) may have major effects on cardiac function. The resultant depolarization rapidly causes **accommodation** in the atria as a result of the more or less complete inactivation of the rapidly inactivated inward sodium current; the atria become inexcitable. The ventricles may show increased excitability as a result of depolarization before they too become inexcitable as a result of accommodation. The effects of the calcium concentration on membrane excitability are considered earlier in this chapter.

Now you see why most textbooks omit a detailed discussion of the cardiac action potential! You have come a long way if you were able to follow this presentation.

Smooth Muscle

Smooth muscle is a difficult tissue to study. Individual cells are typically only 2 to 5 μm in diameter and about 100 μm long; thus they are close to the limit for successful microelectrode penetration. In addition, there are marked differences in properties between smooth muscles

HINTS

51. Where V is not equal to E_{Cl}, it follows that chloride is not passively distributed by the membrane potential. If chloride is not passively distributed, then its distribution must depend on energy-consuming processes, for example, a "metabolic pump."

52. If E_{Cl} is *less negative* than V, the concentration difference for chloride must be less than if distribution were purely passive. Since $[Cl^-]_o$ is the serum chloride level, which is independently regulated, it follows that $[Cl^-]_i$ must be higher than would have been expected. *This could occur only if Cl^- ions are transported inward across the membrane.*

144

from different anatomic locations in the same species and between smooth muscles from the same location in different species (e.g., the catecholamine sensitivity of uterine muscles in the rat, rabbit, and human). Nevertheless, some generalizations can be made.

There are two main types of smooth muscle, unitary (or *visceral*) and multiunit [31, p. 1174].

1. **Unitary smooth muscle** shows spontaneous pacemaker activity, which is then regulated by incoming nerves. Thus it is more like heart muscle than skeletal muscle. However, the pacemaker regions are not fixed, as in heart muscle, but appear to move through the tissue as local excitability varies with nervous input or in response to tissue stretch. (Examples are intestinal, uterine, and ureteral muscles.)
2. **Multiunit smooth muscles** normally do not show spontaneous activity. They are activated by external stimuli of nervous or humoral origin. (Examples are the ciliary muscle of the eye, the vascular smooth muscle of the larger vessels, the iris muscle, and the vas deferens.)

Except for the greater instability of the membrane potential in unitary smooth muscle, the mechanisms of the resting and action potentials appear to be similar in both types.

Data for an example of unitary muscle (teniae coli in the guinea pig) and a multiunit muscle (vas deferens in the guinea pig) are shown in Table 6-3. Notice that the resting membrane potential V is markedly different from E_{Cl} in both cases. A change in external ion concentrations shows that the resting potential is most sensitive to change in $[K^+]_o$ but also responds to change in $[Na^+]_o$ and $[Cl^-]_o$. Therefore, we conclude that g_K is greater than either g_{Na} or g_{Cl} but that **all three ions contribute to the resting membrane potential.**

Question: What other conclusion not mentioned above is implied by the observation that V is not equal to E_{Cl}? (Hint 51↑)

Question: Which way is chloride being pumped in this tissue? (Hint 52↑)

Although the action potential in smooth muscle is slow (of about 50-ms duration), repolarization takes place at a rate similar to that of the depolarization phase, without the plateau or negative after-potential seen in cardiac or skeletal muscle (Fig. 6-45). However, while the "overshoot" is Na-dependent, in the sense that the action potential finally disappears in a zero $[Na^+]_o$ bathing solution, it does *not* show the intimate dependence on $[Na^+]_o$ that is a characteristic of the overshoot in nerve and skeletal muscle membranes. The overshoot is more clearly dependent on $[Ca^{2+}]_o$. And it has been argued that much of, if not all, **the inward current of the action potential is carried by Ca^{2+} ions.**

Table 6-3. Ionic Concentrations and Physiological Parameters of Smooth Muscle Cells

	Teniae Coli	Vas Deferens
$[K^+]_i$	164	158
$[Na^+]_i$	19	28
$[Cl^-]_i$	55	57
E_K	−89	−88
E_{Na}	+52	+42
E_{Cl}	−24	−23
V	−55	−57
Overshoot	+7	+11

Values are milliequivalents per liter or millivolts.
Source: Data from R. Casteels. The Relation between the Membrane Potential and the Ion Distribution in Smooth Muscle Cells. In E. Bülbring, A. F. Brading, A. W. Jones, and T. Tonita (Eds.), *Smooth Muscle.* London: Edward Arnold, 1970.

Fig. 6-45. (*A*) Conducted action potential in cardiac muscle; (*B*) conducted action potential in unitary smooth muscle. Note the shorter duration and similarity between rates of depolarization and repolarization in smooth muscle action potential.

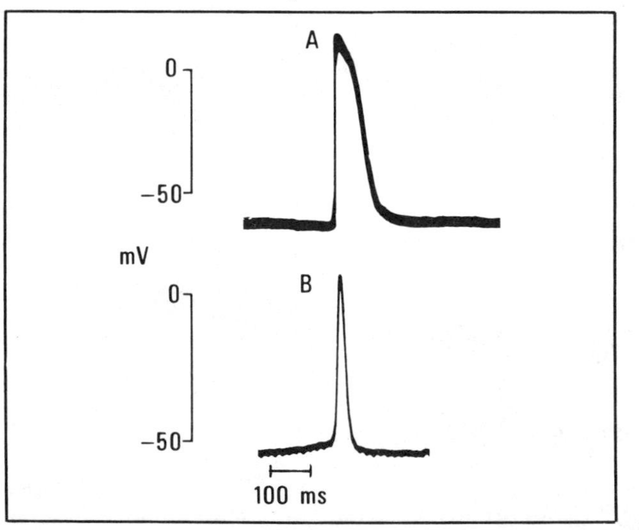

(From J. M. Marshall, Regulation of activity in uterine smooth muscle, *Physiol. Rev.* 42[Suppl. 5]:213, 1962.)

144

Spontaneously active smooth muscles show a slow depolarization ("prepotential") leading to the threshold for the action potential, just as in the pacemaker regions of cardiac muscle (see Fig. 6-46). In some cases, slow oscillations of membrane potential occur without the action potential threshold being reached. Such oscillations can be explained on the basis of **depolarization due to K inactivation followed by repolarization due to a combination of Na inactivation (i.e., accommodation) and K reactivation.** 1

An ingenious hypothesis has been put forward recently. This depends on the observations that a Ca^{2+} extrusion pump is continuously active in the cell membrane and that g_K rises when $[Ca^{2+}]_i$ *increases* and falls when $[Ca^{2+}]_i$ *decreases*. Imagine a cell at its resting potential. Now, $[Ca^{2+}]_i$ falls as a result of continued pump action, and g_K falls. But a reduction in g_K with constant g_{Na} depolarizes the cell. Depolarization increases g_{Ca}, so $[Ca^{2+}]_i$ tends to rise. As $[Ca^{2+}]_i$ rises, g_K rises, the cell repolarizes, and g_{Ca} falls. Now the pump starts to reduce $[Ca^{2+}]_i$, and so on. 2

Smooth muscle membrane potentials may be affected by hormone levels. A good example is the effect of estrogen and progesterone on uterine smooth muscle [31, pp. 1185–1186]. 3

The membrane potential in uterine muscle cells in the immature uterus is only about −35 mV, and the cells are inexcitable (presumably because of Na inactivation). In the estrogen-dominated uterus, the membrane potential increases to about −50 mV and spontaneous activity is observed. In the progesterone-dominated uterus, the membrane becomes hyper- 4

Fig. 6-46. Spontaneous electrical activity in individual smooth muscle cells of teniae coli of guinea pig colon. (*A*) Pacemaker type; (*B*) sinusoidal waves with action potentials on the rising phases; (*C*) sinusoidal waves with action potentials on falling phases; (*D*) mixture of pacemaker, oscillatory, and conducted action potentials.

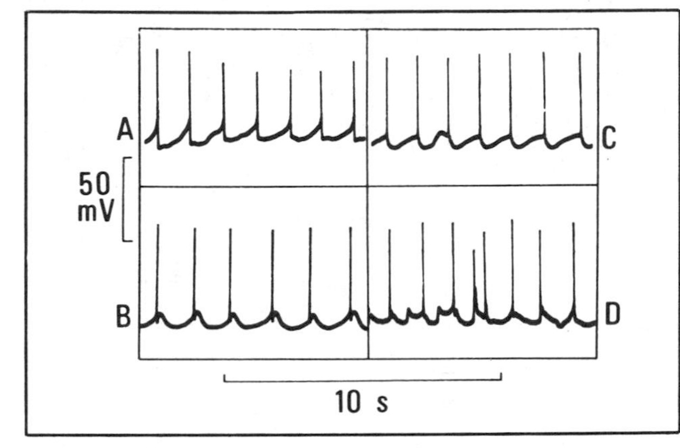

(From E. Bülbring, Physiology and pharmacology of intestinal smooth muscle, *Lect. Sci. Basis Med.* 7:374, 1957.)

HINTS

50. They are if you wait long enough. Normally, however, the gain in the g_K loop is too low for this to be noticeable, except in the pacemaker regions.
53. Clearly, g_K is greater in the progesterone-dominated uterus. We hope that you reasoned this one out before you peeked!
54. Whenever $dV/dt = 0$.
55. (*a*) S; (*b*) B; (*c*) B; (*d*) A; (*e*) N; (*f*) B.
56. Doubling of $[K^+]_o$, presuming this happened acutely, would halve the potassium gradient and very substantially depolarize the cell—quite a sufficient depolarization to make the cell almost inexcitable. Threshold would *increase* markedly.
57. Whenever dV/dt is zero if the situation is not complicated by any external current flowing. In the presence of such an external current, $I_{por} = I_{ext}$ when $dV/dt = 0$ (see page 116).
58. Voluntary hyperventilation can produce an entirely sufficient alkalosis to induce *hypocalcemic tetanus* in a subject with normal total serum calcium levels. As we see later, this tetanus is produced by *motor neuron* hyperexcitability rather than hyperexcitability of the muscle cells themselves. Nevertheless, the reduced ionized calcium levels in the serum will result in a decrease in the threshold in muscle cells even if this decrease is not sufficient to trigger spontaneous activity.

146

polarized to −65 mV and all spontaneous activity ceases (see Fig. 6-47). Similar fluctuations of membrane potential have been seen in the rat at different stages of pregnancy (the uterus is progesterone-dominated until just before parturition, at which point the estrogen level rises sufficiently to outweigh the influence of progesterone; the uterus then becomes spontaneously active). In the pregnant rat, changes in the membrane potential have been found to be associated with change in g_K rather than change in $[K^+]_i$, $[Na^+]_i$, or $[Cl^-]_i$. 1

QUESTION: Is g_K higher in the progesterone-dominated or estrogen-dominated muscle? (Hint 53↑) 2

CLINICAL ESTIMATION OF CHANGE IN MEMBRANE EXCITABILITY

The classic, major, and potentially life-threatening effect of ion concentration on membrane excitability is that produced by hyperkalemia. 3

Not all cells are similarly affected by a similar degree of hyperkalemia. Thus an elevation of serum K^+, which severely *depresses* the excitability of the cardiac pacemaker cells in the atrium, may markedly *increase* the excitability of cardiac muscle cells in the walls of the ventricles! 4

The general rule in any given clinical situation is to *think:* First, how might that situation affect the resting potential? Second, how might such a change in resting potential affect excitability? Third, might that situation more directly affect cell excitability? 5

In spite of the complexity of the relationship between serum ion concentrations and threshold depolarization, certain general conclusions can be drawn:

1. **Changes in serum potassium concentration** (and chloride, if g_{Cl} is high) **primarily affect the steady-state resting potential** (which may subsequently affect excitability). Changes in chloride are relatively unimportant in their direct effects on membrane excitability.
2. **Changes in serum calcium and magnesium primarily affect the threshold** (by affecting the change in I_{Na} for a given change in membrane potential).
3. **Changes in serum pH primarily affect the concentrations of ionized serum calcium and magnesium** (which, in turn, affect the threshold).
4. Changes in serum sodium have profound effects on extracellular fluid volume. While hyponatremia and hypernatremia might be expected to affect both resting potential and threshold, such effects appear first in the central nervous system, leading to confusional states and later to coma rather than to any life-threatening emergency as a result of changes in the excitability of membranes of peripheral nerves or of cardiac, smooth, or skeletal muscles. 6

Fig. 6-47. Effects of estrogen and progesterone on membrane potential of uterine smooth muscle from ovariectomized rats. The units are millivolts.

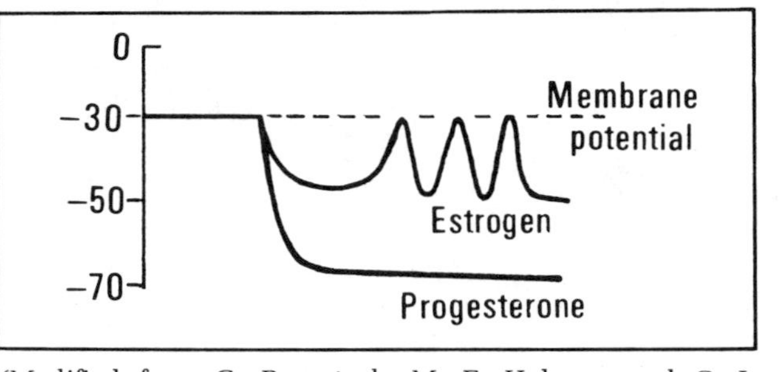

(Modified from G. Burnstock, M. E. Holman, and C. L. Prosser, Electrophysiology of smooth muscle, *Physiol. Rev.* 43:482, 1963.)

146

147

When the effect is primarily on resting potential, remember that **hyperpolarization typically reduces membrane excitability,** whereas **depolarization first increases but then markedly decreases the excitability of cell membranes.**

When the effect is primarily on threshold, remember that **any factor that increases ΔI_{Na} for given ΔV will increase excitability** (and vice versa).

EXAM QUESTIONS

1. Under what circumstances is I_{cap} zero? (Hint 54↑)

2. Mark (*a*) through (*f*) with one of these symbols:

   A = applies to events initiated by an above-threshold stimulus
   S = applies to events initiated by a subthreshold stimulus
   B = applies to both the above
   N = applies to neither of the above

   a. Is proportional to stimulus strength.
   b. Has period of increased G_{Na}.
   c. Has period of increased g_K.
   d. Has absolute refractory period.
   e. Peak of response has a fixed latency from start of stimulus, for various stimulus strengths.
   f. Following this response there is a period of decreased excitability.

   See Hint 55.↑

3. When is I_{por} zero? (Hint 57↑)

4. If $[K^+]_o$ is doubled, would you predict that threshold would be increased or decreased in a skeletal muscle cell? (Hint 56↑)

5. A medical student voluntarily hyperventilates until she feels "swimming." Under these circumstances, what change, if any, might you expect in the skeletal muscle threshold? (Hint 58↑)

147

149

Propagated Action Potential

150

You have traveled down a long, winding road to reach this point in your understanding of membrane potentials, their mechanisms, and their intricacies. Rest not, but continue with us a bit farther. Summon your strength and energies, for though this road does not end here, you may soon sense both the invigoration that comes from step-by-step mastery of a difficult climb and the realization that from your new vantage point you can now enjoy a panoramic view unavailable at lower levels. 1

For those who feel that the preceding paragraph is out of place in scientific writing, we can only say that we wish our emotions to have only the smallest effects on our scientific judgments, but recognize that emotion necessarily plays a large part in our appreciation of the insights thus gained! Why should we hide from you, who have followed us this far, that we find the complex mechanism of the propagated action potential vastly satisfying aesthetically, that this mechanism has for us the intrinsic beauty of a mountain waterfall or of a sunset flashed across a stormy sea? But our vision is not something readily available to the passive viewer; it requires intense, imaginative effort. If these chapters have seemed like a long hike down a winding trail, we can only hope that you, too, find the reward we have found: a stunning panorama of nature at work in her own realm. 2

PROPAGATION OF ACTION POTENTIAL

For reasons that will become apparent gradually, it is a bit easier to describe axonal conduction in myelinated axons. The most obvious advantage is that we can use the nodes as mental "markers" along the distance of the axon. Figure 7-1 shows the sequence of events that might occur if simultaneous recording from 14 nodes were possible. The action potential starts at the top and moves down the axon. Node 1 is the first node to be depolarized; in turn, it depolarizes the nodes ahead (2, 3, and so on), which sequentially generate action potentials, which, in turn, depolarize nodes farther down the axon. By putting a ruler or the edge of a piece of paper vertically on the figure, you can see what is happening simultaneously at all the nodes. For example, at time *A*, node 1 is just past the peak of its action potential, node 2 is near the peak, and node 3 is just about to reach threshold. At time *B*, node 1 is about to finish its repolarization, node 8 is just reaching the peak of its action potential, and node 9 is about to reach threshold. At time *C*, the first three nodes have returned to resting levels, and the peak of the action potential of node 4 is just being reached. Obviously, **since it takes longer to recover from the peak of the action potential than to reach it, at any one time most of the nodes are in the repolarization phase.** 3

Actually, even though it may seem that a lot of nodes are shown in Fig. 7-1, this is only a diagram and does not really indicate how many nodes may be simultaneously active in a typical myelinated axon. 4

QUESTION: If you assume that an action potential lasts only 1 ms, that action potentials are conducted at a rate of 50 mm/ms (or 50 m/s), and that nodes of Ranvier are 5

Fig. 7-1. Theoretical simultaneous recordings from 14 nodes of Ranvier during passage of action potential along the axon (top to bottom).

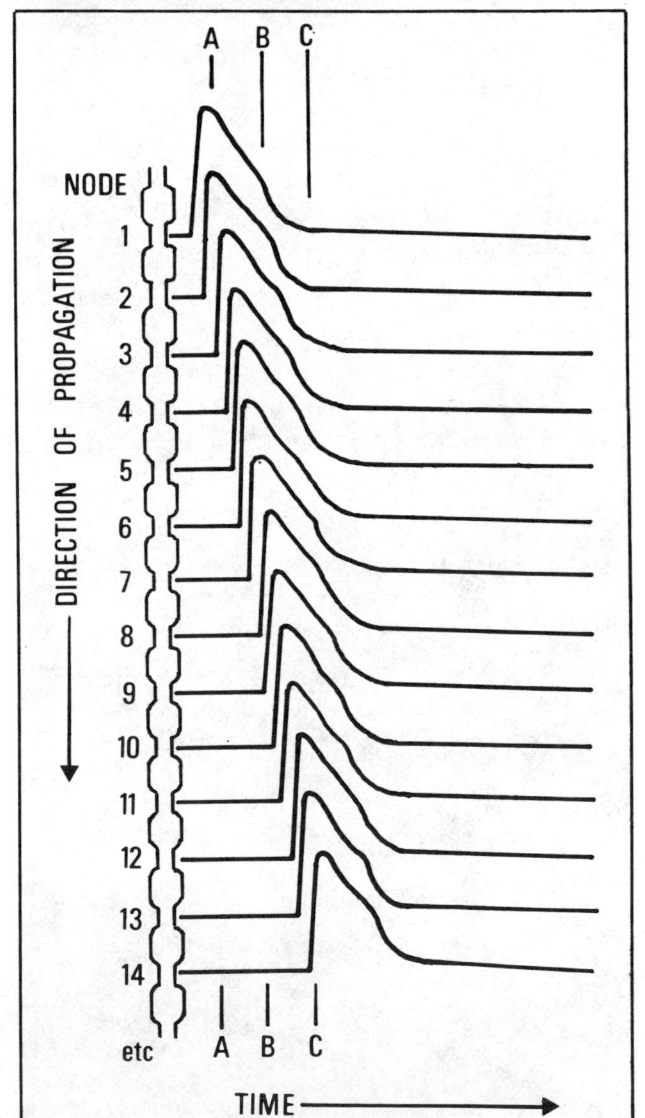

150

about 1 mm apart (all reasonable values), how many nodes are depolarized at one time during the propagation of one action potential? (Hint 1↓)
1

Now, you already know a lot about the detailed mechanism of the action potential in any single region, so what remains to be explained is how the action potential propagates down the axon from region to region (e.g., from node to node). You could probably guess at this mechanism if you think carefully about the consequences of removing that curious (and unnatural) condition that we applied to the action potential throughout Chap. 6.
2

QUESTION: What *was* the special condition applied to the action potential in Chap. 6? Come on now, didn't you realize how important that was? (See Hint 3.↓)
3

QUESTION: So, what happens when we remove that condition? (See Hint 2.↓)
4

Longitudinal currents flowing in **local circuits** between one membrane area and another provide the essential mechanism by which propagation of the action potential occurs. But how does one area arrive at a different potential than another area? It does so by the usual method: capacitative current changes the charge on the membrane capacitor. For simplicity, let's imagine the flow of current between two nodes, node 1 in a part of the membrane that is depolarized during an action potential and node 2 in a resting membrane region in front of that advancing action potential.
5

Figure 7-2 shows two such nodes. The inward I_{por} at node 1 during the action potential drives an outward I_{cap} at *both* node 1 and node 2. Since outward I_{cap} depolarizes, it is obvious that both nodes are depolarized (node 1 more than node 2, you will soon discover).
6

The equivalent circuit diagram is shown in Fig. 7-3, where you can see that at node 1, I_{por} passes out through I_{cap}, but some of the current also passes along the axon to give an outward I_{cap} at node 2.
7

You may have thought from Fig. 7-2 that the action potential at node 1 depolarizes only node 2. Actually, the depolarization extends over many nodes, but it causes less and less effect as a given node moves farther from the depolarized node. Or, to say it in another way, the effectiveness of an action potential in causing depolarization drops off with distance. We attempt to show this in Fig. 7-4, where you can see that more current (from the battery at node 1) passes through node 2 than passes through node 3; in turn, more current passes through node 3 than passes through node 4; etc. You already know of many physical processes that similarly diminish with distance: a light appears fainter as the distance between it and an observer increases; the farther you are from a sound, the fainter it seems.
8

9 QUESTION: Does this rule hold for social and psychological forces as well? (Hint 5↓)

You should note, in the preceding description, that current flowing in through node 1 is flowing out at other nodes (depolarizing them). Because of this current flow, node 1 acts as an "external" current source as far as nodes 2, 3, and 4 are concerned. Recall from Chap. 6 that **when an "ex-**
10

Fig. 7-2. Inward I_{por} at node 1 during the action potential leads to outward I_{cap} at both node 1 and node 2.

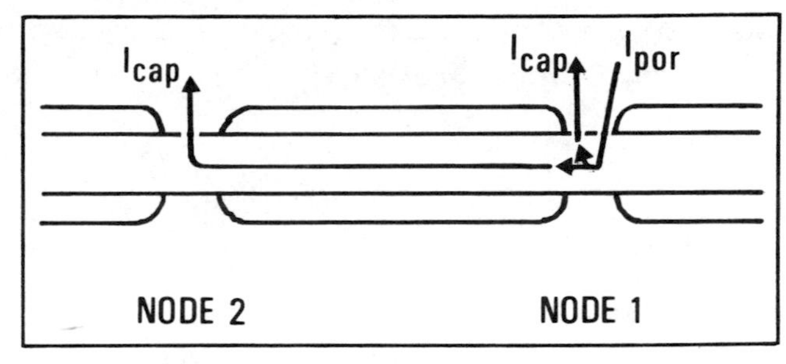

Fig. 7-3. Equivalent circuit for Fig. 7-2; R_i and R_o are resistances of intracellular and extracellular fluids, respectively.

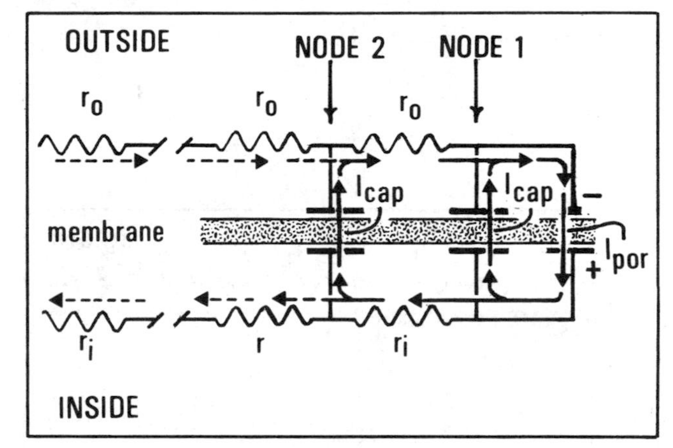

ternal" current source is involved the inward and outward currents through any given (single) node are not equal. 1

Let's describe this idea in greater detail. In Fig. 7-4, it is presumed that a net inward pore current occurs at node 1. The circuit has to be completed; induced current must flow out across the membrane, and **the total current summed over the whole membrane must be zero at all times.** However, unlike the situations we considered previously, the induced current will not leave entirely at node 1. Quite the contrary. **Some part of this induced current must flow out across every single part of the cell membrane!** Thus the outward I_{cap} at this node must be smaller than its inward I_{por} *by just that amount lost to the adjacent nodes.* At the same time, the adjacent nodes become "coupled" to node 1 *to just the extent that they share in the induced outward current spreading from node 1.* We can link these two conclusions by introducing the concept of **coupling current I_{cpl}**, by defining I_{cpl} as that component of transmembrane current (at a given region) lost to, or gained from, other membrane regions. Thus, for any given node,

2 $$I_{cpl} = I_{por} + I_{cap} \qquad \text{Eq. 7-1}$$

Coupling current provides a measure of the extent to which any region either acts as an external current source or is affected by external currents generated elsewhere in the membrane. 3

We introduce the term I_{cpl}, rather than use the previously defined term I_{ext}, to make clear that in this instance the extra current comes from (or flows to) **another part of the cell membrane.** Recall that I_{ext} was defined as current from an external (i.e., non-biological) electrical circuit. 4

Of course, since all current entering the cell must leave it, you can see that the following complete, correct generalization applies in the case of Fig. 7-4: **the sum of all inward currents throughout the membrane equals the sum of all outward currents throughout the membrane.** That is,

5 $$\Sigma I_i + \Sigma I_o = I_{mem} = 0 \qquad \text{Eq. 7-2}$$

The reason why the current is greater at nodes closer to the battery than at those farther away should not be hard to understand if you look carefully at Fig. 7-4. The current through node 2 must pass through $r_i + r_o$, while that through node 3 must pass through $2r_i + 2r_o$. Now, since the battery E is the same for both circuits, the 6

Fig. 7-4. Current flow from action potential at node 1 to adjacent nodes. The farther away from node 1, the less the current.

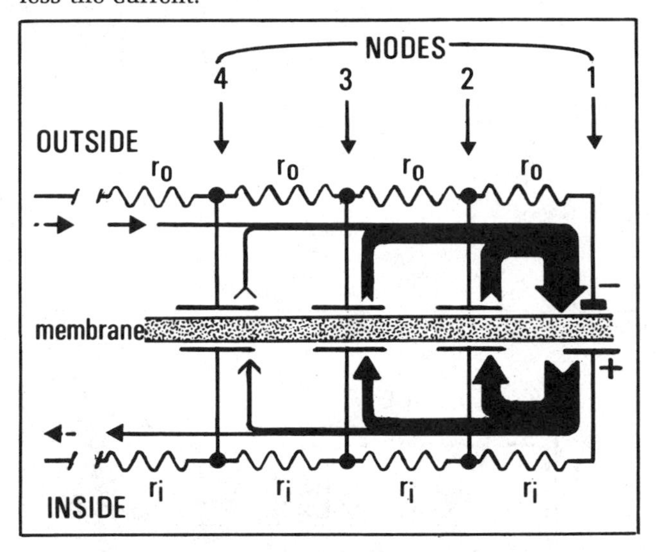

current must be inversely related to the resistance in each circuit, just as it is in Ohm's law:

$$E = IR \qquad \text{Eq. 7-3}$$

Since the circuit through node 3 has a greater resistance, the current must be *less* than that through node 2.

1

Of course, the capacitor will be charging as the current flows, so Ohm's law does not hold directly. However, the comparison between the two circuits still holds in that the current remains inversely related to the dynamic resistance (impedance).

2

To summarize: An action potential propagates by depolarizing the resting membrane area(s) ahead of it. The inward I_{por} in the rapidly depolarizing membrane region (which is generating the rising phase of the action potential) exceeds the outward I_{cap} in that region, producing an inward I_{cpl}. This inward I_{cpl} acts as a current source driving local circuits, which then produce outward (depolarizing) I_{cpl} in resting membrane regions ahead of the advancing action potential. As these previously resting regions become depolarized beyond threshold, in turn, they begin to generate the inward I_{cpl}, which will drive new local circuits to depolarize membrane areas farther down the axon. In this way, the action potential passes over the axon as a grass fire traveling across a dry prairie. (In both cases, note that the rate of spread is determined by the efficacy of the coupling process.)

3

COUPLING AND UNCOUPLING

Before we analyze in greater detail the current flow during the propagated action potential, it may be helpful to give you a little more practice with the concept of coupling between different

4

HINTS

1. 50. Pretty amazing. What? That's 2 inches!
2. All parts of the membrane surface are **not *necessarily*** at the same potential at the same time! O.K., true, but you can go even further than that. What will happen if two adjacent membrane areas are not at the same potential? (See Hint 4.↓)
3. We called it a *spaceclamp*. It ensures that all parts of the membrane are at the same potential at the same time.
5. Consider psychological distance, e.g., the smaller the room, the better the party, *or* the greater the distance from suffering, the less the concern. Make up your own!

154

membrane areas. In the absence of a space clamp, each membrane area contributes a current I_{cpl} to local circuits along the membrane whenever I_{por} and I_{cap} do not sum to zero. Thus we noted that

1 $$I_{cpl} = I_{por} + I_{cap} \quad \text{Eq. 7-1}$$

Advanced students will note that this equation is the same as that previously introduced for I_{mem}. You should be aware that we depart here from orthodox terminology by differentiating between I_{mem} (which we define as the integrated sum of the ionic and capacitative currents **across all membrane areas**) and I_{cpl} (which we define as the sum of the ionic and capacitative currents **at a specific point in the membrane**). We believe you will find the distinction helpful. 2

3 QUESTION: Is it true that $I_{mem} = I_{cpl}$ in a space-clamped axon? (Hint 6↓)

We indicated in the last section that I_{por} in the active (action potential) area of the membrane spreads out to become an outward I_{cap} not only in the active region, but also in resting areas. This was shown in Figs. 7-3 and 7-4, where I_{por} in node 1 spreads to other nodes. Thus, for node 1, it must be that

4 $$|I_{por}| > |I_{cap}| \quad \text{Eq. 7-4}$$

QUESTION: At node 1, under these conditions, there must be a *net current*. Is this coupling current inward or outward? (Hint 9↓) 5

QUESTION: What is the sign of I_{cpl} in the region of node 1 under these conditions? (Hint 8↓) 6

7 QUESTION: Under what conditions can I_{cpl} be positive? (Hint 7↓)

QUESTION: What is the sign of I_{cpl} in a resting region coupled to a depolarized region? (Hint 10↓) 8

We have seen that coupling currents are a consequence of nonuniformity of membrane potential along the cell surface. By contributing to local circuits they affect the potential in surrounding membrane areas. Whenever a coupling current is generated at one area, the local circuit must be completed across all other areas of the cell membrane. And this must be true for every membrane area that we consider. How could one possibly make sense out of such an intricate mass of interacting circuits? 9

154

155

Fortunately,

1. When it is evaluated at each point, I_{cpl} is necessarily the net current developed at that point after summation of all intrinsic and extrinsic currents. Let's see why. All *extrinsic* currents must be distributed between I_{por} and I_{cap}, and so must all *intrinsic* currents. Thus, by the time we have evaluated I_{por} and I_{cap} at one point, we have already taken into account all interactions between that point and the adjacent membrane regions.

2. All local-circuit currents are summed algebraically in the "axial current" flowing up or down the axon core. Thus the *net* magnitude and direction of all coupling currents in a given region of the axon can be evaluated from the magnitude and direction of the axial current. Like many other important concepts, this one is not easy to grasp immediately. Figure 7-5 shows a diagrammatic array of five adjacent nodes. Inward I_{cpl} at nodes 4 and 5 contributes to axial current flow, while nodes 1 and 2 derive their net outward coupling current from the axial current. By evaluating the changes in axial current I_{axial}, we can clearly see that nodes 4 and 5 are depolarizing nodes 1 and 2. 1

The possibility that nodes 1 and 2 were depolarized by action of other nodes outside the picture and to the left can be eliminated by studying the axial current. 2

QUESTION: What is the single fundamental difference between I_{axial} and all the other currents mentioned in this book? (Hint 11↓) 3

Now we can proceed to more detailed analysis of the currents within the propagated action potential. 4

Fig. 7-5. Axial current in relation to coupling currents at individual nodes.

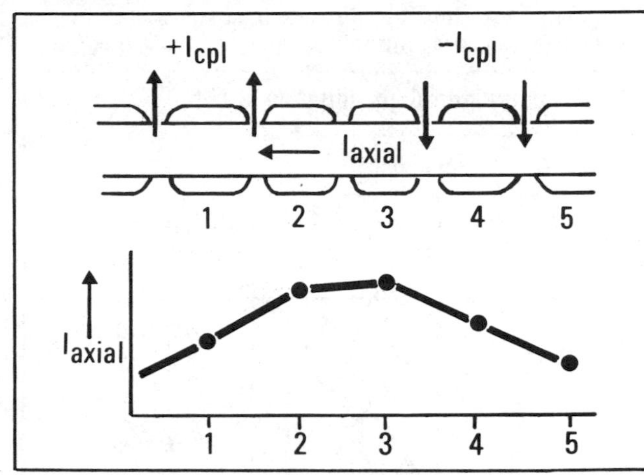

ANALYSIS OF CURRENTS IN PROPAGATED ACTION POTENTIAL

The following analysis depends on some powerful generalizations that we do not derive here. Complete derivations and further detailed explanation are provided in Chapter 8. We provide here an accurate, but nonquantitative, analysis of the propagated action potential. 5

HINT

4. **Longitudinal currents will flow** between one membrane area and the other. Think of it this way: The potential difference between one area and another constitutes a battery, which necessarily drives a current. Current flows from the one area, along the inside of the axon, out at the other area, and back along the outside of the axon to complete the *local circuit*.

155

156

First, nothing about the propagated action potential contradicts the fact that membrane potential is the potential resulting from separation of charge across the membrane capacitance. Thus it remains true that

$$I_{cap} = C_m \frac{dV}{dt} \qquad \text{Eq. 7-5}$$

Second, since, in the propagated action potential, $I_{por} + I_{cap} \neq 0$, how can we obtain I_{por}? Fortunately, it is possible to show (pages 194 to 195) that I_{cpl} must be a function of the second derivative of voltage with respect to time. If d^2V/dt^2 is evaluated throughout the action potential, then I_{cpl} can be calculated. Then I_{por} is given by

$$I_{por} = I_{cap} - I_{cpl}$$

You need to know only C_m, the axon diameter, the interval resistivity, and the conduction velocity. The necessary equations are given in Table 8-1.

Finally, I_{axial} can be obtained directly from potential measurements; I_{axial} is proportional (as I_{cap}) to the first derivative of voltage with respect to time. Figure 7-6 shows a propagated action potential analyzed by this technique into its respective currents. Although it is substantially more complex than the nonpropagated membrane action potential analyzed in Chap. 6, you will soon see that it is not as impenetrable as it may appear at first. Careful study of this figure can provide some fascinating insights as well as verifying some of the intuitive conclusions we reached:

1. At the foot of the action potential, resting membrane regions are depolarized by outwardly directed coupling currents (delivered by axial current flow from more depolarized membrane areas).
2. I_{axial} flows *forward* during the rising phase of the action potential, reaching a maximum where dV/dt is a maximum. This maximum (like the maximum in Fig. 7-5) divides "source" and "sink" regions in the sense that a source contributes I_{cpl}, whereas a sink utilizes I_{cpl} to initiate or increase its depolarization.
3. I_{cpl} is inward throughout the upper part of the action potential.
4. The pore currents are essentially similar here to those of the membrane action potential and are, like them, the result of changes in g_{Na} and g_K. Although the pores themselves respond only to change in membrane potential, just as we saw in Chap. 6, there are some differences in the timing of I_{por} relative to dV/dt that are typical of the propagated action potential.

Fig. 7-6. Analysis of currents along axon in propagated action potential. Note particularly how I_{cpl} is formed by interaction of I_{por} and I_{cap}. Note also "source" and "sink" regions of I_{cpl} and their relationship to direction and magnitude of axial current. (Although this analysis was carried out here for squid giant axon, the form of the currents in myelinated nerve is similar.)

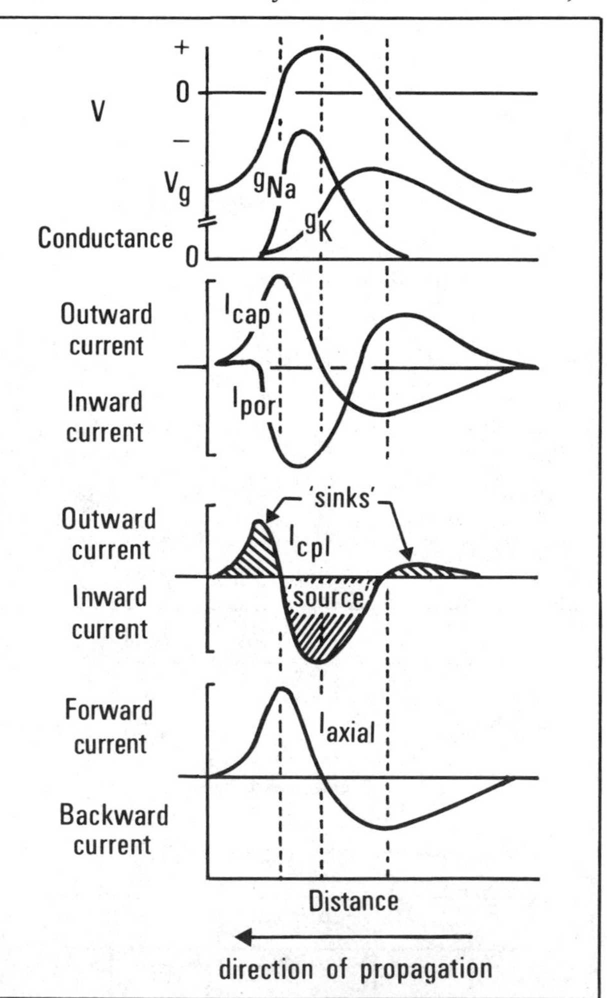

(Modified from D. Noble, Applications of Hodgkin-Huxley equations to excitable tissues, *Physiol. Rev.* 46:1, 1966.)

156

157

a. *dV/dt falls to zero while I_{por} is still inwardly directed!* However surprising this may seem initially, we should have expected it. After all, outward I_{cap} is less than inward I_{por} by the amount lost to coupling currents. So I_{cap} reaches zero when $I_{por} = I_{cpl}$.

QUESTION: Would this affect the size of the overshoot? (Hint 12↓)

b. *I_{por} is zero when I_{cap} is already negative.* What causes inward I_{cap} when I_{por} is zero? The easiest way to think of this is to realize that the difference in potential between the top and bottom of the falling phase must necessarily drive a *backward* axial current. That backward current pulls in current across the membrane capacitance.

c. *I_{por} becomes positive only well after the start of the falling phase.* The rate of recovery increases as this positive I_{por} starts to drive the inward capacitative current.

We now see the validity of our earlier description of the mechanism of propagation. When the foot of the propagated action potential reaches a resting node, that node is depolarized as a result of axial currents, which run far ahead of the region in which they are generated. However, as each node becomes depolarized to threshold, it, too, starts to generate the coupling currents required for further propagation.

MYELINATED AND UNMYELINATED AXONS

You should realize that while we discussed propagation primarily in terms of the nodes of myelinated axons, everything said here could have been said equally of adjacent regions on the surface of an unmyelinated axon. For a given diameter, myelinated axons conduct much more rapidly than unmyelinated ones. Although we delay detailed discussions of the more subtle determinants of conduction velocity until Chap. 8, some simple points should be made at this stage.

It is readily apparent that rapid conduction requires a rapid spread of axial current as far as possible into the region ahead of the advancing action potential. Such current spread will be more effective when axial resistance is low, and in all axons, conduction velocity increases with increasing fiber diameter. (Remember that the larger the diameter of a con-

HINTS

6. Of course it is! It is virtually a definition of the space clamp that $I_{cpl} = 0$ at all times.
7. I_{cpl} is positive in a *hyperpolarizing* region where $|+I_{por}| > |-I_{cap}|$ and in a *depolarizing* region where either $|+I_{cap}| > |-I_{por}|$ or I_{cap} and I_{por} are both positive.
8. I_{cpl} must be negative since I_{por} is inward (negative); that is, $|-I_{por}| > |+I_{cap}|$.
9. Since I_{por} is inward and greater, the net current must be inward.
10. In this case, where the membrane appears to be depolarized by an external current, both I_{por} and I_{cap} are positive. Obviously, I_{cpl} is positive.
11. I_{por}, I_{cap}, I_{mem}, and I_{cpl} are all *transmembrane currents;* I_{axial} is the only current we mention *that does not cross a cell membrane.*

157

ductor, the lower the resistance.) Unfortunately, large-diameter axons are "expensive" in that they make for large, vulnerable nerves. The uniquely vertebrate solution for this dilemma is the myelinated axon. In such an axon, the internodes have high R_m as a result of the effective insulation of the Schwann cell myelin. Transmission is thus efficient, even though internal resistance is quite high. By contrast, the nodes have low resistance and are specialized to produce the high pore currents that are needed to generate the axial currents for rapid propagation.

It has been found that conduction velocity in myelinated nerves is a direct function of the diameter of the nerve fiber (Fig. 7-7).

The slope of the relationship in Fig. 7-7 is 6 m/s per micrometer of diameter, so you can surprise your friends and impress your professors by being able to guess the diameter of an axon, given its conduction velocity. Thus a 10-μm fiber has a velocity of about 60 m/s.

To make life a little more difficult (as usual), the "conversion factor" just quoted is probably too high for the smallest fibers. Below about 10 m/s, the conversion should be about 3, and it may be that the conversion should change even further with unmyelinated fibers [32, p. 115].

The reasons why conduction velocity should vary with diameter have been the subject of much theoretical concern. The most obvious possibility is that the larger-diameter fibers have a lower internal resistance R_i (remember that the resistance of a conductor is inversely related to its cross-sectional area). If R_i is less, the axial currents will spread farther down the axon. In addition, calculations of the internodal capacitance also suggest that the potential would spread farther and faster in large-diameter axons [18, pp. 54–55].

Rushton has produced some theoretical analyses suggesting that conduction velocity should be linear to fiber diameter in myelinated axons and linear to the square root of diameter in unmyelinated fibers (Fig. 7-8). It is of interest that the smallest myelinated nerve fibers are about 1 μm in diameter just where the two graphs cross [1, p. 64]. Thus, myelinated fibers would seem to occur at about the place where an increase in diameter of an unmyelinated axon would be less effective in increasing conduction velocity.

Stop for a moment and try to imagine how big our heads would have to be if we did not have myelinated axons! We like to think of ourselves as large, fast, and smart. Just remember that all the credit should go to the remote ancestor who invented myelination.

Fig. 7-7. Conductor velocity versus axon diameter for myelinated fibers in kittens and cats. Slope of the line is 6 m/s per 1 = μm diameter.

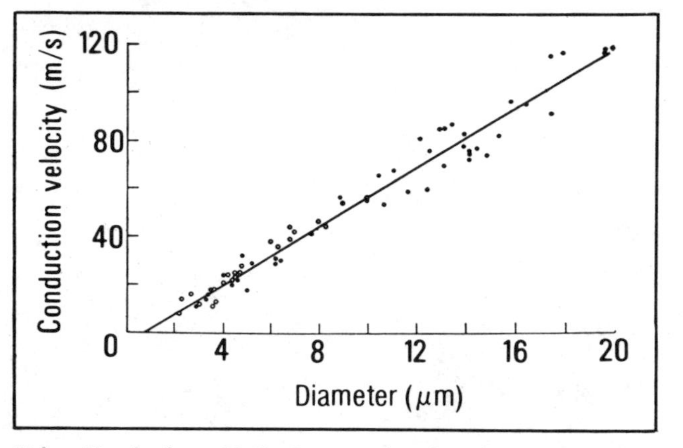

(After Hursh, from H. S. Gasser, The classification of nerve fibers, *Ohio J. Sci.* 41:145, 1941.)

Fig. 7-8. Rushton's theoretical curves for relation of fiber diameter to velocity in myelinated and unmyelinated fibers.

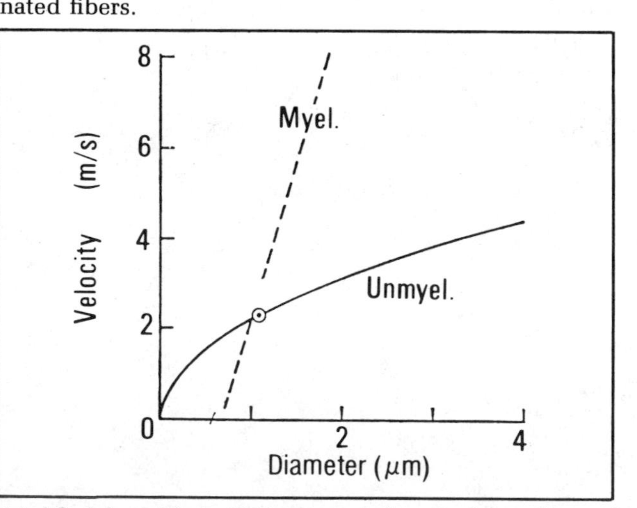

(Modified from W. A. H. Rushton, A theory of the effects of fibre size in medullated nerve, *J. Physiol.* [*Lond.*] 115:101, 1951.)

159

CONDUCTION IN CARDIAC AND SMOOTH MUSCLES

Conduction in both cardiac and smooth muscle, as in skeletal muscle, shows many general similarities to the situation in unmyelinated axons. There is one interesting difference, however, which puts these two muscle types in a class by themselves: **In both cardiac and smooth muscle, some spread of current normally occurs between one cell and the next.** In *cardiac muscle*, this coupling between cells allows an action potential to be propagated throughout the tissue [30, p. 45; 64, p. 139]. In *unitary smooth muscle*, action potentials may be propagated for considerable distances [31, p. 1174; 64, p. 143]. In *multiunit smooth muscle*, coupling is not adequate to permit propagation of an action potential from cell to cell [31, p. 1174; 64, p. 144].

1

Before the days of the electron microscope, it was convenient to think of these tissues as *syncytia* (i.e., as composed of joined or fused cells), and it was thought that the **intercalated disks** in cardiac muscle were apparent rather than real boundaries within the syncytial tissue. However, the electron miscoscope, which showed membranes as a double line, made clear that these double ("unit") membranes were continuous across the intercalated disks and thus established the separate identity of the individual cells in cardiac muscle. However, careful electrophysiological studies, first in cardiac muscle and later in smooth muscle, made it equally clear that very low-resistance connections did exist between adjacent cells in these tissues. Fortunately, the situation was somewhat resolved by the discovery in electromicrographs of regions in which the two adjacent double membranes seemed to become fused for a short distance. These regions were called **close** or **gap junctions,** and it seemed reasonable to suppose that they might be the sites of the low-resistance **electrotonic** connections between cells.

2

In cardiac muscle, the gap junctions are seen as areas of apparent membrane fusion surrounding the intercalated disks. In smooth muscles, normally some gap junctions are present, but considerable variation is seen with regard to the area of contact between adjacent cells. In vascular and other multiunit smooth muscles, small, simple gap junctions may be found. By contrast, in unitary visceral smooth muscle, gap junctions may entirely cover the large, peglike invaginations commonly seen between adjacent cells. As much as 5 *percent of the total cell surface* may consist of gap junction membrane.

3

HINT

12. Why not? In fact, the overshoot is typically 5 to 10 mV lower in propagated action potentials than in a membrane action potential.

159

160

Measurements of the coupling resistance between adjacent cells, together with rough estimates of the *area* of gap junctions from electromicrographs, made it possible to calculate the resistance of gap junction membrane. Such calculations yield values little higher than the resistivity of myoplasm!

1

The coupling resistance remains considerably greater than r_i, since the area of the gap junctions is relatively small when compared with the cross-sectional area of the muscle cell.

2

If ions can pass so readily from cell to cell, as these low-resistance measurements indicated, one might suspect that relatively large holes would be involved. One investigator recently showed that not only do small anions and cations pass readily across gap junctions, but also large dye molecules such as *neutral red* and *Procion yellow* pass readily across gap junctions. These dyes are unable to cross normal cell membranes.

3

Thus, it is not surprising that tangential electron microscope sections through gap junctions show a complex hexagonal pattern consisting, apparently, of an array of tubes about 20 Å long crossing from cell to cell, with extracellular space between them. Aspects of this pattern also can be seen in very thin cross-sections.

4

So the old view of these tissues as syncytia was not all that far wrong, was it? Gap junctions are found also between epithelial cells and gland cells. In both cases, these have been shown to be low-resistance pathways! The functional significance of such junctions as well as the extent to which cells may be metabolically linked (or controlled?) via gap junctions is an entirely open question, and this is currently an exciting area of research.

5

Certain tumor cells show little or no electrical coupling, whereas normal cells from the same tissue are effectively coupled.

6

Notice that the distance over which an action potential spreads through cardiac or visceral smooth muscle is determined by the degree of coupling at each cell boundary. Each cell is a separate "relay station" (equivalent to a node of Ranvier); each cell boundary is a point at which coupling may fail if r_i between cells is so high that sufficient electrotonic spread cannot occur to depolarize the next cell to threshold. In *cardiac muscle*, coupling is always adequate under normal circumstances. But in *visceral smooth muscle*, spread of the action potential usually is limited. (One can imagine that the farther the action potential spreads, the greater its chances of meeting inadequate coupling.)

7

160

161

INJURY POTENTIALS AND CABLE PROPERTIES

1 When a nerve is crushed, it shows a steady *extracellular* potential, which is maximum at the site of crush and drops off with distance in the unaffected nerve.

2 Figure 7-9 shows the **steady potentials** observed along a nerve when the end of the nerve is crushed. The potentials are recorded extracellularly and hence are negative relative to a distant point on the outside of the axon. Presumably, in the area of crush, the membrane is a bit more permeable than usual (it probably has some large holes torn in it!), so that the potential at that point is about midway between that of the inside and that of the outside. In any event, this potential extends a considerable distance down the axon, along healthy nerve.

3 Injury potentials are important historically because they were among the first lines of evidence that the inside of cells were negative and that the magnitude was in the millivolt range. Since the injury potential is steady, it has the space distribution of the "electrotonic potential" of the "cable properties" of the axon.

4 You see, an axon is like one of the early transoceanic telegraph cables: There is an internal conductor (metal in the cable, axoplasm in the nerve), insulation (cable covering, axonal membrane), and an external conductor (seawater, extracellular fluid). Now, if a steady voltage is applied at one end of such a cable, virtually nothing comes out the other end (if the cable is long enough)! The reason is diagrammed in Fig. 7-10, where r_o = resistance outside, r_i = resistance inside, and r_m = resistance of the membrane.

5 Consider the voltages across the various resistors r_m. The farther from the battery, the lower the voltage. Why? It is not hard to see. Compare the circuits through which current must pass to go through the r_m nearest the battery, and the circuit to go through the r_m farthest from the battery. The current passing through the farthest r_m must pass through many more r_o's and r_i's than the current going through the nearest r_m. Just apply Ohm's law, $E = IR$, where the E of the battery is constant. Obviously, there is greater resistance around the longer circuit, so the current will be less. If a smaller current passes through a given r_m, the voltage must be less, as indicated by Ohm's law ($E = IR$), where I and R are given.

The drop-off of potential as a function of distance from the battery can be expressed by the following equation (in which E_0 is the initial potential and E_x is the potential at a distance of x centimeters from that point):

6 $$E_x = E_0 e^{-x/\lambda} \qquad \text{Eq. 7-6}$$

Fig. 7-9. Externally recorded injury potentials along a nerve.

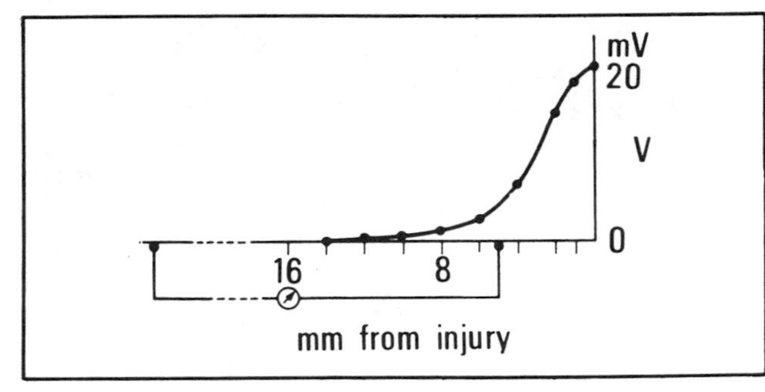

Fig. 7-10. Resistive network for cable properties of axon or transoceanic cable.

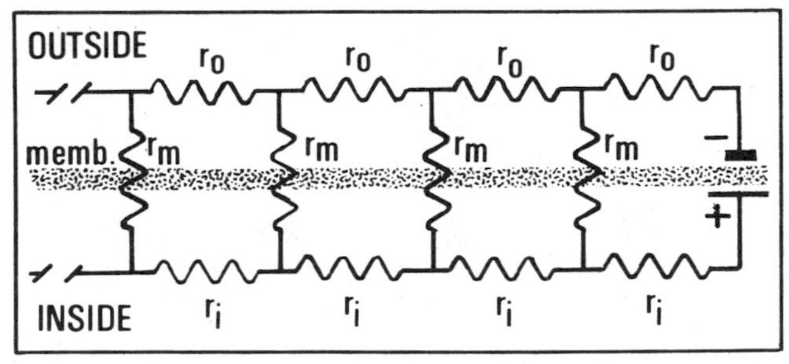

(Modified from R. Lorente de Nó, Correlation of nerve activity with polarization phenomena, *Harvey Lect.* 42:43, 1947.)

161

162

where

$$\lambda = \sqrt{\frac{r_m}{r_o + r_i}} \qquad \text{Eq. 7-7}$$

and

λ = dimensionless "space constant" (lambda) defines distance at which voltage falls to $1/e$ (that is, 37 percent) of its initial value (How can a dimensionless constant represent a distance? Distance is represented as multiples of the unit length chosen for the resistance measurements.)

r_m = resistance of unit length of membrane and myelin (Ω)

r_o = resistance externally (outside) over unit length (Ω)

r_i = resistance internally over unit length (Ω)

1 (For further discussion of the dimensions of these variables, see Chap. 8.)

The form of this relationship is shown in Fig. 7-11. Of course, the actual values of r_o, r_i, and r_m are different for a nerve and for a submarine cable. For the cable, lambda turns out to be miles (!); for the nerve, millimeters. The point is this: **The applied voltage is dissipated by the resistor network and will be undetectable at the other end of the cable unless it is "boosted" at intervals** (i.e., unless extra energy is supplied). For the telegraph cable, booster amplifiers are added at about 40-mi intervals; in the nerve, such boosters are called nodes of Ranvier and occur about each millimeter! The analogy is quite correct; **each node boosts the depolarization that it receives from the previous node into a full action potential!** 2

Incidentally, you might wonder how the resistance of the internodal myelin changes the results of Eq. 7-7. When the electrical resistance and capacitance of myelin are measured, they are found to be equal to those of the squid giant axon, *for each layer* of the wrapping! (Hence it seems likely that each layer of myelin is a Schwann cell membrane [18, pp. 52–53].) The resistance of the entire myelin sheath is about 10 million times that of Ringer's solution—an adequate insulator! Thus, when myelinated and unmyelinated axons are compared, r_m is much larger (per unit length) in the myelinated axon. Thus λ also is larger in the myelinated than in the unmyelinated axon, which partly accounts for the faster conduction velocity of myelinated axons (see Chap. 8). 3

Fig. 7-11. Drop-off of membrane potential with distance (see circuit of Fig. 7-10 and Eqs. 7-6 and 7-7).

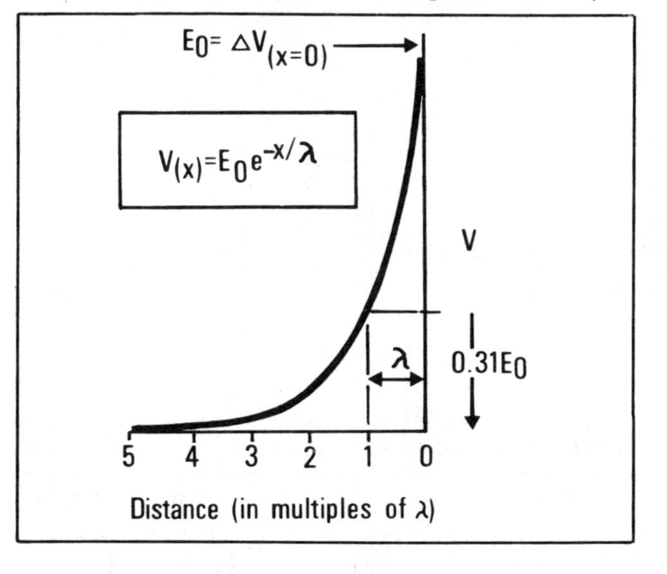

162

163

We're sure you noticed that the membrane capacitance is missing in this analysis. Since the potentials are *steady*, there is no I_{cap}, since $dV/dt = 0$. Hence we can simplify to pure resistors, as in Fig. 7-10 and Eq. 7-7.
1

2 Lambda and electrotonic spread are dealt with in greater depth in Chap. 8.

THRESHOLD AND THE PROPAGATED ACTION POTENTIAL

By now, you should thoroughly understand our earlier statement that "the action potential must be regenerated within each portion of the membrane surface over which the potential travels." The typical propagated action potential is an all-or-nothing phenomenon in which the magnitude of the response in each portion of the membrane is controlled by local conditions. But what happens if those "local conditions" result in an action potential too small to trigger an adequate response in the next membrane region?
3

As we saw in Chap. 6 (pages 123 to 124), a stimulus that barely reaches threshold results in a long "latent period" before the action potential appears. It takes a stimulus that rises substantially above threshold to produce a smoothly rising action potential in which the stimulus is not clearly distinguishable from the subsequent response. Thus, the concept of threshold is not so easily applied in the propagated action potential (except at its initial point of stimulation by, for example, a microelectrode). Each membrane region must be supplied with a **greater-than-threshold** stimulating coupling current, if the action potential is to propagate smoothly and with constant velocity.
4

Typically, the point during a propagated action potential at which $|I_{por}| = |I_{cap}|$ is about halfway up the rising phase of the action potential and some 20 mV more positive than the threshold for initial stimulation of the same axon.
5

So what happens when a given membrane region does *not* receive a sufficient greater-than-threshold stimulus? If this were to happen, say, at a short series of nodes that received current from an injured region, the conduction velocity would fall drastically in the affected region, but would return to normal again in the subsequent normal membrane.
6

QUESTION: So what would you expect in the region just approaching a nerve terminal, where the nerve branches and narrows (raising axial resistance) while at the same time the proportion of electrically excitable sodium pores in the membrane steadily decreases toward the terminal region? (Hint 13↓)
7

The reliability of axonal conduction is based on the all-or-nothing aspects of the action potential, where the magnitude of the response is not a function of the stimulus strength, but depends on only the local membrane "conditions."
8

163

Rushton has expressed the idea beautifully in the following way [53, pp. 169–170]:

"The verbal content of a telephone message may be transmitted exactly by telegraphing it in Morse code, but the conversational overtones, the excitement, or the indignation will be lost, and we cannot help feeling that communication with our fellows would indeed be drab if it had to be restricted to a series of clicks. Yet a series of 'clicks' is in fact all that our brains ever receive—at any rate from nerves. What necessity underlies this impoverishment in the potentialities of communication?

"A telephone wire can transmit a message because conduction along the wire is good and in other directions is bad, so nearly all the electricity flows down the wire and may thus be detected at a distance. But nerves are 'wires' made not of metal but of dilute salt solution, with a resistance of some 25 megohms per millimeter. An ordinary telephone wire [with a resistance of 25 megohms] would stretch across a continent, and after traveling so far the signal would need 'boosting.' Nerve signals . . . need boosting every 1 mm, and in fact the [nodes of Ranvier occur at about this interval and have been shown electrophysiologically to be 'boosting stations.'] So along the stretch of nerve that runs between the fingers and the spinal cord in man there are some 800 nodes or boosting stations. Now if the telephone message is to preserve its quality, each node should restore to the signal exactly what was lost in traveling along the cable from the last node. Suppose that restoration was not perfect but only 99-per-cent complete at each node. Then after 800 nodes the signal would be reduced to $(0.99)^{800} = 1/3000$ of its original size. If on the other hand restoration was overdone by 1 per-cent, the signal would reach 3000 times its initial size—or in practice the system would grossly overload and hence become saturated. So the minutest fluctuation from a perfect restoration will result in a signal that is either saturated or missing (if the response of a node is a function of its depolarization!). Amplitude modulation of signal is impossible: the only reliable code is by unit (saturated) change."

1

STIMULATION WITH EXTRACELLULAR ELECTRODES

So far, we assumed that the **outward** depolarizing current was initiated by either **inward** movement of charge through membrane pores or passing current into the cell through a microelectrode.

2

These situations are essentially identical in that in both cases, charge is injected into the cell, and the circuit must be completed by outward flow of current across the cell membrane.

3

165

But you do *not* need a microelectrode to stimulate a nerve or muscle cell! Entirely adequate stimulation can be achieved with extracellular electrodes placed near the cell surface. **The requirement for stimulation remains simply that current must flow outward across the membrane.** 1

Figure 7-12 shows two external electrodes placed near a nerve and some of the electrical current paths between these electrodes. One path is directly through the external solution (r_0) from **anode** (plus electrode) to **cathode** (minus electrode). Another path involves going through the nodal membrane of node 1, along the interior of the axon, and out through node 2. Under these conditions, **node 2 will be depolarized by the outward current, while node 1 will be hyperpolarized by the inward current.** So, remember: 2

When a nerve is stimulated by external electrodes, stimulation occurs at the cathode and hyperpolarization occurs at the anode. 3

You may remember an exception to the general rule that stimulation occurs at the cathode happens in *anodal break excitation.* (This phenomenon is described in Chap. 6 and under "Accommodation and Break Excitation.") Note also that in Fig. 7-12 the path for stimulating current is longer than and thus has greater resistance than the direct external path *even before* there is any appreciable flow of current through the membrane resistance. In fact, a vast majority of the current will never enter the axons at all. Under these circumstances, in order to pass enough current to stimulate, it is necessary to use *volts,* rather than the millivolt stimulations you might have expected would be enough to change the membrane potential by 20 to 30 mV (to threshold). 4

COMPOUND ACTION POTENTIALS

Up to now, all the potentials described have been those of single cells. That is, the transmembrane potential has been shown. Of course, this requires that an electrode be placed intracellularly, a technique which is relatively recent and certainly not simple, or easy. So, now we take up the subject of recording the electrical activity of nerves **extracellularly.** 5

An example of a compound action potential recorded extracellularly from a nerve is shown in Fig. 7-13. The recording arrangement is shown in the lower part of the figure. After the start of the oscilloscope sweep, a brief electrical stimulus is delivered to electrodes 1 and 2, which starts action potentials from the cathode, electrode 2. On the recording, the stimulus 6

Fig. 7-12. Stimulation by extracellular electrodes. (The "nodes" diagrammed here would not, in reality, be adjacent, because so close an electrode placement would make r_0 too small.)

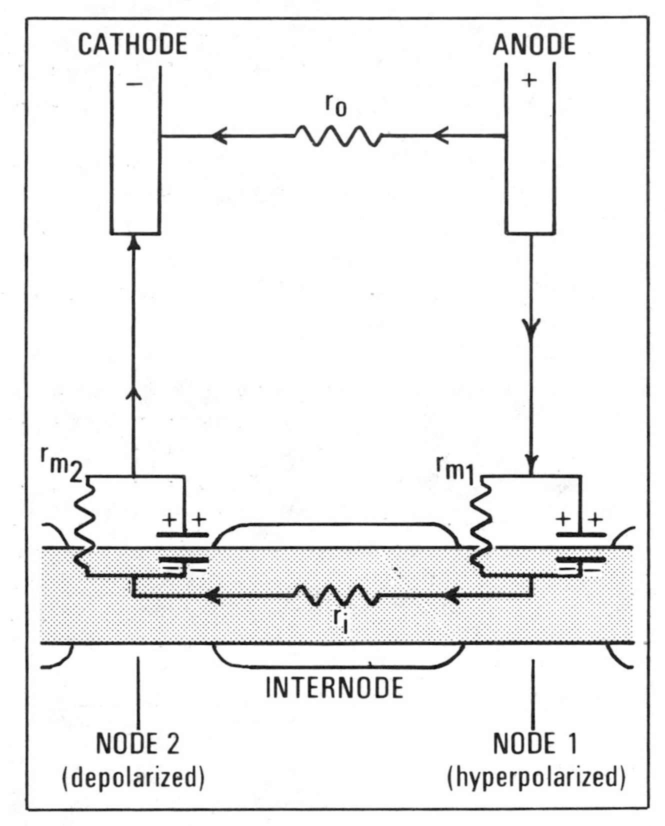

HINT

13. The action potential should decrease in size and slow to a halt.

165

166

gives a small "shock artifact," labeled S. After a latency (*L*) that is necessary for the action potentials to propagate down the nerve, the voltage change between electrodes 3 and 4 shows a monophasic action potential of about 1-ms duration. The area between electrodes 3 and 4 has been crushed, so that the action potential cannot reach electrode 4. [1]

Now, there are two important differences between this extracellular recording and the intracellular recordings you are familiar with: the amplitude of the potential is small, often less than 1 mV, and the amplifier (symbolized by the circle with an arrow in it) is arranged so that an upward deflection of the oscilloscope screen means that electrode 3 is negative with respect to electrode 4. [2]

QUESTION: Why do neurophysiologists show recordings with negative upward when that is not the usual convention in science? (Hint 14↓) [3]

The potential is small because it is recorded extracellularly across r_o. In fact, r_o must be increased artificially by putting the nerve in an insulating medium (air, mineral oil) in order to record even these potentials. If the nerve is in a conducting medium (such as the body), special techniques may be required to obtain satisfactory recordings. Because the electrodes are extracellular, the potential difference is between two points on the nerve. Because of this arrangement, the *action* potential appears as a wave of negativity, as recorded by extracellular electrodes, even though the action potential is positive when recorded across the membrane (i.e., intracellularly). [4]

If the nerve is *not* crushed between the electrodes, a **diphasic** potential is recorded. Since the potential measurement must be made between two electrodes (and assuming that the nerve is not crushed), both electrodes will detect the wave of negativity, at different times. The recorded deflection (Fig. 7-14) will first move upward as the action potential reaches electrode 3 (Fig. 7-14*A*) and then move downward as the action potential reaches electrode 4 (Fig. 7-14*B*). [5]

This occurs because the voltage measurement must take the **difference** between two points, electrodes 3 and 4. If electrode 3 is more negative than electrode 4 (or 4 more positive than 3), then the deflection will be upward. When the reverse is true, the deflection is downward. [6]

If an area between the electrodes is crushed, the action potential cannot propagate into the area of electrode 4, and so **changes observed in a monophasic recording (Fig. 7-13) can be completely ascribed to what is happening at electrode 3.** [7]

Fig. 7-13. Compound action potential recorded from whole nerve, as indicated below.

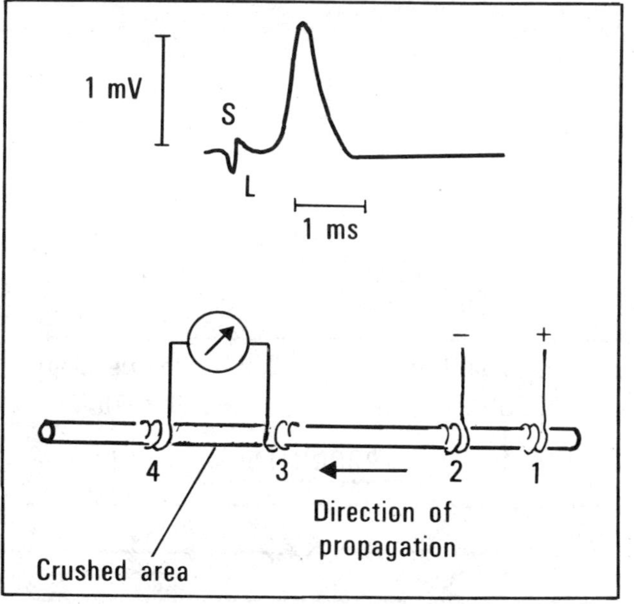

(Modified from S. Ochs, *Elements of Neurophysiology*. New York: Wiley, 1965.)

166

167

1 The extracellular recording of many axons means that they all can contribute to the potential observed. As the stimulus strength is increased, more and more axons are activated, so the potential becomes larger and larger (Fig. 7-15).

2 QUESTION: Isn't Fig. 7-15 evidence against the all-or-nothing law? (Hint 15↓)

3 QUESTION: Why do the axons have different thresholds, so that a stimulus fires some of but not all the axons? (Hint 16↓)

4 The time from the moment at which the stimulus is delivered at electrode 2 to the time when the action potential reaches electrode 3 is called the **latency.** Where both the distance and the time taken to traverse that distance are known, the conduction velocity can be calculated.

5 QUESTION: If electrodes 2 and 3 are 50 mm apart, what will be the latency observed if the fastest fibers conduct at 50 m/s? (Hint 17↓)

CLASSIFICATION OF AXONS

6 **A nerve trunk contains axons of many different diameters and different conduction velocities.** Thus, when all the axons in a trunk are stimulated, the faster-conducting axons get to the recording electrodes first, and a rather "messy" compound potential is recorded (Fig. 7-16).

7 QUESTION: The fiber spectrum of the nerve of Fig. 7-16 (see the inset) shows a large number of fibers with a diameter less than 4 μm, with a lesser peak about 10 μm. Which of these peaks is responsible for the large peak of the compound action potential shown in Fig. 7-16? (Hint 18↓)

The discovery of the compound potential gave physiologists many happy hours as they tried to catalog the different "bumps." They finally ended with a complicated classification scheme, shown in Table 7-1. The function of the fibers contributing to the various humps was unknown at first—it was even unknown whether the fibers were motor or sensory—so the alphabetical classification came to be a mixed classification, based partly on conduction velocity and partly on anatomic function. Subsequently, a classification scheme based on Roman numerals came into use specifically for the sensory fibers, so that the letters came to be used more or less for motor fibers. The differences in letters and numerals have some slight use in talking about specific fibers (as you will see in Chap. 13).
8

9 The various fiber types differ in not only their conduction velocities, but also many other aspects. **The smaller-diameter fibers have higher thresholds in terms of the stimulus strength necessary to make them fire.**

Fig. 7-14. Diphasic recording as an action potential (AP) propagates right to left past two electrodes. (*A*) The AP is at and passing the first electrode. (*B*) The AP is at and passing the second electrode.

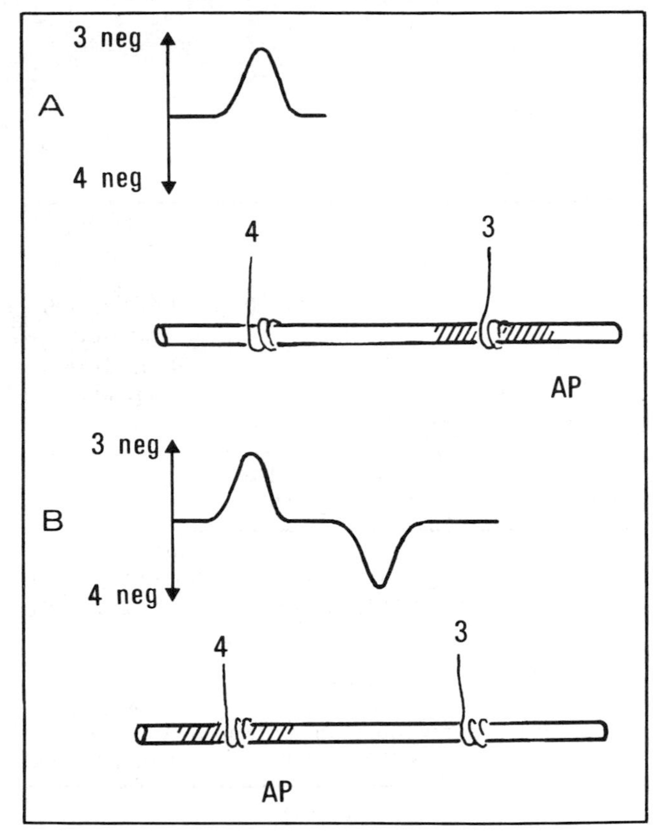

Fig. 7-15. Monophasic compound action potentials with increasing stimulus strength.

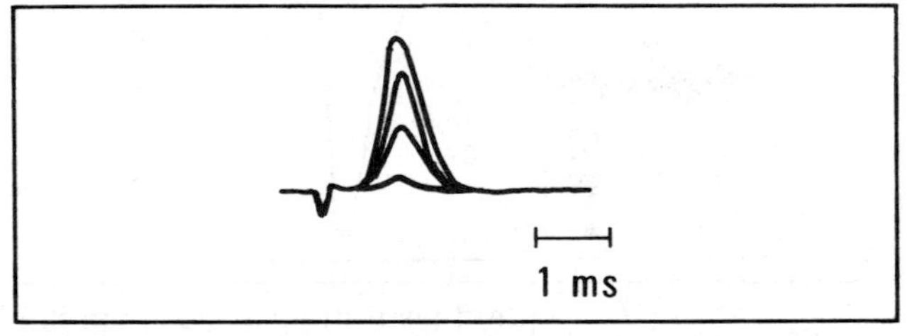

(Modified from S. Ochs, *Elements of Neurophysiology*. New York: Wiley, 1965.)

167

Table 7-1. Classifications of Fiber Types

Fiber Type: Mixed Motor and Sensory	Fiber Type: Sensory	Function	Fiber Diameter (μm)	Conduction Velocity (ms)
A α	Ia	Proprioception; muscle spindle sense; motor to muscle	12–20	70–120
	Ib	Tendon proprioception		
β	II	Touch, pressure	5–12	30–70
γ		Motor-to-muscle spindles	3–6	15–30
δ	III	Pain, temperature	2–5	12–30
B		Preganglionic autonomic	<3	3–15
C Dorsal root	IV	Pain, reflex responses	0.4–1.2	0.5–2
Sympathetic		Postganglionic sympathetics	0.3–1.3	0.7–2.3

This makes it easy first to stimulate the large alpha fibers alone and then to add the beta fibers by stimulating more strongly, etc. (compare Fig. 7-13 with Fig. 7-16). **The highest-threshold axons are those of the C fibers** (see Table 7-1), which may require stimuli 10 to 50 times higher than those of the large myelinated axons. 1

The higher threshold of the smaller fibers follows again from the higher R_i of the smaller fibers. Since R_i is in series with the membrane capacitances that must be depolarized, a high R_i can reduce the amount of current that a given voltage at the stimulating electrodes can move across the membrane capacitance, through the inside of the cell (R_i), and out another segment of the membrane capacitance. 2

Of great importance is the differential sensitivity of fibers of different diameters to local anesthesia. 3

The C fibers are the most easily "narcotized," which explains the clinical observation that when a local anesthetic is applied to a nerve, pain will disappear before motor function. 4

By contrast, the large fibers are more susceptible to block that is due to prolonged pressure (such as falling deeply asleep in an awkward position), which can give rise to a situation where the motor fibers are interfered with, while pain is still perceived. The small C fibers are more resistant to hypoxia, however, as anyone knows whose foot has "fallen asleep"! These observations are summarized in Table 7-2. 5

The lower volume-to-surface-area ratio of unmyelinated fibers means that they do not have as much "concentration reserve" to absorb the effects of ion movements during the action potential. For this reason, the ion pumps of C fibers cannot sustain as many 6

Fig. 7-16. Compound action potential for nerve showing fiber spectrum in inset graph.

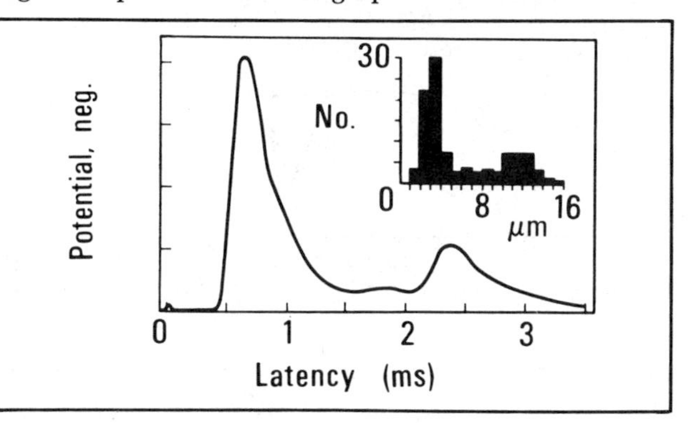

(Modified from H. S. Gasser, Pain-producing impulses in peripheral nerves, *Res. Publ. Assoc. Res. Nerv. Ment. Dis.* 23:44, 1943.)

Table 7-2. Relative Susceptibility of Mammalian A, B, and C Nerve Fibers to Conduction Block Produced by Various Agents

	Most Susceptible	Intermediate	Least Susceptible
Sensitivity to hypoxia	B	A	C
Sensitivity to pressure	A	B	C
Sensitivity to cocaine and local anesthetics	C	B	A

169

action potentials as those of A fibers after the pumps have been poisoned by drugs or slowed by being out of the body (during an in vitro experiment).

CLINICAL ESTIMATION OF CONDUCTION VELOCITY

Earlier we pointed out some simple, rule-of-thumb concepts to assist in estimation of basic membrane function in clinical circumstances. This is much more difficult to do for conduction velocity, so perhaps it is fortunate that conduction velocities can be measured directly in peripheral nerves.

Breakdown of the myelin sheath occurs under some circumstances in peripheral nerves. Such peripheral demyelination leads to a reduction in conduction velocity.

HINTS

14. Some say that it is because neurophysiologists don't know which end is up. Probably that is not true. Others suggest that a neurophysiologist will always arrange the polarity so that the baseline is at the bottom for purely Freudian reasons. Further evaluation of *that* suggestion is beyond the scope of this book.
15. No. The height of a compound action potential is a function of the number of axons firing at any one time under the electrode. Each axon gives an all-or-nothing response, but when many hundreds are summed, the summed potential can be different sizes, depending on the number of axons contributing to the total potential (e.g., Fig. 7-15).
16. Fiber diameters may vary (see next section in text), but even if all the axons have the same diameter, some are farther away from the stimulating electrodes than others in the nerve trunk. (Currents in a volume conductor are smaller at a distance than up close.) Therefore, at a given stimulus voltage, the axons on the nerve trunk surface are stimulated more strongly than those deeper in the trunk.
17. 1 ms. This sort of computation is easiest if you work in millimeters and milliseconds, since 1 m/s = 1 mm/1 ms.
18. The small peak at 10 μm gives the large peak at 0.5 ms! This one fools a lot of students (especially those who are tired, nearing the end of the chapter) who forget that the larger diameter is faster, hence will have the *shortest* latency. From this figure you can now guess that large fibers also generate larger external potentials than smaller fibers.
19. Complete demyelination would reduce the nerve to an unmyelinated state and remove the increase in conduction velocity brought about by myelination. So R_m falls, C_m rises, and the axial current flow falls. Obviously, the velocity must decrease. Such complete demyelination would not be common, although changes in conduction velocity as a result of partial demyelination are readily detected in affected persons.
20. (*a*) Outward current; (*b*) membrane depolarization; (*c*) a doubling of g_{Na}.
21. (*b*), (*d*).
22. This is included as an example of the slightly unfair, semiquantitative question that can be relied on to induce panic in many students! Keep your wits about you, and don't be thrown by the apparent precision required. The logic is simple, providing you remember the effects of pH on free ionized calcium concentration (see Chap. 6): Increasing pH decreases $[Ca]_o$ and increases excitability, which should *slightly* increase conduction velocity. Since the normal range of ulnar nerve conduction velocities is about 50 to 60 m/s, you could hardly expect changes greater than 15 to 20 percent.
23. (a) False; (b) false; (c) false; (d) false.
24. (*a*). All the other answers are clearly incorrect, and now that you *really understand* what is meant by (*a*), it's perfectly O.K. to go ahead and accept it as the best in a bad bunch!

169

170

1 QUESTION: Why would you expect this? (Hint 19↑)

Conduction velocity also may be affected by any means that might alter dV/dt in the rising phase of the action potential. For example, both hypercalcemia and hyponatremia might be expected to reduce dV/dt and thus decrease conduction velocity. Such effects on conduction velocity are likely to be insignificant, compared with the importance of these same parameters in control of muscle, heart, and neuron excitability. While we see that **any factor that affects membrane excitability also affects conduction velocity,** changes in conduction velocity may not be the most noticeable or prominent signs of such changes in excitability. 2

The heart may provide an interesting exception to this general rule, since changes in conduction velocity through the AV node have been suggested as a mechanism for generation of certain abnormal ventricular rhythms. 3

In summary, here are some general rules to remember:

1. Myelinated fibers conduct more rapidly than unmyelinated fibers.
2. Larger-diameter fibers (of either type) conduct more rapidly than smaller-diameter fibers.
3. Demyelination decreases conduction velocity.
4. Any factor that affects membrane excitability also affects conduction velocity. 4

EXAM QUESTIONS

1. In each of the following pairs, which occurs first when a single node of Ranvier is being approached by a naturally occurring action potential?
   (*a*) Inward current; outward current
   (*b*) Membrane depolarization; membrane hyperpolarization
   (*c*) A doubling of g_K; a doubling of g_{Na}
   (Hint 20↑) 5
2. When two unmyelinated axons of similar diameter are compared, which of the following would you expect to have the higher conduction velocity?
   (*a*) The axon with higher membrane capacitance
   (*b*) The axon with higher *resting* membrane resistance
   (*c*) The axon with higher action potential
   (*d*) The axon with higher maximum dV/dt on the rising phase of the action potential
   (Hint 21↑) 6

170

171

3. A medical student is asked to hyperventilate while conduction velocity is measured in the large myelinated fibers of her ulnar nerve. Which of the following possible responses would you expect after several minutes of maximum voluntary ventilation?
   (*a*) Marked increase in conduction velocity (increase >10 m/s)
   (*b*) Slight increase in conduction velocity (increase <10 m/s)
   (*c*) Slight decrease in conduction velocity (decrease <10 m/s)
   (*d*) Marked decrease in conduction velocity (decrease >10 m/s)
   (See Hint 22.↑)

1

4. *Mark each statement true or false.* Two stimulating electrodes, *A* and *B*, and two recording electrodes, *C* and *D*, are placed (in the order indicated) along a frog's sciatic nerve. The latency of the biphasic response observed at electrodes *C* and *D* will be
   (*a*) Increased if the anode is at *A*.
   (*b*) Decreased if the anode is at *B*.
   (*c*) Increased if the cathode is at *B*.
   (*d*) Decreased if the cathode is at *A*.
   (See Hint 23.↑)

2

5. *Pick the single most appropriate answer.* Saltatory conduction is faster than conduction in an unmyelinated nerve of similar diameter because
   (*a*) The action potential jumps from node to node.
   (*b*) The length constant is shorter in myelinated axons.
   (*c*) The time constant is longer in myelinated axons.
   (*d*) The membrane resistance is smaller in the internode regions of a myelinated axon than it is in an unmyelinated axon.
   (See Hint 24.↑)

3

171

173

Advanced Topics: Transient Potentials

In this chapter, we seek a quantitative basis for the more intuitive approach to the action potential provided in Chaps. 6 and 7. **We recognize that this quantitative approach is substantially beyond the needs of many students,** and thus most of this chapter is equivalent in complexity to material which appears at the third and fourth levels in other chapters. 1

The chapter is divided into four major sections. Section I deals with the propagated action potential and factors affecting conduction velocity; section II deals with advanced topics arising from Chap. 6, the membrane action potential; section III covers a more quantitative treatment of the cable equations, material presented only briefly in Chap. 7; section IV is a brief, do-it-yourself guide to some of the major experimental techniques of electrophysiology. 2

These **sections** are **arranged in decreasing relevance to the general reader, but increasing relevance to the more experimentally inclined.** If you are interested in the experimental details, we hope you will read this chapter with pleasure. But we will certainly understand if the more general reader turns to Chap. 9 (with relief?) after the first few pages! 3

I: CONDUCTION VELOCITY AND PROPAGATED ACTION POTENTIAL

The propagated action potential and membrane action potential show only slight differences in shape as a result of local current spread, and it remains just as true in the propagated action potential that I_{cap} can be obtained directly by measurement of dV/dt. There is, however, an interesting twist. 4

In a uniform axon, in which conduction velocity is constant, the abscissa always can be directly, and simply, converted from time to distance. Thus, the shape of the action potential is the same whether it is plotted against time or against distance along the membrane. In the first case, one "stands" at a single place in the membrane to watch the change of voltage with time; in the second case, it is as if one could photograph the whole action potential as it is spread out over the membrane surface at any given instant (see Fig. 8-1). 5

QUESTION: What is the conduction velocity of the action potential shown in Fig. 8-1? (Hint 1↓) 6

The reasoning used is as follows: If the conduction velocity is constant and if the shape of the action potential is the same at all points, then all parts of the wave must be propagated at the same velocity θ. Where $\theta = x/t$, then,

$$\frac{dV}{dt} = \theta \frac{dV}{dx} \qquad \text{Eq. 8-1}$$ 7

Fig. 8-1. Form of propagated action potential plotted against (*A*) time (in milliseconds), (*B*) distance along axon (in millimeters).

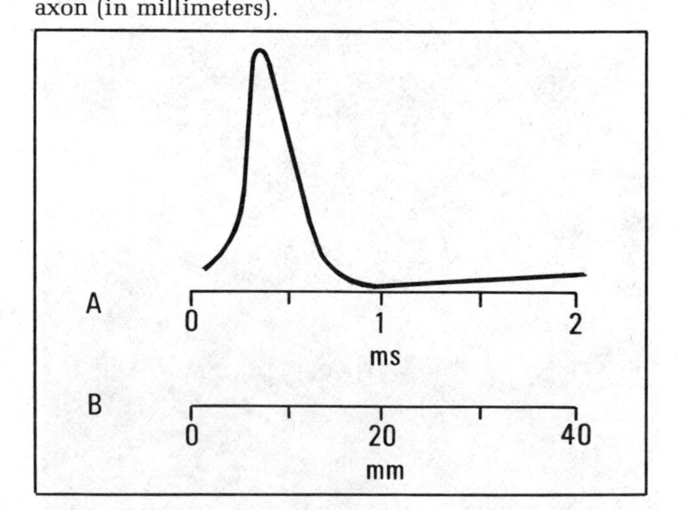

or,

$$\frac{dV}{dx} = \frac{1}{\theta}\frac{dV}{dt}$$ 1

An important point here is that **the action potential is spread out over 4 to 6 cm of membrane surface in the squid giant axon!** Even the depolarized region is about 2 cm long. (Of course, this may not come as a surprise to *you*, but we have noticed that many students would have guessed 2 mm or less, rather than 2 cm, before reading this section.) 2

IA: Factors Affecting Conduction Velocity

Over an evolutionary time scale, there is a clear payoff for certain neurally regulated responses (particularly "escape reactions") being as fast as possible. With each advance in the reaction rate of the prey, the predator must increase his or her speed, and vice versa. Since the ultimate determinant of speed may well be the conduction velocity of motor or sensory axons, or both, it is not surprising that so many invertebrates possess "giant fibers" to mediate escape reactions. Nor is it surprising that conduction velocities vary widely, within any given species, between these fastest-conducting axons and the slower-conducting axons involved in the more vegetative functions of the animal in question. What factors determine conduction velocity? 3

Look at Fig. 8-1. Can you see that the *smaller* the dV/dx and the *larger* the dV/dt, the *greater* the conduction velocity? Figure 8-2 compares two hypothetical action potentials from two different axons. The action potentials are both the same height. Since dV/dt is the same in both, which has the greater conduction velocity? Notice that in both records, the peak of the action potential occurs 1.0 ms from the first discernible depolarization. 4

QUESTION: What are the conduction velocities in these two axons? (Hint 2↓) 5

Figure 8-3 compares two hypothetical action potentials from two different axons. The action potentials are both the same height, and dV/dx is also the same for both. But what about dV/dt? In the upper record, dV/dt must be higher, since the peak of the action potential is reached within 0.5 ms as compared with about 1.0 ms for the lower axon. 6

QUESTION: What are the conduction velocities for these axons? (Hint 3↓) 7

If you think carefully about what you have just seen, you will have grasped an obvious, but nevertheless significant, generalization: **Conduction velocity increases as *dV/dx* decreases** or **as *dV/dt* increases.** But how are these parameters related to the more readily recognizable properties of the membrane? 8

Fig. 8-2. Two action potentials with same height and same dV/dt, but with different conduction velocities. (*A*) Time scale; (*B*) distance scale.

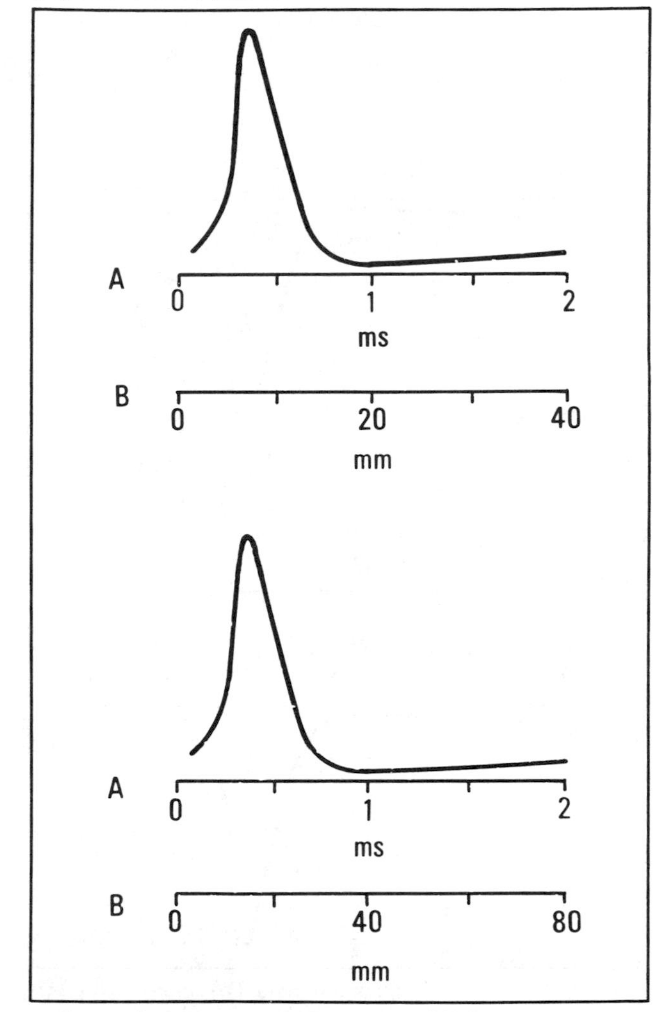

176

dV/dx. This parameter is determined primarily by the spread of axial current. Although axial-current spread is dependent on many factors (and is affected by the reactivity of the membrane conductances to ΔV), nevertheless, one of the most important determinants is the ratio of membrane resistance to internal axial resistance. The larger the ratio, the greater the distance of spread, the smaller *dV/dx*, and the greater the conduction velocity. Think of it this way: You want to fill a bucket from a leaky garden hose. The smaller the leak (higher r_m) and the larger the diameter of the hose (lower r_i), the faster the bucket will fill (i.e., the greater the axial current). 1

dV/dt. This parameter depends on two factors: a low membrane time constant and a high reactivity of the voltage-dependent sodium conductance. Thus, the lower the membrane capacitance and the greater the rate of increase of I_{Na}, the greater *dV/dt*. 2

Unfortunately, there are some rather difficult teleological tradeoffs to be made here. For example, if *dV/dx* is low by virtue of high r_m, then *dV/dt* may be slow since g_{Na} cannot increase rapidly without changing r_m. Or, if *dV/dt* is high because of a highly reactive g_{Na} system, then *dV/dx* will be increased as a result of the low r_m. The escape from this "Catch 22" is to decrease *dV/dx* by **increasing the diameter of the axon.** 3

It has been noted experimentally that conduction velocity is directly related to axon diameter in a wide range of both unmyelinated axons and myelinated axons (although the slope of the relationship is quite different in these two types of axons) (see Figs. 7-7 and 7-8). 4

Since axial resistance falls in relation to the square of the radius, whereas membrane resistance is related linearly to radius, a substantial gain occurs with increased diameter. Nevertheless, the squid giant axon, with its 1-mm-diameter axon, is clearly pushing this process to its limits to achieve a 20-m/s conduction velocity. Fortunately (for us), the vertebrates found a better way, as we noted in Chap. 7 and describe next. 5

IB: Conduction in Myelinated Axons

The problem, remember, was how to get r_m low enough to permit high *dV/dt* without increasing *dV/dx* and reducing the distance over which significant coupling can occur. The answer lies in specialization: design a high-resistance, low-capacitance membrane that will have a reasonable time constant and charge at low-current densities; then insert intermittent "nodes" specialized to generate high inward currents with high *dV/dt*. It's cunning, it should work, and it does! 6

At the risk of repeating what we said in Chap. 7, let's run through that a little more slowly. 7

In the internode regions, the many turns of the myelin sheath tremendously increase the resistance to current passed outward across them. Simultaneously, each turn of the myelin sheath further separates the plates of the membrane capacitor, thus 8

Fig. 8-3. Two action potentials with same height and same *dV/dx*, but with different conduction velocities. (*A*) Time scale; (*B*) distance scale.

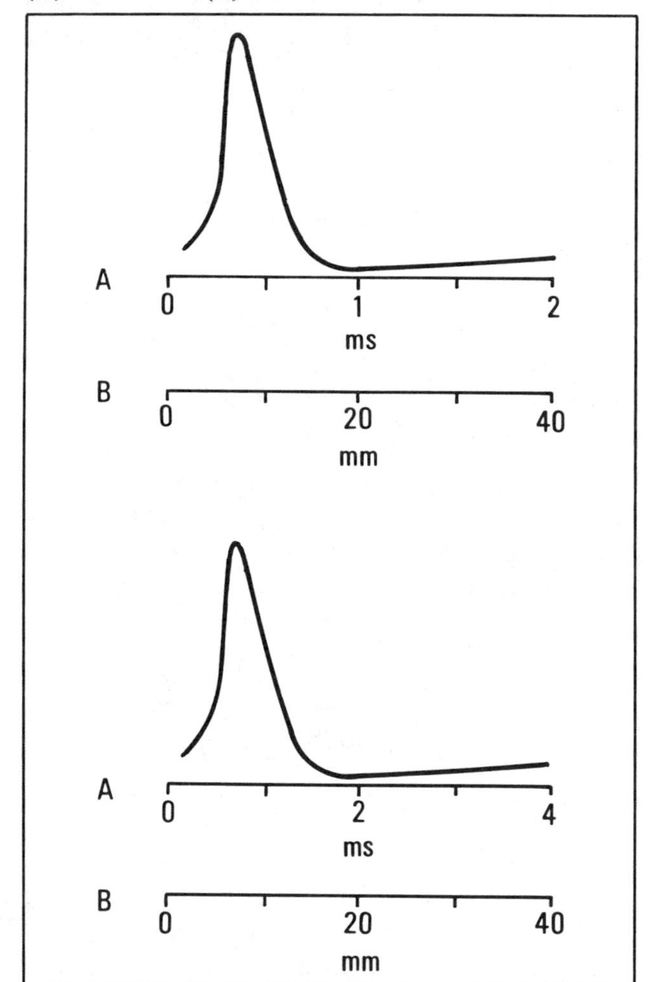

176

177

reducing the effective capacitance. The reciprocal nature of these changes leaves the membrane time constant almost unchanged, although the current density required to produce a given change of voltage will, of course, be drastically reduced. The high membrane resistance and low capacitance both serve to increase the distance over which effective coupling can occur; in fact, entirely adequate coupling for extremely rapid conduction can be achieved with axons of 20-μm diameter. Thus, a high axial resistance is tolerable when it appears in association with very high r_m.

At the nodes, the virtual absence of the myelin sheath removes the added resistance found throughout the internodes. Additionally, the nodal membranes contain the highest concentrations of electrically excitable sodium pores yet found (4000 as compared with 400 per 1 μm^2 in the unmyelinated squid giant axon or 20 to 100 per 1 μm^2 in more typical unmyelinated axons). Clearly, nodal membrane is specialized to generate high inward current and high dV/dt.

But even in these sophisticated axons, there are some difficult tradeoffs to be made. One might think that internode length should be as great as possible, so that dV/dx would be as low as possible. However, then pressure or damage at a single node would block conduction of the entire axon. In fact, most myelinated nerves seem to be set up so that normal conduction can take place across at least two, if not three, blocked nodes. (Typical internode lengths are 0.25 to 1.0 mm.)

Conduction in myelinated nerves is often termed *saltatory*, in that the ionic currents are truly generated at discrete, separate "point sources" scattered along the membrane surface. Many students seem to think that *saltatory conduction* means that the action potential occurs at one node, then jumps to the next, and so on down the axon. This is absolutely *not* the case, as we repeatedly stated in Chap. 7. No, the power of myelination is in the specialization of function between node and internode. And it *is* true that *current* passes from node to node. As a result, vertebrates have achieved conduction velocities close to 5 times greater than those of squid giant axons, with myelinated axons less than one-fortieth of their diameter. Even allowing for extracellular space, you can fit about 1000 such axons into the space occupied by a single squid giant axon.

In myelinated axons, just as in unmyelinated axons, conduction velocity increases as a function of axon diameter. After all, it is just as important to be able to reduce the resistance to axial current flow. However, the specialization of function between current-generating regions (the nodes) and current-transmitting regions (the inter-

HINTS

1. The conduction velocity can be obtained by comparing the two scales. Here it is 20 mm/ms, or 20 m/s, which is about average for a squid axon at 18°C.
2. 20 m/s for the upper axon; 40 m/s for the lower axon.
3. 20 m/s for the upper axon; 10 m/s for the lower axon.

177

nodes) tremendously alters the constant of proportionality involved in this relationship.

Experimental observations show that the effect of diameter on the conduction velocity of myelinated axons is even greater than for unmyelinated axons, so that

$$\theta \propto d^{1.5}$$

Question: Can you think of one possible reason for the additional steepness of this effect of diameter on conduction velocity? (Hint 4↓)

IC: Conduction in Branching Axon Systems

It has long been noted experimentally that action potentials initiated in a small side branch may not propagate back into the main axon, whereas similar action potentials initiated in the main axon can propagate effectively into the side branch. Such an experimental situation is shown in Fig. 8-4. Stimulation of the main axon at either electrode 1 or 2 causes an action potential to pass electrodes 3 and 4. By contrast, when stimulation occurs at electrode 3, an action potential passes electrode 4 at the base of the side branch, but fails to propagate into the main axon. Even more curious, however, is the fact that if stimulation is applied at electrode 4, an action potential is observed at all three other electrodes! Thus, it cannot be that there is some one-way valve, or rectifier, that prevents current flow back into the main axon.

Although this observation was baffling initially, it turns out to have a simple geometric solution, as shown in Fig. 8-5. When an action potential initiated at A (corresponding to electrode 3 in Fig. 8-4) reaches point B, that small area of membrane surrounding point B would be required to depolarize the much larger-capacitance and lower-resistance membrane of the main axon. But dV/dt will be slow at C, and there simply may not be sufficient current generated by the small branch membrane (B) to depolarize the main axon to threshold.

Obviously, this problem does not arise at every point at which an axon branches. It arises only where there is a substantial disparity in size between the main axon and side branch and where there is an *abrupt* transition in axon diameter.

You should note that no problem exists when an action potential passes from the main axon into the side branch, since the larger membrane at C (Fig. 8-5) would clearly be capable of depolarizing the smaller-diameter membrane (with lower capacitance and higher resistance) at B.

If, however, a *stimulus* is applied at point B (electrode 4 of Fig. 8-4), then the additional resources of the stimulator are sufficient to depolarize the main axon to

Fig. 8-4. Four electrode placements in branching axon. Electrodes 1 and 2 are in the main axon; electrodes 3 and 4 in the side branch.

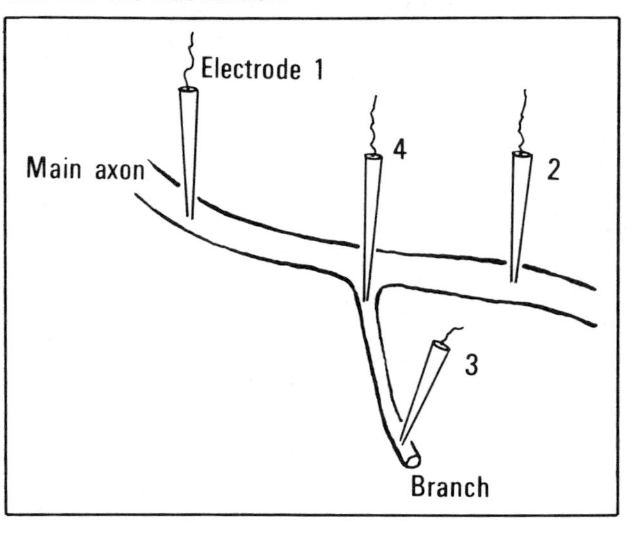

Fig. 8-5. A modification of Fig. 8-4 to suggest situation that occurs when action potential in side branch attempts to propagate into main axon.

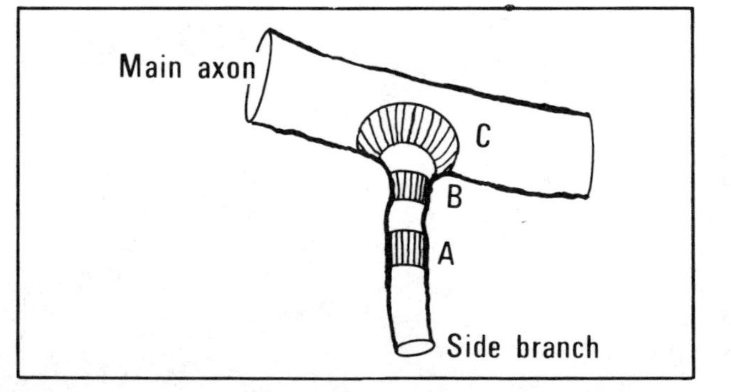

1 threshold, and action potentials can propagate to the other electrode sites.

The problem just described applies equally to myelinated and unmyelinated axons. It is no surprise that myelinated axons always branch at nodes (or, if you prefer to think of it in this way, myelinated axons always have nodes at their branching 2 points). Think about that for a moment, and then check Hint 5.↓

ID: Conduction in Skeletal Muscle

Conduction in skeletal muscle fibers closely resembles conduction in a large, unmyelinated nerve axon. Vertebrate skeletal muscle fibers typically range from, say, 20 to 150 μm in diameter and thus are large by comparison with all but the giant nerve fibers of invertebrates. The conduction velocities of skeletal muscle fibers generally seem appropriate to unmyelinated fibers in this size range. However, they have one 3 interesting property that deserves some discussion.

We noted that most unmyelinated nerve membranes have a capacitance very close to 1 $\mu F/cm^2$. In muscle, the observed capacitance rises to 8 or 10 $\mu F/cm^2$! It appears that this is due not to some curious difference in the intrinsic capacitance of muscle fiber membrane, but to its complexly infolded form (discussed in detail in Chap. 10). Simply, the surface area of the membrane to be depolarized is some 8 to 10 times greater than would be the case in a cylindrical axon of comparable diameter. In the absence of any other compensatory differences, the higher capacitance of muscle membrane would slow dV/dt, since more current would be required to achieve a given ΔV. Thus it is not surprising that measurements of the sodium pore density in 4 frog sartorius muscle indicate about 400 pores/μm^2.

5 Question: Why would a high Na^+ pore density help in this situation? (Hint 6↓)

As a result, θ is about 2 m/s in frog sartorius muscle and varies from less than 1 m/s to 6 at least 10 m/s in mammalian skeletal muscle.

II: MEMBRANE ACTION POTENTIAL

In Chap. 6, we glossed over two very significant points: discussion of the voltage-clamp technique and quantitative description of voltage-dependent conductances as functions of both voltage and time. We now discuss these points and further show how such an analy- 7 sis can be utilized to synthesize the space-clamped membrane action potential.

IIA: Voltage-Clamp Technique

In the normal course of events, sodium and potassium conductances change as functions of both voltage and time. Furthermore, these changes of conductance themselves initiate further changes in membrane potential. Detailed experimental study of voltage-dependent conductances scarcely would have been possible without development of an experimental technique that allowed more complete separation of these variables. The method used by 8 Hodgkin and Huxley, despite its technical elaboration, is conceptually simple: If the membrane potential could be "clamped" constant at a chosen level, then any resultant changes in conductance could be viewed solely as functions of **time.** The effects of **voltage** on conductance could be obtained later by comparing the responses determined at different, chosen, clamped potentials. But how can you hold the membrane potential constant while ionic currents are changing by as much as a thousandfold? How is it possible to "clamp" a membrane potential?

1

Avoiding, for the moment, the technical aspects of this problem, can you think of the *theoretical* requirement for maintenance of constant membrane potential? After the potential has been changed from its resting level to the desired new value, how can it be changed at that point? Clearly, we want dV/dt to be zero. But what does that necessitate? Think for a moment, and then check Hint 8. ↓

2

Note, in Fig. 8-6, that if I_{por} is inward and I_{ext} is in the direction shown, and if the magnitude of I_{ext} can be so rapidly adjusted that V remains constant, then there can be no capacitative current across the axon membrane. To maintain the voltage clamp, all that is required is a feedback amplifier capable of passing current in either direction in the external circuit, such that V is held constant. Under these circumstances, I_{cap} will always be zero, while I_{por} will be equal to I_{ext}. Clearly, if I_{ext} is monitored, then I_{por} will always be known. Thus, the voltage-clamp circuit provides a moment-by-moment measure of the ionic currents flowing across the axon membrane.

3

Actually, I_{ext} gives the current (in amperes) in the external circuit; this can be corrected to I_{por} (in amperes per square centimeter) when the effective membrane area is known.

4

You probably have realized that any effective voltage-clamp system is likely to be substantially more complex than the arrangement illustrated in Fig. 8-6. Even the more complex system in Fig. 8-7 omits the "guard" electrodes often used to reduce edge effects near the ends of the space-clamped region.

5

So how do we change membrane potential during the initial stepwise ΔV? A command voltage is applied across terminals of the clamp amplifier. Then I_{ext} must change rapidly to charge the membrane capacitor to reach the new command voltage. A brief I_{cap} will flow (sometimes called the **capacitative transient**) while the membrane is charging; afterward, I_{cap} will be zero and current flow in the external circuit will be determined solely by the ionic currents across the axon membrane. A diagrammatic version of a typical voltage-clamp record is shown in Fig. 8-8. Before we go on, let's see whether you can answer some simple questions based on this figure.

6

QUESTION: Why is the capacitative transient shown as an outward current? (Hint 11 ↓)

7

Question: What is the *initial current*, and why is it outward in direction? (Hint 10 ↓)

8

Fig. 8-6. Simplified diagram to explain basic principle of voltage-clamp method.

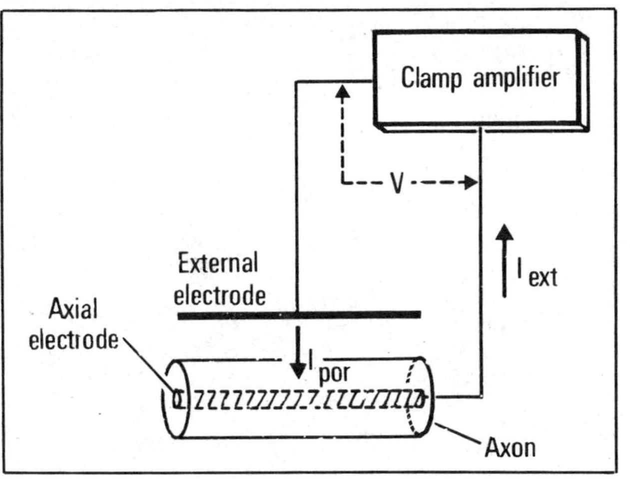

Fig. 8-7. A more complex "simplified diagram" of a typical voltage-clamp apparatus.

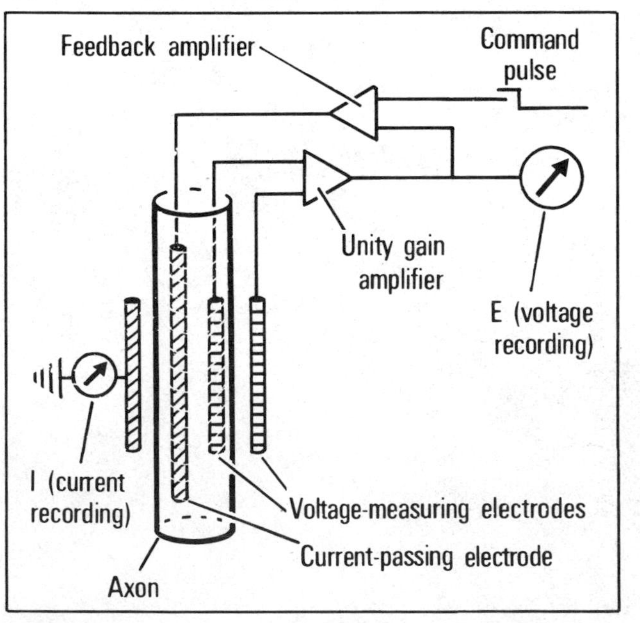

(Modified from T. H. Bullock, R. Orkand, and A. Grinnell, *Introduction to Nervous Systems.* San Francisco: Freeman, 1977.)

181

Question: What are these "early inward" and "delayed outward" currents? Aren't these just I_{Na} and I_K, respectively? (Hint 13↓)

We recommend that you try to identify all the components just described in some more typical record from the research literature. Note, however, that we exaggerated the size of the initial current in Fig. 8-8 (for didactic purposes) and that the capacitative transient is blanked out in most published records.

We make no particular apology for not offering here a more definitive treatment of the technical aspects of the voltage clamp. Interested students should consult Cole [5, 6] and Sjodin [56].

Fig. 8-8. Diagrammatic representation of a typical response to voltage-clamp step depolarization in electrically excitable membrane.

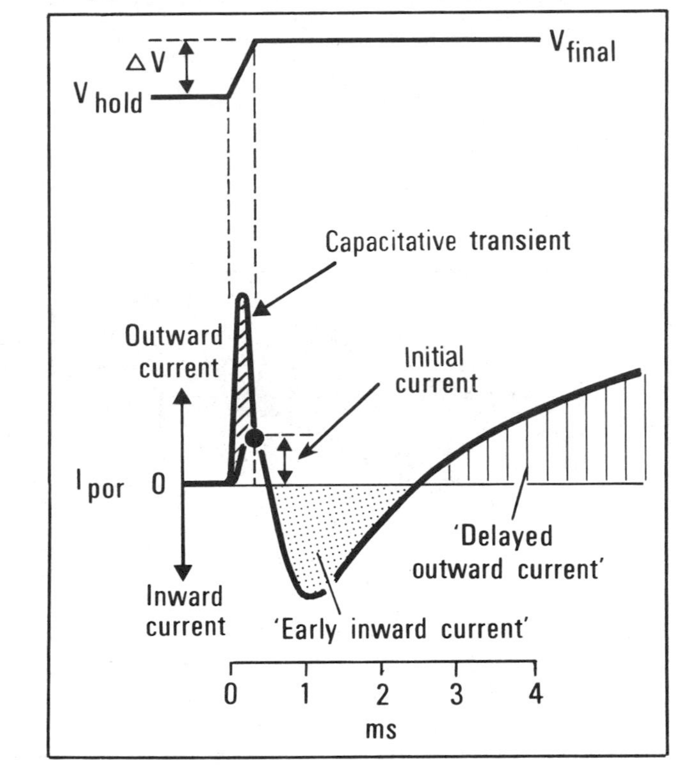

IIB: Separation of Ionic Currents

Although we have seen how to measure total I_{por} by using the voltage-clamp technique, clearly, we do not yet have our data in their most useful form. The next step must be the further analysis of I_{por} into its ion-specific component currents.

In their analysis of the squid axon membrane, Hodgkin and Huxley were able to consider I_{por} as made up of two time- and voltage-dependent components, I_{Na} and I_K, together with a small, time independent *leakage current* I_{leak} representing the sum of all other membrane conductances.

QUESTION: What would you guess to be the principal ionic current involved in I_{leak} in the squid axon? (Hint 14↓)

The method used for analysis of I_{por} into its components was as follows:

1. Separation of I_{Na}. When does an ionic current fall to zero no matter how large the conductance may be? When the driving force is zero, right? Thus if the early inward current were really carried by Na^+ ions, it would be reduced to zero when the membrane potential was clamped to E_{Na}. And if the membrane were clamped to a potential more positive than E_{Na}, then the early inward current would reverse its sign to flow *outward* across the membrane. That the early inward current does reverse when V is more positive than the estimated E_{Na} was the first indication that this current was probably due to Na^+ ions.

HINTS

4. In myelinated axons, an increase in diameter will increase axial current flow and hence increase the initial dV/dt at the next node. However, such an axon could now tolerate a longer internode length, thus decreasing dV/dx. It seems likely, therefore, that the interaction between axon diameter and internode length introduces the power function.

5. If you think of the nodes as booster stations inserted into a transmission line to maintain signal strength, doesn't it make sense to place a booster station at every branch point? We think so.

6. Remember, $\tau_m = R_m C_m$. When C_m is increased, the time constant can remain unchanged if R_m falls. Increasing the Na^+ pore density will increase g_{Na} and reduce R_m. Or we can look at it in a different way; see Hint 7.↓

181

182

Now let's reduce $[Na^+]_o$ to 30 percent of its normal value, substituting with impermeable choline ions to maintain osmolarity; E_{Na} will become more negative, and so will the "reversal potential" for the early inward current.

1

Where early inward current is purely Na^+-dependent, this reversal potential should move 59 mV (at 20°C) for a 10-fold change in $[Na^+]_o$. A smaller change than this would occur if other ions made any substantial contribution to this current. (The 59 mV follows directly from the constant in the Nernst equation.)

2

Similarly, the change in magnitude of the early inward current at other potentials should appropriately reflect the imposed ΔE_{Na}. In Hodgkin and Huxley's experiments, the changes observed in 30% and 10% sodium solutions were exactly those that would have been predicted from the assumption that the early inward current was carried almost entirely by sodium ions. Since that hypothesis has been confirmed, now it is possible to calculate the magnitude of I_{Na}, as shown in Fig. 8-9. For a voltage step at which $V_{final} = E_{Na}$, I_{Na} must be zero, since $V - E_{Na} = 0$. It follows that I_{por} under these circumstances is $I_K + I_{leak}$. The difference between that result and I_{por} found when the experiment is repeated in a normal $[Na^+]_o$ medium, therefore, must be I_{Na}. Typical experimental curves are shown in the upper part of Fig. 8-9; the lower part gives the resulting analysis of the currents.

3

This method would become laborious if a new external solution had to be used for each voltage step. In practice, Hodgkin and Huxley used only the normal and 30% and 10% normal Na^+ solutions, extrapolating from these values for intermediate values of V_{final}. Separation of ionic currents can now be achieved by pharmacological methods. External tetrodotoxin (at appropriate concentration) completely blocks I_{Na} without affecting I_K. Similarly, internal perfusion of axons, using solutions in which cesium has been substituted for potassium, effectively blocks all current through potassium pores with only minimal effects on I_{Na}.

4

2. Estimation of I_{leak}. The next problem was clearly to separate I_{leak} from $I_K + I_{leak}$. The method used was, again, simple and direct. When choline is completely substituted for sodium in the bathing medium, I_{Na} can be ignored at potentials near resting potential. Under these circumstances, the main voltage-dependent current should be I_K. Hodgkin and Huxley found that the delayed outward current had a reversal potential 12 to 14 mV more negative than the resting potential, which coincided well with other estimates of E_K in these axons. Since I_K should be zero at this reversal potential, the current required to hold the membrane at that potential must be I_{leak}.

5

Repeating this observation after change of E_K (by increase of $[K^+]_o$) gives an equivalent estimate of I_{leak} at a different membrane potential. Thus sufficient data are obtained to solve for the two unknowns, g_{leak} and E_{leak}, from simultaneous equations of the form

6

Fig. 8-9. Separation of I_{Na} from $I_K + I_{leak}$. See subsequent paragraphs for estimation of I_{leak} and hence separation of I_K.

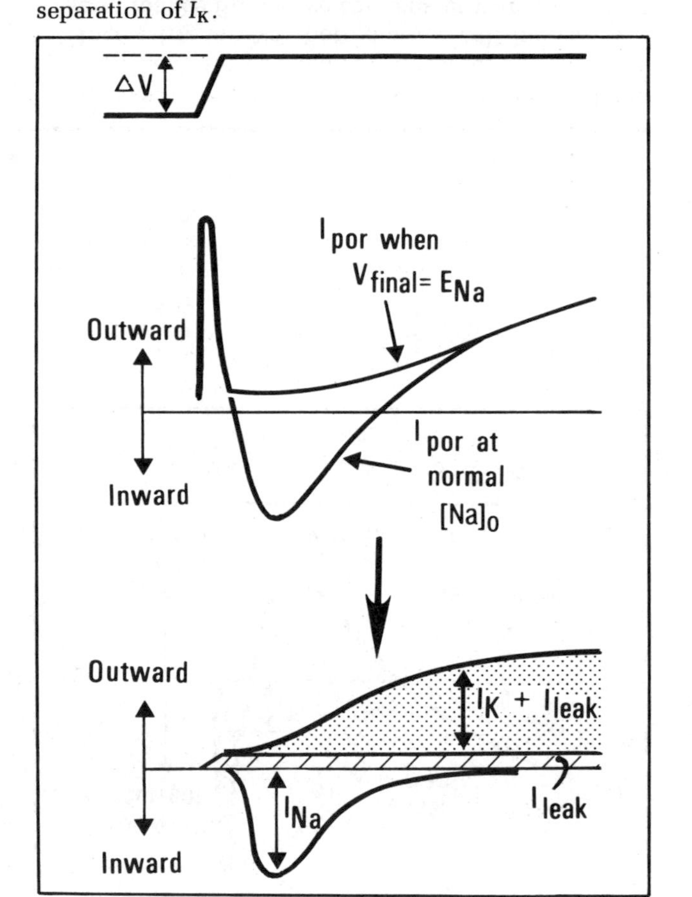

182

183

1 $$I_{leak} = g_{leak}(V - E_{leak}) \quad \text{Eq. 8-2}$$

Since g_{leak} is a non-voltage-dependent conductance that does not change with time, it is a simple matter to calculate I_{leak} for any V_{final} and subtract this from $I_K + I_{leak}$ to obtain the true value of I_K alone. 2

Question: How could this method of calculation of I_{leak} be compatible with our earlier suggestion that perhaps K^+ ions contribute to the leakage current? (See Hint 15.↓) 3

3. Calculation of g_{Na} and g_K. Now that I_{Na} and I_K can be obtained separately as functions of time, and by remembering that the driving force for an ionic current must remain constant throughout a given voltage-clamp step, conductances can be obtained directly from

$$g_{Na} = \frac{I_{Na}}{V - E_{Na}}$$

and

$$g_K = \frac{I_K}{V - E_K}$$

It was by this method that the plots of g_{Na} and g_K as functions of time, shown as Figs. 6-15 and 6-16, were obtained. 4

HINTS

7. For given dV/dt, greater dQ/dt is required when C_m is increased. The higher the pore density, the greater the density of inward ionic current, hence the higher the available density of the outward capacitative current.
8. Doesn't it mean that I_{cap} must be zero at all times? But if I_{por} is changing rapidly, how *can* I_{cap} be zero? Recall that $I_{mem} = I_{por} + I_{cap} = I_{ext}$. (Eq. 6-11), and then turn to Hint 9.↓
10. Before the voltage-dependent conductances have had time to change, an "initial current" appears. This is the current that would flow across the *resting* membrane conductances, but at the new voltage V_{final}. Why is it outward? See Hint 12↓
11. Because it takes an outward I_{cap} to depolarize the membrane.
13. Certainly, we would like to believe that the early inward current is carried by Na^+ ions, while the delayed outward current is K^+-mediated. But until we manage to complete the proof, a certain integrity is maintained by not prejudging the issue. And in some tissues (e.g., barnacle muscles), the early inward currents are carried by calcium rather than sodium ions.
14. Almost certainly I_{Cl}, wouldn't you think? However, g_{Cl} is very small in the squid axon, and thus it has been suggested that a substantial part of the leakage current is provided by K^+ ions passing through non-voltage-sensitive pores. Do we really mean to suggest that a separate, small population of pores exists that may not be voltage-sensitive? Why not?

183

184

IIC: Mathematical Description of Voltage-Dependent Conductances

To complete the task that Hodgkin and Huxley had set for themselves within their 1952 series of papers [19–23], they had to demonstrate that both the membrane and propagated action potentials could be synthesized from the voltage and time dependence of g_{Na} and g_K, **as measured by the voltage-clamp technique.** Such a tour de force would prove not only that the method allowed *qualitative* demonstration that action potentials result from voltage-dependent changes in sodium and potassium conductance, but also that the experimental methods were sufficiently accurate to provide detailed *quantitative* information as to the magnitude and time course of the changes that take place. Therefore, they sought the simplest mathematical expression that would yield accurate representation of the voltage and time dependence of these conductances.

1

1. Mathematical description of g_K. Remember that the rise in g_K after depolarization follows an S-shaped curve, whereas the fall of g_K after rapid repolarization seems to be a simple exponential process. Bearing this in mind, Hodgkin and Huxley suggested that $g_{K(t)}$ could be modeled effectively as the fourth power of a first-order exponential equation:

$$g_{K(t)} = \bar{g}_K(n_t)^4 \quad \text{Eq. 8-3}$$

where $\bar{g}_K$ is a scaling factor representing the maximum possible potassium conductance and n_t is a simple exponential function of time, such as would be the case in a simple reaction of the form

$$n' \underset{\beta_n}{\overset{\alpha_n}{\rightleftharpoons}} n \quad \text{Reaction 8-1}$$

where $n + n' = 1.0$ and α_n and β_n are the rate constants of the two unidirectional reactions. Now, if α_n and β_n are considered to be instantaneous functions of voltage (but independent of time), then voltage dependence will be included in any equation defining n as a function of time, in spite of its apparently simple, first-order form. When n_0, α_{n_0}, β_{n_0} are the values for $t = 0$, when α_n and β_n are the values of the voltage-sensitive rate constants immediately after the voltage step, and when n_∞ is the value of n at $t = \infty$, the required equations are

$$dn/dt = n'\alpha_n - n\beta_n = (1 - n)\alpha_n - n\beta_n \quad \text{Eq. 8-4}$$

and

2

$$n_t = n_\alpha - (n_\alpha - n_0)e^{-t/\tau_n} \quad \text{Eq. 8-5}$$

184

185

where

$$n_0 = \frac{\alpha_{n_0}}{\alpha_{n_0} + \beta_{n_0}} \qquad \text{Eq. 8-6}$$

$$n_\infty = \frac{\alpha_n}{\alpha_n + \beta_n} \qquad \text{Eq. 8-7}$$

$$\tau_n = \frac{1}{\alpha_n + \beta_n} \qquad \text{Eq. 8-8}$$

1

However forbidding these equations may seem at first, they are sufficiently simple to become old friends with amazing rapidity as soon as you start to use them in the description of g_K. For illustrative purposes, they may even be given a physical basis if you assume that the opening of each potassium pore is governed by four such independent "n particles," where all particles must be in the n position, rather than the n' position, for the pore to open. The probability of such an occurrence is clearly n^4. Thus the pore opening varies as $d(n^4)/dt$, whereas the pore closing will vary as $4\,dn/dt$. 2

Values of α_n and β_n for best fit at each V_{final} were obtained from a set of voltage-clamp data points such as those shown in Fig. 8-10. Then empirical equations could be generated to describe the form of α_n and β_n as functions of voltage. 3

At this point, the advanced student should attempt to gain some familiarity with the equations noted (Eqs. 8-4 through 8-8). We suggest that you attempt to recal- 4

Fig. 8-10. Rise of g_K associated with different depolarizations. The circles are experimental points, and the lines are from analytic solutions.

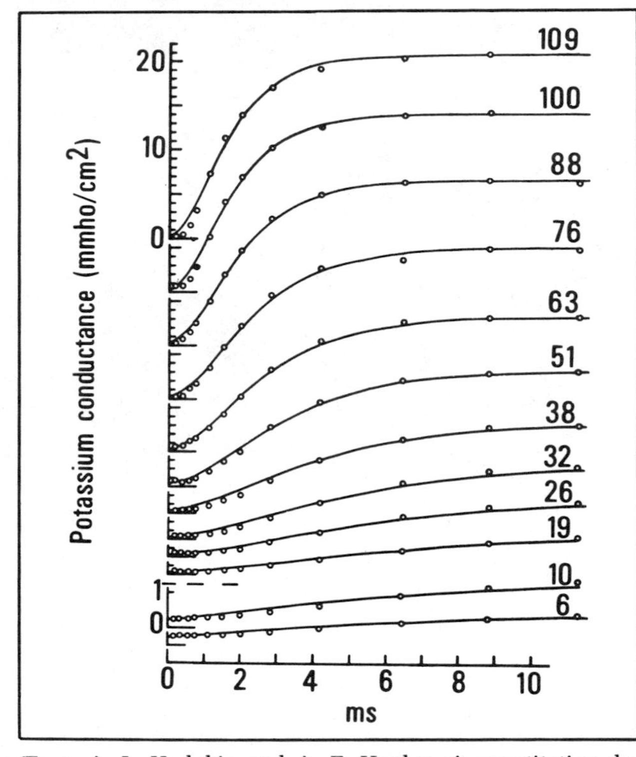

(From A. L. Hodgkin and A. F. Huxley, A quantitative description of membrane current and its application to conduction and excitation in nerve, *J. Physiol.* [*Lond.*] 117:500, 1952.)

HINTS

9. Obviously, if I_{cap} is to be zero, then I_{ext} must be equal and opposite to I_{por}. If you got that, you will now have figured out, for yourself, both the power and the simplicity of the voltage-clamp technique! Let's return to the main text to see how this works out.
12. Because $|V - E_K|$ is increased, whereas $|V - E_{Na}|$ is reduced. Thus, $I_K > I_{Na}$ and net I_{por} is outward.
15. What if this channel's Na-K selectivity ratio gave it an E_i close to that of E_{Cl}? Then it would pass current at $V = E_K$ and contribute to I_{leak}. This suggestion is sufficiently reasonable that it could be disproved only by accurate quantitative studies.

185

culate at least some of the points for the smooth curve in the 100-mV depolarizing step shown in Fig. 8-10. The necessary values for this task are

$\bar{g}_K = 24.31$ mmho/cm^2

$n_0 = 0.315$

$\alpha_n = 0.866$ ms^{-1}

$\beta_n = 0.043$ ms^{-1}

Start by calculating the 1.0-ms value for g_K. You will need to use Eq. 8-8 to get τ_n; then use Eq. 8-7 to obtain n_∞. Now substitute into Eq. 8-5 to find n_t; then raise this to the fourth power, and substitute into Eq. 8-3. Check your answers against Fig. 8-10. You will see that the correct value at 1.0 ms is about 5.7 mmho/cm^2. (If you had trouble obtaining this answer, turn to Hint 19↓.)

1

2. Mathematical description of g_{Na}. Fortunately, it was possible to describe g_{Na} by a method very similar to that used for g_K. Here $\bar{g}_{Na}$ is used again as a scaling factor equivalent to $\bar{g}_K$, while m is the *activation function* (in place of the n used to describe potassium activation). It was noted that Na activation could be described adequately by m^3 rather than as the fourth power of the exponential function. More importantly, however, it was necessary to introduce an additional *inactivation function* represented by the simple h variable. The equation for $g_{Na(t)}$ becomes

$$g_{Na(t)} = \bar{g}_{Na}(m_t)^3 h_t \qquad \text{Eq. 8-9}$$

where m increases but h decreases following a depolarizing step. Although m and h move in opposite directions, the equations describing the m and h reactions apparently are identical. The difference is introduced (as we see later) in the voltage dependence of the rate constants; i.e., while α_m *increases* in depolarization, α_h *falls*, etc. The reactions are

$$m' \underset{\beta_m}{\overset{\alpha_m}{\rightleftharpoons}} m \qquad \text{Reaction 8-2}$$

and

$$h' \underset{\beta_h}{\overset{\alpha_h}{\rightleftharpoons}} h \qquad \text{Reaction 8-3}$$

Following the same terminology as for the n variable yields

2

187

$$m_t = m_\infty - (m_\infty - m_0)e^{-t/\tau_m}$$ Eq. 8-10

$$m_0 = \frac{\alpha_{m_0}}{\alpha_{m_0} + \beta_{m_0}}$$ Eq. 8-11

$$m_\infty = \frac{\alpha_m}{\alpha_m + \beta_m}$$ Eq. 8-12

$$\tau_m = \frac{1}{\alpha_m + \beta_m}$$ Eq. 8-13

while

$$h_t = h_\infty - (h_\infty - h_0)e^{-t/\tau_h}$$ Eq. 8-14

$$h_0 = \frac{\alpha_{h_0}}{\alpha_{h_0} + \beta_{h_0}}$$ Eq. 8-15

1 $$h_\infty = \frac{\alpha_h}{\alpha_h + \beta_h}$$ Eq. 8-16

2 QUESTION: What is the equation for τ_h? (Hint 16↓)

3 QUESTION: Given α_h, β_h, and h_t, how would you evaluate dh/dt? (Hint 17↓)

As before, we recommend that advanced students carry out some additional calculations based on these equations. Why not evaluate g_{Na} for the 100-mV depolarization at $t = 0.5$ and $t = 1.0$ ms? Assume $\bar{g}_{Na} = 70.7$ mmho/cm^2.

$m_0 = 0.042$

$\alpha_m = 6.2\ \text{ms}^{-1}$

$\beta_m = 0.02\ \text{ms}^{-1}$

$h_0 = 0.608$

$\alpha_h = 0$ (or, $< 0.01\ \text{ms}^{-1}$)

$\beta_h = 1.50\ \text{ms}^{-1}$

4 Check your answers against Fig. 8-11.

Fig. 8-11. Changes of g_{Na} associated with different depolarizations. The circles are experimental observations, and the lines are from analytic solutions.

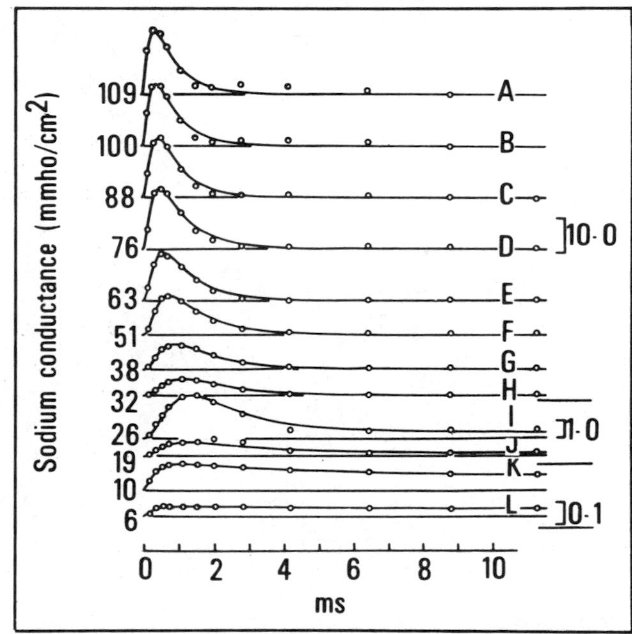

(From A. L. Hodgkin and A. F. Huxley, A quantitative description of membrane current and its application to conduction and excitation in nerve, *J. Physiol.* [*Lond.*] 117:500, 1952.)

187

188

Question: If V_{final} is +40 mV for this same 100-mV depolarization and if $E_K = -75$ mV, $E_{Na} = +55$ mV, $E_{leak} = -50$ mV, and $g_{leak} = 0.3$ mmho/cm^2, what is I_{por} 1.0 ms after the start of the depolarizing step? (Take g_K from your calculation on page 186.) See Hint 18.↓

1

IID: Hodgkin and Huxley Revisited

It is difficult to believe that students could have followed us this far without wanting to satisfy their curiosity by checking out the classic sequence of papers published by Hodgkin and Huxley in 1952 [19–23] (and reprinted in Cooke and Lipkin's excellent volume [7]). *You are strongly encouraged to read these important papers,* and you should instantly feel at home among ideas that have been both the stimulus and the source material for much of this book. Nevertheless, be on your guard, since **there are important differences in terminology between their papers and standard usage followed here.** The most noteworthy differences follow.

2

1. The sign of the transmembrane potential. Designated E, the membrane potential is positive at rest and becomes negative when depolarized beyond zero. Thus E_K is +72 and E_{Na} is −55 mV in this terminology. This is the sign of potentials seen when the intracellular axial wire electrode is grounded.

3

2. The sign convention for direction of transmembrane currents. This is also reversed in their terminology such that inward currents are positive and outward currents negative. Their voltage-clamp current records thus appear upside down by comparison with those appearing in the more recent literature (and in this book).

4

NOTE: Since both membrane potential and currents are reversed in sign, the ionic Ohm's law equations appear spuriously familiar and are given as (for example)

$$I_{Na} = g_{Na}(E - E_{Na})$$

5

3. The use of V. In their terminology, V is used to indicate differences from resting potential. Thus depolarization is shown as an increasingly *negative* value of V, with hyperpolarization indicated by *positive* values of V. Thus if resting potential is −60 mV and E_K is −72 mV in our notation, this appears in their papers as $E = +60$, $V = 0$ at resting potential and $E_K = +72$ mV while $V_K = +12$ mV.

6

QUESTION: If V_{leak} is given as -13 ± 1 mV in their notation, what would this translate to in the notation used in this book? (Hint 21↓)

7

The confusion is so appalling in this example that **we shall use V_{HH} throughout the remainder of this chapter to distinguish their V from ours!**

8

4. The use of I_m. This usage can be somewhat confusing since they do not distinguish, as we have, between I_{mem} and I_{cpl}. However, don't let these problems discourage you. Forewarned is forearmed. With this terminology mastered, you should find their papers thoroughly rewarding.

9

188

189

IIE: Synthesis of Membrane Action Potential

We commented (pages 184 to 187) on the extreme theoretical significance of Hodgkin and Huxley's successful synthesis of the membrane action potential from the data obtained in voltage-clamp experiments.

All the equations required to synthesize the membrane action potential have been presented *except for the crucial empirical equations indicating the voltage dependence of* α_m, β_m, α_h, β_h, α_n, *and* β_n. The equations used by Hodgkin and Huxley are the following:

$$\alpha_m = \frac{0.1(V_{HH} + 25)}{e^{(V_{HH} + 25)/10} - 1} \qquad \text{Eq. 8-17}$$

$$\beta_m = 4e^{V_{HH}/18} \qquad \text{Eq. 8-18}$$

$$\alpha_h = 0.07e^{V_{HH}/20} \qquad \text{Eq. 8-19}$$

$$\beta_h = \frac{1}{e^{(V_{HH} + 30)/10} + 1} \qquad \text{Eq. 8-20}$$

$$\alpha_n = \frac{0.01(V_{HH} + 10)}{e^{(V_{HH} + 10)/10} - 1} \qquad \text{Eq. 8-21}$$

$$\beta_n = 0.125e^{V_{HH}/80} \qquad \text{Eq. 8-22}$$

HINTS

16. By analogy with Eq. 8-13, this must be

$$\tau_h = \frac{1}{\alpha_h + \beta_h}$$

17. By analogy with Eq. 8-4, you may conclude, correctly, that dh/dt in a reaction such as Reaction 8-3 is

$$dh/dt = (1 - h_t)\,\alpha_h - h_t\beta_h$$

19.

$$\tau_n = \frac{1}{\alpha_n + \beta_n} = 1.10 \text{ ms}$$

$$n = \frac{\alpha_n}{\alpha_n + \beta_n} = 0.953$$

$$n_t = 0.953 - (0.953 - 0.315)e^{-1/1.1} = 0.696$$

$$(n_t)^4 = 0.235$$

$$g_{K(t)} = 5.705 \text{ mmho/cm}^2$$

189

190

The basic equation used in synthesis of the space-clamped membrane action potential is

$$I_{mem} = I_{cap} + I_{por} = C_m \frac{dV}{dt} + I_{por} = 0$$

Thus,

$$-dV/dt = \frac{I_{por}}{C_m}$$

which can be expanded to

$$-dV/dt = \frac{\bar{g}_K n^4(V - E_K) + \bar{g}_{Na} m^3 h\,(V - E_{Na}) + \bar{g}_{leak}(V - E_{leak})}{C_m} \qquad \text{Eq. 8-23}$$

This equation, together with the following three equations, provides a total of four simultaneous equations for use in the integration procedure:

$$dn/dt = \alpha_n(1 - n) - \beta_n n$$

$$dm/dt = \alpha_m(1 - m) - \beta_m m$$

$$dh/dt = \alpha_h(1 - h) - \beta_h h$$

The numerical integration procedures are not described further here. Advanced students are referred to the original paper by Hodgkin and Huxley [22] and to a more recent description (together with a FORTRAN IV computer program) by Palti [42, 43]. Note the excellent simulation achieved by this method, as shown in Fig. 8-12.

Fig. 8-12. Upper graph: the membrane action potential computed by Hodgkin and Huxley. Lower graph: the membrane action potential actually recorded in squid giant axon. Notice that the time scales have been slightly distorted to compensate for temperature differences between the simulation (calculated for 18.5°C) and the experimental record (obtained at 20.5°C).

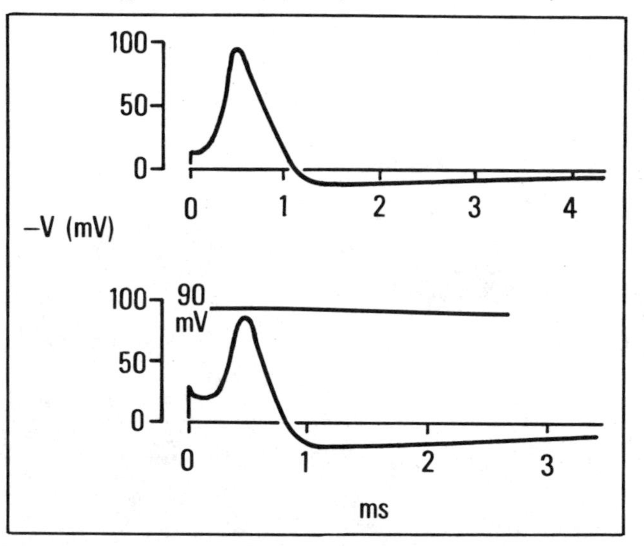

(From A. L. Hodgkin and A. F. Huxley, A quantitative description of membrane current and its application to conduction and excitation in nerve, *J. Physiol.* [Lond.] 117:500, 1952.)

IIF: Gating Currents

In the 30 years since its initial introduction, no more than minor modifications have been required in the mathematical description of voltage-dependent conductances of the squid axon to extend this formulation to cover a wide range of other electrically excitable membranes. This rather general applicability has suggested to many that the equations used might represent something more substantial than a simple curve-fitting exercise—that these equations might be describing the behavior of actual components within the membrane responsible for the control of conductance. Perhaps there really *are* four identical "gates" con-

190

191

trolling the opening of each potassium pore. Perhaps there really *are* three *m* gates and one *h* gate controlling the sodium pore opening, as the equations suggest. It is certainly tempting to imagine that *m*, *n*, and *h* states represent, for example, short forms of the gating molecules. By contrast, *m'*, *n'*, and *h'* might represent long forms of the same molecules that close off the pores to permeating ions by protruding into the pore lumen.

Whether the gating process involves conformational changes in large molecules or actual translocations of mobile "gating particles," it seems highly likely that the moving component of the gate must carry substantial charge. Otherwise, why would their movements be affected by the voltage applied across the membrane?

But if charged gating particles are moving in the plane of the applied voltage field (i.e., across the membrane), shouldn't this movement be expected to give rise to a detectable *gating current*? This current would be largest in the case of the fast-moving *m* gates. Thus any detectable gating current might be expected to show a time course equivalent to that of dm/dt.

Using a voltage-clamped squid axon under experimental circumstances that prevented any normal ionic currents through the membrane pores, investigators have detected very small currents with a time course similar to that expected for dm/dt. The existence of these gating currents has since been confirmed by other experimenters; the correspondence between dm/dt and gating current time course has been evaluated carefully. While it is probably too early to conclude that there are, indeed, three such gating particles per sodium pore, it remains possible that this might be the case, and little doubt remains that these currents are generated by the *voltage sensors*—those charged particles whose movements control the voltage-dependent opening and closing of sodium channels.

Why, then, have components of gating current related to *n* and *h* particles remained undetected? First let's consider *h*: τ_h typically is around 3 τ_m for large depolarizations, while there could only be one-third as many *h* particles. Thus the expected "*h* contribution" would be about one-tenth the size of the "*m* contribution" to gating currents. Similarly, τ_n is slow, and potassium pores are less numerous than sodium pores, to the extent that the *n* contribution also might be at the limits of detectability by present techniques. In spite of the tremendous

HINTS

18. Don't quit now! However amazing it may seem, you can perform this calculation easily by using the values provided for the variables of the *m*, *h*, and *n* systems on page 184. Try it; then turn to Hint 20.↓

21. By assuming $V_{HH} = 0$ when our $V = -60$ mV, $E_{leak} = -47$ mV.

191

technical difficulties involved in making accurate gating-current measurements, it should now be clear why this has become an exciting frontier in current membrane research.

III: CABLE EQUATIONS AND PROPAGATED ACTION POTENTIAL

In Chap. 7, we faced a problem that seemed to baffle an entire generation of textbook writers, that of providing a reasonably *simple* description of the spread of current (and change of potential) along cell membranes. The problem is that the spread of current along even uniform resistive-capacitative cable insulation is too complex for easy presentation, although this must be far simpler than the nonuniform situations encountered most often in excitable cells. To avoid what has seemed to us a resultant overemphasis on the concept of λ, the *space constant*, we considered spread of axial current in only the most general and nonquantitative manner. Not only does λ neglect the capacitative properties of the membrane (which we all agree are crucially important), but also it requires a *uniform* resistive array and thus is almost completely inapplicable in any situation where conductance changes are occurring.

However, as we hinted in Chap. 7, axial current *can* be defined rigorously and quantitatively. Now we attempt a more complete description of current spread along cell membranes, with particular attention to analysis and synthesis of the propagated action potential.

IIIA: Advanced Treatment of Cable Properties

The initial requirement is to be more explicit with respect to the effects of membrane capacitance on electrotonic spread of potential away from a single point source of applied current. Note that the simple exponential decay of potential seen in Fig. 8-13/7-11 occurs only when the system is in steady state, i.e., when $dV/dt = 0$ at each point within the array of circuit elements diagrammed in Fig. 8-14. This can occur only when I_{cap} is zero across each of the capacitors in the array.

In the more interesting, and more typically encountered, situation in which dV/dt is *not* zero at each point, the membrane capacitance must be taken into account. It is not difficult to intuit the general form of the effect of these capacitors on spread of potential. Since the capacitor provides an initial, low-resistance bypass for membrane resistance, more current (than otherwise expected) will flow out close to the point of origin of the depolarization, and less will be left for more distant membrane areas. Thus dV/dt will be faster than expected close to the point of origin and slower elsewhere along the membrane (see part *A* of Fig. 8-15). Similarly, the recovery of resting potential after I_{ext} is turned off also will be affected by membrane capacitance (see part *B* of Fig. 8-15). Here, too, dV/dt is fastest at the point of origin of the depolarization and slower in more distant parts of the array.

Note that while parts *A* and *B* of Fig. 8-15 seem curiously *dissimilar* at first, they are exactly reciprocal. In both cases, dV/dt falls off with distance from the point of

Fig. 8-13/7-11. Drop-off of membrane potential with distance from an applied voltage.

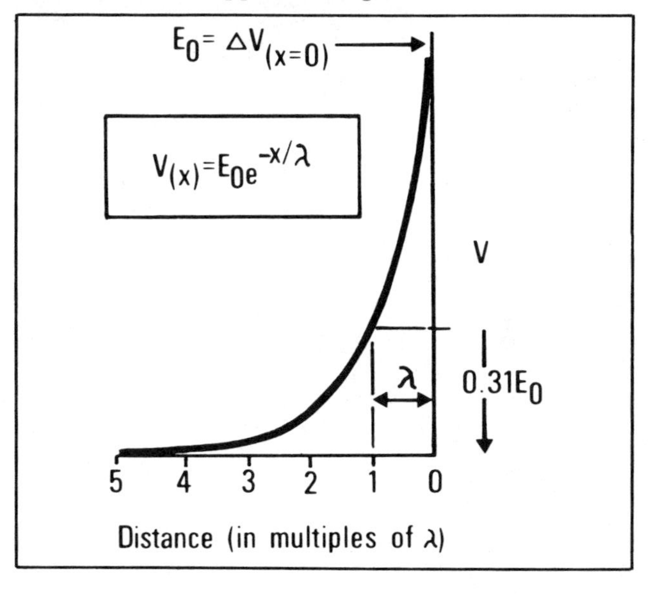

Fig. 8-14. The combined resistive and capacitative model of the cable properties.

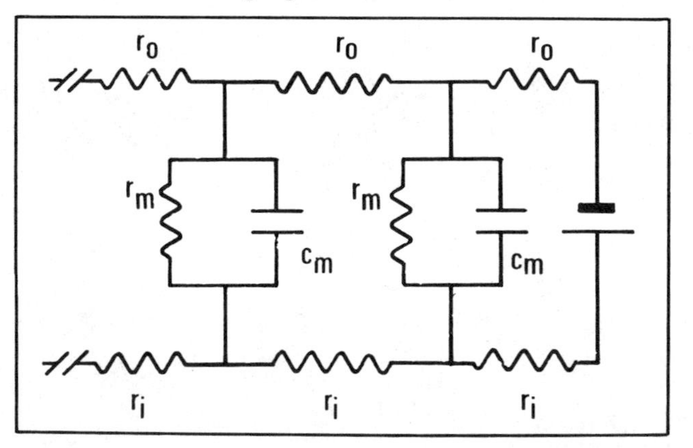

193

origin of the depolarization. Plots of I_{cap} against distance, rather than ΔV against distance, would have been symmetrical for both turn-on and turn-off of I_{ext}.

But the problem need not be handled at the merely intuitive level. Clearly, the curves of Fig. 8-15 represent solutions of an equation yielding ΔV as a function of ΔV_0, λ, and τ (the expected RC time constant for the uniform array). Unfortunately, the equation used by Hodgkin and Rushton is too complex for full description here. The interested reader should consult their classic 1946 paper in its original form [24] or in Cooke and Lipkin [7].

Despite the mathematical subtlety of the methods used to generate Fig. 8-15, these equations presume an infinite array of uniform circuit elements. That is, r_m is assumed to be constant and identical in every element of the array. We know that this condition does not apply during the action potential. Thus when Hodgkin and Huxley later needed to consider longitudinal current flow within the axon (in order to synthesize the propagated action potential), they could not use those equations in their final form. Instead, they returned to an earlier point in the derivation by Hodgkin and Rushton, where it was noted that the current across the membrane I_{cpl}, at any point distant from the point of origin of the depolarization, would be given by the *change* in axial current ($I_{axial_1} - I_{axial_2}$) occurring as it passes that point. This concept is presented diagrammatically in Fig. 8-16, where it can be readily seen that

Fig. 8-15. Changes in membrane potential with time (t) and distance from the point of current injection (computed on the assumption of a uniform, electrically inexcitable membrane). (*A*) Current on. (*B*) Current off.

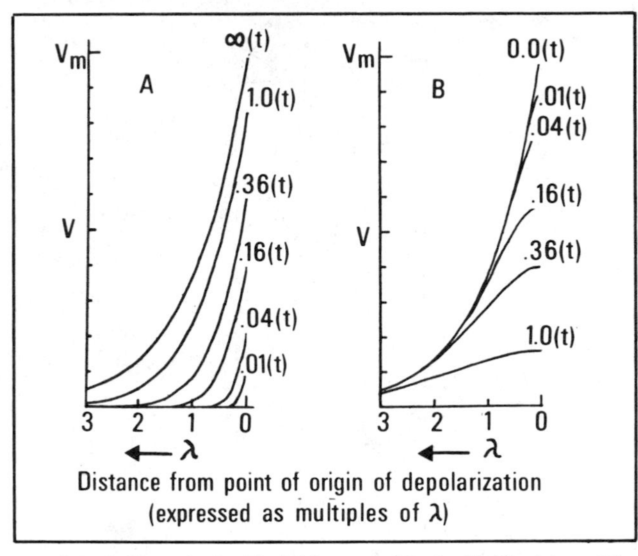

(Modified from A. L. Hodgkin and W. A. H. Rushton, The electrical constants of a crustacean nerve fibre, *Proc. R. Soc. Lond.* [*Biol.*] 133:444, 1946.)

HINT

20. $\tau_m = 0.161$ ms $\quad m_\infty = 0.997 \quad m_t = 0.995$

$(m_t)^3 = 0.985$

$\tau_h = 0.667$ ms $\quad h_\infty = 0 \quad h_t = 0.136$

$g_{Na(t)} = 9.448$ mmho/cm²

$I_K = 5.705[40 - (-75)] = 5.705(115)\ \mu A/cm^2$
$= +656\ \mu A/cm^2$

$I_{Na} = 9.448[40 - (+55)] = 9.448(-15)\ \mu A/cm^2$
$= -141.7\ \mu A/cm^2$

$I_{leak} = 0.3[40 - (-50)] = 0.3(90)\ \mu A/cm^2$
$= +27\ \mu A/cm^2$

$I_{por} = 656 + 27 - 142 = +541\ \mu A/cm^2$

Incidentally, this answer is in good agreement with Hodgkin and Huxley's estimates of the value of I_{por} taken from the 100-mV step, curve I, in Fig. 11, page 521 in one of their 1952 papers [22].

Fig. 8-16. Relationship between axial current flow and current across the membrane in a short section of axon of length Δx. See text for further explanation of symbols used.

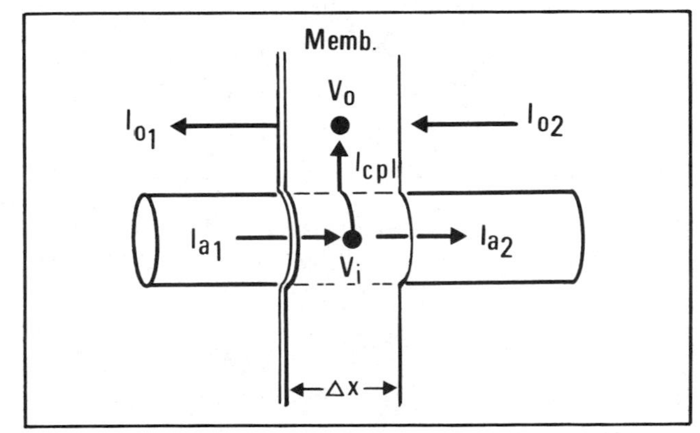

193

the internal and external longitudinal currents must be equal in magnitude at every point (since this is a closed circuit). Thus $I_{axial_1} = I_{o_1}$, $I_{axial_2} = I_{o_2}$, and necessarily $dI_o/dx = -dI_{axial}/dx$. Thus it must follow that

$$I_{cpl} = dI_o/dx = -dI_{axial}/dx \qquad \text{Eq. 8-24}$$

Now, if the axoplasmic resistance is r_i and the extracellular resistance along the membrane surface is r_o, then at the point in question (since $\Delta V = IR$)

$$\frac{dV_i}{dx} = -r_i I_{axial} \qquad \text{Eq. 8-25}$$

and

$$\frac{dV_o}{dx} = r_o I_o \qquad \text{Eq. 8-26}$$

where V_i and V_o are the potentials inside and outside the axon, respectively, with respect to a distant ground. Thus the transmembrane potential V must be the difference $V_i - V_o$, and since we already know that $I_{axial} = I_o$ at each such point,

$$\frac{dV}{dx} = -I_{axial}(r_i + r_o) \qquad \text{Eq. 8-27}$$

Note that this equation permitted us to calculate axial current in Chap. 7. Rearranging the equation, and assuming a constant velocity of propagation for the action potential, now gives the following very useful form:

$$I_{axial} = \frac{1}{(r_i + r_o)\theta} \frac{dV}{dt} \qquad \text{Eq. 8-28}$$

Returning to our previous train of thought, we may note that differentiating Eq. 8-27 gives

$$\frac{d^2V}{dx^2} = \frac{-dI_{axial}}{dx}(r_i + r_o) \qquad \text{Eq. 8-29}$$

Fig. 8-17/7-6. Transmembrane and axial currents during a computer-synthesized propagated action potential.

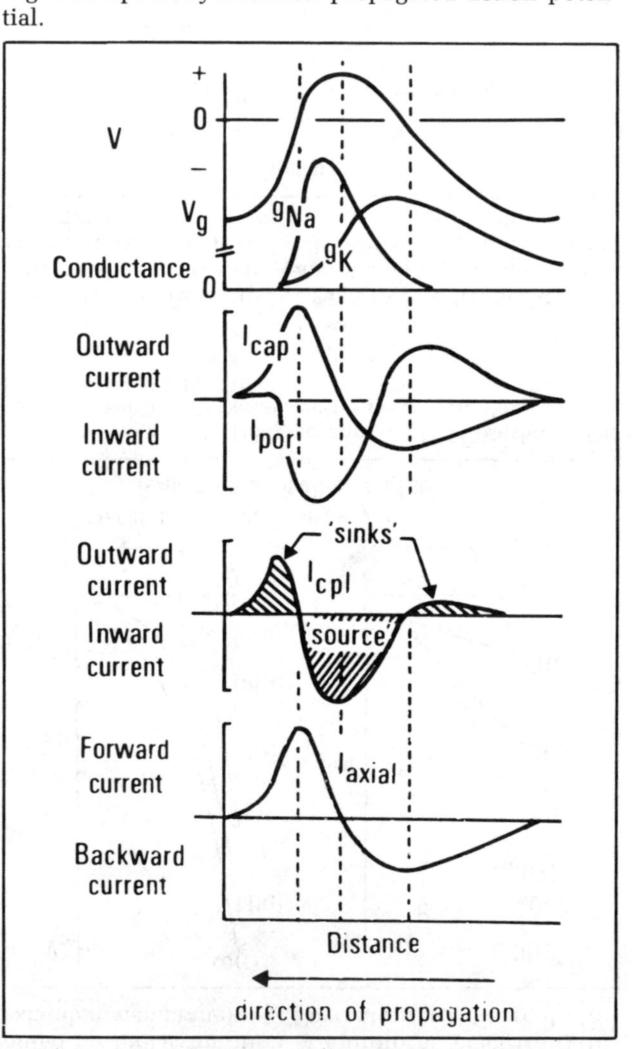

(Modified from D. Noble, Applications of Hodgkin-Huxley equations to excitable tissues, *Physiol. Rev.* 46:1, 1966.)

Substituting into Eq. 8-24 and remembering that I_{cpl} has both ionic and capacitative components, we get

$$I_{cpl} = \frac{d^2V}{dx^2}\,\frac{1}{r_i + r_o} = I_{por} + I_{cap} \qquad \text{Eq. 8-30}$$

The membrane current at a given point (i.e., coupling current) is seen to be proportional to the second derivative of the membrane potential. Notice that this equation includes none of the assumptions that were so troubling before. It does not matter that r_m be variable. Nothing is said as to how I_{cpl} should be apportioned between r_m and C_m, that is, between I_{por} and I_{cap}. All that is stated is that **total current across the membrane, at a point distant from the point source, is proportional to the second differential of the membrane potential at that point.**

Since r_o is negligible under typical experimental conditions (although not necessarily in nature), the equation can be further simplified:

$$I_{cpl} = \frac{d^2V}{dx^2}\,\frac{1}{r_i}$$

Or, where I_{cpl} has units of membrane current density and where constant conduction velocity is assumed,

$$I_{cpl} = \frac{d^2V}{dx^2}\,\frac{a}{2R_i} = \frac{d^2V}{dt^2}\,\frac{a}{2R_i\theta} \qquad \text{Eq. 8-31}$$

where R_i is the specific axoplasm resistance and a is the axon radius. Hodgkin and Huxley utilized this form in their version of Eq. 8-34 (described in the next section).

Before continuing with the *synthesis* of the propagated action potential, we should point out that this analysis also included the major equations required for the *analysis* of the propagated action potential, shown in Fig. 8-17/7-6. These equations are grouped in Table 8-1.

Table 8-1. Equations Required for Analysis of Propagated Action Potential

Equation	Units
$I_{cap} = C_m \frac{dV}{dt}$	A/cm^2
$I_{axial} = \frac{1}{\theta r_i}\frac{dV}{dt}$	A/cm
$I_{cpl} = \frac{1}{\theta^2 r_i}\frac{d^2V}{dt^2}$	A/cm^2
$I_{por} = I_{cap} - I_{cpl}$	A/cm^2

Remember, r_i is the resistance, *in ohms*, of a 1-cm length of axon; θ must be in *centimeters*, for the same unit of time as you pick for *dV/dt*.

IIIB: Synthesis of Propagated Action Potential

As we noted in Chap. 7, the challenge in both analyzing and synthesizing the propagated action potential is to take appropriate account of the longitudinal spread of current along the axon resulting from those local circuits that must invariably exist in the

196

non-space-clamped axon. How much current will flow into a given membrane region from a neighboring point source? How much current will flow away from an active point source into surrounding membrane areas?

Hodgkin and Huxley faced this problem in the equation that we have just derived, Eq. 8-30. Expanding both I_{por} and I_{cap} in that equation, they achieved the following partial differential equation in space and time:

$$\frac{1}{r_i + r_o}\frac{\partial^2 V}{\partial x^2} = C_m \frac{\partial V}{\partial t} + \bar{g}_K n^4 (V - E_K) + \bar{g}_{Na} m^3 h (V - E_{Na}) + \bar{g}_{leak}(V - E_{leak}) \qquad \text{Eq. 8-32}$$

To avoid the difficulties inherent in solution of partial differential equations, they noted that all parts of the action potential must propagate at the same velocity (normally), since the shape of the action potential does not change as it passes along the axon. It follows, therefore, that if θ is the velocity of conduction, then

$$\partial^2 V/\partial x^2 = \frac{1}{\theta^2}\frac{\partial^2 V}{\partial t^2} \qquad \text{Eq. 8-33}$$

Substituting Eq. 8-33 into 8-32 gives the following ordinary differential equation:

$$\frac{d^2V}{dt^2} = \theta^2 (r_i + r_o)[C_m \frac{dV}{dt} + \bar{g}_K n^4 (V - E_K) + \bar{g}_{Na} m^3 h (V - E_{Na}) + \bar{g}_{leak}(V - E_{leak})] \qquad \text{Eq. 8-34}$$

Then this equation can be solved by standard iterative methods.

The methods used are similar to those for the membrane action potential. See Hodgkin and Huxley [22] and Palti [42, 43] for further discussion of these techniques.

Notice, however, that θ appears as an unknown in Eq. 8-34. Nevertheless, as we saw in section I of this chapter, θ also must be dependent on C_m, on the reactivity of the voltage-dependent conductances, and on $\bar{g}_K$, $\bar{g}_{Na}$, and so on. In the absence of any simple analytic equation giving θ in terms of all these parameters, the standard procedure has been to guess at what seems a reasonable value. If the guess is too high, V does not return to resting potential, but takes off to infinity. Similarly, if the guess is too low, V tends to minus infinity. Thus it is a simple matter to adjust θ to that single value, sufficiently consistent with the other predetermined membrane properties, that permits V to return to V_s. The results obtained by solution of Eq. 8-34, for the data obtained by Hodgkin and Huxley, are shown in Figs. 8-18, 8-19, and 8-20.

196

197

Fig. 8-18. Upper graph: the computed solution of Hodgkin's and Huxley's equations for the propagated action potential at 18.5°C. Lower graph: the experimentally recorded propagated action potential in squid giant axon (also at 18.5°C).

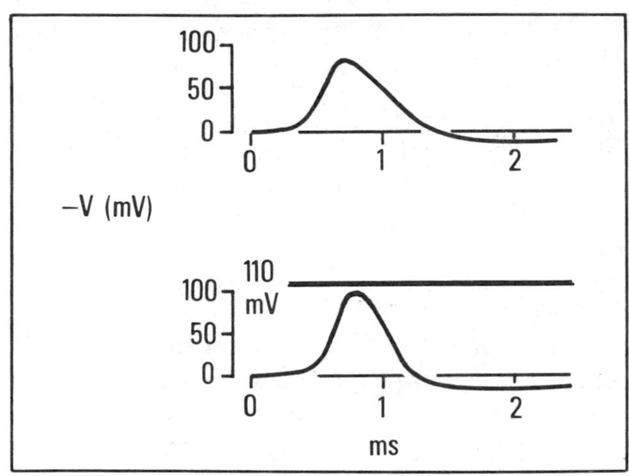

(From A. L. Hodgkin and A. F. Huxley. A quantitative description of membrane current and its application to conduction and excitation in nerve, *J. Physiol.* [*Lond.*] 117:500, 1952.)

Fig. 8-19. Computed conductance changes occurring in the simulated propagated action potential shown in Fig. 8-18.

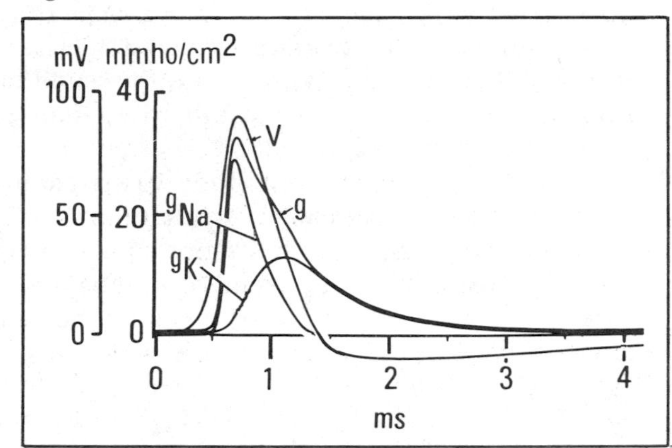

(From A. L. Hodgkin and A. F. Huxley, A quantitative description of membrane current and its application to conduction and excitation in nerve, *J. Physiol.* [*Lond.*] 117:500, 1952.)

Fig. 8-20. Computed transmembrane currents during the simulated propagated action potential shown in Fig. 8-18. (Remember that I_{axial} would have the same form as the capacitative current shown here.)

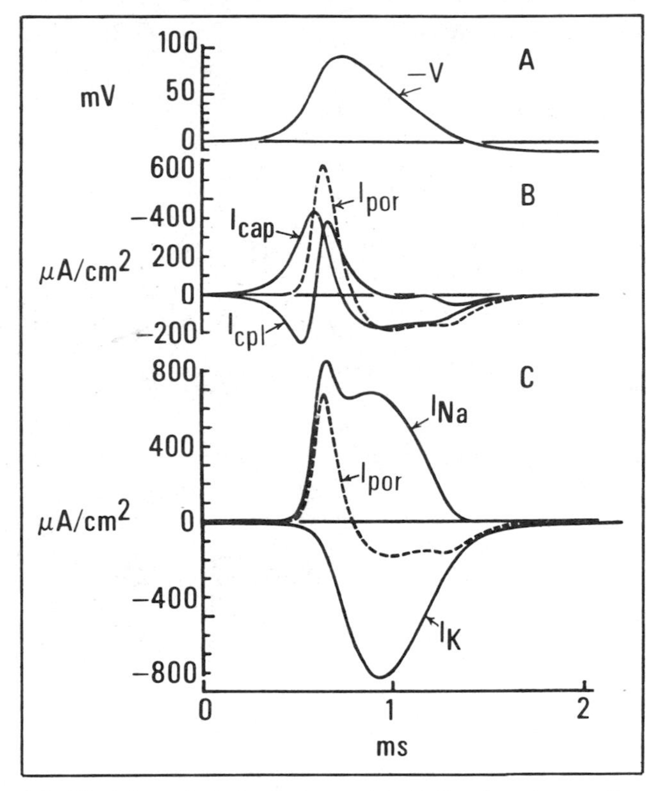

(Modified from A. L. Hodgkin and A. F. Huxley, A quantitative description of membrane current and its application to conduction and excitation in nerve, *J. Physiol.* [*Lond.*] 117:500, 1952.)

197

198

Equation 8-34 has been used by a substantial number of different investigators over the last 30 years, and impressive fits have been obtained between the predicted and the observed action potentials, thresholds, conduction velocities, etc. Nevertheless, you may find one aspect of this equation disturbing. You will remember that Eq. 8-34 was obtained, essentially, by expansion of Eq. 8-30. And Eq. 8-30 was derived by consideration of local circuits at some distance from the point source. Thus Eq. 8-30 appeared with the limitation that it did *not* apply at the point source itself. How can an equation with this kind of limitation be applied to those parts of the action potential in which the membrane region in question is clearly acting as a point source of current (i.e., where I_{por} and I_{cap} have *opposite signs*)? Could Hodgkin and Huxley have made so curious an error?

1

The resolution of this problem lies in Fig. 8-21. Notice that, at the microscopic level, the pores themselves are the point sources of current. Between these point sources lie the relatively large areas of the capacitative membrane. Now, we are concerned with only the potential across this capacitor. Thus, at the level of Fig. 8-21, we see that the membrane in which we are most interested must *always* be distant from the point source of current! The limitation is avoided, and the equation is seen to apply to membranes at all points throughout the action potential.

2

Fig. 8-21. Internal view of a 1000 Å square of squid axon membrane depolarized beyond threshold. Note that the open pore area here is only some 30 to 45 $Å^2$. (We believe this to be a high estimate of the open pore area *at any one time*, since a maximum of four pores might be expected here.)

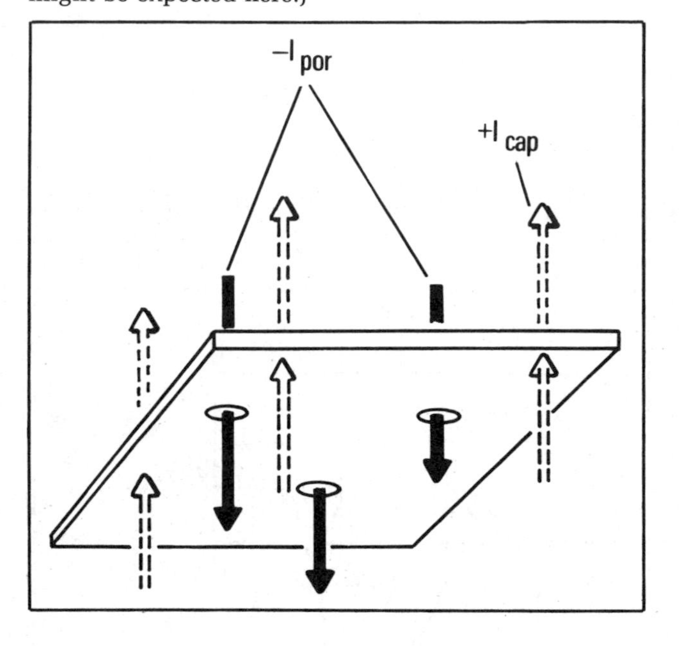

IIIC: Concept of Coupling

In this chapter, we make relatively little reference to the concept of "coupling" between adjacent membrane areas. This concept is introduced in Chap. 7 to direct attention to just exactly those time-dependent aspects of the spread of axial currents that now have been covered in considerable detail and along more classic lines. Nevertheless, this concept remains interesting, particularly for the questions that it poses.

3

While it is clearly true that coupling results from currents flowing in local circuits, and while it is equally clear that these currents are summed in the axial currents flowing at a given point, it is not easy to give a simple response to the following paradox: How can axial currents flow over distances substantially greater than the λ calculated for resting, let alone active, membranes? Are internally generated currents capable of flowing over distances far greater than externally applied currents?

4

A typical value of λ might be 1.5 mm at rest. Since $\lambda = \sqrt{r_m/r_i}$, if r_m fell by even a factor of 100 during the action potential, λ should fall by a factor of 10, to be about 0.15 mm. And yet axial current apparently enters as $-I_{cpl}$ near the peak of the action potential and exits *centimeters* away near the foot (as $+I_{cpl}$). How can this be?

5

198

199

This paradox is answered by careful consideration of the nature of coupling currents. In Chap. 7, we point out that I_{cpl}, at a point, is always the *net* current. Now imagine several points close together on the membrane. Here I_{cpl} from point *A* contributes primarily to I_{cap} at point *B*; I_{cpl} at *B* contributes to I_{cap} at *C*; etc. But axial current builds up steadily, although each intrinsic circuit seems completed close to the site at which it is generated. We see that while each membrane area obeys the cable equations, this does not prevent the *appearance* that summed currents travel over much longer distances. The reality is that these *long-range currents are being regenerated within each part of the membrane over which they seem to travel.* Only at the very foot of the propagated action potential, where no local response has had time to occur, is the spread of potential finally limited to that predicted by the cable equations. 1

It is, therefore, our contention that the more fully you understand the material of this chapter, the more useful coupling becomes as an intuitive, first approach to the mechanisms of interaction between membrane areas within the propagated action potential. However, be aware that this concept is difficult to define in any truly rigorous sense. One term that has proved useful in describing electrotonic coupling between cells (via gap functions—see Chap. 7) is the **coupling ratio.** If depolarization of cell *A* causes a depolarization in cell *B*, the degree of coupling seen can be expressed as

$$\text{Coupling ratio} = \frac{\Delta V_B}{\Delta V_A}$$

A coupling ratio of 0.8 or higher would indicate very close coupling between these cells. Lower ratios are more usual, even when synchronous action potentials indicate tight functional connection between one cell and another. 2

Unfortunately, this concept does not help us analyze or define coupling between adjacent membrane regions during the action potential. Without apologizing further for the complexities of nature, we continue to utilize this concept, bearing in mind that its major utility may be to force us to think more clearly about the form of the local circuits we invoke so readily. 3

IIID: Decremental Conduction and Electrotonic Spread

We defined electrotonic spread as the spread of potential that would occur across a membrane, from a point source of current, if that membrane were purely "passive," in the sense that it did not show voltage-dependent conductance changes. Obviously, such a spread of potential should be accurately represented by the classic cable equations as applied to uniform systems. 4

We noted, however, that even quite small depolarizations are sufficient to initiate at least a small *local response.* We merely wish to point out that **wherever a local response occurs, axial currents will spread, as in the action potential, over distances greater than predicted for the uniform membrane.** Such greater-than-predicted 5

199

200

spread seems to occur between dendrite and cell soma, as well as along the long dendrites of some sensory cells.

In the extreme instance where the local response approaches the size of an action potential, but fails to produce all-or-nothing propagation, this process is readily recognizable and is called **decremental conduction.** This term correctly indicates that the mechanisms involved are essentially those of the propagated action potential. Confusion arises only when the local response is small. **Neither the term *electrotonic spread* nor any analysis depending on the concept of λ as a determinant of such spread should ever be utilized where any local response at all is suspected.**

IV: BRIEF GUIDE TO SOME BASIC MEASUREMENTS

The approach taken in this book has been consistently theoretical rather than experimental or historical, and we make no particular apologies for that decision. It seems appropriate, however, to offer at the very least some theoretical comments as to how the parameters we so glibly discuss (V, R_m, C_m, τ_m, etc.) actually are measured in real membranes.

IVA: Measurement of Membrane Potential

You just insert the microelectrode into the cell and record its membrane potential, right? Right, and yet there is a little more than this that you should know. What you are recording is the potential generated across the cell membrane. Thus the membrane is represented as a charged capacitance in Fig. 8-22. The two resistors in series are the microelectrode itself R_L and the recording system R_R. Unfortunately, the resistance of a microelectrode with a tip small enough to penetrate a cell without damage is likely to be in the range of 10 to 30 × 10^6 Ω. Since the circuit of Fig. 8-22 is clearly a "voltage-divider" circuit, what does this tell us about the input impedance required of the recording system if we want accuracy to within 1 percent? Well, R_R has to be greater than 3 × 10^9 Ω. In the early days of microelectrode recording, such input impedances were extremely difficult to achieve. Fortunately, recent advances in technology and circuit design have made high-impedance amplifiers relatively inexpensive.

The circuit of Fig. 8-22 is considerably oversimplified. In real systems, a number of other batteries appear in series with the cell membrane, which may provide major distortion of the apparent membrane potential. These are the junction potentials between (1) the indifferent electrode and the extracellular solution, (2) the solution filling the microelectrode and the intracellular medium, and (3) the wire inserted into the open end of the microelectrode and the filling solution. Typically, (1) and (2) can be made equal and opposite by using the same substance for both electrodes (usually platinum or silver-silver chloride). The *tip potential* of the microelectrode can be made as small as possible by choice of an appropriate filling solution and then can be "backed off" to zero within the recording system.

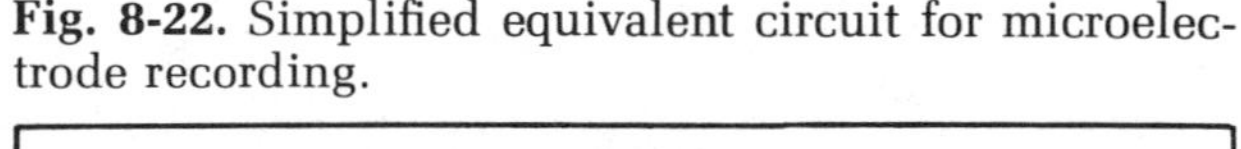

Fig. 8-22. Simplified equivalent circuit for microelectrode recording.

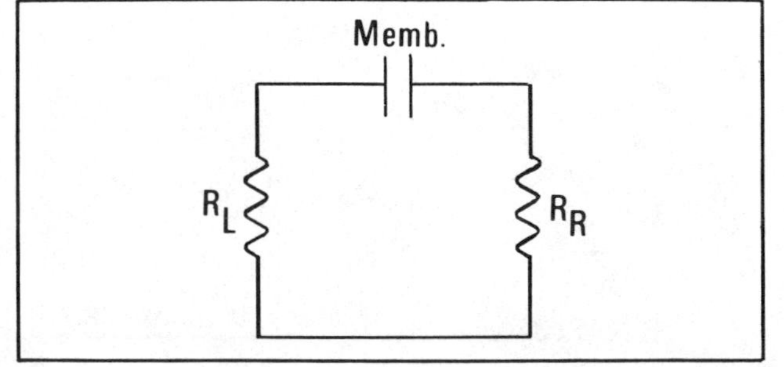

200

201

And then there are other problems, such as possible changes in tip potential between extracellular and intracellular media, neutralization of capacitance in the shielded leads between microelectrode and amplifier, etc. The determined student should read, for example, the relevant chapters in Lavallée, Schanne, and Hébert [29].

What does a microelectrode "see"? There are at least two different answers to this question, based on the geometry of typical cell types.

1. Spherical cells. In all but the very biggest cells, it is usually safe to assume that the electrode tip lands up somewhere near the middle of the cell. (Try sticking a pin into a Ping-Pong ball if you don't believe this!) Thus the electrode will look equally at all events on the cell surface, whatever their location. But this does not mean that discrete changes in membrane potential may not occur. It just means that you will *see* them as if they occurred simultaneously all over the cell surface.

2. Cylindrical cells. Once again, we assume that the electrode tip ends up near the center of the cylinder. Therefore it sees all events equally within a torus around the circumference of the cylinder. But the torus for 20-20 vision is very narrow. Events occurring any appreciable distance away will be seen only insofar as that piece of membrane is "coupled" to the narrow torus surrounding the electrode penetration site. Such coupling will be more effective for slow potential changes, less effective for rapid changes in potential, etc.

IVB: Current-Passing Microelectrode

Most of the more interesting measurements require you to pass current across cell membranes, either to stimulate an action potential or to investigate membrane resistance and capacitance. Almost universally now, such measurements are made by using **constant-current stimulators,** since it is the current across the cell membrane that you wish to control.

This is in contrast to the older work done by using extracellular electrodes, where **constant-voltage sources** were used as stimulators. You should be quite suspicious of stimulators; lack of proper understanding of their performance characteristics can be a major pitfall for a beginning student.

As in the case of the recording electrode, the current-passing microelectrode is profoundly affected by cell geometry:

1. Spherical cells. It is reasonable to assume that current density is uniform over the entire membrane surface. This assumption normally will be valid unless there are large membrane areas with an unexpectedly low resistance. Such areas would take more than their fair share of the current.

2. Cylindrical cells. Current density will not be uniform, but rather will be spread out along the membrane, in the manner predicted by Eq. 8-30, on either side of the narrow torus surrounding the electrode site.

201

202

Even with accurate constant-current stimulators, there are hazards to torment the unwary experimenter. In stimulating, the major hazard is likely to be polarization, occurring between the wire electrode and the filling solution within the microelectrode. Fortunately, most cells are small enough that the currents required also are small. Under these circumstances, normal Ag-AgCl electrodes are sufficient to prevent polarization. A more subtle hazard results from unwitting injection of K^+ ions into the cell during prolonged depolarizing pulses, with resultant change in E_K (or in E_{Cl} during hyperpolarization). Forewarned is forearmed. You should always calculate the expected ion shifts when prolonged currents are applied to a cell.

Additionally, you must be aware that while "cylindrical cells" may just possibly be nearly cylindrical, very few neurons are truly spherical! So current densities are unlikely to be truly uniform, and it is wise to become a little cynical whenever the spherical approximation is invoked.

IVC: Measurement of Membrane Electrical Properties

The first step in measurement of cell membrane resistance is always to obtain R_{eff}, the **effective resistance,** and this is often *all* that need be, or can be, measured. Now R_{eff} is defined as the resistance that would be calculated by Ohm's law, where I is the applied current at the electrode site and V is the observed steady change in membrane potential at the same point. Just this simple measurement may be enough to indicate that, for example, a given toxin exerts its effect by changing the sodium conductance of the resting membrane: change in R_{eff} was observed in normal saline medium, but not in a sodium-free, choline-substituted medium. Any further analysis requires measurements concerning membrane area, axoplasmic resistivity, etc., which cannot be made except under the most favorable circumstances.

However, even to obtain R_{eff}, clearly it is necessary to both stimulate and record from the same cell. There are two ways in which this can be done:

1. The single-electrode technique. Either by use of the simple Wheatstone bridge circuit or by sophisticated, feedback-compensated versions of the same concept, it has been possible to stimulate and record through the same microelectrode (see Fig. 8-23). This method has several advantages: It yields R_{eff} directly, since ΔV is recorded at the same point as I_{ext} is applied; it is not necessary to get more than one electrode into the same cell; and problems arising from measurement of interelectrode spacing are eliminated. However, certain disadvantages are associated with accurate balancing of the bridge circuit, particularly where large changes in R_m occur as a result of the applied stimulus.

2. The double-microelectrode technique. Separate electrodes are used for passing current and for recording membrane potential. This method has the tremendous advantage of electronic simplicity and reliability. However, two electrodes must be

Fig. 8-23. Simple Wheatstone bridge circuit. If R_3 is adjusted such that $R_3/(R_1 + R_m) = R_1/R_2$, then $V_a = V_b$ since the voltage drop will be the same across both arms of the voltage-divider circuit. The stimulus is then not seen by the recording circuit.

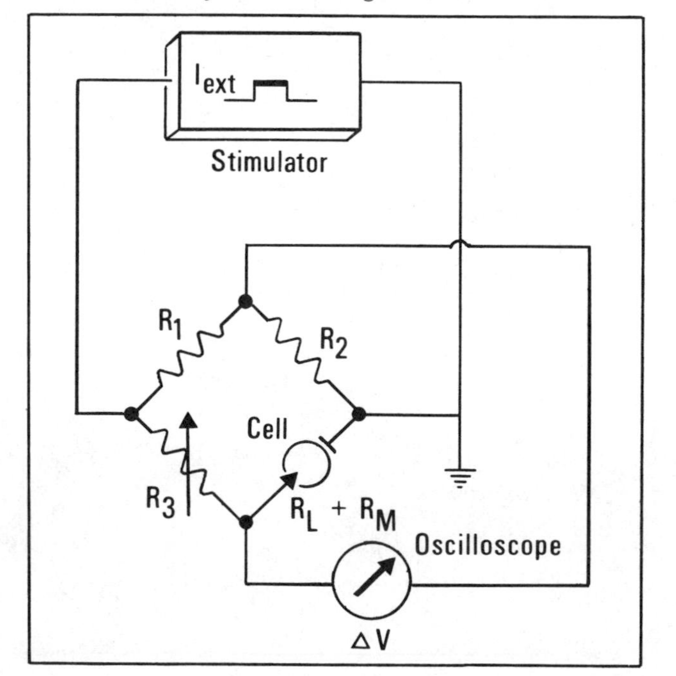

202

203

placed in the same cell (often no easy task), and (at least in cylindrical cells) accurate measurements must be made of the interelectrode spacing.

In the previous two subsections, we noted that differences are encountered, both in recording and in passing current through "spherical" as opposed to "cylindrical," cell types. So it is not surprising that the electrical properties of the cell are determined differently for each cell type.

1. **Determination of the electrical properties of spherical cells.** Because of the special properties of the spherical cell, electrode spacing need not be monitored, and results obtained by both single- and double-electrode techniques are analyzed in the same way. In such a cell, the change of membrane potential ΔV_t after *onset* of either a hyperpolarizing or depolarizing current pulse should follow the following equation:

$$\Delta V_t = \Delta V_f(1 - e^{-t/(R_m C_m)}) \qquad \text{Eq. 8-35}$$

Here ΔV_f is the final value of ΔV. Similarly, when the pulse is turned off,

$$\Delta V_t = \Delta V_o \, e^{-t/(R_m C_m)} \qquad \text{Eq. 8-36}$$

Thus the membrane time constant $\tau_m = R_m C_m$ can be obtained directly from the time taken for the membrane to charge to 0.63 of ΔV_f or discharge to 0.37 of its initial ΔV_o. And, where ΔV_f is known, R_{eff} can be calculated directly (in ohms) as

$$R_{eff} = \frac{\Delta V_f}{I_{ext}} \qquad \text{Eq. 8-37}$$

To go further than this, you need a reasonably accurate estimate of cell diameter d. Clearly, R_m, the specific resistance in ohms times square centimeters, must be given by

$$R_m = R_{eff}(\text{surface area}) = R_{eff}\pi d^2 \qquad \text{Eq. 8-38}$$

Finally, where both τ_m and R_m are known, C_m must be

$$C_m = \frac{\tau_m}{R_m} \qquad \text{Eq. 8-39}$$

203

204

Obviously, you are going to have to work through these calculations at least once before you realize just how easy it is!

1

Question: In a spherical cell 20 μm in diameter, a current of 0.05×10^{-9} A is sufficient to hyperpolarize the cell by 10 mV. If the time constant was 2.5 ms for both turn-on and turn-off of the current pulse, obtain values of R_m (in ohms times square centimeters) and C_m (in microfarads per square centimeter). (See Hint 22.↓)

2

2. Determination of the electrical properties of cylindrical cells by the single-microelectrode method. Although R_{eff} may be obtained directly from I_{ext} and the observed ΔV_f, just as in Eq. 8-37, τ_m cannot be estimated by the same method as in spherical cells. With hyperpolarizing currents, or in other circumstances in which no marked change in R_m is expected along the membrane surface, the tables provided by Hodgkin and Rushton [24] can be used. When the electrode spacing is zero, the true τ_m (that is, R_mC_m) is the time at which ΔV_t reaches 84.3 percent of ΔV_f. The only remaining problem is to obtain R_m from R_{eff}. Unfortunately, this requires that both the diameter of the cylinder and the specific resistivity of the axoplasm be known:

$$R_m = \frac{\pi^2 d^3 R_{eff}^2}{R_i} \qquad \text{Eq. 8-40}$$

Where R_i is in ohms times centimeters and R_{eff} is in ohms, eq. 8-40 gives R_m, the specific membrane resistance, in ohms times square centimeters. Note, however, that accurate measurement of cell diameter is extremely important, since d appears cubed in this equation! Similarly, the axoplasm resistivity normally must be assumed; direct measurements of axoplasm resistivity are quite rare in the literature.

3

QUESTION: Where d is estimated to be 80 ± 5 μm in a frog muscle fiber, R_i is estimated at 250 ± 10 $\Omega \cdot$cm, and you expect resting values of R_m to range from 2.5 to 4 k$\Omega \cdot$cm^2 in this muscle, what would be the range of external currents required to produce a 10-mV hyperpolarization at the site of current injection? (See Hint 24.↓)

4

3. Determination of the electrical properties of cylindrical cells by using two microelectrodes. In the simplest possible method, two microelectrodes are merely inserted from opposite sides of the cylinder into the same "torus" at zero electrode spacing. The only advantage gained here is the possible increase in accuracy that may result when one no longer has to fight the vagaries of the Wheatstone bridge

5

204

205

circuit. Otherwise, the method of calculation would be exactly as described for the single microelectrode. However, with the second electrode, additional information may be gained that may substantially reduce the uncertainty of the single-electrode method, provided that the cell is of sufficient diameter for reliable multiple penetrations.

The method involves moving the recording electrode to two (or preferably more) different points at different spacings from the current-passing electrode. At each site ΔV_f is carefully measured and plotted against the electrode spacing on a semilog plot (see Fig. 8-24). The resulting points should yield a straight line according to the following equation (taken from Hodgkin and Rushton [24]):

$$\Delta V_{f(x)} = \frac{1}{2} I_{ext} \sqrt{r_m r_i}\, e^{-x/\lambda} \qquad \text{Eq. 8-41}$$

Here x is the interelectrode spacing, r_m and r_i are the membrane and axial resistances (in ohms), respectively, of the cylindrical cell, and $\Delta V_{f(x)}$ is the final change of potential observed at spacing x.

If this straight line is extrapolated back to its origin, $\Delta V_{f(x=0)}$ is obtained. Then this permits λ, the distance at which $\Delta V_{f(x)} = (1/e)\, \Delta V_{f(x=0)}$ to be read off directly

Fig. 8-24. Graphic method for estimation of λ. See text for further details of method.

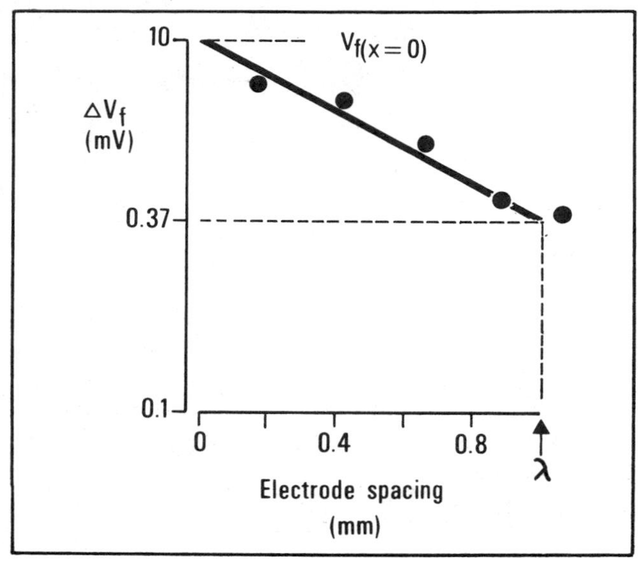

HINTS

23. $R_{eff} = 2 \times 10^8\ \Omega$
$R_m = 2.5 \times 10^3\ \Omega \cdot cm^2$
$C_m = 1 \times 10^{-6}\ \mu F/cm^2$

25. The low estimate is

$$R_{eff} = \sqrt{\frac{R_m R_i}{\pi^2 d^3}} = 5.0 \times 10^5\ \Omega$$

$$I_{ext} = \frac{\Delta V_f}{R_{eff}} = 2.0 \times 10^{-8}\ A$$

The high estimate is

$$R_{eff} = 3.15 \times 10^5\ \Omega$$

$$I_{ext} = 3.2 \times 10^{-8}\ A$$

Notice that these estimates are not as far apart as we might have feared. Nevertheless, this current is about 1000 times greater than was needed to produce a similar hyperpolarization in a spherical cell of one-fourth the diameter—even though R_m was about the same in both instances.

26. $\lambda(l) = 2\pi(d/2)^2\ R_{eff}/R_i = 1.7$ mm. And, just for interest, $\lambda(l)$ would have been 1.5 mm for the lower estimate of R_{eff}.

205

206

from the same line (see Fig. 8-24). Now, R_{eff} may be calculated directly as $\Delta V_{f(x=0)}/I_{ext}$. But if we set $x = 0$ in Eq. 8-41 and solve for R_{eff}, we obtain

$$R_{eff} = \frac{1}{2}\sqrt{r_m r_i} \qquad \text{Eq. 8-42}$$

and you will remember that

$$\lambda = \sqrt{\frac{r_m}{r_i}} \qquad \text{Eq. 8-43}$$

Since λ is properly a dimensionless number (it has units of multiples of the "characteristic length"), these simultaneous equations may be solved to yield

$$r_i = \frac{2R_{eff}}{\lambda} \qquad r_m = 2R_{eff}\lambda$$

From the cable geometry (where the diameter d is known),

$$R_i = 2\pi\left(\frac{d}{2}\right)^2 \frac{R_{eff}}{\lambda l} \qquad \text{Eq. 8-44}$$

$$R_m = 2\pi d R_{eff}\lambda l \qquad \text{Eq. 8-45}$$

1 and where l is the characteristic length of axon to which λ, r_m, and r_i are all related.

Usually, l is omitted from these equations, since it is normal to select 1 cm as the characteristic length. Unfortunately, this makes dimensional nonsense out of these equations and has unnecessarily confused many people (including the authors of this text). You should note that Fig. 8-24 actually estimates $\lambda(l)$ rather than λ. Do we need to be this pedantic? Why not! 2

Now, τ_m could be obtained from any of the responses used to construct Fig. 8-24. It is only necessary to go to table 1 of Hodgkin and Rushton [24], noting the relevant electrode spacing as a fraction of λ, and read off the percentage of $\Delta V_{f(x)}$ that corresponds to the true $R_m C_m$ time constant. Then C_m can be obtained, once R_m has been calculated from Eq. 8-45. 3

206

207

Question: For the muscle fiber described in the question on page 204, what is the value of $\lambda(l)$ appropriate to the higher R_{eff} value? Use Eq. 8-44. See Hint 26.↑

1

Before you rush out to measure any cell membrane resistances, there is one nasty problem. The quantity **$1/R_m$ is equal to the sum of the ion-specific conductances only if there has been no "local response."** This caveat is so commonly ignored that you may have difficulty believing that it is true. But it is, as you will soon prove. So remember, resting R_m should always be measured in hyperpolarization if you suspect that the cell shows voltage-dependent conductance changes. Otherwise, there is nothing but the voltage clamp to allow you accurate estimates of membrane conductance. Here is how to prove it to yourself:

2

1. Imagine a simple cell in which $g_m = g_{Na} + g_K$. Pick values for these conductances and for E_{Na} and E_K. Select a depolarizing stimulus-current density, and calculate ΔV from g_m. Now substitute that value of V into the ionic Ohm's law equations. To your delight, everything will come out right; the currents are exactly as expected.
2. Now assume that g_{Na} doubled during depolarization. We want to know what happens to ΔV, so we calculate the expected new value, substitute into the ionic Ohm's law equations, calculate the currents, and . . . *bleah!* It doesn't work out! Why not? Well, the membrane generated an inward I_{Na}, which then increased I_{cap} to give a *bigger* depolarization than the expected ΔV—the local response. The "observed g_m" is no longer equal to the sum of the conductances. . . . Convinced?

3

IVD: Current-Voltage Plots

One of the most thoroughly satisfying ways to present data from investigations of membrane electrical properties is by means of *current-voltage plots*. So we feel obliged to mention these plots and to warn you against pitfalls in their interpretation.

4

Data presented on crossed current-voltage axes can come from two sources: from **voltage-clamp experiments** in which voltage is the independent variable (and so should be represented on the abscissa), while current, the dependent variable, ap-

5

HINTS

22. Calculate R_{eff} from Eq. 8-37. Calculate R_m from Eq. 8-38. Calculate C_m from Eq. 8-39. See Hint 23.↑

24. The higher R_{eff}, the smaller I_{ext} for given ΔV_f. We know from Eq. 8-40 that $R_{eff} = \sqrt{R_m R_i / \pi^2 d^3}$. Therefore, it follows that for the *low* estimate of I_{ext}, use $d = 75\ \mu m$, $R_i = 260\ \Omega \cdot cm$, and $R_m = 4\ k\Omega \cdot cm^2$. For the *high* estimate of I_{ext}, use $d = 85\ \mu m$, $R_i = 240\ \Omega \cdot cm$, and $R_m = 2.5\ k\Omega \cdot cm^2$. See Hint 25.↑

207

208

pears on the ordinate; and from **current-clamp experiments** in which a constant-current source is used. In the second case, the current axis should be the abscissa, and the resulting voltage change will be the dependent variable.

A pitfall for the unwary is to confuse data from the one source with data from the other. (This is not as difficult as one might hope that it should be, since authors have sometimes been careless with their choice of axes; one occasionally discovers current-clamp data presented on voltage-clamp axes.)

Figure 8-25 shows, in diagrammatic form, a typical plot from voltage-clamp data. The solid line represents the peak currents of the early part of the record (traditionally called I_{peak}). Typically, these peaks will be I_{Na}-dependent. Note that this curve crosses the zero current point three times: at resting potential, at the threshold, and finally when the early inward current reverses its direction at E_{Na}. Thereafter this curve approaches a line of maximum slope, the slope of $g_{Na_{max}}$. Two definitions of conductance may be extracted from this plot. The first is the familiar form obtained from the relationship of I to V at given point on the curve. This is called the **chord conductance** and is always positive. Then there is the **slope conductance,** which can be calculated from the *slope,* dI/dV, at any point on the curve. It is a general observation that **the slope conductance is always negative over a region of potential within which all-or-nothing membrane responses can be obtained.**

When the peak delayed current (often known as I_{ss}, that is, $I_{steady\ state}$) is plotted on the same axes, we realize that the region between the two curves defines the area in which time-dependent changes can occur. In fact, this area could be filled in by the family of curves that would be obtained at different times following the early inward current peak and preceding the later peak of the delayed currents. Notice that the delayed currents also tend toward a maximum slope, the slope of $g_{K_{max}}$, but that there is no negative conductance region for the delayed currents.

Figure 8-26 shows the very different appearance of the curve that might have been obtained for the same membrane in a current-clamp situation (i.e., a space-clamped membrane exposed to a constant-current stimulus source). Looking first at the curve for peak response, note that on the hyperpolarizing side of resting potential, the curve is steep, reflecting the high resistance of the resting membrane. The curve tends to become steeper, however, as threshold is approached and reflects an "infinite" resistance in the region between threshold and E_{Na}. (This "infinite R_m" region is equivalent in significance to the negative-conductance region in the voltage-clamp plots.) The membrane cannot easily be pushed beyond E_{Na}, however, and the reduced resistance of the fully activated membrane now becomes apparent.

When very long-lasting current pulses are utilized, the "delayed response" to steady depolarization can be investigated. This response reflects maintained g_K, and the curve changes with time in membranes that show slow inactivation of potassium conductance.

Fig. 8-25. Typical current-voltage plot prepared from voltage-clamp data. See text for full description.

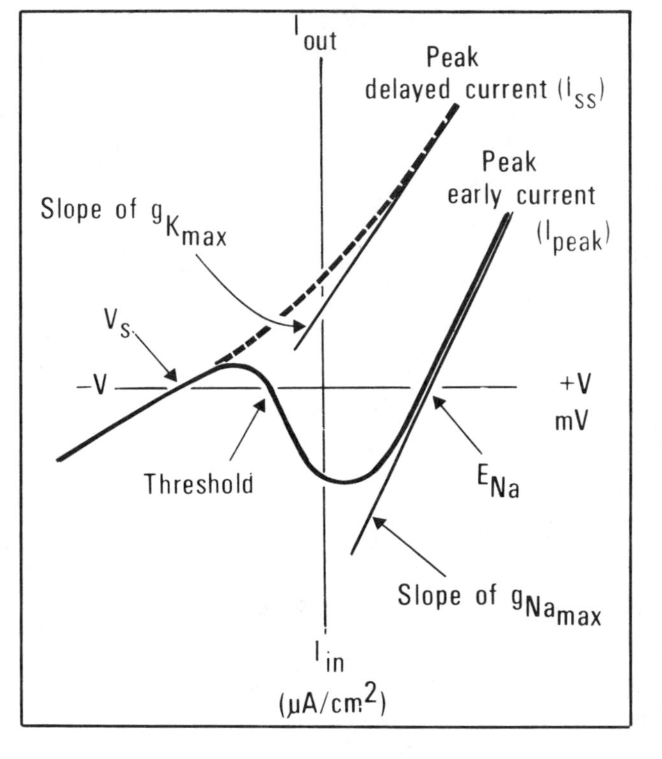

208

209

These curves make an excellent method by which the curious and unusual responses of some membranes can be presented. Membranes are known in which a second negative conductance region appears during hyperpolarization as a result of increased g_K under these conditions. As you might expect, large hyperpolarizing stimuli induce hyperpolarizing responses in such membranes.

Any reader wishing to look further into this aspect of membrane behavior is referred to the fascinating treasure trove of unusual membranes described by Grundfest [17].

So, here you are at the end of our presentation of the axonal action potential. Can *you* imagine the whole panorama of the propagated action potential? Or did it fall apart into a sequence of disconnected details? If so, try again. Start off with the rising phase (we admit the falling phase is only lightly sketched in these chapters), and see whether you can picture it, with its complexly circulating currents and changing conductances. As you gain insight, you may touch the esthetics of a complex science, and you can guess at what it must have been like to have been among the first to gain these insights.

Fig. 8-26. Typical current-voltage plot prepared from current-clamp data. See text for full description.

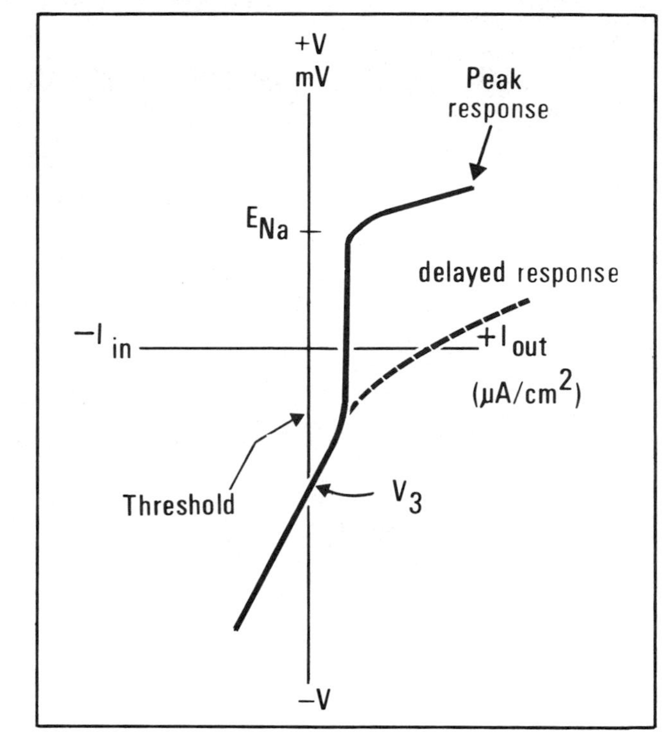

209

211

Generator and Receptor Potentials

INTRODUCTION TO SENSORY MECHANISMS

1 Let us briefly review the sequence of ideas presented so far.

The cells of the body are negative inside (relative to the outside). **This intracellular potential is the potential on the membrane capacitance,** the magnitude of which is determined by the amount of charge asymmetry occurring on either side of the membrane (capacitance). The charge asymmetry is brought about by the net fluxes of ions being moved by electrical and concentration forces. The **membrane conductances** to specific ions **determine the relative movement of the ions** and hence the net flux. 2

At the Nernst equilibrium potential for a single ion, the net flux is zero. That is, the fluxes are equal in both directions (in one direction as a result of the electrical force, in the other direction as a result of the concentration force). **When the membrane potential is at the Nernst equilibrium potential for a given ion, the current carried by that ion species is zero.** 3

At the resting potential, the total current across the membrane is zero. However, for all ions that are *not* in equilibrium at this potential, some current will flow. To keep the cell in steady state, such ion movements must be counteracted by ion "pumps." These pumps require metabolic energy, but serve to maintain relatively constant ionic gradients across cell membranes. 4

Changes in membrane potential occur when a change in the conductance for specific ions results in a net flow of ionic current through the membrane pores, generating an opposite current across the membrane capacitance. In axons, conductance changes are brought about by electrical depolarization of the membrane. If the depolarization is small enough to be below threshold, the conductance changes are small and proportional to the depolarization. However, if the depolarization exceeds threshold, then an all-or-nothing action potential is produced (by an all-or-nothing, maximal change in Na^+ conductance, followed by changes in K^+ conductance). **As a result of longitudinal current flow, the action potential in one region depolarizes the adjacent membrane areas past threshold** by the mechanism of an inward current flow at the "active" region, which then results in a net outward flow across the membrane capacitance of "inactive" regions. In this way the action potential propagates down the length of an axon. 5

Now, we warned you that electrical depolarization is *not* the only mechanism by which conductance changes can be initiated. In fact, **the entire process of sensation depends on the existence of membrane regions that are insensitive to electrical depolarization but markedly sensitive to a variety of forms of incident stimuli.** 6

The goal of this chapter is to discuss how sensory neurons accomplish the task of changing membrane conductance in response to the appropriate incoming stimuli. Thus, we are dealing with the "short-range transmission" portion of the sensory neuron, as shown in Fig. 9-1/2-13. 7

Table 9-1 shows some examples of the variety of **stimuli** that **may produce** conductance changes and hence **depolarization (or hyperpolarization) of specialized sensory mem-** 8

Fig. 9-1/2-13. Short-range transmission in sensory neurons is the subject of this chapter.

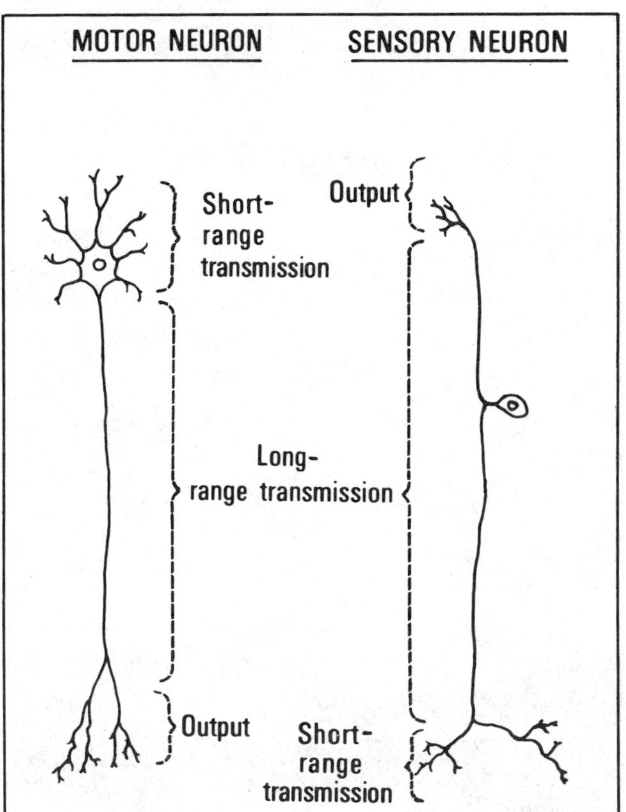

(Modified from E. L. House and B. Pansky, *A Functional Approach to Neuroanatomy* [2d Ed.]. New York: McGraw-Hill, 1967.)

213

Table 9-1. Classification of Receptors by Form of Energy to Which They Respond at Lowest Stimulus Intensity

Incident Stimulus	Examples of Receptor Types and Function Served	Intermediate (Transduction) Mechanism
Mechanical force	Mechanoreceptors Touch pressure in skin and subcutaneous tissues; both organized and free endings Position sense and kinesthesia: mechanoreceptors of joints and vestibular receptors of inner ear Mechanoreceptors of cochlea, serving hearing Stretch receptors of muscle and tendon, which do not serve conscious sensation directly Visceral pressure receptors: receptors in carotid sinus, right atrium, intestines (?), bladder, trachea and bronchi, some of which may reach consciousness	Unknown; possibilities are: (1) Change in static properties of nerve endings, e.g., resistance, capacitance (2) Specific or nonspecific change in membrane conductances (3) Intermediate release of specific chemical agent and chemoreception at nerve endings
Light	Photoreceptors of eye, serving vision	Photochemical transduction, leading to conductance changes in receptor (can have hyperpolarizing or depolarizing effect depending on species)
Heat	Thermoreceptors, separately for warmth and cold	Unknown (by regulation of chemical reaction that influences state of nerve ending?)
Chemicals in solution	Chemoreceptors, separately for taste and smell Osmoreceptors, which do not reach consciousness Carotid body receptors (pO_2, pH, pCO_2), which do not reach consciousness	Uncertain, probably excitation of receptor cell or nerve ending by specific chemical combination, leading to change in conductance
Extremes of mechanical force, heat, cold, some chemicals	Nociceptors, serving pain	Incipient or actual destruction of tissue cells (release of intracellular substance, exciting nerve ending?) *or* massive, prolonged stimulation of endings listed above?

Modified from V. B. Mountcastle, Sensory Receptors and Neural Encoding: Introduction to Sensory Processes. In V. B. Mountcastle (Ed.), *Medical Physiology*, 14th Ed. St. Louis: Mosby, 1980. Chap. 11.

213

214

branes. The process by which one form of energy (the incident stimulus) is converted to another (the ionic currents that produce membrane depolarization) is called **transduction.** While the details of the transduction mechanism are not *fully* understood in any known sensory receptor, some of the basic mechanisms are discussed in this chapter both as an introduction to the study of sensory systems and because these ideas turn out to be very important in Chap. 10. Since it is such an important concept, let us repeat: ***transduction* means to transform from one energy to another.** (You are a transducer of chocolate bars into body heat!) Typically, the cell that is specialized to transduce light is found to be *relatively* insensitive to other energies, e.g., gentle force. This is the basis of the separations of the left-hand column of Table 9-1.

1

QUESTION: If a cell responds to mechanical deformation, what energy has been transduced? (Hint 1↓)

2

Since, with few exceptions, a given receptor seems to transduce one form of energy much better than others, it has been postulated that **only a specific sensation can be elicited by activation of a specialized nerve ending or its CNS connections.**

3

This is sometimes known as **Müller's law of specific energies.** It seems to hold well, except possibly for some aspects of pain and a few receptors that have dual sensitivities.

4

In some ways, biology has tried unsuccessfully to emulate the supposedly more rigorous fields of physical science. Thus there is a tendency to try to establish "laws" or "doctrines," as if these ideas were as immutable as Newton's laws of motion (which, come to think of it, weren't so immutable after all!). It is surprising how many biological "laws" initially were hypotheses based on relatively little direct experimental support, e.g., Starling's law, Dale's principle, or the Weber-Fechner law. In general, be cautious in accepting as gospel anything in biology labeled in this fashion! Shall we call this the "Jewett-Rayner law"?

5

Considerable anatomic research has been devoted to attempting to correlate the many anatomic structures found in skin (Ruffini endings, Krause's end bulbs, etc.) with specific sensations. This analysis has been called into question by several observations.

6

The strongest observation is that several sensations can be elicited from the cornea—touch, pain, warmth, cold—even though only bare nerve endings are present. In addition, attempts at correlating sensation "spots" on the skin with the endings in the skin (under the "spot") have failed to show the "expected" correlations with previous anatomic classifications. At present, we may conclude that specialized endings are not a *necessary* structure for a specific sensory transduction, although they certainly may play a part in the way the ending behaves physiologically (e.g., the Pacinian corpuscle, described later in this chapter).

7

214

215

SENSORY CODE

Since **the entire sensory input to the CNS** (and to consciousness) **must be carried in the form of all-or-nothing action potentials,** one may well wonder how this information is "coded." 1

Such a question has yet to be completely answered for *any* system! However, it *is* clear that **the frequency of firing may be an important code** and that **sudden changes in frequency** (even from inactivity, i.e., zero frequency) **may transmit information** on important, sudden changes in the modality sensed. Moreover, both these codings (frequency and rate of change in frequency) can be transmitted by the same axon (see the next section)! It is certainly possible that various sensory systems use different types of coding and that the coding may vary in different parts of the same system. 2

The latter case will be shown shortly for the Pacinian corpuscle and for the stretch receptors, where **the response of the receptor sensory end organ is amplitude-modulated, while the transmission in the axon is frequency-modulated!** 3

Other, more complex codes have been sought, such as changes in variance around a mean and autocorrelation functions [4, pp. 351–352]. 4

An important philosophical point is that frequency modulation of an all-or-nothing signal is *not* a digital function! Some have likened the action potential to the all-or-nothing logic modes in digital computers and concluded that the brain is like a digital computer. However, **if frequency is the code, the interimpulse interval is the transmitter of the message,** and since the interval can have an almost unlimited number of states, **the information actually is in analog form.** Thus, in this case, the brain should be considered a very complex, simultaneously acting "megacircuit" analog computer. However, since it has been shown that any computable function can be accomplished on a digital computer (Turing machine), there is no fundamental limitation preventing a digital computer from imitating an analog machine, and hence the brain. (In this view, we take a basically "rationalist" position, and will not admit of any "vitalism" in the guise that there is *necessarily* something special in "ionic" computers of salt water and proteins, compared with electronic computers of metal and semiconductors.) 5

When one approaches the problem of sensory coding from the viewpoint of psychophysics, one relates the magnitude of the stimulus to subjective sensation. For many sensory systems, there is a logarithmic relationship between absolute stimulus intensity and the perceived increments in sensation magnitude that is called the **Weber-Fechner law.** 6

215

216

Expressed in mathematical terms, this law says that if I is the insensity of the stimulus and S is the amount of sensation, then

$$S = a \log I + b \qquad \text{Eq. 9-1}$$

where a and b are constants [1, p. 322; 52, p. 338; 61, pp. 1653–1656]. Thus, if a and b are each 1, then a 10-fold increase in stimulus intensity leads to only a doubling of the sensation magnitude. In this way, **sensation is seen to cover a large range of stimulus intensities.** In some sensory systems, it can be similarly shown that the firing frequency of the sensory neuron shows a logarithmic relationship to stimulus intensity. It will remain a question to be answered later in this chapter how this logarithmic transformation occurs; but note that in these cases, one of the qualities of perception has been determined largely in the PNS, before the information reaches the CNS.

Recent studies suggest that many sensations do not fit the logarithmic relationship of the Weber-Fechner law, but can be described by Stevens' "power law" in the form

$$S = aI^n \qquad \text{Eq. 9-2}$$

where the exponent n differs in various sensations [61, pp. 1657–1659]. Measured exponents have ranged from 0.33 for brightness discrimination to 3.5 for perception of electric shock, but several sensations have an exponent near unity [58, p. 13; 61, p. 1659]. Thus, one can conclude that in a neuronal sensory system, there can be transformations in the CNS that also affect sensation. Furthermore, not all sensory endings behave according to the logarithmic relation.

The supervening of the Weber-Fechner law by Stevens' power law should be supportive in your mind of the Jewett-Rayner law, page 214.

Some of the observations of psychophysics are found, as this chapter unfolds, to be due to the properties of the sensory endings, while others must be ascribable to CNS properties.

ADAPTATION, SLOW AND FAST

We now return to the question of how information about a stimulus is transmitted by the sensory neurons. One very common feature of sensory coding (the firing pattern in sensory axons) is adaptation. **Adaptation is a reduced response to a sustained constant stimulus.** The extremes of adaptation come in two forms, slow and fast (see Fig. 9-2).

Those nerve fibers that show slow adaptation experience relatively little change in firing rate during a prolonged stimulus (e.g., the muscle spindle, Fig. 9-2). However, fibers that

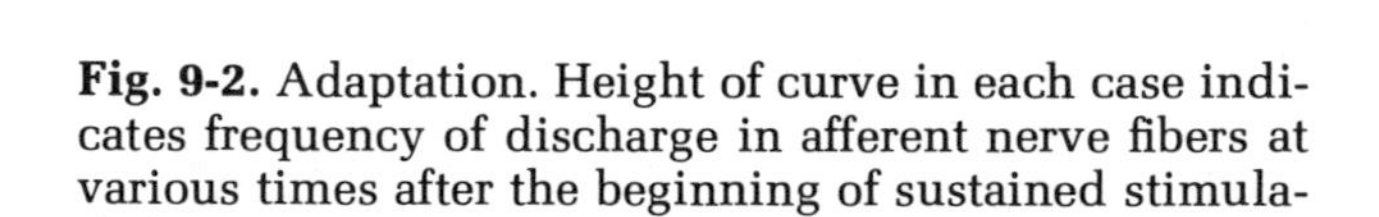

Fig. 9-2. Adaptation. Height of curve in each case indicates frequency of discharge in afferent nerve fibers at various times after the beginning of sustained stimulation.

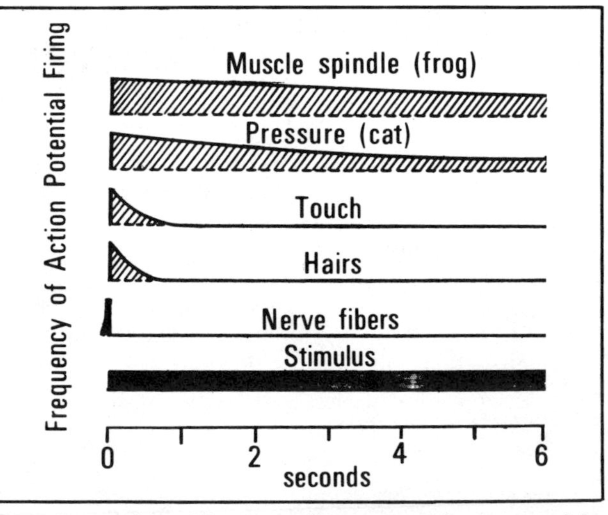

(From E. D. Adrian, *The Basis of Sensation: The Action of the Sense Organs.* New York: Hafner, 1964.)

216

show fast adaptation exhibit marked changes in firing rate during the stimulus (e.g., touch and hairs in Fig. 9-2). You are already familiar with an extreme form of adaptation—the single firing of an axon to a prolonged electrical stimulus (shown as nerve fibers in Fig. 9-2)—which we ascribed to accommodation in Chap. 7. It is easy to confuse the two: **accommodation is a change in threshold to a prolonged stimulus.** Thus, in some cases, adaptation can be due to accommodation (as in the nerve fiber). In this chapter, you learn *other* mechanisms for adaptation. 1

While vertebrate axons accommodate very rapidly to an electrical stimulus, this does not occur in some invertebrate axons, which will produce a train of impulses when they are continuously depolarized [16, p. 128]. 2

A further confusion arises in using these terms relative to the visual system, where accommodation (as defined here) is called "adaptation," while the term *accommodation* refers to changes in the focus of the lens! 3

A little reflection on Fig. 9-2 will reveal that the fibers showing one type of adaptation (so-called static receptors) will be good at transmitting information about the steady intensity of the sensed quality, while fibers showing the other type of adaptation (so-called phasic receptors) will be able to transmit information only when conditions are changing. 4

QUESTION: Which is which? (Hint 2↓) 5

Fast adaptation can be considered as providing information on the first derivative of the function sensed. Thus, a **fast-adapting fiber sensitive to displacement probably is coding velocity** (i.e., the rate of change in displacement). 6

In Chap. 13, you learn about the Ia muscle spindle fibers that primarily code velocity changes by rapid adaptation. Some may even transmit information on acceleration (!). 7

GENERATOR POTENTIALS

What mechanisms are responsible for initiating the patterns of impulses seen in the sensory axon? Let's move "out" along the sensory neuron to its very end to see where these action potentials arise. 8

HINT

1. Static or kinetic energy of motion has been transduced to electrical potential energy (change in charge on the membrane capacitance).

218

The simplest sensory system is one in which the specialized sensory membrane is located at the tip of a sensory axon (physiologically, this is one form of "input" zone, as described in Chap. 2). Depolarization of the sensory membrane as a result of the incident stimulus, the *generator potential*, sets up longitudinal current flow within the sensory axon (refer to Fig. 9-3/7-3 and imagine that node 1 is the depolarized sensory region) that may be sufficient to initiate an action potential at the nearest node of Ranvier (see node 2 in Fig. 9-3/7-3). 1

QUESTION: On what basis can you hypothesize that the mechanism by which the sensory ending activates the axon is that of an electrical potential? (Hint 3↓) 2

We define the **generator potential** as any change in membrane potential in the distal end of a sensory neuron that arises as a direct consequence of the initial sensory transduction process. In some specialized cases, considered later (e.g., the retina), the initial sensory transduction occurs in a separate "receptor cell" that lacks an axon. In such cases, the term **receptor potential** can be introduced to make clear that **sensory action potentials cannot result directly from the initial transduction process.** However, in the examples we consider most extensively, no separate receptor cell is involved, and action potentials may be initiated directly from the transduction process via the intermediary action of the generator potential. 3

QUESTION: Where there is no separate receptor cell, isn't the generator potential also a receptor potential? (Hint 4↓) 4

Before we come to grips with the ionic mechanisms of generator potentials in the next section, we need to offer some further insights into the behavior of a "typical" receptor. 5

PACINIAN CORPUSCLE

Much valuable information has been obtained from a detailed study of the **Pacinian** (pah-chin-ee-an) **corpuscle.** In isolated preparations, **it is possible to record both the generator potential and the axonal action potential excited by this receptor.** 6

Pacinian corpuscles, which are found throughout the body in the mesentery, intermuscular septa, joint capsules, subcutaneous tissues, etc., are made up of lamellae surrounding an unmyelinated axon terminal (Fig. 9-4). A myelinated axon leaves the "onion," the first node occurring within the lamellations. By clever experimental methods, it is possible to record externally, from the isolated preparation, both the generator potentials from the unmyelinated axon terminal and the action potentials from the myelinated axon, simultaneously, while rapid, controlled mechanical stimuli are administered. 7

For a description of the techniques, see Mountcastle [36, p. 1350] and Walsh [60, p. 50]. The methods involve use of fine electrodes and sometimes a sucrose gap. Stimulation is by an energized piezoelectric crystal. 8

Fig. 9-3/7-3. Circuit diagram for myelinated axon when node 1 is undergoing depolarization and node 2 is near resting level; r_o and r_i are resistances of extracellular and intracellular fluids, respectively. The same circuitry holds when "node 1" is instead the membrane creating the generator potential.

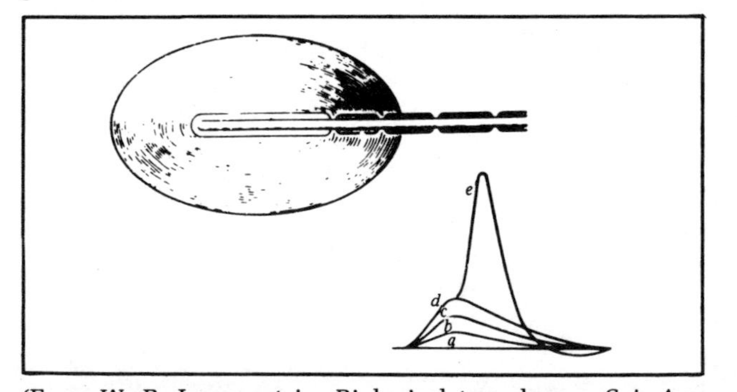

Fig. 9-4. Responses (both generator potential and axonal action potential) in isolated, intact Pacinian corpuscle to increasing magnitudes (*a* to *e*) of brief pressure.

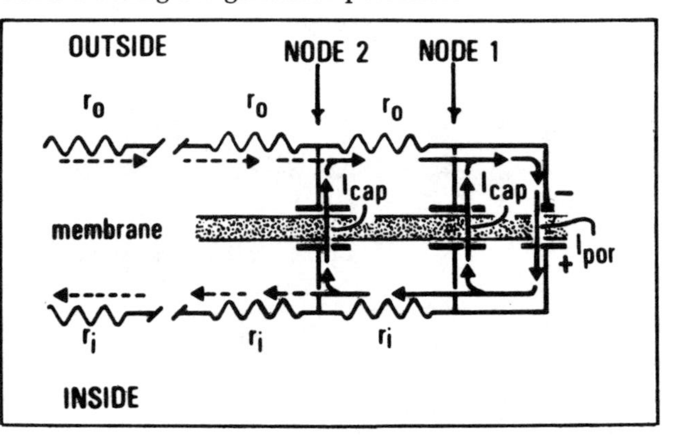

(From W. R. Loewenstein, Biological transducers, *Sci. Am.* 203:98, 1960.)

218

219

As can be seen in Fig. 9-4, **increasing stimulus strength causes an increase in the graded generator potential. If the generator potential from the unmyelinated portion is sufficiently great, depolarization of the first node of Ranvier occurs,** and an action potential also is recorded.

1

Note that since these recordings were made by special external electrodes, the potentials observed are not the transmembrane potential, but are proportional to it. The potential recorded approximates the potential at the first node of Ranvier, which is depolarized by electrotonic spread of the generator potential from the unmyelinated ending. That is, the inward current flow at the ending leads to an outward capacitative current that depolarizes not only the ending itself, but also the first node.

2

The basis for these ideas is shown in the following experiments: the lamellations can be removed without changing the responses of the organ to brief stimuli (Fig. 9-5). Furthermore, when the first node of Ranvier (at the arrow in Fig. 9-6) is blocked with a local anesthetic, the graded generator potentials are unaffected (Fig. 9-6). Finally (not shown), it is possible to stimulate at various distances from the recording point and show that the magnitude of the generator potential diminishes with distance—a longitudinal spread of potential much the same as you studied in the axon (Chap. 7).

3

Question: A stimulus strength barely above threshold was used to give rise to an action potential in Fig. 9-5. After a local anesthetic was applied at the first node, the same stimulus failed to initiate an action potential (Fig. 9-6). Why was a weak stimulus strength chosen? (Hint 5↓)

4

If just the generator potential is taken into account, the amplitude of the potential is related to the stimulus strength, as shown in Fig. 9-7. The graded aspect of the potential is clearly apparent.

5

The potential is quite linear over the range 0 to 5 of stimulus strength (Fig. 9-7). This could be the "normal" operating range of the corpuscle. Reasons for the lesser slope above stimulus strength 5 are given below.

6

Fig. 9-5. This is the same as Fig. 9-4 except nerve ending has been exposed by dissecting away the lamellar covering.

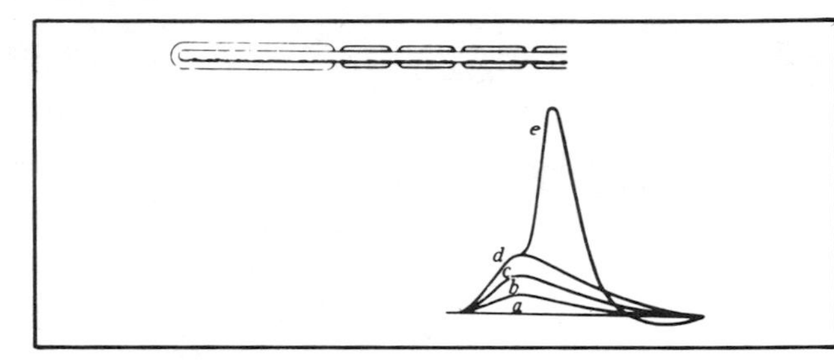

(From W. R. Loewenstein, Biological transducers, *Sci. Am.* 203:98, 1960.

Fig. 9-6. This is the same as Fig. 9-5 except first node of Ranvier has been pharmacologically blocked. Note that generator potentials are unchanged, but action potential from axon cannot be elicited.

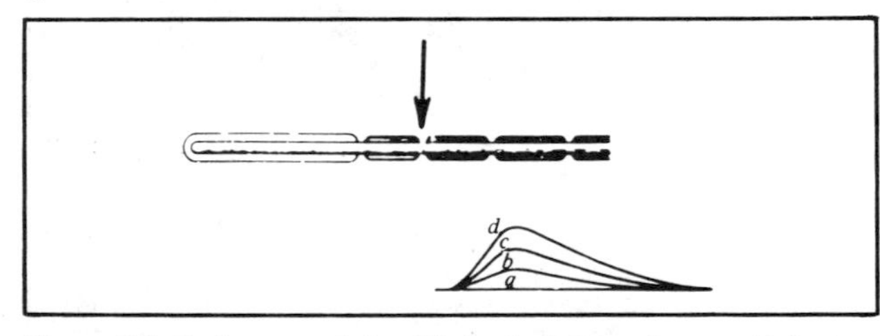

(From W. R. Loewenstein, Biological transducers, *Sci. Am.* 203:98, 1960.

HINT

2. **Static receptors** show slow adaptation (if any); **phasic receptors** show fast adaptation.

219

Note that the rate of rise in the generator potential also shows two slopes relative to stimulus (Fig. 9-7, bottom), but there is less difference between the slopes than in the case of the peak amplitude. 1

The amplitudes must be expressed as a percentage of maximum since only external electrodes are used, with the absolute potentials obtained being influenced by experimental conditions. 2

IONIC MECHANISMS OF GENERATOR POTENTIAL

Not all generator potentials have the same ionic mechanism. All that is required to signal the presence of a given stimulus is that there be an appropriate change in the firing rate of the sensory axon. (Thus it is theoretically possible to imagine a hyperpolarizing generator potential that would *reduce* the discharge rate of a spontaneously active sensory axon.) Fortunately, in the case of the Pacinian corpuscle and the crayfish stretch receptor, **the ionic mechanism seems straightforward: a "nonspecific" conductance increase affects at least Na^+, K^+, and possibly Cl^- ions.** 3

Question: How can an increase in conductance of both Na^+ and K^+ lead to depolarization of the membrane? (Hint 7↓) 4

The depolarization by changes in conductance of Na^+ and K^+ is readily understood if you consider the case in which g_{Na} and g_K both increase by the same absolute amount. Since g_K is much larger than g_{Na} in the resting state, g_K will be changed less (percentagewise), compared with the relatively larger change in g_{Na}. For example, assume that $g_K = 100$ and $g_{Na} = 1$ in the resting state. Then if we increase both conductances by 1, we have doubled g_{Na}, but increased g_K by *only* 1 percent. Thus, a depolarization would be the result of this change, since **the ratio of g_{Na} to g_K changes as an equal increment is added to each of the resting values.** 5

Question: Is the change in V, which is due to the combined conductance changes in Na^+ and K^+, larger or smaller than what would have happened if g_{Na} had changed the same amount, without a change in g_K? (Hint 9↓) 6

Question: Does an increment in g_{Cl}, if E_{Cl} is at the resting level, affect how much the increments in g_{Na} and g_K depolarize the membrane? (Hint 10↓) 7

An equal increment in g_{Na} and g_K, as we have been describing, **is very easy to hypothesize if you imagine that a given mechanical deflection of a part of the membrane acts to open the same number of closed Na^+ and K^+ pores.** Thus, in the resting state, if we assume that each pore has the same conductance, there are 100 open K^+ pores and only 1 open Na^+ 8

Fig. 9-7. Generator potential and rate of rise as related to stimulus strength in Pacinian corpuscle.

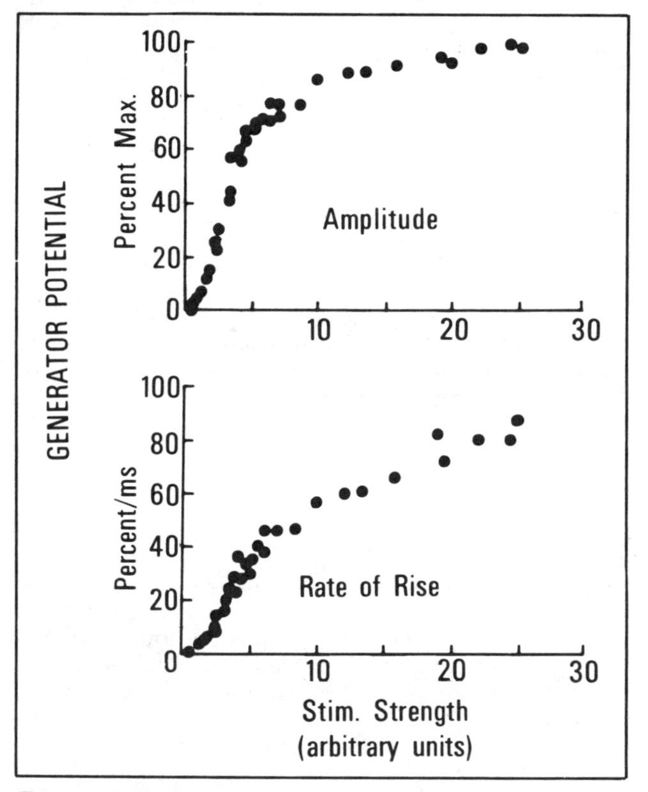

(From J. A. B. Gray and M. Sato, Properties of the receptor potential in Pacinian corpuscles, *J. Physiol.* [Lond.] 122:610, 1953.)

pore. Then, as the membrane is deformed, this leads to an opening of 25 Na^+ and 25 K^+ pores. The net result is a change in the g_K/g_{Na} ratio, which depolarizes the membrane (Fig. 9-8). If the stimulus were great enough to bend an adjacent part of the membrane, then an additional 25 pores of each might be opened, changing the conductance of that localized region still further, etc. You will see in Chap. 10 that **a very similar explanation applies to transmission at the myoneural junction, where the conductance change is due to a chemical, rather than to mechanical, deformation.** 1

Since the increment of conductance appears to be rather similar for all three ions, it should not be surprising that the generator potential seems to be most sensitive to change in $[Na^+]_o$. 2

QUESTION: Why might you expect this? (Hint 12↓) 3

If the voltage-clamp technique could be applied successfully to these small endings, it would be possible to determine the ionic contributions more exactly. One interesting idea that has been developed concerns Pacinian corpuscles. Mountcastle [36, p. 1351] postulates that the unmyelinated ending has a large number of pores (or small regions of pores) that are affected by stretch in an all-or-nothing fashion. Thus the graded response would be made up of large numbers of such responses in the same manner that the "graded" compound action potential comprises all-or-nothing firings from individual axons. This sort of mechanism also would explain why the amplitude of the generator potential finally flattens out at higher deformations (Fig. 9-7) as the number of individual units activated approaches the maximum available. This view is completely consistent with what we presented concerning pores in this chapter and in Chap. 6. 4

The conductance changes brought about by mechanical deformation of the membrane are markedly dependent on temperature, whereas the electrically excitable nodes of Ranvier do not show as great a dependence [36, p. 1352]. Should this be taken as evidence that the Pacinian corpuscle senses temperature 5

Fig. 9-8. Ratio of open to closed pores in resting state. As membrane of sensory ending is stimulated, the number of additional Na^+ and K^+ pores opening is diagrammed by movement of heavy black line to the right. As this happens, ratio of Na^+ to K^+ pores varies.

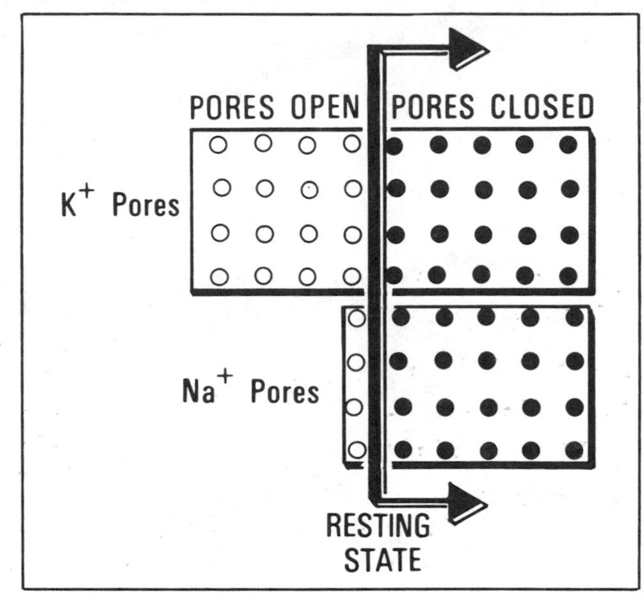

HINTS

3. The axon is **electrically excitable;** i.e., its permeability changes are initiated by depolarization (Chap. 6).

4. No terminology is perfect! The question just posed has caused considerable confusion, and often the two terms are used quite indiscriminately, even in the research literature. We prefer to simplify matters by relatively rigid adherence to the definition just provided. Thus, our answer must be: "No, in the absence of a separate receptor cell, there is no receptor potential."

5. If the stimulus were stronger, the generator potential might "jump over" the blocked region and make the interpretation of the results more difficult! Can you say this in more technical language? (See Hint 6.↓)

as well as pressure? The answer has been nicely settled by showing that temperature changes do not give any generator potentials, so the ending would seem to be affected by temperature—but only if pressure, the primary stimulus, also is applied.
1

INITIATION OF ACTION POTENTIAL

At the first node of Ranvier of the Pacinian corpuscle, **the action potential is initiated.** This is readily explained since the depolarizing generator potential drives axial currents along the axon.
2

QUESTION: Since mechanical deformation of the unmyelinated ending leads to a net inward current at that point (by increasing g_{Na}, g_K, and g_{Cl}), what is the result in terms of current at the first node of Ranvier? (Hint 13↓)
3

TEMPORAL AND SPATIAL SUMMATIONS

Clearly, the adequate stimulus for the Pacinian corpuscle seems to be mechanical deformation of the unmyelinated ending. **The graded electrical response of the unmyelinated ending outlasts the stimulus, allowing temporal summation** (Fig. 9-9) to occur under appropriate conditions.
4

Fig. 9-9. Temporal summation of mechanical stimuli in generator potentials of Pacinian corpuscle. Mechanical stimuli are at short horizontal marks.

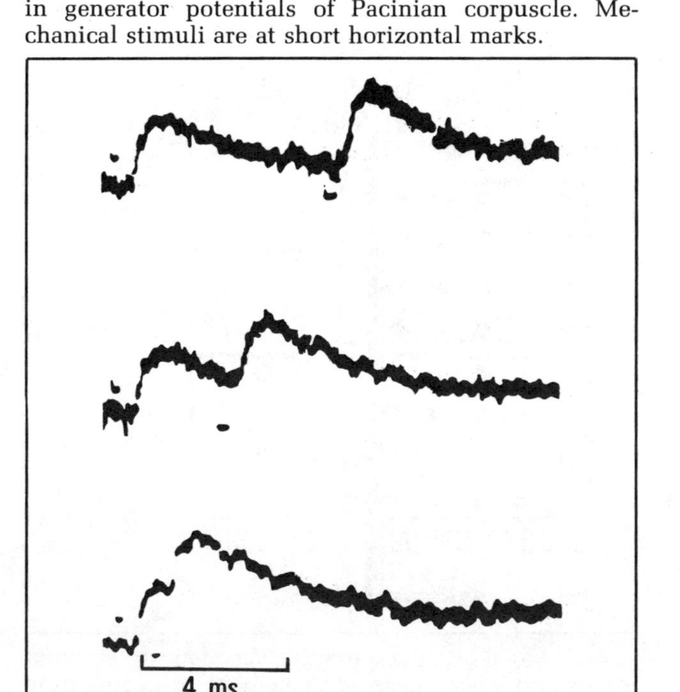

(From J. A. B. Gray and M. Sato, Properties of the receptor potential in Pacinian corpuscles, *J. Physiol.* [*Lond.*] 122:610, 1953.)

The two responses shown in Fig. 9-9 clearly sum. So you can see that the response elicited by a long series of brief, light deformations might easily be different from that of a single, brief, large deformation.
5

6 Question: Why is no action potential shown in Fig. 9-9? (Hint 15↓)

Experiments in which the exposed, unmyelinated portion of the ending was stimulated systematically at various places show that each point seems to develop an independent depolarization, which spreads axially to the first node of Ranvier. This has two important consequences. First, **spatial summation from separate parts of the ending** (at the first node) undoubtedly **occurs.** Second, this fact suggests that the **unmyelinated portion is *not* electrically excitable.**
7

With regard to spatial summation, presumably this occurs frequently in the ending as the outer lamellae are distorted by a mechanical movement and various parts of the unmyelinated ending are deformed.
8

The complex manner in which the lamellae interact mechanically has been studied by clever microscopic methods. The relative movements of the inner and outer lamellae differ depending on the type of displacement (slow or rapid, short or prolonged) [14, p. 55; 16, p. 138]. Presumably, the mechanical properties of the interacting lamellae cause the "off" response to a prolonged stimulus, as shown in part *A* of Fig. 9-10, by deformation at two separate locations. The response with the lamellae removed is clearly different and shows no off response (part *B* of Fig. 9-10).
9

The function of the lamellae can be likened to a mechanical high-pass filter. That is, only the higher frequencies are transmitted to the innermost lamella, and hence to the unmyelinated ending. This also can be shown electrophysiologically, by studying the response (action potential frequency) as a function of stimulus frequency (mechanical deformation). The corpuscle can transmit information (over the axon) about vibrations between 40 and 1000 Hz (cycles per second), but the maximum sensitivity (least required mechanical displacement) occurs at about 300 Hz [46, p. 109].

All membranes are sensitive to mechanical deformation, but the unmyelinated ending is much more sensitive than, for example, the node of Ranvier in the same preparation. The ending can be activated by as little as 0.2 μm of movement [46, p. 98]. (It's important for biologists to realize that study of biological systems brings one right down to the dimensions studied in physics, for 0.2 μm = 2000 Å.)

The idea of an electrically inexcitable membrane is one of the most important in this chapter.

It would certainly seem from Fig. 9-6 that **the action potential occurs at the node of Ranvier** and hence **not at the unmyelinated ending.** Spread of depolarization along the unmyelinated ending appears at least superficially similar to true electrotonic spread. This lends further credence to the idea that this portion of the membrane is entirely **electrically inexcitable;** that is, *the membrane permeabilities are changed not by depolarization of the membrane, but by mechanical deformation.* This would be very useful, since it would increase the range over which summations could occur at the first node of Ranvier. (It would also increase the range of frequencies over which the axon could be activated, as described later in this chapter.)

Fig. 9-10. Generator potentials recorded from Pacinian corpuscle (*A*) before and (*B*) after removal of lamellae. Note both that with lamellae present (*A*) the generator potential is not sustained, even though movement is, and that the bare membrane is capable of sustained depolarization (*B*).

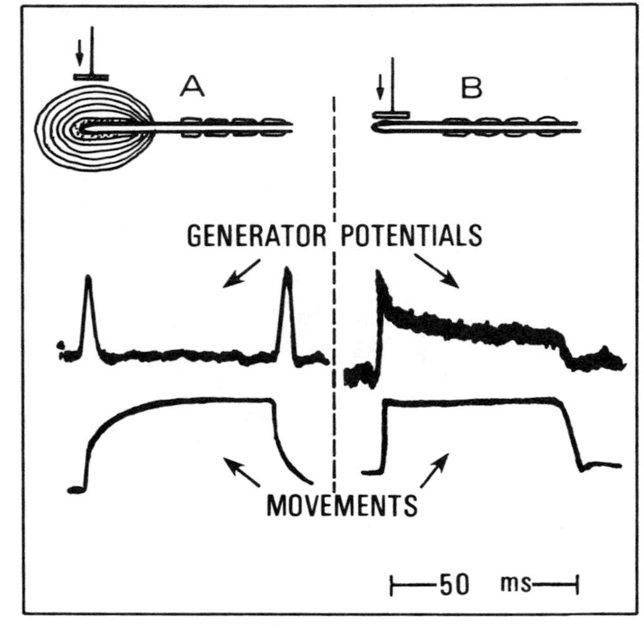

(Modified from W. R. Loewenstein and M. Mendelson, Components of adaptation in a Pacinian corpuscle, *J. Physiol.* [*Lond.*] 177:377, 1965.)

HINTS

6. If the generator potential is strong enough, the electrotonic depolarization at the second node may be sufficient to cause an action potential, even if the first node is rendered inactive by a local anesthetic.
7. The *ratio* of g_{Na} to g_K must change as both increase, e.g., by the same amount. This one is easy for those who remember the consequences of the steady-state equation in Chap. 4. Otherwise, go to Hint 8.↓
9. Smaller. Those with rusty memories, on to Hint 11.↓
10. It does! To get this, you have to go back to the steady-state equation and plug in the values. In words, the larger the Cl^- conductance, even when $V = E_{Cl}$, the larger the "shunting" effect, whereby a given depolarizing current is carried by movement of Cl^- through the membrane pores rather than by a current through the membrane capacitance. *Note:* This question is at the third level and is a "Question," not a "QUESTION." Cl^- plays a role in reducing effects at several synaptic locations (see Chaps. 10 and 11).
12. In resting membranes, the driving force on the sodium ion normally is considerably higher than on K^+ or Cl^-, while g_{Na} is much less than either g_K or g_{Cl}. An equal *increment* of conductance thus will be a proportionately greater change in I_{Na} than for I_K or I_{Cl}. So the size of the generator potential will be markedly dependent on the inward movement of sodium ions, and hence on the magnitude of the Na^+ gradient.
14. Depolarizes. If you had to look, you must be pretty rusty and need to go back to Chaps. 7 and 4 (and do not collect $200!).

Actually, in some, but not all, experiments an antidromic impulse has been able to cause an action potential in the ending [36, p. 1352]. Thus, according to one theory, the ending *is* partially electrically excitable, but with a very high threshold!

1

If this were the case, the generator potential would spread to the first node by *decremental conduction,* in that locally generated coupling currents would increase the apparent length constant of the unmyelinated membrane. (See Chap. 8 for further discussion of this concept.)

2

This whole argument is not trivial, since some of the mechanisms found here may apply to the functioning of dendrites (where even less is known).

3

We speculate that there may be some physiological difference between two membranes, one of which could maintain a range of graded responses without an all-or-nothing action potential and the other of which is repeatedly "wiped clean" by antidromic activation of the ending from the firing of the first node of Ranvier. In the second case, there might be a greater overall flux of ions—and there may be effects on the subsequent rate of depolarization of the node after each action potential (as described later in this chapter). Furthermore, an all-or-nothing response must have a refractory period (see Chap. 6), which could affect the reactivation of both the ending and the node—so we'll leave it to one of you to figure it out. Please send us a reprint when you prove your theory!

4

WHY PACINIAN CORPUSCLE ADAPTS RAPIDLY

Prolonged deformation of the Pacinian corpuscle will give only one or two action potentials, so the response is **rapidly adapting.** Why? There are only three plausible reasons: mechanical properties of the lamellae, changes in the generator potential, or accommodation at the first node.

5

1. Mechanical properties of the lamellae. The lamellae move relative to one another in such a way that a steady displacement applied to the surface of the corpuscle is transmitted to the center for only a brief time. This can be shown by recording the generator potential to a prolonged deformation (part *A* in Fig. 9-10). Note that the generator potential decays to the prestimulus level in about 10 ms. (There is also a response at the cessation of the deformation, presumably resulting from mechanical idiosyncrasies in the corpuscle—you can ignore it for now.) Obviously, such a decay means that prolonged deformation would have little chance of stimulating the node over a prolonged period.

6

2. Decreasing generator potential. Even if the mechanical properties of the lamellae are removed as a factor, the unmyelinated ending itself shows a decrease in generator potential (for unknown reasons) during a prolonged (that is, 50 ms!) deformation (part *B* of Fig.

7

9-10). Possibly this mechanism contributes to the rapid decline in the potential when the lamellae are present (part *A* of Fig. 9-10).

3. Accommodation at the first node. By applying steady electrical currents, it is possible to show that the first node of Ranvier (which *is* electrically excitable) accommodates rapidly (i.e., it changes its threshold for firing). Therefore, even **if the generator potential were constant, the organ would show rapid adaptation in its response,** although **the combined effects of all three mechanisms undoubtedly make the rate of adaptation greater** (faster adaptation) than if only one of the mechanisms were operating.

SLOW ADAPTATION IN OTHER RECEPTORS

How can we continue without pointing out that there is a marked difference between the receptors that adapt rapidly and those that adapt slowly? Obviously, **the generator potentials of a slowly adapting sense organ must remain relatively constant in response to a prolonged stimulus.**

The response of the photoreceptor element in the lateral eye of the horseshoe crab, *Limulus* (when exposed to 20-s illuminations of increasing intensities), is shown in Fig. 9-11. The *intracellular* recording technique detects the transmembrane responses, showing both the generator potential and action potentials at the same time. **The generator potential is easily seen as the prolonged displacement of the membrane potential from the baseline.** The action-potential spikes (just small, vertical lines at such a slow sweep speed) are superimposed on the generator potential. Note that prolonged stimulation gives very little diminution in either the generator potential or the firing rate. The latter indicates that this receptor is of the slowly adapting type.

Fig. 9-11. Records 1, 2, and 3 taken at increasing light intensities.

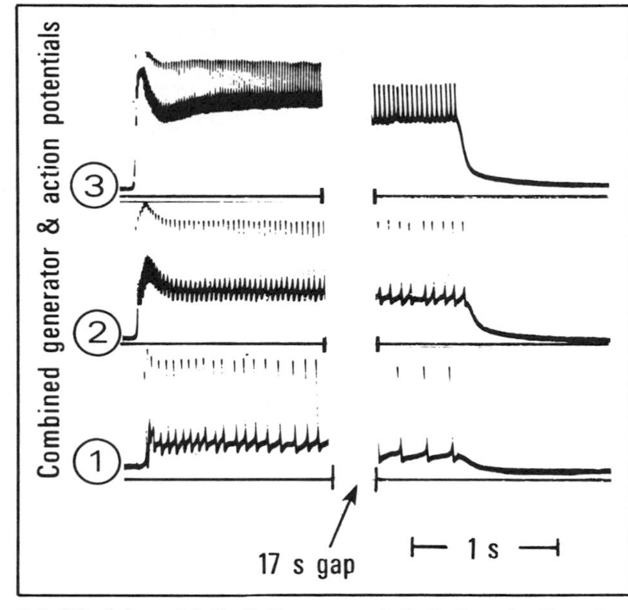

(Modified from M. G. F. Fuortes and G. F. Poggio, Transient responses to sudden illumination in cells of the eye of *Limulus*, *J. Gen. Physiol.* 46:435, 1963.)

HINTS

8. Recall from Chap. 4 (or go back if this is unclear to you) that if you take the conductance equation in the resting state,

$$V = \frac{+40g_{Na} - 100g_K}{g_{Na} + g_K}$$

and then increase the g's in the *same proportion*, there is *no change* in *V*. So, **to depolarize** the membrane (a change in *V* toward zero), **the changes in g_{Na} and g_K must not be proportional.** How they both change is now revealed—if you go back to the main text!

11. The action of an increase in g_K is to oppose the membrane change induced by g_{Na}, so that if g_K were removed from the picture, the membrane potential would change more. If this is unclear, go back to Chap. 4 and play around with changes in the steady-state equation some more.

13. There must be a net outward current across the capacitance of the first node. This, in turn, either depolarizes or hyperpolarizes the nodal membranes. Which? (Hint 14↑)

15. Three possibilities come to mind: First, the generator potential, although large in the terminal region, is still below threshold for the first node of Ranvier. Second, in this experiment, the first node(s) may be rendered inactive by local anesthetic, in order to reveal the summation more easily. Third, for didactic purposes, they've "made it simple." In this case, the first and third possibilities are *not* correct.

1 QUESTION: Does the first node in this preparation accommodate much? (Hint 16↓)

Note that the generator potential shown in Fig. 9-11 is graded, increasing with increasing intensity. But did you also notice that the **spike height** (the distance from the bottoms to the tops of the action potentials) **diminishes with increased firing** rate? This is **probably due to changing ionic concentrations at the higher rates.** "*Wait a minute!*" we hear you say. "How come you spent all that time in Chap. 6 getting us to learn that there aren't concentration changes with the action potential, and now you say there are! How are you going to get out of this?" It's easy, but why not try to figure it out yourself before turning to Hint 17?↓ 2

Another reason for the decrease in spike height during rapid trains of impulses is that the action potentials may fall in the relative refractory period of the preceding action potential. Residual sodium inactivation may not be completely removed between each action potential, leading to a slow reduction in maximum g_{Na}, and hence in maximum I_{Na} with a consequent reduction in the overshoot. 3

Another example of a slowly adapting receptor is the crayfish stretch receptor, a preparation that has been extensively studied because of its large size and ready accessibility to microelectrode recording. Intracellular recordings can be obtained by microelectrodes, which show the relationship of the generator potential to the action potentials. 4

Figure 9-12 shows the responses to two different amounts of stretch. If the pull is weak, then only a generator potential is recorded (which is below threshold for the action potential). **The generator potential is maintained throughout the period of stretch.** If the pull is harder, then the generator potential is above threshold for action potentials and a train of impulses of quite uniform frequency is generated (lower recording, Fig. 9-12). Note that, just as in the Pacinian corpuscle, **the membrane giving rise to the generator potential is not the same membrane that gives rise to the action potential.** 5

The changes in the firing rate and the generator potential when the stretch is increased and decreased in steps are shown in Fig. 9-13. The records are segments from a continuous recording in which the degree of stretch was progressively increased and then progressively decreased. The increased firing rate as the generator potential increases is readily seen. At the end of the top line, the firing rate is so high that the action potentials blur together at the slow sweep speed. 6

QUESTION: Figure 9-13 shows the action-potential height decreasing with increasing firing rate! What can explain this? (Hint 18↓) 7

The sort of data obtainable from the crayfish stretch receptor can be shown in graphic form (Fig. 9-14). Note that the generator potential is linearly related to the length of the "stretched" receptor (upper graph). The firing rate (impulses per second) is also linearly proportional to the length (lower graph). 8

227

Fig. 9-12. Intracellular recording from cell depolarized by crustacean stretch receptor, at two different amounts of stretch. The first shows maintained generator potential at a stretch below threshold for this cell; the second recording is above threshold and shows both generator-potential and action-potential spikes.

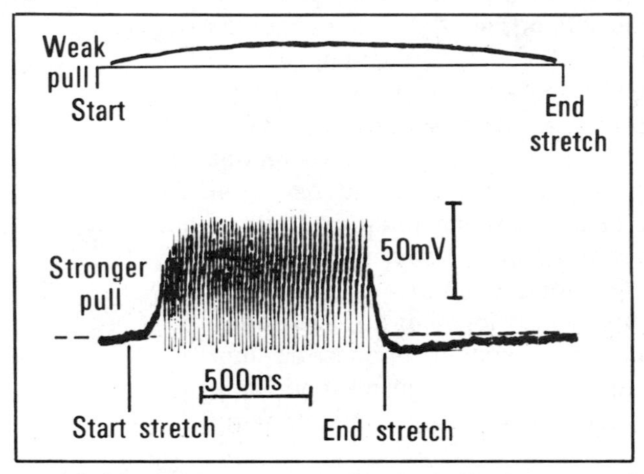

(Modified from C. Eyzaguirre and S. W. Kuffler. Processes of excitation in the dendrites and in the soma of single isolated sensory nerve cells of the lobster and crayfish, *J. Gen. Physiol.* 39:87, 1955.)

Fig. 9-13. Segments from continuous recording from same type of cell as in Fig. 9-12.

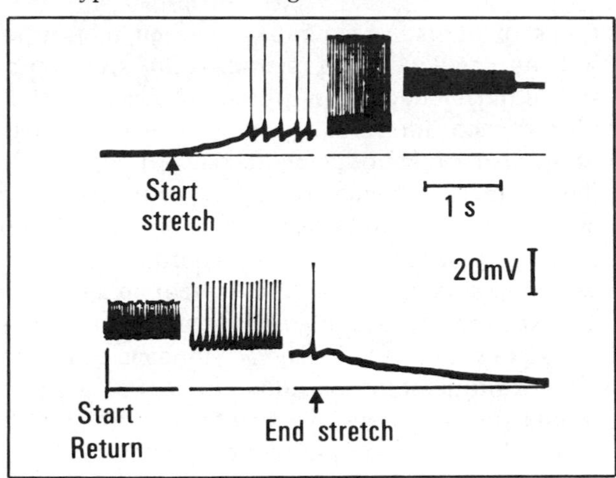

(From C. Eyzaguirre and S. W. Kuffler, Processes of excitation in the dendrites and in the soma of single isolated sensory nerve cells of the lobster and crayfish, *J. Gen. Physiol.* 39:87, 1955.)

Fig. 9-14. Relation between muscle length and size of generator potential (above) and impulse frequency (below) in crayfish stretch receptor. Squares and circles indicate values in two different preparations.

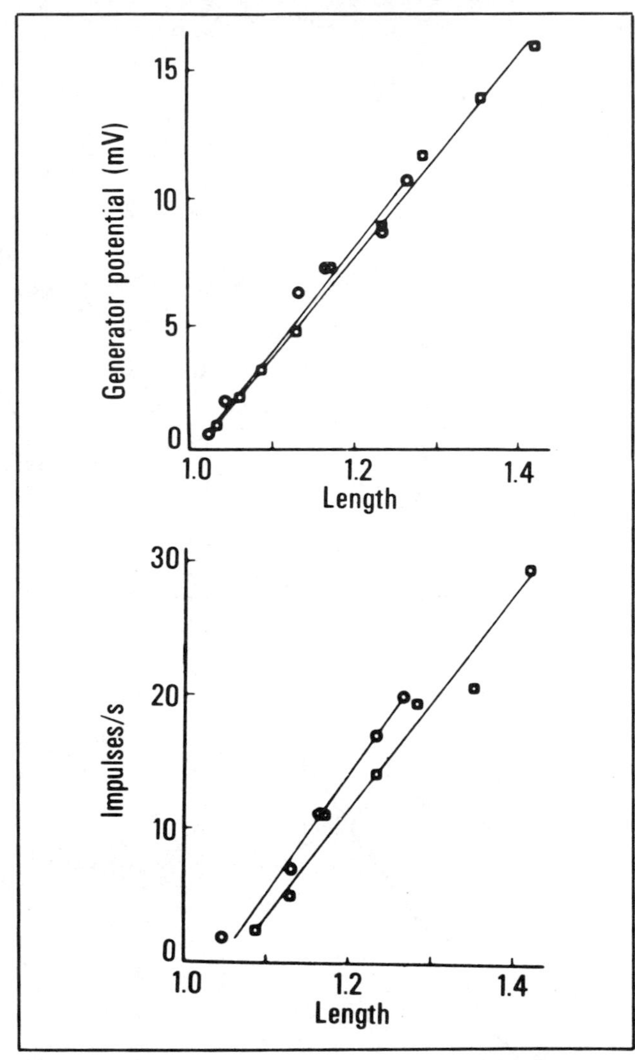

(From C. A. Terzuolo, and Y. Washizu, Relation between stimulus strength, generator potential, and impulse frequency in stretch receptor of crustacea, *J. Neurophysiol.* 25:56, 1962.)

227

228

1 QUESTION: What simple-minded conclusion can you draw from these two graphs regarding the relationship of the generator potential to the firing rate? (Hint 19↓)

2 The same simple-minded conclusion can be drawn with respect to the data from the frog muscle spindle (Fig. 9-15).

HOW THE MAGNITUDE OF THE GENERATOR POTENTIAL AFFECTS THE FIRING RATE

3 Despite all that has been said, we still have not described how **an increased depolarization of the membrane of the first node gives an increased firing rate,** as in Fig. 9-15.

4 Question: Will the relative refractory period explain it? (Hint 20↓)

A good possibility seems to be that as the generator potential becomes larger, the rate of depolarization by the local response of the membrane increases. You have seen the marked changes in latency that occur with varying strengths of stimuli (Chap. 6). Presumably, the same mechanisms used in Chap. 6 to explain the differences in latency of the action potentials when stimulus strength is varied also occur when the generator potential is the "stimulus." For example, in the crayfish stretch receptor, shown in Fig. 9-16, note that as the firing rate increases because of an increased stimulus, the membrane potential approaches the "takeoff point" (threshold) more and more rapidly. 5

Finally, even where the receptor region is essentially inexcitable electrically, the action potential of the first node must spread electrotonically (or decrementally) back into the receptor region. During the rising phase of the spike, this effect would further depolarize the receptor. But in the falling phase (and for some time into the positive after-potential), the generator potential must be reduced. While such effects (like relative refractoriness) have a rather short time course, the complex interactions among these multiple mechanisms seem sufficient to account for observed receptor behavior in fast-adapting neurons, such as the Pacinian corpuscle. **Slow-adapting neurons** may show very slow firing rates, which can hardly be regulated by any of the mechanisms previously discussed. However, typically such neurons show regular, repetitive firing rates even in the absence of any obvious stimulus (i.e., at "resting length" in the case of stretch receptors). It seems probable that this "resting discharge" results from spontaneous activity at the first node, generated by mechanisms more similar to those that produce repetitive firing in the pacemakers of heart and smooth muscle. Clearly, these slow, cyclical excitability changes then can be modified by depolarizing currents (*on* response) or hyperpolarizing currents (*off* response) from the generator potential in the receptor region. Such a system may be extremely sensitive to small changes in generator potential. 6

Fig. 9-15. Relation between size of receptor/generator potential and frequency of impulse discharge in frog muscle spindle.

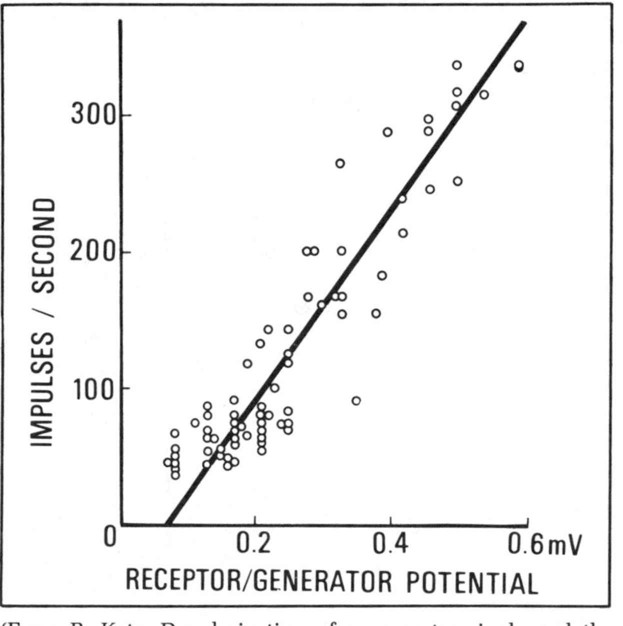

(From B. Katz. Depolarization of sensory terminals and the initiation of impulses in the muscle spindle, *J. Physiol.* [*Lond.*] 111:261, 1950.)

228

229

A LAST LOOK AT THE PACINIAN CORPUSCLE

1 Let us sum up the main points of the sensory physiology of the Pacinian corpuscle (and, as well, the invertebrate and amphibian stretch receptor).

2 Mechanical deformation of the terminal (input) portion of the sensory neuron leads to changes in the conductance of that electrically inexcitable membrane by increments in g_{Na}, g_K, and possibly g_{Cl}. This leads to an inward current flow (a generator potential) in this region, which then spreads to the electrically excitable part of the axon (the first node of Ranvier in myelinated fibers). At this point, a net outward capacitative current brings the node to threshold and results in a train of impulses whose rate is proportional to the magnitude of the generator potential. The generator potential is variable, depending on the magnitude of the "stimulus," i.e., the deformation of the membrane, possibly because the mechanically sensitive parts of the membrane act independently, thus providing a form of spatial summation.

3 The net result is that when the sensory ending undergoes mechanical deformation, a train of impulses is generated. The information transmitted is modified by both mechanical and physiological factors, so that rapid adaptation occurs in the case of the Pacinian corpuscle (while there is little adaptation of stretch receptors).

4 This adaptation makes the information sent to the CNS depend markedly on the rate of change of the membrane deformation. Thus the Pacinian corpuscle is well suited to sense vibration above 50 Hz.

RECRUITMENT

5 Up to now, we considered only single receptors and their axons. **If the CNS responds,** as is likely, **to the number of impulses it receives** in a given sensory modality, then another method (in addition to variation of firing rates of individual neurons) can be used to transmit information about stimulus magnitude: **fibers with higher thresholds can be recruited, so that the number of impulses reaching the CNS is increased as the stimulus intensity increases.**

6 It is a normal and characteristic finding that **a population of sensory fibers will have widely differing thresholds to their own specific stimulus modality.** The most sensitive sensory endings have the lowest thresholds and transmit information about faint stimuli.

Fig. 9-16. Discharge (firing) threshold at different levels of stretch. A slowly adapting stretch receptor cell is impaled with intracellular microelectrode. The resting potential is about 70 mV. Top recording, near threshold stretch: discharge frequency about 5 impulses per second. Middle recording, greater steady stretch: regular discharge at about 16 impulses per second. Bottom recording, stretch further increased: discharge at 40 impulses per second. Horizontal line shows approximate firing level. This level occurs at point of inflection between slowly developing depolarization and abrupt takeoff of the spike potential.

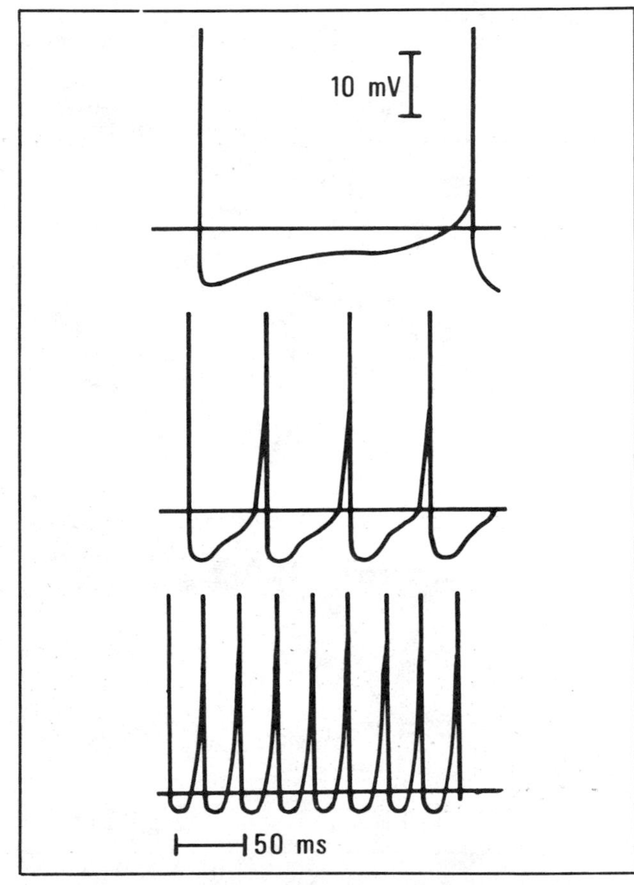

(Modified from C. Eyzaguirre, and S. W. Kuffler, Processes of excitation in the dendrites and in the soma of single isolated sensory nerve cells of the lobster and crayfish, *J. Gen. Physiol.* 39:87, 1955.)

HINTS

16. If it did, you wouldn't see the continued firing during the prolonged stimulus! Note that in Fig. 9-11 you are actually seeing the potential (the generator potential) that would cause accommodation if it did occur (which it doesn't).
17. The explanations of Chap. 6 applied to a *single* action potential, but not always to trains of them! Sneaky, eh?
18. This was just explained with regard to Fig. 9-11. Turn back to Hint 17.

229

230

Higher-threshold endings transmit information about stronger stimuli. In this way, **a population of neurons can transmit information over a greater range of stimulus intensities than a single neuron can.**

1

Interestingly, in some sensory organs the cells tend to fall into just two types: low-threshold and high-threshold. This is true in the eye, the ear, and the skin, even without considering pain as a high-threshold "touch"! By means of lateral inhibition (covered in Chap. 11) such bimodal distributions can be arranged to have a very large dynamic range.

2

ANOTHER MECHANORECEPTOR: THE COCHLEA

As shown in Table 9-1, **mechanoreceptors subserve many different senses, with a variety of specialized anatomic structures.**

3

The most highly specialized mechanoreceptor is the organ of Corti in the cochlea, which, together with the other auditory structures of the ear, detects rapid variations in air pressure (i.e., sound waves).

4

(A complete description of the mechanisms of hearing is beyond the scope of this book. We present only those aspects of cochlear physiology that relate to the main ideas of this chapter. For a broader view, consult a standard textbook of physiology.)

5

Sound waves are detected by mechanical movement of hair cells in the cochlea (inner ear) after the ear drum, moved by the sound waves, transmits the movement through a sequence of bones in the middle ear to the fluid in one compartment in the cochlea, the scala tympani. The fluid, in turn, moves the flexible basilar membrane. **As the basilar membrane is displaced, the processes of the hair cells (the sensory end organs) are bent by the motion of the basilar membrane relative to the tectorial membrane,** which has a bony attachment on one side (Fig. 9-17). **This movement causes, by unknown means** (just as unknown as in the case of the Pacinian corpuscle), **receptor potentials in the hair cells, which, in turn, depolarize the axons of the next neurons, the bipolar cells of the spiral ganglion.** [The spiral ganglion bipolar cells are similar to the dorsal root cells of the somatosensory system in that they have processes that extend both centrally and peripherally (Chap. 2).] The central processes of the spiral ganglion cells, on passing to the CNS, make up the auditory nerve (VIII cranial nerve).

6

Before going into further detail, we should point out some unusual features of this anatomy. There are **two kinds of fluid in the cochlea, the perilymph and the endolymph.** As Fig. 9-18 shows, the perilymph occupies all spaces except the scala media, which contains endolymph and hence is also called the *endolymphatic space*. Now, **the ionic compositions of endolymph and perilymph are markedly different** (Table 9-2). **The high concentration of K^+ in the endolymph** should be noted, being in this way similar to intracellular fluid, but having, in addition, **a high Cl^- content** (Table 9-2). Now, it is of interest that **the electrical potential of the endolymphatic space is +80 mV** (Fig. 9-18).

7

Fig. 9-17. Movement of tunnel of Corti and tectorial membrane, based on descriptions by von Békésy. Shearing action between two stiff structures (tectorial membrane and reticular lamina) bends processes of hair cells.

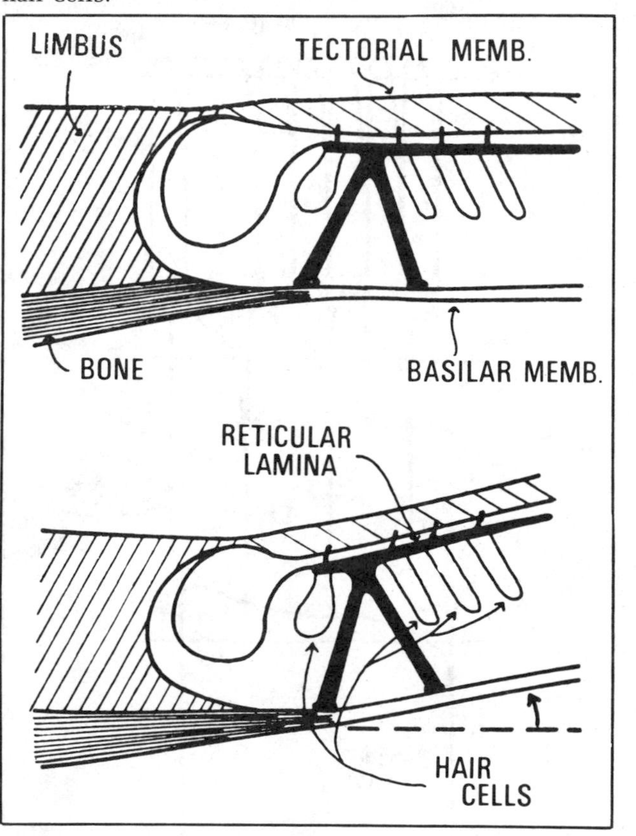

(From H. Davis. Initiation of Nerve Impulses in Cochlea and Other Mechano-Receptors, in T. H. Bullock (Ed.), *Physiological Triggers and Discontinuous Rate Processes*. Washington, D.C.: American Physiological Society, 1957.)

230

231

1 QUESTION: Knowing two facts—the +80-mV potential and the high K^+ concentration in the scala media (endolymphatic space)—can you conclude that the +80-mV potential is due to the K^+ concentration gradient? (Hint 22↓)

2 The cells of the stria vascularis have been implicated as being the source of both K^+ and (by an active pump) the positive potential. Note that in the absence of such information, you should not assume that K^+ is significant in this situation.

3 Note, further, that the tunnel of Corti, shown in Fig. 9-18 between the inner and outer hair cells, contains perilymph.

4 QUESTION: What would be the consequence if the tunnel of Corti contained endolymph? (Hint 21↓)

5 **The unusual position of the hair cells** can be appreciated now: They are **surrounded by perilymph, except at the portion under the tectorial membrane** (Fig. 9-17) **where the hair cells are moved.** At this point, they are exposed to a high K^+ concentration with a potential of about 155 mV across their apical membranes.

6 Question: How did we calculate the 155-mV transmembrane potential? (Hint 25↓)

Fig. 9-18. Distribution of positive endocochlear potential, which is positive (+) relative to the inferior perilymph (0). The minus signs inside the cells indicate that they are negative relative to the perilymph. The tunnel of Corti is the unlabeled dotted-line area in the center of the figure, between the inner and outer hair cells.

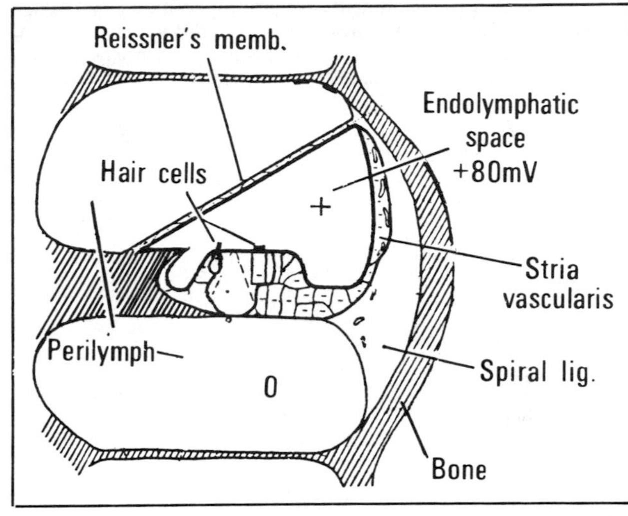

(Modified from I. Tasaki, H. Davis, and D. H. Eldredge, Exploration of cochlear potentials in guinea pig with a microelectrode, *J. Acoust. Soc. Am.* 26:765, 1954.)

Table 9-2. Composition of Endolymph, Perilymph, and Spinal Fluid

	Endo-lymph	Peri-lymph	Spinal Fluid
Potassium (mEq/L)	144.4	4.8	4.2
Sodium (mEq/L)	15.8	150.3	152
Chloride (mEq/L)	107.1	121.5	122.4
Protein (mg/100 mL)	15	50	21

Data from C. A. Smith, O. H. Lowry, and M. L. Wu, The electrolytes of the labyrinthine fluids. *Laryngoscope* 64:141–153, 1954.

HINTS

19. The firing rate must be linearly related to the amount of depolarization that is due to the generator potential. This is a usual finding, even when the stimulus required to produce the generator potentials is nonlinear (e.g., logarithmic).

20. If you thought so, you were in good company! Adrian, one of the early, important workers on sensory endings, long ago suggested this as a possible mechanism. The axon would be activated by a suprathreshold generator potential, which would fire the first node. When the excitability (decreased in the relative refractory period) returned sufficiently, the generator potential (assumed to be still continuing) would be strong enough to fire the node again. The stronger the generator potential, the sooner after the first action potential the node would fire again; hence, the larger the generator potential, the faster the firing rate. Unfortunately, the times over which the relative refractory period can be detected are much *too short* to account for the usual frequencies of firings observed. This mechanism may work at the very highest firing rates (e.g., greater than 300 impulses per second), but it is hard to imagine it working satisfactorily at 10 impulses per second, at which rate many sensory nerve fibers show a linear relationship of generator potential to firing rate (Fig. 9-14).

231

The 155 mV is due to a −75-mV potential of the hair cells relative to the perilymph (Fig. 9-18) combined with the +80-mV potential of the endolymph relative to the perilymph. The net result of the two potentials is the very unusual 155 mV across the membrane, obviously much different from the usual −75 to −90 mV for most cell membranes.

The consequences of this unusual situation have been studied in a squid giant axon that was placed under similar conditions of external K^+ and an additional external positive potential. **Under these conditions, the membrane becomes exquisitely sensitive to mechanical vibration, firing action potentials in response to a finger tap on the experiment table!** Thus, we can hypothesize that these conditions play a role in the unusual mechanical sensitivity of the hair cells, which can be stimulated by movements of the basilar membrane (at the threshold of hearing) and which are about 10^{-7} μm, that is, about one-tenth the diameter of a hydrogen atom [60, p. 261].

The range of basilar membrane movements that the hair cells can detect is increased by having two groups of cells with different sensitivities, the inner hair cells and the outer hair cells. The inner hair cells, which are fewer, **have a higher threshold than the outer hair cells.** Thus, they tend to be stimulated by larger movements (louder sounds). The outer hair cells are more sensitive, but cannot respond to the larger movements.

This mechanism (of using two groups of receptors, one more sensitive than the other) is quite common in sensory systems, as previously mentioned. The mechanism of interaction between these cell types is discussed in Chap. 11.

The generator potentials connected with hearing can be recorded with electrodes in or near the cochlea: the cochlear microphonic potential (so named because it so nicely follows the sound waves that it can act as a microphone, if amplified and put into a loudspeaker).

Figure 9-19 shows the remarkable similarity between the sound stimulus and the cochlear microphonic (generator) potential.

The transformation from sound waves to neural impulses obviously depends on the movements of the basilar membrane. These movements are complex, but cannot account for the transformation observed at the level of the axons of the spiral ganglion cells. One must postulate that there is interaction of neural elements, as described in Chap. 11.

SPECIALIZED RECEPTORS IN VERTEBRATE RETINA

Another transducer that is highly specialized and highly sensitive is the retina. **The photosensitive cells of the retina transduce light energy (photons) into electrical signals.** The sensitivity is so great that **a single photon can activate a single receptor!** Certainly, this is a remarkable genetic adaptation, showing the great importance of this transduction to the survival of the animal.

Fig. 9-19. Microphonic electrical recordings from basal turn of guinea pig cochlea in response to sound waves of various frequencies. Note that basal turn responds to all frequencies as demanded by traveling-wave theory, and amplitude of sound waves need not be altered greatly (10 dB) to give equal responses.

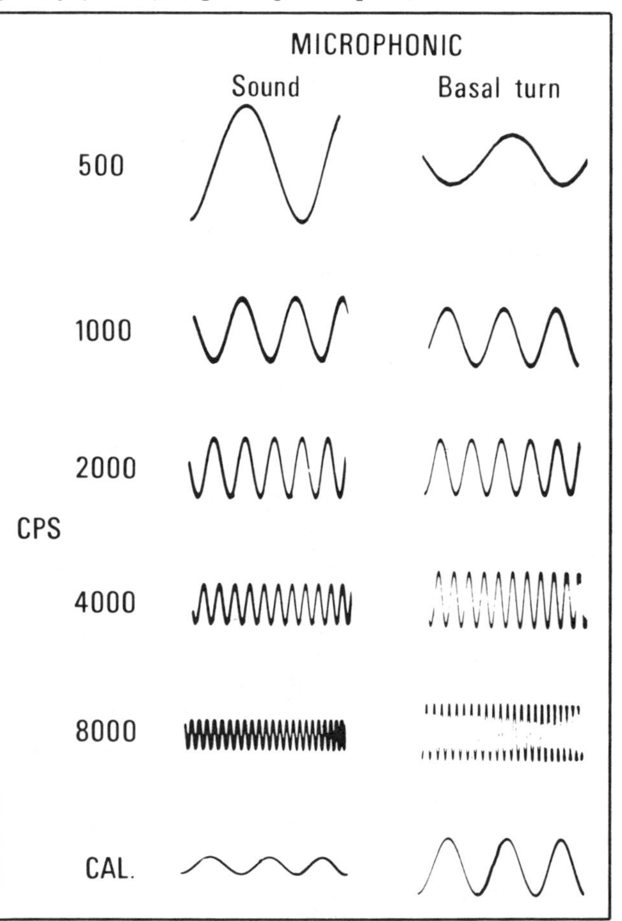

(From I. Tasaki. Nerve impulses in individual auditory nerve fibers of guinea pig, *J. Neurophysiol.* 17:97, 1954. Courtesy of Charles C Thomas, Publisher, Springfield, Ill.)

233

Earlier in this chapter, a few recordings from invertebrate photoreceptors are shown. **In this section, we deal exclusively with the vertebrate retina.** In many respects, the vertebrate retina is unique, with respect to not only other types of sensory receptors, but also other types of photoreceptors in the animal kingdom. For example, in many invertebrates, the light reaches the photoreceptors rather directly, whereas in the vertebrate retina, **light passes through the cells of the retina before reaching the photoreceptors,** as shown in Fig. 9-20. In Fig. 9-20, the arrow indicates the direction in which light travels in reaching the photoreceptor layer—the layer of rods and cones.

1

The significance of the synaptic connections shown in Fig. 9-20 is discussed in Chap. 11. At this point, we concentrate on the mechanisms of the photoreceptors: the rods and cones.

2

Cones are sensitive to color, whereas rods detect all wavelengths of light. An enlarged portion of a rod is seen in Fig. 9-21, which shows that the photoreceptive end of the rod consists of repeated enfoldings of the cell membrane in the form of disks. In this "outer segment" region, the cell absorbs light energy in a "visual pigment," rhodopsin.

3

As you might expect, the absorption of light energy changes the cell membrane conductance. Now comes the unexpected part: g_{Na} **decreases when light is absorbed!**

4

The resting state of this cell is, itself, highly unusual. The g_{Na} of the outer segment is not as small as usual (Chap. 4), which leads to a larger-than-usual inward I_{Na} in this region when the receptor is in its resting state (i.e., in the dark). This *inward* I_{Na} is balanced by an *outward* potassium current, just as in the normal resting cells considered in Chap. 4. However, because the sodium and potassium currents occur primarily in different regions of the photoreceptor cell (*inward* I_{Na} in the *outer* segment, *outward* I_K in the *inner* segment), a steady current flows along the inside and outside of the cell between these two regions.

5

Fig. 9-20. Synaptic connections in primate retina, as seen by electron microscopy. C, cones; R, rods; H, horizontal cells; MB, midget bipolar cells; FB, flat bipolar cells; RB, rod bipolar cells; A, amacrine cells; MG, midget ganglion cells; DG, diffuse ganglion cells.

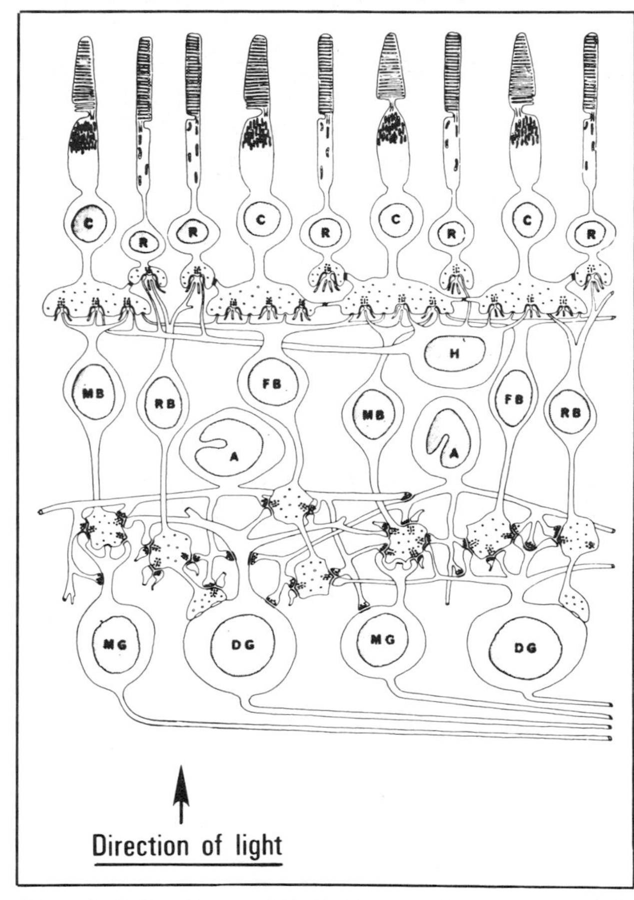

(From J. E. Dowling and B. B. Boycott, Organization of the primate retina: Electron microscopy, *Proc. R. Soc. Lond.* [*Biol.*] 166:80, 1966.)

HINTS

21. The axons of the spiral ganglion cells, which run out to innervate the hair cells (through the fluid of the tunnel of Corti), wouldn't function very well. If you don't know why, then go, go, go to Hint 24.↓
22. No! Passive diffusion of K^+ cannot be the mechanism of the potential! For the reasoning, see Hint 23.↓
25. See the next paragraph in the text—we just thought you would enjoy figuring it out for yourself—very likely you'll remember it better if you do.
26. (1) The stimulus passes through the nervous system before reaching the sensory ending. (2) There is a high resting g_{Na}. (3) A stimulus causes a decrease in g_{Na}. (4) A stimulus leads to hyperpolarization of the membrane. (5) Hyperpolarization affects the release of synaptic transmitter. (6) There is no action potential in the first two cells.

While you ponder this list, also realize that the world is actually upside down, since the simple single lens of the eye inverts the image!

233

234

This current is called the *dark current* because it is largest in the dark and is reduced by illumination.

Transduction of the incident light causes a *fall* in g_{Na} in the outer segment. Thus I_{Na} is reduced in this region. So I_K will be larger than I_{Na}, tending to hyperpolarize the cell until the potassium driving force is reduced to a point at which the ionic currents are, once again, in balance. **The membrane hyperpolarizes in response to a light stimulus, and the dark current falls** (since both I_{Na} and I_K are reduced during steady illumination). **This hyperpolarization occurs across the entire length of the cell, whose entire membrane seems electrically inexcitable.** At the synaptic connection with the bipolar cells, this hyperpolarization leads to reduction in release of transmitter substance, which hyperpolarizes the bipolar cell by turning off the release of a depolarizing transmitter substance. However, this cell also appears to be electrically inexcitable, and potential charges spread passively along the small bipolar cells. Only when hyperpolarization of the bipolar cells results in depolarization of the next cell—the ganglion cell—is an electrically excitable membrane encountered, and an action potential is generated.

Thus, in the retina, the receptor potential of the rods and cones is clearly separated from the generator potential in the ganglion cell, with the distinction between receptor and generator potentials being as defined earlier in this chapter.

The small size of the receptors and bipolar cells is completely compatible with relatively passive spread of potential (without need for a propagated action potential). By contrast, the action potential is necessary in the case of the ganglion cell, whose axons make up the optic nerve, transmitting information from the retina to the CNS. Recall from Chap. 2 that only when distances between cells become large is it necessary for the specialized mechanism of the action potential to be used.

Further description of the interactions of the cells in the retina, as affecting retinal function, is postponed until Chap. 11. By that time, you will have a much better idea of how synapses operate, which makes explanations of the interactions of the cells much easier to understand.

The sensitivity of the receptors is determined by the photopigment.

In the case of the cones, the photopigments show a restricted sensitivity, being primarily sensitive to light of one color. There are three "primary" colors, which correspond to the sensitivities of three types of cones. The cones are mainly concentrated at the point of the retina that light strikes when one fixates on a small spot of light. The peripheral areas of vision are supplied mainly with rods, which do not show color selectivity.

Fig. 9-21. Structure of mammalian rod, as seen by electron microscopy. OS, outer segment; CC, connecting cilium; IS, inner segment; C_1 and C_2, centrioles. Transverse sections through (*a*) connecting cilium and (*b*) centriole are shown at the right. rs, rod sacs; cf, ciliary filaments; cm, cell membrane; mi, mitochondrion; er, endoplasmic reticulum.

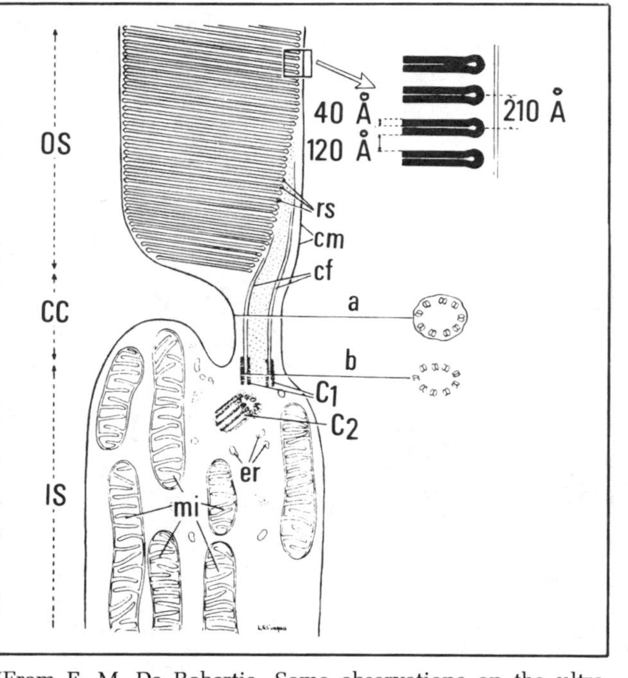

(From E. M. De Robertis, Some observations on the ultrastructure and morphogenesis of photoreceptors, *J. Gen. Physiol.* 43 [Suppl. 2]:1, 1960.)

234

235

The sensitivity of the visual receptors varies with the amount of light to which they have been exposed recently (within the last few seconds or minutes). Thus they maintain their sensitivity above the average illumination that they have experienced.

1

This feature is called adaptation, but note that the meaning in this context is completely different from its use elsewhere in this book and in neurophysiology, where adaptation refers to a change in firing rate to a constant stimulus. (It is especially ironic, as well as confusing, but since neither the receptors nor the bipolar cells have action potentials, their "adaptation" cannot be related to changes in firing frequency!)

2

When receptors are exposed to light (during *light adaptation*, the visual pigment bleaches (becoming white) and is less responsive. Thus, more light is needed to bring about activation of the receptor (the receptor is less responsive). In this way, very large ranges of illumination can be handled—the range of sensitivity of the rods can be as great as 10 orders of magnitude.

3

The adjustment of the retina to the amount of light is the equivalent of having an automatic camera that adjusts the film speed to the amount of light, rather than limiting the amount of light reaching the film, as in the usual camera.

4

QUESTION: In how many ways is the visual system of vertebrates unusual compared with other neural systems? (Hint 26↑)

5

CHEMICAL TRANSDUCTION: TASTE AND SMELL

We complete this chapter with a brief mention of sensory receptors of a completely different type—those sensitive to chemicals. **For both taste and smell, the sense organs distinguish different chemicals in solution.**

6

Even though in the case of smell the chemicals are airborne, still, before they can be sensed, they pass into solution in the moist mucous membranes lining the nose. Similarly, substances not in solution are not tasted on the tongue.

7

---

23. After all, if a high K^+ concentration and passive diffusion *can* account for the scala media potential, as in the resting membrane potential, then the scala media should be *negative*, not positive, inside. Hence we must postulate some other mechanism (an active pump?) as the basis for a *positive* potential, with a high internal K^+.

24. They would be continuously depolarized by the high K^+ surrounding them. If this puzzles you, then you really have blanked out completely on Chap. 4, and you had better have some coffee and compose yourself—after all, you can't have gotten this far without having learned what K^+ is all about!

235

236

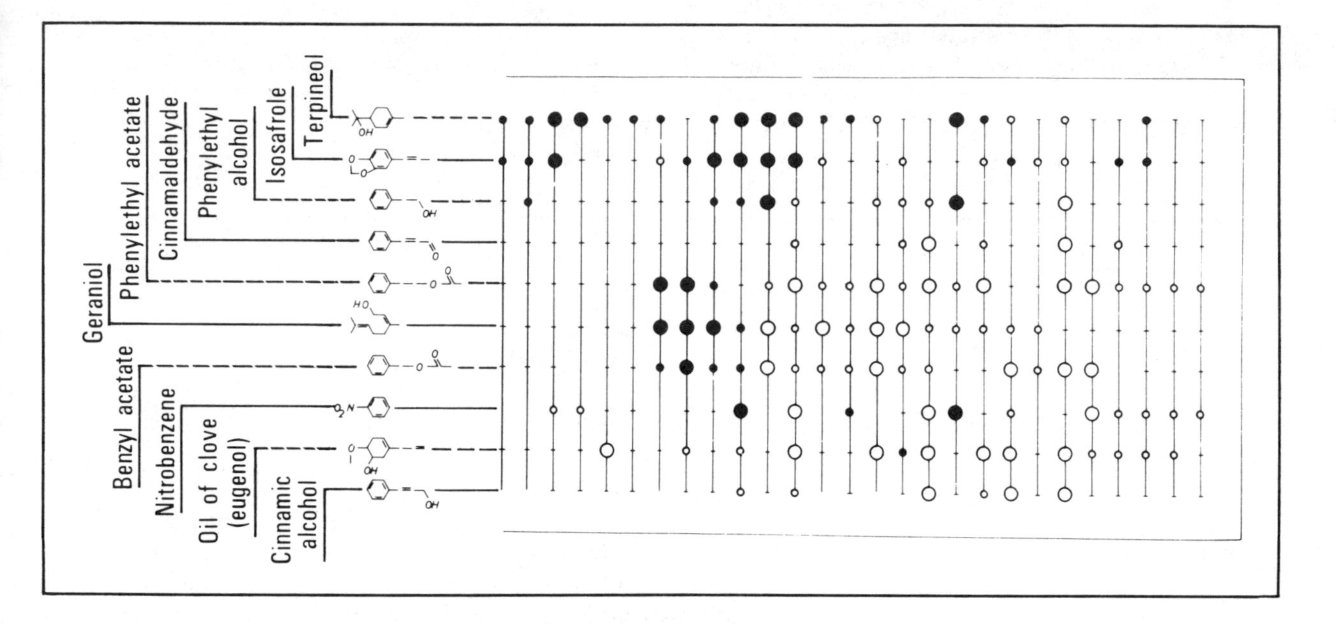

Fig. 9-22. Olfactory responses of 27 sensilla basiconica in the moth *Antherea pernyi* to 10 different compounds. Filled circles indicate excitation, open circles indicate inhibition; sizes of circles reflect extent of change in nerve impulse frequency. Small, horizontal lines: no effect.

(From J. Boeckh, K. E. Kaissling, and D. Schneider, Insect olfactory receptors, *Cold Spring Harbor Symp. Quant. Biol.* 30:263, 1965.)

The mechanism by which various chemicals can cause different reactions in these receptors is not known. However, in some cases, generator potentials can be measured, so that **presumably there is a change in membrane conductance** (probably to at least Na^+ and possibly to other ions as well) **that is proportional to the amount of stimulating substance.** Of course, hyperpolarizing as well as depolarizing responses can be imagined. Thus, many of the general principles that you learned relative to other receptors are also applicable to these receptors. 1

It may be that not only does the ability to discriminate among different chemicals involve selective responses on the part of a single type of receptor, but also **the selectivity possible in the chemical senses may depend on the pattern of responses of a number of different receptors,** each with a different selectivity. This idea is illustrated with respect to olfaction in the moth in Fig. 9-22. Note that the various receptors have different patterns of response, some excited, others inhibited. Conceivably, many chemicals are recognized only by their combined effects on several different receptors. Thus, **a multichannel sensory system may be able to distinguish different stimuli that any single receptor alone could not.** 2

The preceding generalization may be more interesting after you learn about lateral inhibition in Chap. 11. 3

236

237

Myoneural Junction and Muscle Function

MYONEURAL JUNCTION OF SKELETAL MUSCLE

Slowly (but surely) you are learning the different mechanisms involved in the workings of the nervous system. You can see from Fig. 10-1/2-14 how far you have come. We now discuss the "output" side of the nervous system—the means by which nerves control skeletal muscles.

The junction between nerve and muscle—the myoneural junction—is important not only because of its role in producing bodily movement, but also because it is a synapse that is quite well understood. A **synapse** is the point of functional contact between excitable cells (in the nervous system and the neuromuscular system).

There are various reasons why we know so much about the myoneural junction. Historically, muscular contractions were the first recognized "biological amplifiers," i.e., the first means of studying how nerves work. Obviously, the myoneural junction was part of the system and so was also an object of study. In addition, the myoneural junction is larger than most synapses, which makes experimental methods easier to apply. (For this same reason, we know lots about the squid giant axon.)

In contrast to **electrical synapses** (e.g., the low-resistance gap junctions seen in cardiac and smooth muscles in Chap. 7), the **myoneural junction will serve as the prototype of the chemical synapse,** which is a connection between cells where information is transmitted by means of special chemicals called synaptic transmitters. For **the myoneural junction, the synaptic transmitter is acetylcholine** (ACh) (Fig. 10-2).

Proof of the existence of such chemical transmitters dates to the classic experiments of Loewi, who in 1921 showed that when an isolated heart was slowed by electrical stimulation of the vagus nerves, a second isolated heart also was slowed if the fluid perfusing the second heart came from the outflow from the first (neurally slowed) heart. The causative agent was subsequently identified as ACh. The history of the discovery has an amusing sidelight. Loewi recalls that he awoke one night with an idea for an interesting experiment. He thought he might try it in the morning, and he fell asleep again. In the morning, he could remember only that he had had an interesting idea, but could not remember the experiment. Some nights later, he again awoke with the same idea, but this time he was not going to let it escape him, for he got up, went to the lab, and did the experiment directly! He was later to say that it seemed such a strange and foolish experiment that he probably could not have thought of it in the daytime!

Chemical transmission implies changes in postsynaptic membrane conductances in response to a synaptic transmitter. Therefore, with this subject we complete the introduction

Fig. 10-1/2-14. Sequence of topics covered in this book, showing chapter numbers.

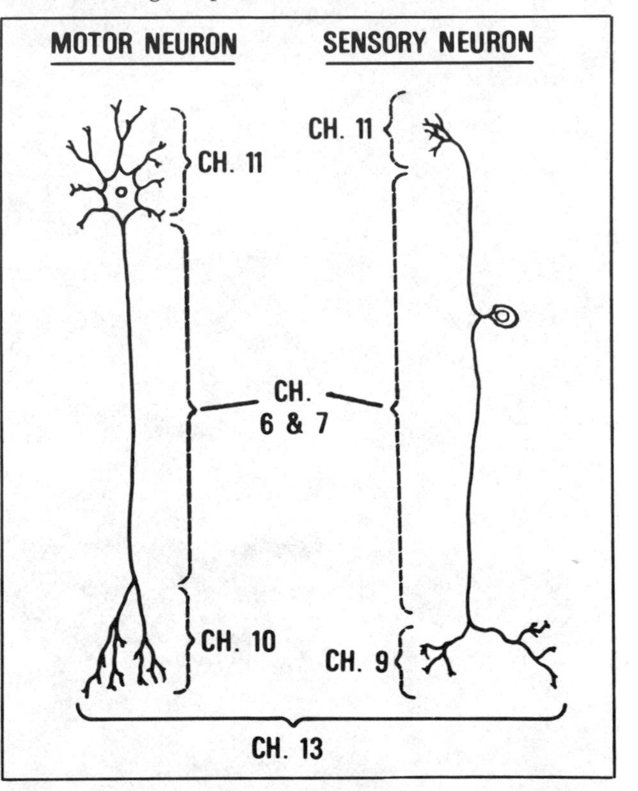

(Modified from E. L. House, and B. Pansky. *A Functional Approach to Neuroanatomy* [2d Ed.]. New York: McGraw-Hill, 1967.)

Fig. 10-2. Chemical structure of ACh.

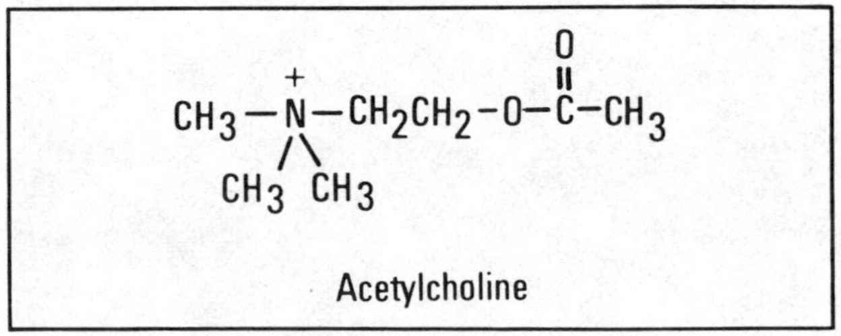

$$CH_3-\overset{+}{N}(CH_3)_2-CH_2CH_2-O-\overset{O}{\overset{\|}{C}}-CH_3$$

Acetylcholine

239

to the **five main types of excitability** (previously mentioned in Chap. 6): electrical excitability (e.g., the axon), mechanical excitability (e.g., the Pacinian corpuscle), photic excitability (e.g., the retina), general chemical excitability (e.g., olfaction), and specific chemical excitability (e.g., the chemical synapse).

1

Chemical transmission at the myoneural junction is an absolute necessity. The currents generated by the nerve endings are too small to depolarize the large, numerous muscle fibers that may be innervated by a single nerve fiber.

2

See Katz [27, pp. 99–104] for the calculations involved. Chemical synaptic transmission at the myoneural junction can be considered as an impedance-matching device, matching the high output-impedance of the nerve to the low input-impedance of the muscle.

3

Some General Properties of Chemical Synapses

The myoneural junction, as seen by the electron microscope, is shown in Fig. 10-3. The important structures are described here.

4

At the **presynaptic terminal,** the nerve ending branches to innervate several muscle fibers. The ending has **mitochondria** and **presynaptic vesicles** near the synaptic regions. The **synaptic cleft** is a 100- to 200-Å region between the two cells that communicates with the extracellular space. The **postsynaptic membrane** is a specialized portion of the muscle fiber membrane with numerous enfoldings, thus it has a large surface area.

5

Transmission across a chemical synapse is unidirectional. Thus at any synapse, one cell membrane is *presynaptic* and the other cell membrane is *postsynaptic,* while the extracellular space between the presynaptic and postsynaptic membranes is referred to as the *synaptic cleft.*

6

The presynaptic and postsynaptic membranes have entirely different functions. **The presynaptic membrane,** like the membranes of a neuroendocrine cell, **releases the transmitter substance when the membrane is depolarized. The postsynaptic membrane is a chemoreceptor that undergoes conductance changes,** and hence a change in membrane potential, **in response to the released transmitter.**

7

Sequence of Events in Neuromuscular Transmission

The sequence of events involved in transmission of information across the myoneural junction is diagrammed in Fig. 10-4.

8

The numbers of the following paragraphs refer to the numbers in Fig. 10-4:

1. An action potential propagates down the axon to the (presynaptic) nerve ending. Note that although the axon is myelinated throughout most of its length, the terminal branches of the ending are unmyelinated (engulfed by only a single Schwann cell).

9

Fig. 10-3. Myoneural junction of frog. (*A*) One portion of the junction. (*B*) General position of endings of motor axon on muscle fiber, showing (*A*) as a small rectangle. (C) Schematic drawing from electron micrographs of longitudinal section through the muscle fiber. 1: Terminal axon membrane. 2: "Basement membrane" partitioning gap between nerve and muscle fiber. 3: Folded postsynaptic membrane of muscle fiber.

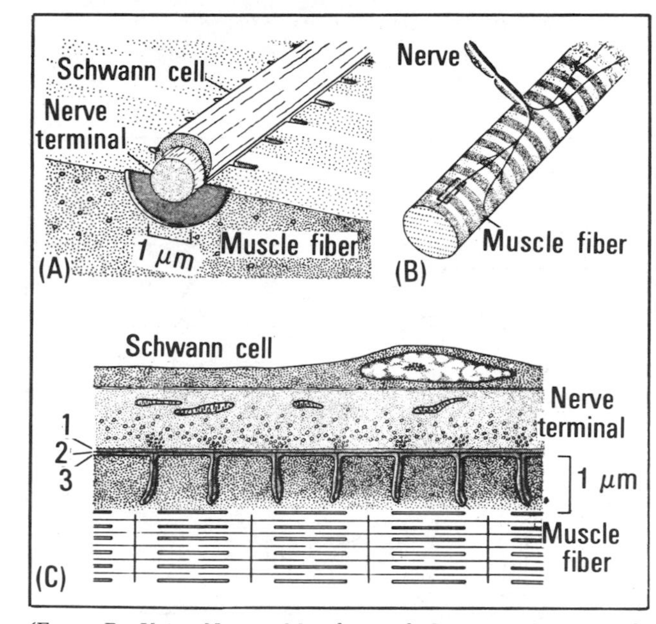

(From B. Katz, *Nerve, Muscle, and Synapse.* New York: McGraw-Hill, 1966.)

239

240

2. The depolarization of the ending by the action potential leads to the release of synaptic transmitter (ACh) into the synaptic cleft.
3. The transmitter diffuses across the (about 200-Å) synaptic cleft.
4. The transmitter reaches and attaches to "receptor sites" on the postsynaptic membrane, which produces a change in membrane conductance and consequent depolarization, called the **end-plate potential** (EPP).

   The **postsynaptic membrane,** like other receptor membranes, **is electrically inexcitable.** In Fig. 10-4, the true EPP is shown by the broken line; this is the potential that would be observed if no action potential occurred in the adjacent membrane areas.
5. The adjacent **electrically excitable** muscle membrane (5 in Fig. 10-4) is electrically coupled to the postsynaptic region, such that it may be brought to threshold of firing by spread of current from the EPP. In detail, **net inward current at the postsynaptic membrane leads to outward capacitative current that not only depolarizes the postsynaptic membrane (giving the EPP), but also spreads along the surface to produce outward capacitative current at adjacent areas of electrically excitable membrane, depolarizing these regions to threshold.**

1 2

There is an interesting problem in terminology here that should be brought to your attention before you become unnecessarily confused. Although the EPP is the potential generated by the transmitter acting on the *postsynaptic* (i.e., muscle) membrane, the *end plate* is a histological term for the platelike *presynaptic* terminal that can be seen with a dissecting microscope under favorable conditions of illumination. The postsynaptic response came to be called the end-plate potential because it was a potential that could *be recorded only in the immediate vicinity of the visible end plate.* Nevertheless, the EPP is recorded *from the muscle,* not from the end plate itself!

3

When the electrically excitable membrane surrounding the postsynaptic region is depolarized to threshold, action potentials propagate in both directions along the muscle fiber. (Note that Fig. 10-4 shows the muscle fiber action potentials propagating to only one side; in actuality, the myoneural junction typically is near the center of the muscle fiber, so that propagation occurs in both directions.) The potential change normally observed beneath the end plate (shown by the solid line in insert 4 of Fig. 10-4) shows the summation of two factors: the EPP (shown by the broken line in this insert) and potentials produced by coupling from action potentials in the adjacent muscle membrane. The main events of this sequence are discussed in greater detail in the following sections.

4

It is entirely legitimate to regard the postsynaptic membrane as a "chemoreceptor" located not at the end, but near the middle of a "sensory axon" (the muscle fiber membrane outside the synaptic area).

5

Fig. 10-4. Sequence of events when a nerve activates a muscle through synaptic transmission across myoneural junction.

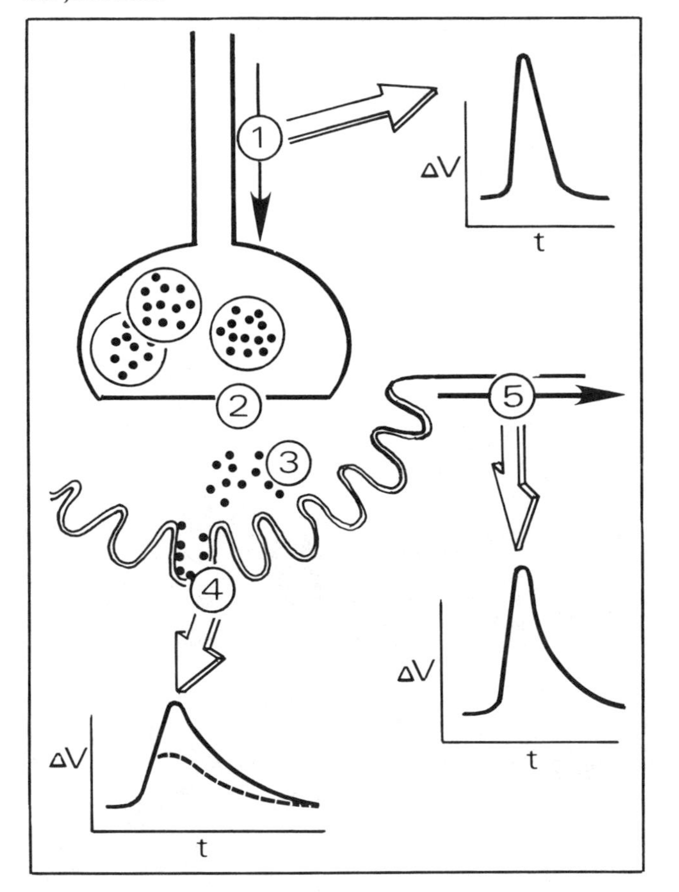

240

241

Thus the arrival of the transmitter causes a depolarization of the postsynaptic membrane (the EPP) that is exactly equivalent to a generator potential. This end-plate potential initiates an action potential that propagates along the muscle fiber surface, just as the generator potential may elicit an action potential in a sensory axon.

Under normal circumstances, transmission is 1:1 at the neuromuscular junction. That is, for each action potential in the presynaptic axon, one, but only one, action potential propagates along the muscle fiber surface. **Only one muscle action potential occurs for every EPP because the EPP decays during the relative refractory period of the muscle membrane action potential.** The rapid decay of the EPP is the result of two processes: the rapid, enzymatic breakdown of ACh by acetylcholinesterase (AChEase), located in high concentrations on and near the postsynaptic membrane, and, somewhat less importantly, diffusion of ACh out of the synaptic cleft.

Biosynthesis of Acetylcholine

At the myoneural junction, all the biochemicals needed for synthesis and breakdown of ACh are present.

The structure of ACh and the biochemical pathways for its synthesis and degradation are shown in Fig. 10-5. Note that the process is entirely cyclic; i.e., all products of reactions are used in the other reactions. (The conversion of ADP to ATP is not diagrammed.) Thus, all the chemicals needed for all the steps are present at the myoneural junction. When the released transmitter is broken down by the extracellular enzyme AChEase, a large proportion of both the choline and the acetate is taken up by the presynaptic ending for resynthesis to ACh within the synaptic vesicles.

Recently it has been shown that the uptake of choline is carrier-mediated, but not energy-dependent. This passive diffusion implies a diffusion gradient with a low intracellular choline concentration. The intracellular choline concentration remains extremely low as a result of ACh synthesis within the presynaptic vesicles.

Transmitter Release

For many years, it was suggested that the mechanism of transmitter release (which had been shown to be "quantal" on the basis of physiological evidence presented in a later part of this chapter: see pages 247–249) might involve individual synaptic vesicles disgorging their contents through the presynaptic membrane into the synaptic cleft in response to membrane depolarization [27, pp. 132–133]. However, clear evidence in support of this hypothesis has been obtained only recently.

Now we know that **synaptic vesicles actually fuse with the presynaptic membrane surface in the central regions of the synaptic area,** losing their contents in a process known as **exocytosis.** Such a mechanism might be expected to produce a steady increase in the total

Fig. 10-5. Summary of reactions involved in synthesis and breakdown of ACh. Acetyl-Co A: acetyl coenzyme A. HS-Co A: reduced coenzyme A.

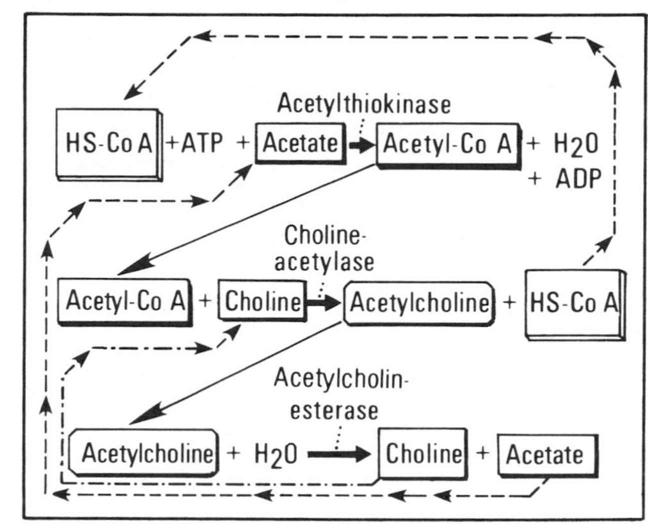

(Modified from W. F. Ganong, *Review of Medical Physiology* [6th Ed.]. Los Altos, Calif.: Lange, 1973.)

241

area of the presynaptic membrane. However, this does not happen because **new synaptic vesicles constantly are being formed by *pinocytosis* at the edges of the synaptic region.** Presumably, these new vesicles fill with ACh as they migrate toward the center of the synapse. Such a process is shown diagrammatically in Fig. 10-6.

This recent evidence, which helped to legitimize this happy hypothesis, was as follows:

1. Rapid stimulation of motor nerves resulted in an increase in the total synaptic membrane area; i.e., the rate of vesicle exocytosis temporarily exceeded the rate of pinocytosis.
2. When an extracellular marker (horseradish peroxidase) was added to the living preparation some time before fixation, pinocytotic vesicles containing the marker were noted at the edges of the synaptic area. These vesicles were otherwise identical to normal synaptic vesicles.

Although it is not yet clear how depolarization brings about the fusion of transmitter-laden vesicles with the surface membrane of the presynaptic cell, it is known that this process is initiated by entry of calcium into the presynaptic terminal.

The number of quanta of **transmitter released** for a given depolarization of the presynaptic terminal **is increased if $[Ca^{2+}]_o$ is increased.** However, transmitter release **is decreased when $[Mg^{2+}]_o$ is increased.** These observations are consistent both with a greater Ca^{2+} influx when the Ca^{2+} gradient is increased and with the suggestion that Mg^{2+} reduces the entry of Ca^{2+} (perhaps by interacting at the same membrane pore) [27, p. 133].

This interaction between Ca^{2+} and Mg^{2+} in the control of transmitter release seems to be a general property of not only myoneural junctions, but also just about every known type of chemical synapse. Note also that **the effects of $[Ca^{2+}]_o$ on transmitter release are opposite to those on membrane excitability** (Chaps. 6 and 8).

Chemistry and Competition in Synaptic Cleft

The cyclic nature of the chemical reactions involved in the synthesis and breakdown of the transmitter has been previously mentioned (Fig. 10-5). The presumed location of some of these reactions is shown in Fig. 10-7.

The location of AChEase at the end-plate region has been ascertained by experiments on denervation hypersensitivity (covered later in this chapter). Also AChEase has been shown by various biochemical methods to be on the outside of the end-plate membrane [39, pp. 1213–1214].

Fig. 10-6. Arrows follow route of maturing synaptic vesicle from its site of formation at edge of synaptic area to its fusion with membrane (and release of contained transmitter).

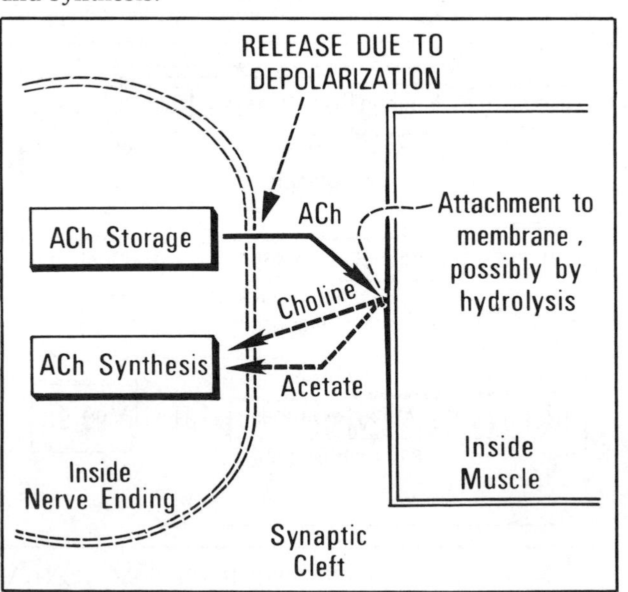

Fig. 10-7. Cycle of ACh storage, release, inactivation, and synthesis.

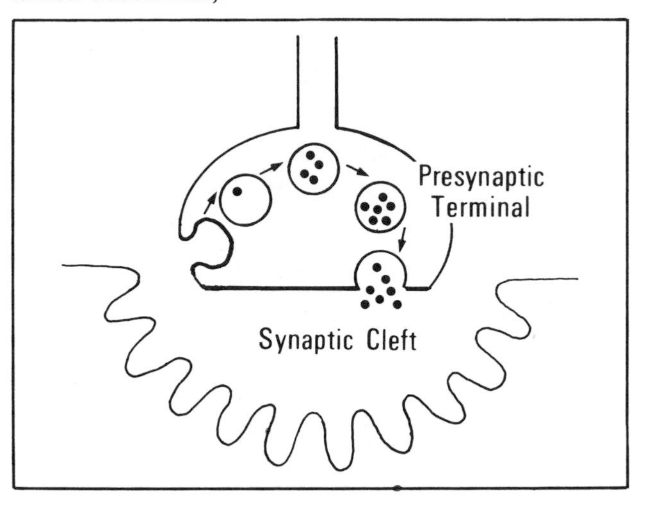

243

The termination of the EPP conductance changes presumably is due to the hydrolysis of ACh into inactive choline. Of course, diffusion away from the synaptic cleft also might account for the shortness of the conductance change. However, drugs that interfere with AChEase activity greatly prolong the EPP (Fig. 10-8). So diffusion alone cannot be too important a factor. Obviously, if diffusion occurred to a great extent, then the cyclic pattern shown in Figs. 10-5 and 10-7 would be interrupted. That some ACh does leak away is suggested by the fact that the bloodstream contains AChEase. As you might imagine, it might not be a good idea for ACh to build up in the blood, and hence in the extracellular space!

The number of ACh molecules that attach to the postsynaptic membrane apparently influences the extent of the conductance changes and the resulting size of the EPP. Curare is a drug that competes with ACh for attachment to the end plate and hence reduces the EPP (Fig. 10-9). The chemical dynamics are those that you would expect from a competition between two chemicals for the same receptor sites.

The detailed action of many drugs that affect neuromuscular transmission is now quite well understood.

Detailed study of the action of drugs is covered in pharmacology. Here the actions are described in order to demonstrate that a knowledge of the physiology of the synapse is necessary for an understanding of the actions of many drugs. (Do not memorize the drug names at this time. What is important is which parts of the biochemical and physiological cycles can be changed by drugs.)

1. **ACh release can be prevented or reduced** by botulinus toxin as well as by decreased Ca^{2+} and/or increased Mg^{2+}.
2. **ACh attachment can be blocked** by prior attachment to the transmitter receptor sites of drugs such as curare and succinylcholine.
3. **Some drugs can mimic ACh** in attaching to the postsynaptic membrane and causing permeability changes (e.g., carbachol, succinylcholine).
4. **Some drugs interfere with AChEase activity,** thus reducing the rate of hydrolysis and prolonging the action of ACh (Fig. 10-8) (e.g., neostigmine, physostigmine).
5. **Choline uptake can be blocked** (e.g., by hemicholinium).
6. Drugs that interfere with ATP production will **reduce synthesis of ACh.**

End-Plate Potential

The EPP does not begin until about 0.25 to 0.50 ms after the nerve ending has depolarized. This time is called the **synaptic delay.**

Synaptic delay is a consistent finding in chemical transmission across synapses. You will encounter it again in Chap. 11.

Fig. 10-8. Intracellular EPP before and after AChEase inhibitor.

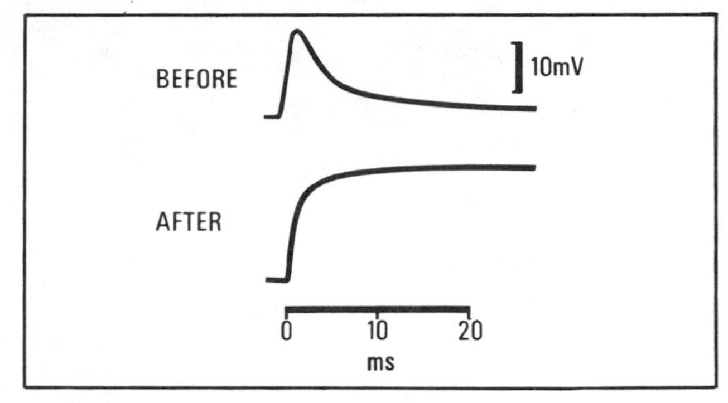

(Modified from P. Fatt and B. Katz. An analysis of the end-plate potential recorded with an intra-cellular electrode, *J. Physiol.* [*Lond.*] 115:320, 1951.)

Fig. 10-9. Recorded EPPs each 3 s during washout of curare from preparation. At seventh record, action potential from adjacent membrane is recorded. Note gradation in heights of EPPs.

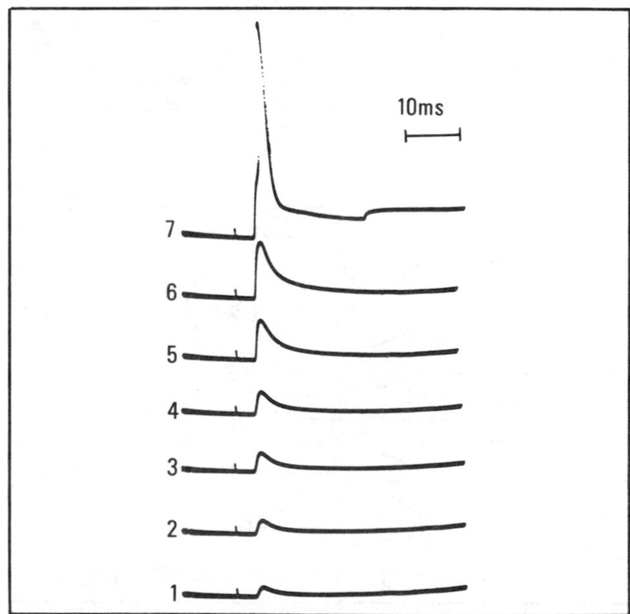

(Modified from W. L. Nastuk, Neuromuscular transmission: Fundamental aspects of the normal process, *Am. J. Med.* 19:663, 1955.)

243

244

Only a small part of the synaptic delay can be ascribed to diffusion across the synaptic cleft. It has been calculated that the diffusion of ACh across the 100- to 200-Å cleft takes only about 0.1 ms. The remaining time probably is consumed in the processes of release of the transmitter and the action of the transmitter in changing the end-plate permeabilities. 1

Since the muscle cell is large and readily penetrated with microelectrodes, one of the earliest experimental techniques used in the study of postsynaptic responses was to hyperpolarize or depolarize the end-plate region and observe the effect on the EPP. It was observed that the EPP becomes larger when the postsynaptic membrane is *hyperpolarized*, but becomes smaller and smaller as the membrane is depolarized (see Fig. 10-10) until a potential is reached at which the EPP "reverses." Further applied depolarization then *increases* the size of the *reversed* EPP. The potential at which no EPP is visible is called the **reversal potential.** 2

The magnitude and polarity of the reversal potential have considerable theoretical significance. Let us imagine, for example, that the EPP were produced by a change in g_{Na} while all other conductances remained at normal resting levels. If the postsynaptic membrane were depolarized toward E_{Na}, then the EPP would become smaller because the driving force is $V - E_{Na}$. Finally, *when V became equal to* E_{Na}, there would be no EPP, however great the increase in g_{Na}. For such a cell, the reversal potential would be equal to E_{Na}. Thus, a knowledge of the reversal potential can reveal the underlying ionic mechanisms (if the concentrations are known). 3

QUESTION: Why would there be no EPP if the membrane potential were at E_{Na}? (Hint 2↓) 4

QUESTION: What would happen to the EPP if the cell were depolarized beyond E_{Na}? (Hint 1↓) 5

In typical myoneural junctions, the reversal potential is between −20 and −5 mV and is not affected by $[Cl]_o$. 6

QUESTION: What does this tell us about the nature of the conductance changes produced by transmitter action? (Hint 3↓) 7

Since the reversal potential is clearly a steady-state condition in which I_{cap} is zero, it is possible to show (by applying the steady-state equation) that the reversal potential V_R is dependent on only the *change of conductance*. The following general equation can be obtained:

$$V_R = \frac{\Sigma(\Delta g_i E_i)}{\Sigma \Delta g_i} \quad \text{Eq. 10-1}$$

8

Fig. 10-10. EPPs in a hypothetical cell with a "reversal potential" of −20 mV.

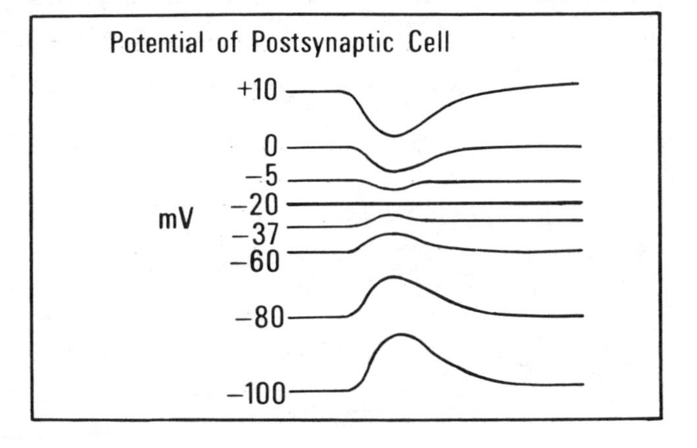

244

245

Here Δg_i is the change in the conductance of the ith ion species, and E_i is its Nernst battery potential.

1

Now, if $V_R = -20$ mV (as in Fig. 10-10), $E_{Na} = +60$, and $E_K = -100$, substituting in Eq. 10-1 gives

$$-20 = \frac{\Delta g_{Na}(60) - \Delta g_K(100)}{\Delta g_{Na} + \Delta g_K} \qquad \text{Eq. 10-1}$$

Rearranging this equation yields

$$\Delta g_K(100) - \Delta g_K(20) = \Delta g_{Na}(60) + \Delta g_{Na}(20) \qquad \text{Eq. 10-2}$$

The final, easy-to-remember result is

2

$$\Delta g_K = \Delta g_{Na} \qquad \text{Eq. 10-3}$$

This result can be interpreted in terms of the numbers of pores that are either closed or open [to a fixed value, just as in the sensory ending of the Pacinian corpuscle (see pages 218 to 220)]. Thus, we can imagine that Fig. 10-11/9-8 shows the relationship of closed to open pores in the resting state. As more transmitter acts, the proportions change, moving the solid vertical line to the right.

3

If a reversal potential at some other value is found, then the changes in conductances are still proportional. For example, if the reversal potential were found in the previous instance to be -40 mV, then $\Delta g_{Na} = 0.6\,\Delta g_K$. Thus Fig. 10-11/9-8 would be changed so that the height of the two pore "blocks" would be in that proportion.

4

Note that the *ratios* shown in the examples would not change with different quantities of transmitter. Thus one can hypothesize that a given quantity of transmitter "opens" a certain number of pores (but always the same proportion of Na^+ to K^+ pores).

5

By skillful extension of the voltage-clamp technique to the study of the EPP in skeletal muscle, Takeuchi and Takeuchi confirmed that the EPP is the result of an equal increment in g_{Na} and g_K. Furthermore, the changes in g_{Na} and g_K show the same time course.

6

QUESTION: If there are equal increments in g_K and g_{Na}, why does the membrane potential change? (Hint 4↓)

7

Fig. 10-11/9-8. Relationship of open to closed pores in resting state. With ACh, changes in conductances increase, so that solid vertical line moves to right, as indicated by arrows.

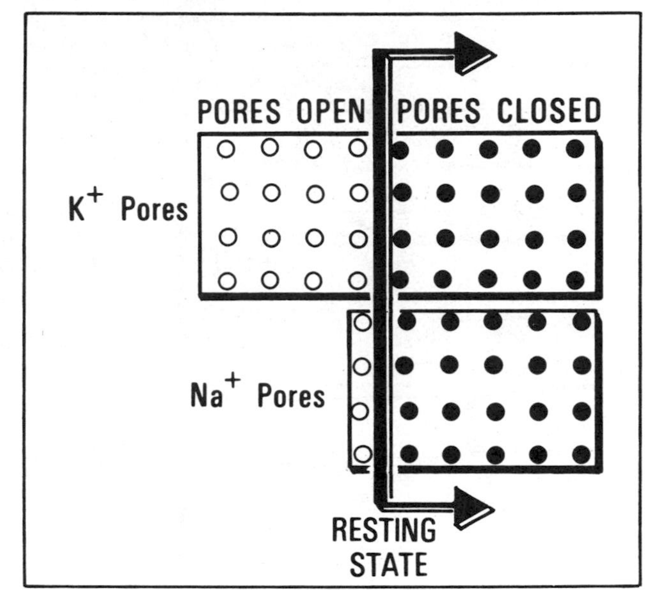

245

246

The change in membrane properties that gives rise to the EPP sometimes is described as *nonspecific* because it is as if *either* an equal number of Na^+ and K^+ pores were opened *or* the pores themselves were equally permeable to both Na^+ and K^+ (that is, nonselective). The term *nonspecific* is quite misleading, however, since it is clear that g_{Cl} and the conductances for many other ions are *not* changed by the action of ACh at the myoneural junction of skeletal muscle. 1

The exact mechanism by which ACh changes the membrane conductances is, of course, unknown. However, it is known that the **acetylcholine acts on the outside of the postsynaptic membrane.** 2

By means of special, double-barreled micropipettes, a small amount of ACh can be ejected at the edge of the synaptic cleft, where it gives rise to an EPP. If the ACh is injected intracellularly into the muscle, however, there is no effect. 3

The conductance of the postsynaptic membrane is not affected by the potential of the membrane. That is, **the postsynaptic membrane is electrically inexcitable.** The conductance changes are brought about only by chemicals (ACh or certain drugs). 4

The evidence is obtained by electrically depolarizing the membrane (in a voltage clamp) and observing that there are no conductance changes [27, p. 126]. 5

Several consequences follow from the electrical inexcitability: (1) No action potential is generated in the postsynaptic region. (2) Thus the response can be graded over a larger range, just as with generator potentials. (3) The responses can then show temporal summation (discussed later in this chapter). (4) There is no refractory period, just as for generator potentials. 6

A good example of the graded nature of the EPP is provided by studying the blocking actions of various concentrations of the ACh-blocking drug curare (see Fig. 10-9). 7

Action Potential of Muscle Fiber

The membrane of the muscle fiber (exclusive of the postsynaptic region) generates an action potential in the same way as in axons. **The membrane adjacent to the postsynaptic region is electrically excitable and is brought to threshold by the spread of the outward capacitative currents along the membrane from the "end-plate region."** 8

The measurement of the spread of the EPP is shown in Fig. 10-12. Actually the EPP generated is sufficiently large that the safety factor is 3 or 4; that is, the EPP is 3 or 4 times as large as is necessary to bring the electrically excitable membrane to threshold. Such a safety factor ensures that a depolarization of the presynaptic axon will *always* lead to depolarization of the muscle (under normal conditions). 9

Fig. 10-12. Transmembrane potential changes produced in curarized muscle fiber by stimulation of motor nerve to muscle. *Abscissa:* Time in milliseconds. *Ordinate:* Change in transmembrane potential in millivolts. Number by each curve is distance (in millimeters) of intracellular recording microelectrode from end-plate region. As distance increases, recorded potential becomes smaller and slower.

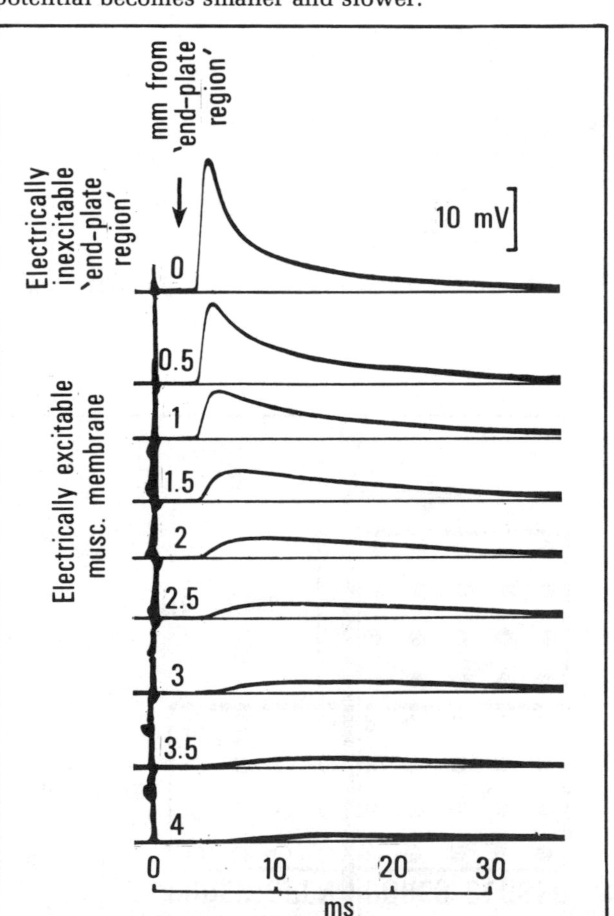

(Modified from P. Fatt and B. Katz. An analysis of the end-plate potential recorded with an intra-cellular electrode, *J. Physiol.* [*Lond.*] 115:320, 1951.)

246

247

Question: In Fig. 10-12, can you guess why the EPP gives no action potential? (Hint 5↓)
1

Electrical depolarization of the motor **nerve ending** (by an intracellular microelectrode) with simultaneous measurement of the end-plate potential confirms what we said at the beginning of this chapter: There is no electrical depolarization of the postsynaptic membrane resulting from current spread from the nerve ending.
2

Figure 10-12, besides showing electronic spread, illustrates **decremental conduction** of a subthreshold response. The classic teaching is that the all-or-nothing action potential propagates while the graded local response does not. This may be incorrect! In Fig. 10-12, note that the start of the membrane depolarization (as well as the peak depolarization) has a greater and greater latency as the distance between the source of the depolarization and the microelectrode increases. Classically, one would ascribe this only to the passive cable properties and, in particular, to the time constant associated with the distributed capacitance along the fiber. However, since we know that for any sizable depolarization there must also be a local response (which makes the depolarization larger than it would have been as a result of the electrotonic spread alone), the small depolarization occurring far away from the end plate shown in Fig. 10-12 must be partly due to the local response of the intervening muscle membrane. Thus, the local response actually "conducts," but it "conducts with decrement" and is, in this way, a self-limiting process. Decremental conduction has been proposed (with heated arguments pro and con) as a mechanism in the functioning of dendrites in the CNS, and here it is appearing right in the classic membrane response to the EPP! (See also ("Decremental Conduction and Electrotonic Spread" in Chap. 8 for further discussion of this concept.)
3

Quantal Miniature End-plate Potentials

A phenomenon of considerable importance in understanding the mechanisms of the storage and release of ACh is the recording of **miniature end-plate potentials** (MEPPs).
4

HINTS

1. The EPP would reverse because both the sign of $V - E_{Na}$ and the direction of I_{Na} would be changed.
2. Recall that $I_{por} = g_{Na}(V - E_{Na})$. Obviously, when $V = E_{Na}$, there can be no I_{por}, hence no I_{cap}, and no change in the charge on the membrane capacitance.
3. Since the reversal potential is substantially less than E_{Na}, some other ion must be involved in the EPP. Since the EPP is not affected by change in external chloride concentration, g_{Cl} is not changed by transmitter action. Do you want to guess that the other ion is K^+?
4. The g_{Na}/g_K ratio changes for different increments, just as you learned in Chap. 9, pages 220 to 221.

247

Figure 10-13 shows MEPPs recorded immediately below the end plate, without stimulation of the nerve. Note that the amplitude of the random fluctuations of the membrane potential is about 0.5 mV. In some preparations, the average rate is about 1 MEPP per second. A careful study of these spontaneous potentials shows that they are of quantal amplitudes. That is, all are multiples of the height of the smallest. The statistical distribution of the larger MEPPs is what you would expect from the chance occurrence of separate quantal events [27, pp. 133–137]. 1

An additional finding of great interest is that **the EPP elicited by nerve stimulation also is a summation of quantal events.** 2

This fact was demonstrated by reducing the amount of $[Ca^{2+}]_o$ and increasing the amount of $[Mg^{2+}]_o$ until the amount of ACh released was so markedly reduced that the amplitude of the EPP evoked by nerve stimulation was seen to be a multiple of the minimal height [27, pp. 133–137]. 3

It has been possible to estimate the number of quanta released by an action potential at the myoneural junction to be between 100 and 400 under normal conditions. 4

The whole idea of quanta in myoneural transmission was unified by the findings (by means of electron microscopy) of numerous vesicles in the presynaptic terminal. These are sufficiently large to contain the 10^4 molecules of ACh calculated to be involved in the minimal quantal event. 5

Thus, **the theory was formed that the vesicles represent the storage sites for ACh before release, that at rest the vesicles sometimes are "accidentally" released** (presumably because of random movement of the vesicles against the nerve terminal membrane), **and that the presynaptic action potential somehow coordinates the sudden release of about 100 vesicles.** 6

The idea that the vesicles may be part of the method of ACh synthesis is supported in part by the finding that fractions (separated by centrifugation) containing the vesicles have high cholineacetylase activity. 7

The same factors that affect the amount of ACh released from the ending as the result of an action potential also influence the spontaneous firing rate of the MEPPs. 8

Thus, the MEPP rate is increased by an increase in $[Ca^{2+}]_o$ and decreased by an increase in $[Mg^{2+}]_o$. However, the **number of quanta released by a presynaptic action potential varies with the height of the presynaptic action potential.** Thus, prior depolarization of the ter- 9

Fig. 10-13. Spontaneous MEPPs recorded intracellularly at frog end plate.

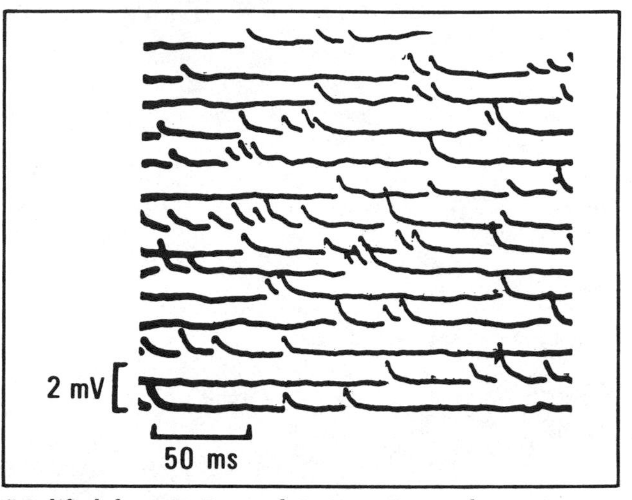

(Modified from P. Fatt and B. Katz. Some observations on biological noise, *Nature* 166:597, 1950.)

249

minal reduces both the height of the presynaptic action potential and the number of quanta released, although such depolarization may increase the rate of MEPPs [27, p. 138].

1

Drugs such as curare, which influence the height of EPP by blocking receptor sites, affect only the size of the MEPPs, not their frequency.

2

Denervation Hypersensitivity

The spontaneous bombardment of the postsynaptic membrane by MEPPs has raised the question of whether such bombardment at the end plate prevents the **denervation hypersensitivity** that **occurs when the nerve innervating the muscle is cut and allowed to degenerate.** Under these conditions, **the muscle becomes exceedingly sensitive to circulating ACh.** It has been shown that the hypersensitivity is due to the whole of the muscle membrane becoming sensitive to ACh.

3

Under normal (innervated) conditions, only the postsynaptic membrane of the myoneural junction is sensitive to ACh.

4

With denervation, the ACh-sensitive end-plate region becomes larger, spreading over the muscle. If only one of the nerves innervating one of those rare muscles with dual innervation is cut, then the hypersensitivity occurs only around the denervated end plate.

5

The sensitivity of the muscle membrane has been tested by using the iontophoretic application of ACh from a micropipette, as described earlier. Drugs that inhibit AChEase do not potentiate the effect of the iontophoretically applied ACh, which suggests that AChEase remains localized at the original end-plate region.

6

Denervation also affects the firing of MEPPs. Immediately after cutting of the nerve, the MEPPs disappear, reappearing after a few days with a low spontaneous rate.

7

There is some evidence that under these conditions, the Schwann cells begin to produce ACh while the axon cylinder undergoes wallerian degeneration.

8

During reinnervation, the MEPPs maintain a slow firing rate even when the new nerve ending has returned to the old synaptic gutters and the muscle membrane surrounding the end plate has returned to its normal sensitivity.

9

HINT

5. The response of the postsynaptic membrane has been decreased by application of curare, so just the EPP can be studied without the added complexity (to the recording) of the action potential.

249

250

From such evidence, it must be concluded both that the area of ACh sensitivity is not determined by ACh secretion itself and that some neurotropic influence from the postsynaptic surface may attract the regrowing nerve fiber. 1

For further details on denervation and reinnervation, see Nastuk and Mountcastle [39, pp. 1225–1226] and Ochs [40, pp. 177–180]. 2

Of course, denervation hypersensitivity may play a role in the uncontrolled, spontaneous, disorganized movements of muscle (fibrillations) that occur when a muscle is denervated as a result of accident or disease. 3

Temporal Summation, Facilitation, and Depression

What would a chapter be without at least one mention of temporal summation? The prolonged EPP, together with the electrical inexcitability of the postsynaptic membrane makes temporal summation easy, if not inevitable! 4

Figure 10-14 shows the postsynaptic response to repetitive stimulation. It is easy to see that temporal summation occurs. The broken lines indicate the potential that would have occurred if the stimuli had stopped at that point. 5

In addition, facilitation during the early period of stimulation is apparent in Fig. 10-14. **Facilitation is the increase in response to a single presynaptic action potential during repetitive stimulation.** 6

Facilitation can be seen by comparing the vertical lines marked on the second through fourth responses, which indicate the heights of the responses that are being summed onto the previous response. During facilitation, the EPP can be 10 times greater than the first response! Experiments have shown that the increase is due to a change in the *number* of quanta released, and not to a change in the *size* of the quanta. Small changes in the size of the presynaptic spike might account for such a change in number. (This is found to be the case in presynaptic inhibition in the spinal cord, Chap. 11.) 7

A period of facilitation may be followed by a period of **depression,** where the responses are less than the first response (see Fig. 10-14). 8

The same comments made above apply here. Namely, the **number of quanta released seems to be the controlled variable,** rather than the size of the quanta. 9

It might be supposed that such a decline is the mechanism of muscle fatigue. 10

Fig. 10-14. Changes in postsynaptic muscle potentials resulting from repetitive stimuli at about 25 stimuli per second.

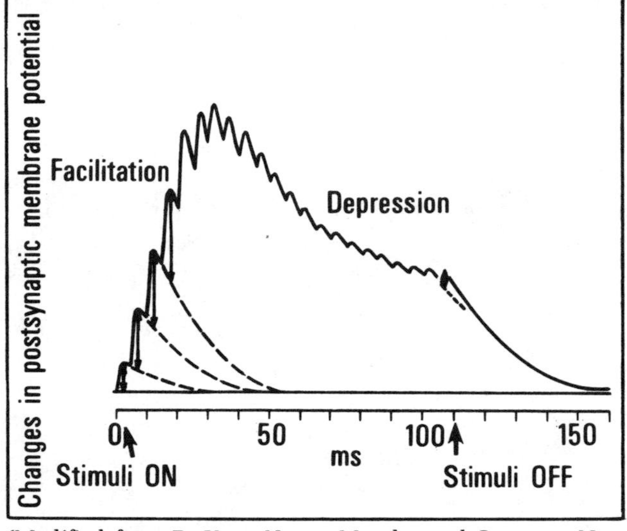

(Modified from B. Katz. *Nerve, Muscle, and Synapse.* New York: McGraw-Hill, 1966.)

250

Indeed, **in isolated nerve-muscle preparations,** if a nerve is stimulated repeatedly, the muscle twitch may finally weaken and disappear, at which time direct stimulation of the muscle can still elicit a contraction. **Under such conditions, a decreased release of ACh quanta at the neuromuscular junction can be demonstrated, which supports the idea of a "block" at the myoneural junction in these experiments. However, this cannot be the cause of fatigue in "intact" humans.** It has been shown that **with failure of maximal voluntary effort as a result of fatigue, direct stimulation of the nerve fails to give a contraction,** even though neuromuscular transmission has occurred, as shown by the presence of muscle *action potentials.* That is, the muscle is unable to respond mechanically even though all the neural transmission mechanisms are still functioning [39, pp. 1222–1223]. Thus, conclusions based on in vitro physiological experiments should be extended to intact humans with some care, no matter how soothing explicatory direct comparison may seem. 1

Myasthenia Gravis

Brief mention should be made of the disease myasthenia gravis, since the manifestations of the pathological process are most prominent at the neuromuscular junction. 2

The patient shows weakness on sustained effort. Experiments show that neuromuscular transmission fails after a period of continuous contraction. The patient is relieved of some symptoms by administration of drugs that inhibit AChEase, thus prolonging the action of the available ACh (Fig. 10-8). Recent evidence suggests that the number of quanta released is normal, but that the size of the quantal depolarization is decreased. 3

Question: Can you conclude from this evidence that the pathological defect acts on the presynaptic side of the synapse? (Hint 6↓) 4

There is now substantial evidence that immunological derangements may play a part in the disease [39, pp. 1226–1227], leading to destruction of ACh receptor sites on the postsynaptic membrane. 5

Muscle Structure and Function

Before we can go further with the story of how muscle contraction occurs, you need to know some of the details of muscle structure at the level shown by the electron microscope. 6

Figure 10-15 shows the molecular structure of striated muscle. **The muscle is made up of fasciculi, each of which is composed of muscle fibers. Each muscle fiber comprises many myofibrils, which in turn contain the myofilaments constructed of the actin and myosin molecules.** The regular arrangement of the actin and myosin causes the striated appearance of the muscle. 7

The myosin molecule is shaped somewhat like a golf club (*M* in Fig. 10-15), and the "golf clubs" are arranged on the filament with the handles all pointing to the center of the 8

Fig. 10-15. Organization of skeletal muscle from gross to molecular levels.

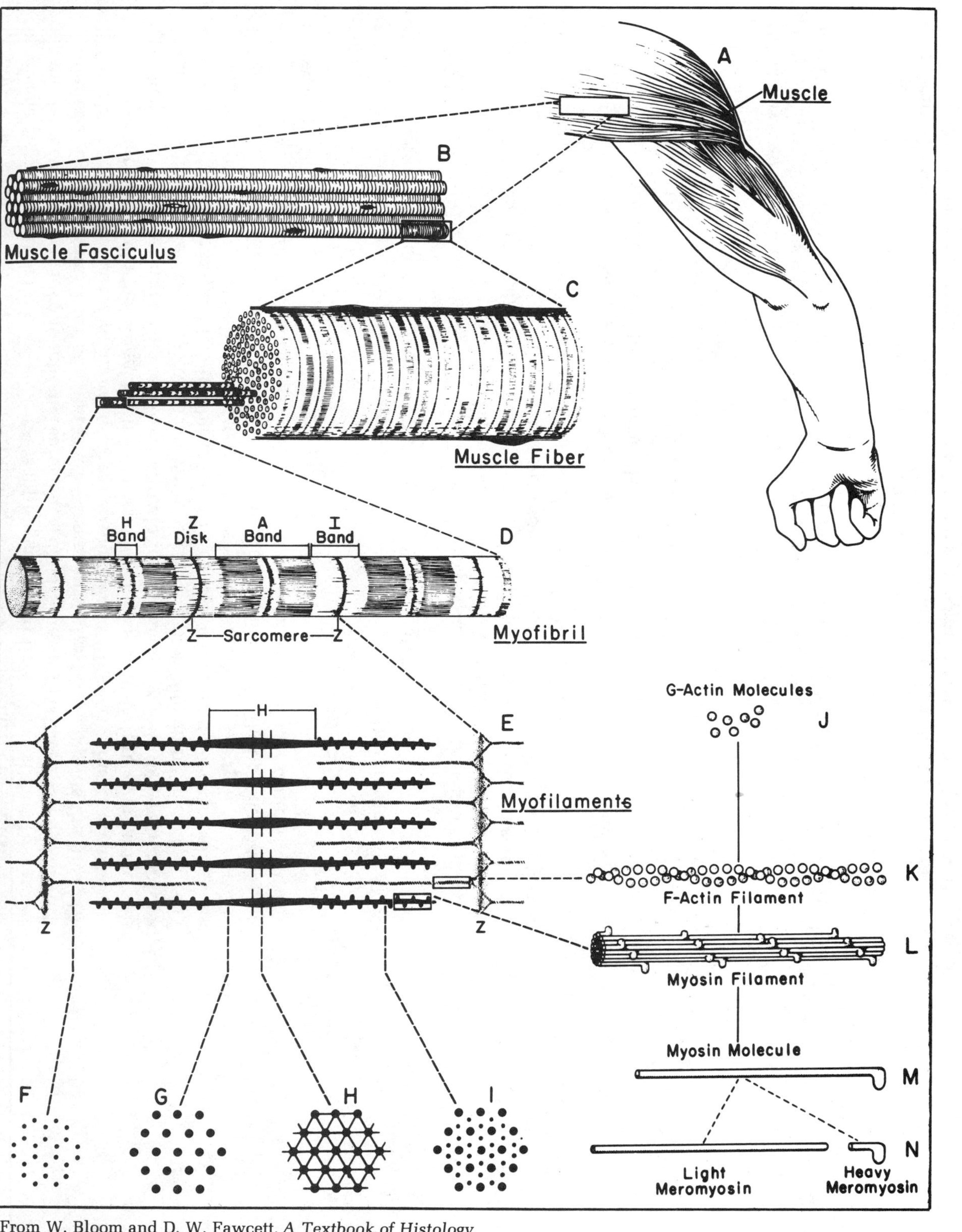

(From W. Bloom and D. W. Fawcett. *A Textbook of Histology* [10th ed.]. Philadelphia: Saunders, 1975.)

253

myosin filament. Thus there is a "blank" area in the H zone where the "handles" overlap, as shown in Fig. 10-15. **The "heads" of the "golf clubs" interact in an unknown manner with the actin molecules. But the heads must have an important role in actin-myosin interaction since ATPase is located on the "heads."** Since energy from ATP is needed for muscle contraction, the presence of ATPase on the "head" (heavy meromyosin) certainly suggests that this is where the "action" is, at least with regard to the energy!

When the muscle contracts, the actin and myosin filaments interact, and the myofilament structures slide into one another as shortening occurs. One way of visualizing this is to realize that the actin myofilaments form a "tunnel" (best seen at *F* in Fig. 10-15) into which the myosin molecules (*G* in Fig. 10-15) can move. A writer (who shall remain nameless) has suggested that the process is best visualized as a centipede (the myosin filament) trying to swim into a piece of wet macaroni (the spaces between the actin filaments). The centipede's "feet" are the "golf-club heads" referred to earlier. To make the picture complete, you must also realize that there are two centipedes tied tail to tail (*H* in Fig. 10-15), swimming in opposite directions! This is all very useful if you want to remember how **maximal muscle tension varies with muscle length,** since the amount of tension depends on how far the centipedes get their heads into the macaroni.

As you can see from Fig. 10-16, there can be no active tension if the muscle length (by stretching) is such that the centipedes cannot get their "feet" into the macaroni. With less stretching the centipede can get a few feet in, and then the amount of active tension is directly and linearly related to the amount of interaction of the actin and myosin (between *A* and *B* in Fig. 10-16). Between *B* and *C* in Fig. 10-16, there is no difference in the tension developed, since even though the centipedes are farther into the macaroni at *C*, they have no more "feet" with which to develop additional grip on the macaroni. Between *C* and *D* of Fig. 10-16, there is less tension (even though the number of centipede "feet" is the same) because the ends of the macaroni are getting tangled up in one another. Finally, between *D* and *E* there is a sharp drop-off in the developed tension for two possible reasons: the centipedes are smashing their heads against the *Z* lines, or the macaroni are crossing so far that a given strand of actin must have "feet" from the two (oppositely directed) centipedes working in opposite directions on it.

Thus the variations of muscle tension that depend on length are nicely explained by the physical interrelationships of the actin and myosin (not to mention centipedes and macaroni).

Fig. 10-16. Isometric tension (active) of maximally stimulated single frog muscle fiber, compared with length and relationships of actin and myosin myofilaments.

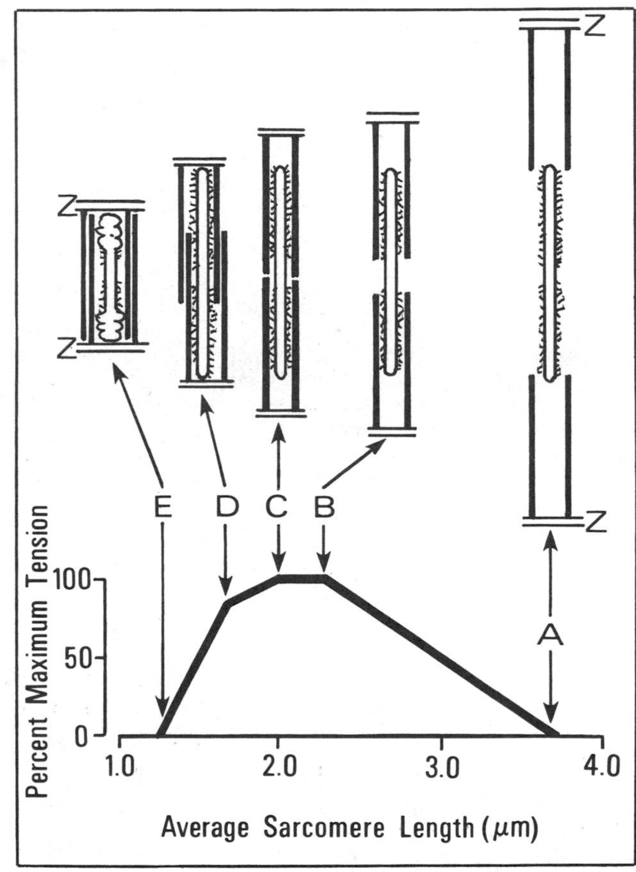

(Modified from A. M. Gordon, A. F. Huxley and F. J. Julian. The variation in isometric tension with sarcomere length in vertebrate muscle fibres, *J. Physiol.* [Lond.] 184:170, 1966.)

HINT

6. No! The size of the response might be due to partial occlusion of the receptor sites (as with curare) or to less transmitter per vesicle.

253

254

(*Note:* In order to avoid giving this book a bad reputation, we would appreciate it if, when discussing this subject with your instructors, you would talk about actin and myosin rather than centipedes with golf-club feet and their tails tied together swimming up pieces of wet macaroni.) 1

The length-versus-tension diagram of maximally stimulated whole muscle is shown in Fig. 10-17. Clearly, as the muscle length is increased, the maximal tension developed by the muscle is increased, but only up to a point. Beyond that point, the amount of active tension declines. 2

This behavior of the whole muscle is readily explained by the results shown in Fig. 10-16. (In Fig. 10-17, the muscle was not lengthened to the point where no tension developed, as was the case in Fig. 10-16.) Remember, however, that this length-tension diagram shows the behavior of the **maximally stimulated whole muscle.** A whole family of such curves could be drawn (all below this curve) if the proportion of activated fibers is altered, if the degree of activation of the contractile mechanism is changed, or if both factors change at the same time. (For further details see the section on length-tension diagrams in Chap. 13.) 3

The **passive tension** developed by the muscle may be **caused by tension in the connective tissue sheaths around the muscle fibers** and not by stretch of the muscle fibers themselves (see the section on the detection of muscle tension in Chap. 13). Nevertheless, the "stiffest" muscle known, in terms of its passive tension, is insect flight muscle, which has no surrounding connective tissue at all! 4

As you will see later, **tensions less than that shown in Fig. 10-17 can be developed by a muscle when the stimulation rate is decreased.** 5

6 First, however, we must finish the story of how the actin and myosin interact.

Intracellular Events in Excitation-Contraction Coupling

The mutual attraction of actin and myosin must be very great indeed, for a "relaxing mechanism" prevents continual interaction from occurring (which would be a waste of energy and a considerable bother as well). 7

The relaxing mechanism has two components: tropomyosin and troponin. The tropomyosin is attached to the actin molecule. (You can remember this if you realize that if it were attached to the myosin, it would be both logical and easy to recall!) **The troponin is attached to the tropomyosin. When tropomyosin and troponin are present** (attached to the actin), **there is no interaction between the actin and myosin, hence no contraction.** However, **the relaxing mechanism can be inhibited by the presence of Ca^{2+}**, which binds to the troponin, causing a conformational change that then **permits the actin and myosin to interact** by means of cross-bridging (we're back to the centipede golf-club feet). The contraction continues for as long as Ca^{2+} is attached to the troponin. So where does the Ca^{2+} 8

Fig. 10-17. Length-tension diagram for skeletal muscle. Passive tension curve measures tension exerted by muscle at each length when it is not stimulated. Total tension curve represents tension developed when muscle contracts isometrically in response to maximal stimulus. Active tension is difference between the two.

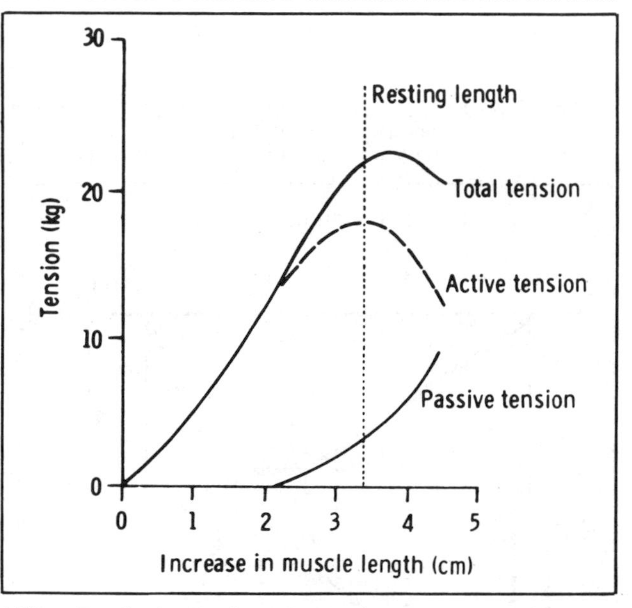

(After Prosthetic Devices Research Project, *Fundamental Studies of Human Locomotion and Other Information Relating to Design of Artificial Limbs*, Berkeley, 1947; from W. F. Ganong. *Review of Medical Physiology* [9th Ed.]. Los Altos, Calif.: Lange, 1979.)

254

usually hide out when the muscle isn't contracting? **The lateral sacs of the sarcoplasmic reticulum are known to contain significant amounts of Ca^{2+}** (see Fig. 10-18).

Then how does the Ca^{2+} get out of the lateral sacs? Apparently, Ca^{2+} is released from the lateral sacs by inward spread of depolarization from the surface membrane action potentials. This depolarization travels from the surface of the muscle fiber down the transverse tubules and lateral sacs that form the triad seen in Fig. 10-18.

Indirect evidence for the role of the transverse tubules in excitation-contraction coupling comes from experiments in which the transverse tubular system has been disrupted by osmotic fluid movements (as shown by electron micrographs). With the transverse tubules disrupted, the outer muscle membrane can still transmit an action potential, but no contraction occurs.

The whole sequence involved in a single muscle twitch can now be listed (Table 10-1).

QUESTION: Which of the events of Table 10-1 are of an all-or-nothing character? (Hint 7↓)

Question: In what sense is the contraction of muscle due to inhibition of an inhibition? (Hint 8↓)

For those who are clinically inclined, some hereditary muscle diseases (the myotonias) are characterized by the difficulty that the patient has in relaxing the muscle after a contraction.

As you might expect, since the amount of Ca^{2+} released is a *graded* phenomenon, **the amount of tension developed by the muscle also is related to the amount of Ca^{2+} binding to the troponin.** Presumably, the amount of Ca^{2+} can be increased by a rapid firing rate of the muscle action potentials.

When a motor nerve is depolarized just once, a single muscle **twitch** occurs. However, if the frequency of stimulation of the nerve is increased, there comes a point at which the muscle does not completely relax between contractions (Fig. 10-19), called **incomplete tetanus.** If the rate of nerve firing is fast enough, then there is no detectable relaxation at all, and the muscle tension is continuous—the **complete tetanus** (Fig. 10-19). Note that tension in a single muscle fiber (as in Fig. 10-19) can be changed by varying the frequency of stimulation of the innervating nerve fiber, presumably because of differing amounts of interaction between actin and myosin, in turn due to different amounts of Ca^{2+} attached to the troponin.

Thus, the following generalization can be seen to bring together the factors controlling muscle tension **in a single fiber: tension is directly related to the amount of interaction between the actin and myosin.** This interaction can be varied by two basic methods. Either the number of

Fig. 10-18. Cutaway drawing showing relations of sarcotubular system to muscle fibrils.

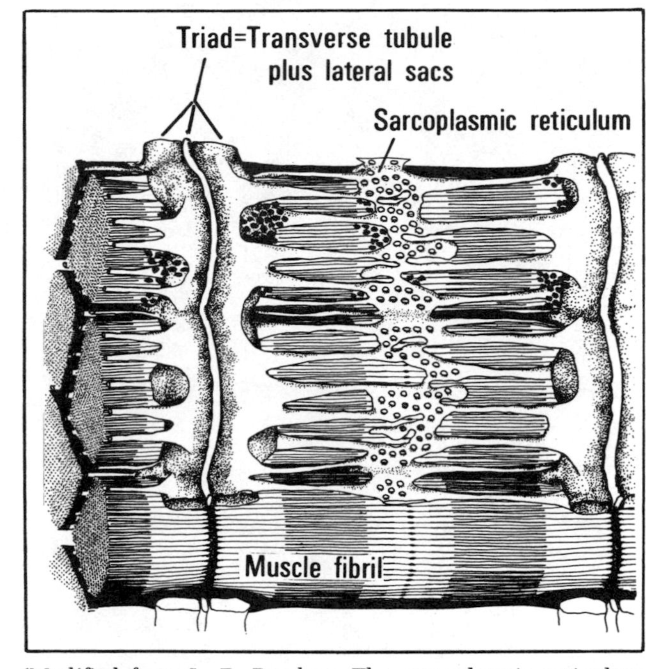

(Modified from L. D. Peachey. The sarcoplasmic reticulum and transverse tubules of the frog sartorius, *J. Cell. Biol.* 25(3):209, 1965. By copyright permission of The Rockefeller University Press.)

256

Table 10-1. Sequence of Events in Single Muscle Twitch

1. Single action potential in the efferent alpha motor neuron.
2. Depolarization of the presynaptic nerve terminals at the myoneural junction.
3. Release from ACh from the presynaptic vesicles, into the synaptic cleft.
4. Diffusion of the ACh across the synaptic cleft.
5. Attachment of the ACh to the outside of the postsynaptic membrane, causing changes in g_K and g_{Na} proportional to the amount of ACh and depolarizing the postsynaptic membrane (electrically inexcitable).
6. Depolarization of the adjacent electrically excitable muscle membrane past threshold, with generation of a propagated action potential up and down the muscle cell membrane.
7. Depolarization passing down the transverse tubules into the depths of the muscle; notice that even in a large muscle fiber the distance to its center is only about 50 μm (that is, approximately one-thirtieth of the length constant of the surface membrane).
8. Release of Ca^{2+} from the lateral sacs of the sarcoplasmic reticulum so that it interacts with the troponin, which is attached to the tropomyosin, which is attached to the actin; thus Ca^{2+} inhibits the relaxing mechanism (troponin and tropomyosin).
9. Active cross-linkages occurring between actin and myosin with development of tension and consumption of ATP energy.
10. With termination of the depolarization of the action potential in the muscle cell membrane and the transverse tubules, "taking up" of the Ca^{2+} back into the lateral sac (by a mechanism requiring "pumping" energy)
11. Prevention of further interaction of actin and myosin by troponin and tropomyosin (the relaxing mechanism).
12. Muscle relaxation.

Fig. 10-19. Muscle tension in single muscle fiber in response to increasing frequency of stimulation. At about three stimuli per second, muscle shows *individual twitches*. At about 10 to 15 stimuli per second, twitches fuse into an incomplete tetanus (relaxation is not complete between contractions). Finally, above 30 stimuli per second, muscle tension becomes constant: the complete tetanic contraction.

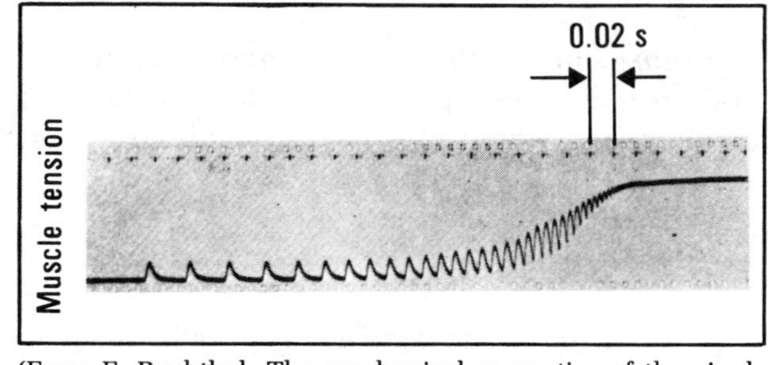

(From F. Buchthal. The mechanical properties of the single striated muscle fibre at rest and during contraction and their structural interpretation. *Dan. Biol. Med.* 17:1, 1942.)

256

centipede feet that can reach the macaroni may be limited by the (stretched) length of the fiber. Or, of those actin-myosin points that can interact as a result of length, tension will develop in only those that are released (by Ca^{2+}) from the relaxing mechanism inhibition that prevents their interaction. Thus, the common factor limiting muscle tension is always the number of interactions that can occur, restricted by either length or absence of Ca^{2+}.

1

Note that muscle force is graded, not all-or-nothing. The physiological mechanisms of the all-or-nothing action potentials can be seen as methods of transmitting information over long distances, either down the axon or up and down the long muscle fiber. Thus the frequency of firing in the axon becomes translated to a quantity of transmitter at the myoneural junction, which in turn determines the frequency of firing in the muscle membrane, which in turn affects the quantity of Ca^{2+} released, which in turn affects the amount of tension developed. In this whole description, remember, the variables mentioned (quantity of transmitter or chemical, frequency of action potential firing) are **graded.** In this way, the actions of the nervous system are similar to those of an analog, rather than a digital, computer. (See further discussion of this issue in "Sensory Code" in Chap. 9.)

2

For whole muscle, muscle force is graded by three means (in addition to muscle length): summation of twitches in an individual fiber to form a tetanus; increase in the average firing rates for the muscle as a whole (subtetanic); and recruitment of additional muscle cells, thus increasing the number of cells firing.

3

You might view the single fiber twitch (at a given length) as a "quantum of contraction." As more and more muscle tension is needed, the following mechanisms can be brought into play. First, more individual fibers can be activated (recruitment). Second, the average firing of the fibers, if they are activated asynchronously, can be increased to give a further gradation of tension, by summation of more "quanta" without any individual fiber going into tetanus. Third, for maximal tensions, an increasing number of fibers, and ultimately the whole muscle, can go into tetanus.

4

The subject of muscle physiology has been sketched here because you need to know the purpose of the whole myoneural junction apparatus. However, many details of muscle action and control

5

HINTS

7. Numbers 1, 2, and 6.

8. The relaxing mechanism inhibits contraction by inhibiting the interaction of actin and myosin. The action of Ca^{2+} is to inhibit the inhibitory action of the relaxing mechanism. Q.E.D.
9. See Table 10-2 for summary.

258

are left for Chap. 13, when you will have more background in sensory processes and spinal cord function. 1

Hypocalcemic Tetany

When blood levels of calcium are decreased, a characteristic clinical picture of hypocalcemic tetany ensues: involuntary, sustained contractions of striated muscles. Let us consider the mechanism of this clinical entity. 2

QUESTION: What are the effects of decreased $[Ca^{2+}]_o$ on motor neuron excitability, neuromuscular transmission, and muscle contraction? (Hint 9↑) 3

From Table 10-2, you can see that, with regard to overall effects on muscle tension, the effects of decreased $[Ca^{2+}]_o$ are contradictory. **While myoneural transmission and muscle contraction are reduced, the axonal excitability is increased to the point of spontaneous firing.** One can only conclude from this, combined with the clinical observations, that **the sustained contractions of hypocalcemia are due primarily to the greater functional effect on the motor neuron.** Thus the overall result is increased muscle tension, even though the tension is less than would occur if the same number of nerve impulses reached a myoneural junction and the interior of a muscle with normal Ca^{2+} concentrations! 4

At first, such differential sensitivity seems unusual or abnormal. Yet a little reflection (and/or experience) will reveal that **most drugs rely on a differential sensitivity between the described effect and the (undesired) side effect.** The situation with Ca^{2+} does not differ in principle. 5

MYONEURAL JUNCTIONS OF CARDIAC AND SMOOTH MUSCLE

You should realize that **the specialized structure of the myoneural junctions discussed so far is peculiar to skeletal muscle.** No such highly organized, one-to-one, synaptic contact is seen in either cardiac or smooth muscle tissues (even in the case of neurogenically controlled, multiunit smooth muscle). 6

In these tissues, the innervating axons usually show a number of **varicosities** along the course of the branching axon terminals. These varicosities are seen as small swellings on the axon in the light microscope. Under the electron microscope, you can see that these swellings contain the vesicles typical of presynaptic terminals. Physiologists may refer to such terminals as *en passant*, meaning that the repeated synaptic contacts along the axon are made as the axon "passes" the muscle fibers (rather than only at the termination of the axon, as in striated muscle). The surrounding "postsynaptic" cells show no apparent specialized synaptic receptor areas. Often the *en passant* endings are localized at the surface of the muscle. Hence although the synaptic gap between the terminal and the 7

Table 10-2. Effects of Decreased $[Ca^{2+}]_o$

Structure	Effect	Page Reference
Motor neuron	Excitability increased to point of spontaneous firing	135–137
Myoneural junction	ACh release diminished	242
Muscle	Decreased strength of contraction (at extremely low levels, sufficient to reduce Ca^{2+} release from lateral sacs)	255–257

258

259

nearest postsynaptic cell may be only a few hundred angstroms, there may exist other cells in the same tissue that may be as much as several millimeters from the nearest nerve terminal! 1

Since this change in the synaptic gap must produce wide variations in the **synaptic delay**, it is not surprising that in these tissues **either the innervating axons merely modulate an ongoing myogenic rhythmicity or,** where contraction is neurogenically regulated, **the rate of change of tension is quite slow** (and the tissues usually are quite thin). 2

Excitation-contraction Coupling in Cardiac Muscle

Cardiac muscle cells possess triads that are similar in all general respects to those found in skeletal muscle. 3

Thus, the same mechanisms for inducing contractions in striated muscle also occur in cardiac muscle. That is, given an action potential in the muscle fiber, the sequence of events is similar. Of course, cardiac muscle differs from striated muscle in two important respects: the inherent rhythmicity (membrane instability—see page 141) and the direct communication between cells, by which the action potential propagates from one muscle cell to the next (page 157). 4

Although some appreciable part of the Ca^{2+} required to raise the concentration of $[Ca^{2+}]_i$ above the threshold for contraction may enter through the surface membrane, clearly by far the greater part of the cell's calcium store is within the longitudinal elements of the sarcoplasmic reticulum system. 5

Excitation-contraction Coupling in Smooth Muscle

It was stated for many years that **smooth muscle cells** did not contain any discernible sarcoplasmic reticulum. Nevertheless, these cells **clearly require the presence of intracellular calcium ions for initiation of contraction.** 6

Since smooth muscle action potentials appear to be related to inward Ca^{2+} movement across the surface membrane and since these cells have a very high surface-to-volume ratio as a result of their small diameter, it appears entirely reasonable that much of, if not all, the calcium required for excitation of the contractile mechanism might come from outside the cell. 7

HINT

10. Check your answer with Table 10-1. Numbers 1, 2, and 6 are all-or-nothing.

259

260

However, it has been known for some years that *most* (if not all) smooth muscles can be made to contract by the appropriate pharmacological agent, **even when completely depolarized by high-$[K^+]_o$ solutions. Such agents also can cause contraction in calcium-free bathing media.** So it was concluded both that smooth muscle cells must contain some intracellular calcium store and that release of calcium from these storage sites can be initiated by the appropriate drug **without change in membrane potential.** 1

So where are these intracellular calcium stores? Recently, investigators showed that a primitive, calcium-containing, sarcoplasmic reticulum system is much *more widely present* in smooth muscles than had previously been realized. Similarly, it has been shown that mitochondria contain rather high internal calcium concentrations, and they are able to pump calcium out of the intracellular fluid. Presumably, one or both of these sites are the missing intracellular calcium stores in smooth muscle. 2

EXAM QUESTION

Outline the sequence of events when a single action potential in a nerve to a muscle fiber leads to a contraction of the muscle. Indicate which of the events are all-or-nothing in character. (Hint 10↑) 3

260

261

Synaptic Transmission and Neuronal Networks

262

As you can see from Fig. 11-1/2-14, you are now ready to understand some of the "inner workings" of the nervous system! This chapter unifies what you have learned, bringing together your prior knowledge, at the same time that you are considering the mechanism in which the CNS unifies and brings together *its* information: synaptic transmission. 1

Here we are concerned mainly with the important ideas that seem to form the basis for understanding significant interactions in the nervous system. Even where the details have not been worked out, our present knowledge is sufficiently rich that we can make reasonable guesses about how the system works at the cellular level. This statement will mean much more to you at the end of the chapter! 2

One of the major ideas developed by Sherrington (before microelectrodes) was *integrative action* of the nervous system. By this he meant the **interaction of the parts of the nervous system in such a way as to provide complex activities that are functionally organized** (for survival, of course). We now know enough to have a good idea of the cellular basis of the interaction and hence of the integrative action. That is what this chapter is about. 3

Several parts of the nervous system have been studied in sufficient detail that many of the mechanisms of neural interaction are quite well understood. These include the mammalian spinal motor neuron, the sympathetic ganglion, the crayfish neural cord, and the nervous system of the leech. We concentrate on the first. 4

Of course, it is sometimes technically easier to do experiments where nature has provided a structure that is less difficult to work with than the mammalian CNS. Thus our knowledge is collected from a variety of species. However, any data about what mechanisms *any* nervous system uses are very valuable in directing thoughts and experiments in some *other* nervous system. The study of invertebrate systems has advanced our knowledge much more rapidly than if we had tried to study only mammals. In this way, "basic" science on subjects that seem far removed from a goal may ultimately shorten the path to that goal! 5

Most of the ideas you learned concerning the neuromuscular junction also apply to synaptic transmission. In this sense, **the myoneural junction can be considered typical of the synapse.** 6

However, **in some ways the myoneural junction is atypical.** First, it has only one presynaptic membrane, which together with the postsynaptic membrane acts as a functional unit *independent of any other synapses*. Second, there is a high safety factor; that is, a firing of the presynaptic cell causes firing of the postsynaptic cell on a 1:1 basis under normal conditions. By contrast, **at the typical synapse, many thousands of presynaptic terminals make contact with each postsynaptic cell. Many of these endings must be more or less synchronously activated to excite the postsynaptic cell.** 7

Fig. 11-1/2-14. Sequences of topics covered in this book, showing chapter numbers.

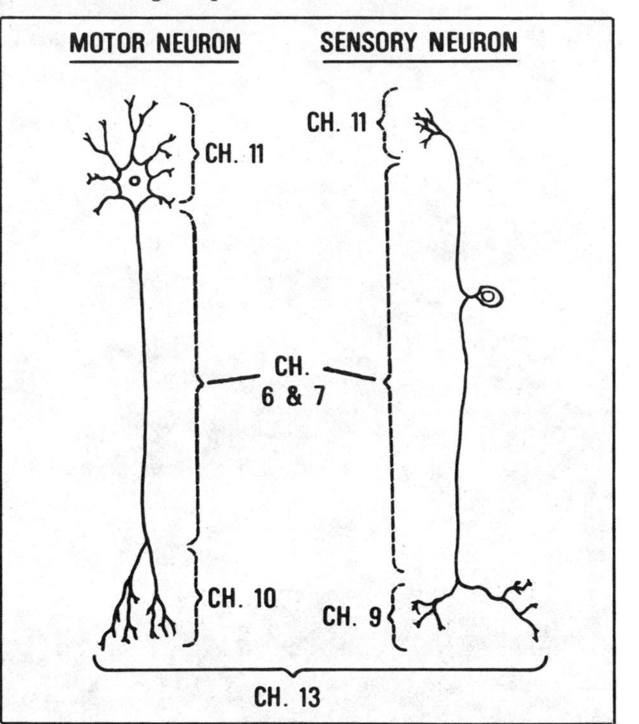

(Modified from E. L. House and B. Pansky. *A Functional Approach to Neuroanatomy* [2d Ed.]. New York: McGraw-Hill, 1967.)

Fig. 11-2. Two types of contact with dendrite (axodendritic synapses).

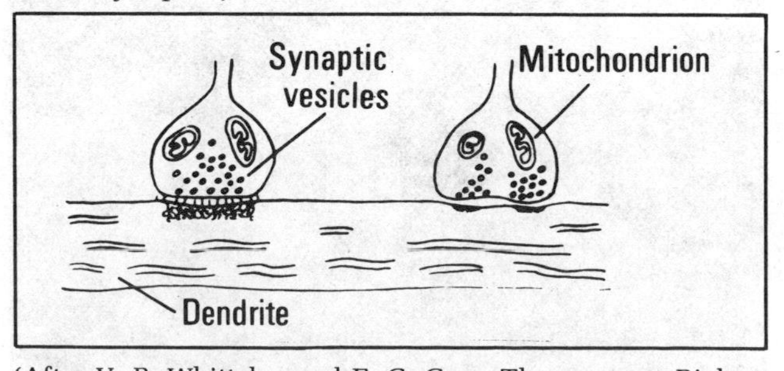

(After V. P. Whittaker and E. G. Gray. The synapse: Biology and morphology, *Br. Med. Bull.* 18:223, 1962; from W. F. Ganong, *Review of Medical Physiology* [6th Ed.]. Los Altos, Calif.: Lange, 1973.)

262

263

ANATOMY

A brief description of the neuroanatomy of synapses (relevant to the physiology) is worthwhile at this point. (For further details, see a neuroanatomy textbook.)

1

Synapses. Electron microscopy has revealed many details of the structure of synapses, a few of which are diagrammed in Figs. 11-2, 11-3, and 11-4. Note that there are a variety of synaptic structures. We do not yet know the significance of these structural differences, but one might guess that *differences in structure may have some influence on function.*

2

While synapses do occur on the cell body (Fig. 11-5), **the greatest number of synapses by far occur on dendrites** (Figs. 11-2 and 11-3). Of course (as a variant of Murphy's Law), we know much more about the less common synapses that end on the cell body (*axosomatic synapses*) than we do about the more common synapses that end on dendrites (*axodendritic synapses*). The variety of dendritic "trees" is large (hinted at in Fig. 11-6; also see Chap. 2). Additionally, axoaxonal synapses are common and provide the anatomic basis for the important concept of *presynaptic inhibition* (see the section "Inhibitory Systems, In Particular" later in this chapter).

3

(For completeness, we also mention dendrodendritic synapses, but please don't ask us what they do!)

4

Axosomatic synapses on the motor neuron are shown in Fig. 11-5. In this chapter, we are concerned mainly with what happens at these axosomatic synapses. Note that *not all the terminal buttons are stained by the method from which this illustration was derived.* It is likely that **most of the surface area of the soma is covered by synapses.**

5

Figure 11-7 diagrams the experimental methods used to study the synaptic activity in the motor neuron. Note that stimuli can be delivered to the *ventral root,* giving direct *antidromic* activation of the neuron. If the *dorsal root* is stimulated, the motor neuron can be activated by *orthodromic* action potentials (synaptically).

6

A further refinement involves orthodromic stimulation of either a cutaneous nerve or nerves from the specific muscles when the ventral root has been cut to prevent antidromic activation.

7

Remember that **orthodromic** means "in the normal direction of action propagation," while **antidromic** means "against the normal direction of propagation." (Thus a car backing up might be said to be traveling antidromically.)

8

Fig. 11-3. Synapse on dendritic spine.

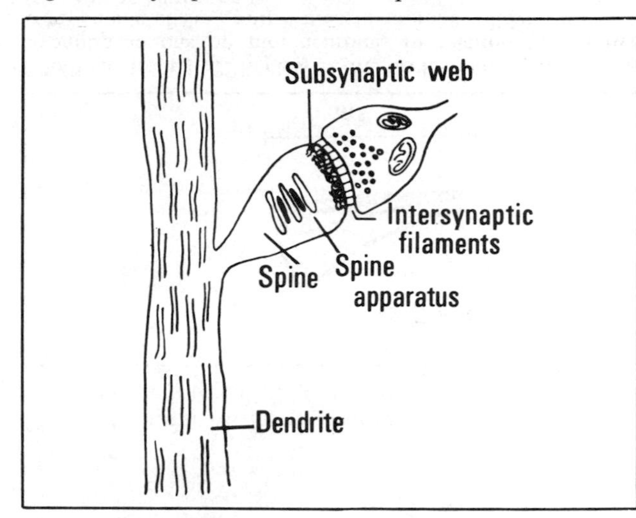

(After V. P. Whittaker and E. G. Gray. The synapse: Biology and morphology, *Br. Med. Bull.* 18:223, 1962; from W. F. Ganong, *Review of Medical Physiology* [6th Ed.]. Los Altos, Calif.: Lange, 1973.)

Fig. 11-4. Axosomatic contacts.

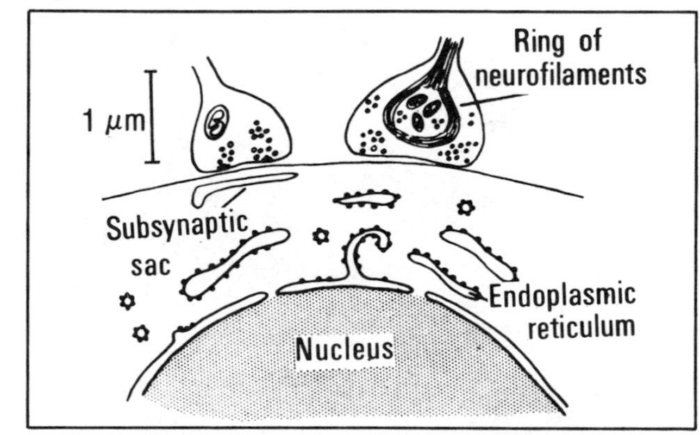

(After V. P. Whittaker and E. G. Gray. The synapse: Biology and morphology, *Br. Med. Bull.* 18:223, 1962; from W. F. Ganong, *Review of Medical Physiology* [6th Ed.]. Los Altos, Calif.: Lange, 1973.)

263

Fig. 11-5. Model of anterior horn cell from lumbar spinal cord of cat. Dark objects are terminal buttons of presynaptic neurons.

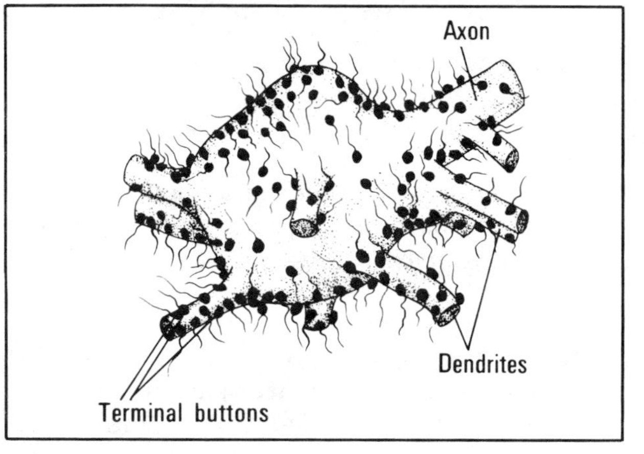

(After R. A. Haggar and M. L. Barr. Quantitative data on the size of synaptic endbulbs in the cat's spinal cord, *J. Comp. Neurol.* 93:17, 1950; from W. F. Ganong, *Review of Medical Physiology* [7th Ed.]. Los Altos, Calif.: Lange, 1975.)

Fig. 11-6. Types of neurons in mammalian nervous system.

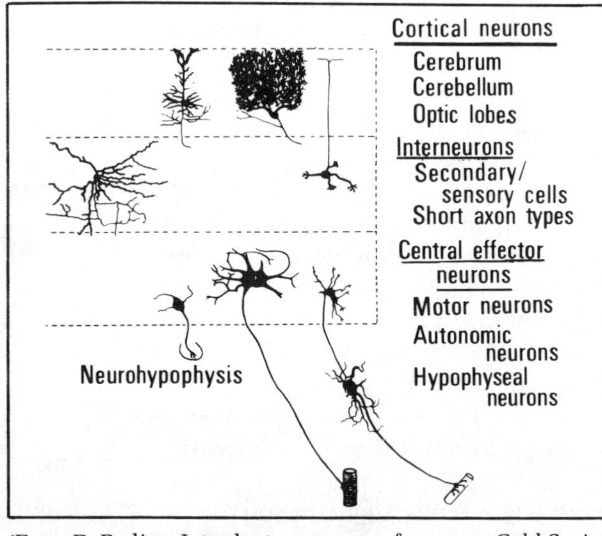

(From D. Bodian, Introductory survey of neurons, *Cold Spring Harbor Symp. Quant. Biol.* 17:1, 1952.)

Fig. 11-7. Arrangement of recording electrode and stimulators for studying synaptic activity in spinal motor neurons in mammals. One stimulator (S_2) is used to produce antidromic impulses for identifying the cell; the other (S_1) is used to produce orthodromic stimulation via reflex pathways.

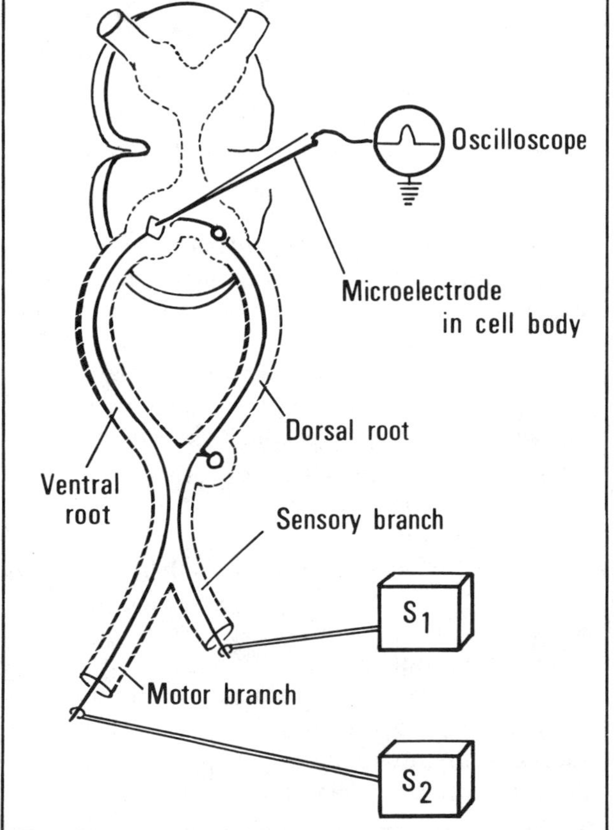

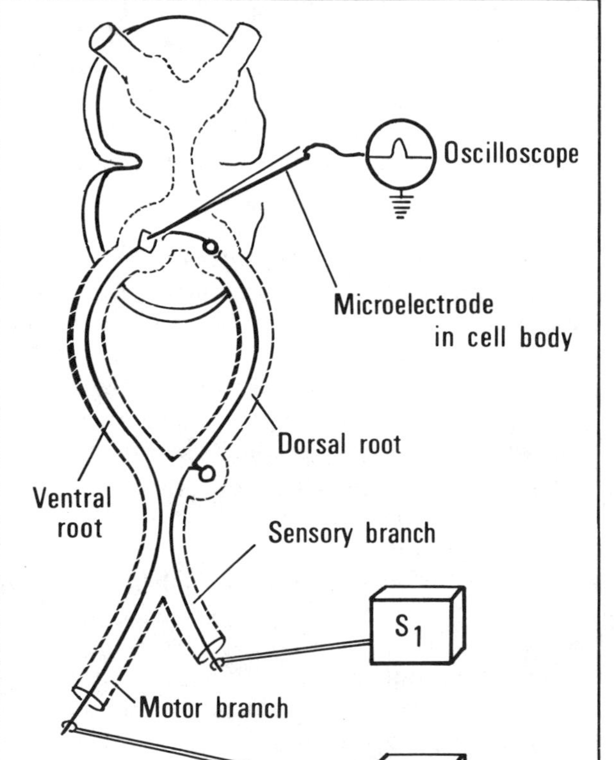

(Modified from W. F. Ganong, *Review of Medical Physiology* [9th Ed.]. Los Altos, Calif.: Lange, 1979.)

Fig. 11-8. Intracellular recordings of EPSPs as number of input fibers to cell is increased.

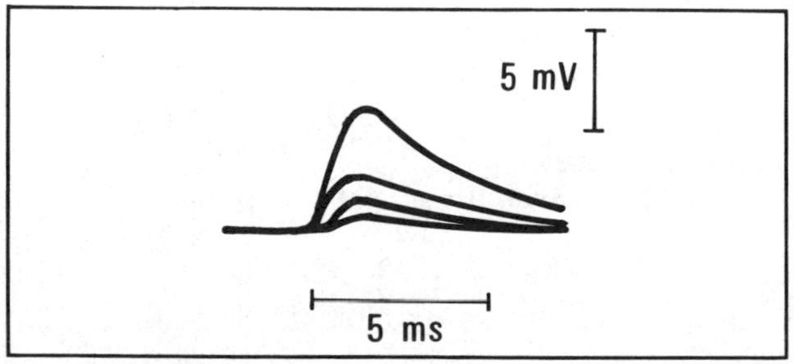

(Modified from J. C. Eccles, R. M. Eccles and A. Lundberg. Synaptic actions on motoneurones in relation to the two components of the group I muscle afferent volley, *J. Physiol.* [Lond.] 136:527, 1957.)

265

EPSP AND IPSP

Orthodromic activation (of appropriate nerve fibers) **causes,** in the motor neuron, **a depolarization known as an *excitatory postsynaptic potential* (EPSP).**

Figure 11-8 shows the superimposed recordings of several EPSPs, the response increases as the number of synaptic inputs to the cell is increased. The response clearly is *graded*, being dependent on the number of simultaneous synaptic inputs to the cell. The depolarization is below threshold, so no action potential occurs. If the number of synaptic inputs is increased still further, then an action potential is also recorded (Fig. 11-9). In Fig. 11-9, note that the top of the action potential is missing as a result of the high oscilloscope gain needed to show clearly the small EPSP. As is explained in greater detail later, **the EPSP and the action potential are generated at two different locations in the motor neuron.**

QUESTION: Where else is it possible to record a depolarizing potential and the action potential it causes at the same time and from the same electrode position? (Hint 1↓)

QUESTION: Why is there a decreasing latency in Fig. 11-9 as the number of synaptic inputs increases? (Hint 2↓)

The same cell can also show **inhibitory postsynaptic potentials (IPSPs)** if other synaptic inputs are activated (Fig. 11-10).

The IPSP is typically a hyperpolarization of the membrane that, like the EPSP, is graded in size by the number of synaptic inputs activated (Fig. 11-10).

Fig. 11-9. Intracellular recordings of EPSP and action potential as number of synaptic inputs is increased (top to bottom).

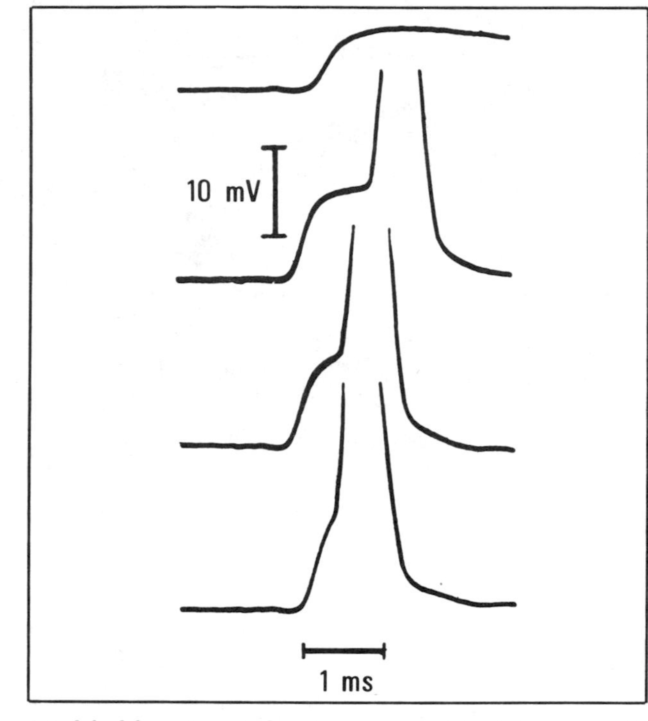

(Modified from J. C. Eccles. Excitatory and inhibitory synaptic action, *Harvey Lect.* 51:1, 1955.)

MECHANISMS OF EPSP AND IPSP

Both the EPSP and the IPSP are postsynaptic responses that result from synaptic transmission. In most cases, the specific chemical transmitter is not known, but it is clear that **in most cases it is not ACh.**

In general, even though most of the synapses in the CNS are probably chemical in nature, relatively few transmitters have been identified with reasonable certainty.

This is not for want of possible candidates: serotonin, GABA, epinephrine, norepinephrine, glutamic acid, glutamate, tyramine, other amino acids, etc.

The full proof that a substance is a transmitter at a given location and for a given response is *very* difficult. Ideally, the following lines of evidence are required to prove that X is a transmitter: (1) X is present at the synapse, in vivo; (2) the enzymes for synthesis and breakdown of X are present at the synapse; (3) there is evidence that X is actually released when transmission occurs; (4) X, when applied experimentally, gives the response measured; and (5) agents that block transmission also block direct application of X.

Fig. 11-10. Intracellular recordings of IPSPs as increasing numbers of synaptic inputs are activated.

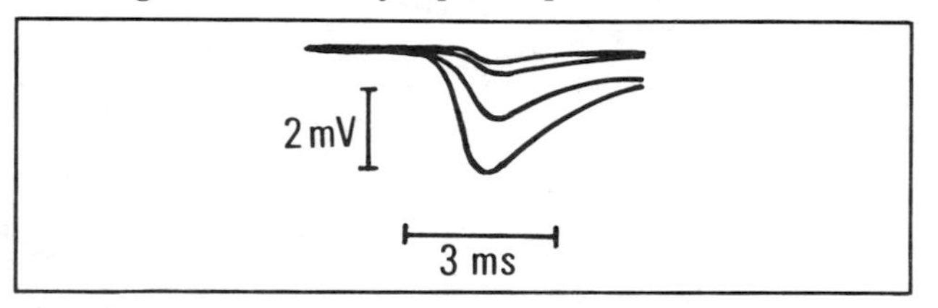

(Modified from J. C. Eccles. The Behaviour of Nerve Cells, in G. E. W. Wolstenholme and C. M. O'Connor [Eds.], *The Neurological Basis of Behaviour* [CIBA Foundation Symposium]. London: Churchill, 1958.)

265

1 Also, in general, there are **different transmitters for the EPSP and the IPSP.**

Theoretically, the same transmitter can have various effects on different membranes. But in a single cell, the effect of an extracellular transmitter (iontophoretically applied) is to either depolarize or hyperpolarize, but not both. 2

3 **Both EPSP and IPSP are characterized by the synaptic delay of a chemical synapse.**

The EPSP synaptic delay is about 0.5 ms (after allowances for conduction velocity and lengths of afferent fibers). However, the *latency* of the IPSP is noticeably longer than that of the EPSP (by about 2 ms in Fig. 11-11). This extra 2 ms is not wholly due to the inhibitory synapse, though. It has been found that the reflex path by which the IPSP is generated involves an *additional synapse,* plus some small, more slowly conducting fibers. When the presynaptic inhibitory endings are stimulated directly, **the synaptic delay of the IPSP is just about 0.5 ms** [40, p. 326]. 4

The duration of action of the transmitters in the EPSP and IPSP has been hard to prove, but the evidence favors a short action (about 2 ms) in each case. The remaining duration (about 7 to 15 ms) is thought to be due to a passive return to the resting membrane potential as a result of resting level conductances (as in the EPP, Chap. 10). 5

Evidence concerning the duration of transmitter action is based on measurements of the time course of the synaptic current. 6

Part of the decay of the EPSP and IPSP also can be related to the electrotonic spread of the potential down the axon and, more importantly, out into the dendritic arborizations. The time course of an *electrical hyperpolarization* is about the same as that of the IPSP [40, pp. 323–324]. However, the time course of decay from an *electrical depolarization* is shorter than that of the EPSP, which argues *against* the hypothesis that since some parts of the soma are electrically excitable (see below), a local response prolongs the EPSP more than the IPSP. 7

8 As might be expected from chemical synapses, **miniature EPSPs (MEPSPs) have been recorded.**

However, it has not been possible to distinguish whether they are produced by random quantal release of transmitter substance from the presynaptic endings or by low-level excitation of afferents causing EPSPs. Either cause would appear as "synaptic noise" on the postsynaptic side of the synapse. 9

Similarly, when relatively few fibers are stimulated peripherally to elicit the EPSP (Fig. 11-12), the variations in heights of EPSPs could be due to variations in threshold of the peripheral axons. 10

Fig. 11-11. Intracellular recording from motor neuron, comparing EPSP from one type of input with IPSP from another type of input.

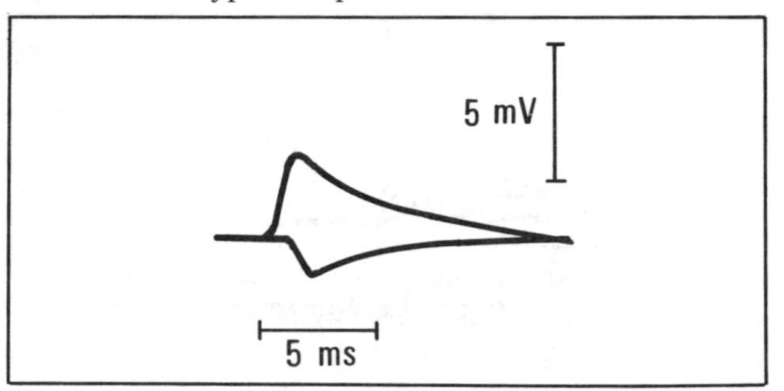

(Modified from J. S. Coombs, J. C. Eccles and P. Fatt. The specific ionic conductances and the ionic movements across the motoneuronal membrane that produce the inhibitory post-synaptic potential, *J. Physiol.* [*Lond.*] 130:326, 1955.)

Fig. 11-12. Intracellular recording of EPSP variations as result of repetition of same stimulus at afferent fibers.

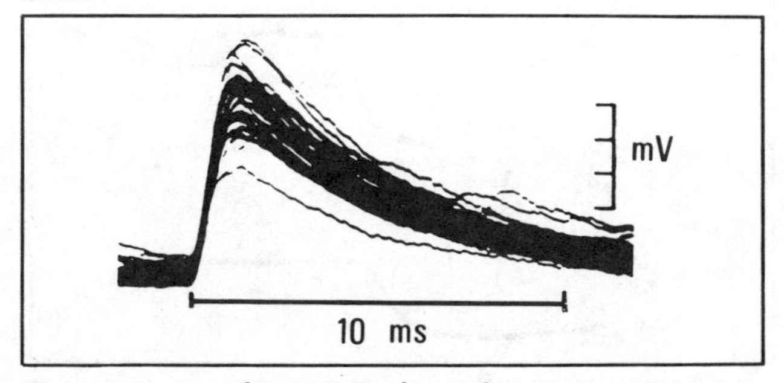

(From J. S. Coombs, J. C. Eccles and P. C. Fatt. Excitatory synaptic action in motoneurones. *J. Physiol.* [*Lond.*] 130:374, 1955.)

267

The ionic basis of the EPSP rests on changes in ionic conductances. (So what's new?) **During the EPSP it seems certain that both g_{Na} and g_K are increased, whereas during the IPSP, g_K and g_{Cl} are increased.** (At last! Chloride does play a role!)

It has not been possible to measure membrane conductances directly even in relatively large motor neurons, let alone in the small interneurons. So synaptic mechanisms usually have been inferred from indirect evidence [12, pp. 49–53].

A classic study of this kind was carried out by Eccles and his coworkers long before the theoretical significance of the reversal potential was fully understood. They concluded that the EPSP must result from a "nonspecific" increase in permeability to both anions and cations since (1) the reversal potential was near 0 mV and (2) the height of the EPSP was affected by local changes in concentration of a wide variety of small anions and cations (see Fig. 11-13). **It has not yet been demonstrated that anions affect the reversal potential for the EPSP.** Eccles himself concluded that his earlier work should be repeated and suggested that many of the ions used may have affected the size of the EPSP by altering resting potential rather than affecting reversal potential [12, p. 53]. In the absence of direct evidence to the contrary, **we presume that the ionic mechanism of the EPSP is similar to that of the EPP.**

In the IPSP, the combined effect of increasing g_K and g_{Cl} may be either to hyperpolarize the cell toward E_K (if $I_K > I_{Cl}$) or to so markedly decrease membrane resistance that small changes in g_{Na} can no longer depolarize the membrane to threshold (where $I_{Cl} > I_K$).

Experiments with injection of various ions into the motor neuron (by iontophoresis) also suggested that the pore that opened during an IPSP might be *nonspecific* in that different ions can pass through it (Fig. 11-13). Of course, **only K^+ and Cl^- are present in vivo in sufficient quantity to cause the ionic currents of the IPSP.** More recent work, however, again suggests that this nonspecificity is illusory. Perhaps the inhibitory transmitter opens two pore types, a relatively specific K^+ pore and a rather less selective anion pore, in approximately equal numbers.

Fig. 11-13. Relative sizes of hydrated ions and *postulated* size of pores for EPSP and IPSP, based on ions that affect the height of postsynaptic potential. The EPSP pores, labeled *E*, were thought to pass all the ions; the IPSP pores, *I*, were considered to pass only those ions that are smaller than Na^+.

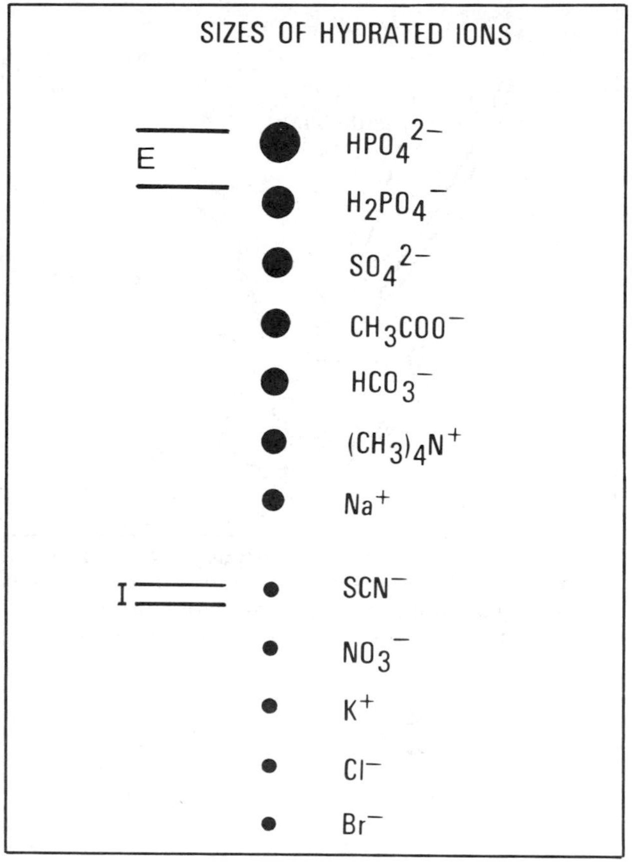

SIZES OF HYDRATED IONS

E

HPO_4^{2-}

$H_2PO_4^-$

SO_4^{2-}

CH_3COO^-

HCO_3^-

$(CH_3)_4N^+$

Na^+

I

SCN^-

NO_3^-

K^+

Cl^-

Br^-

HINTS

1. The generator potential and the action potential from the first node in the Pacinian corpuscle (Chap. 9) and the EPP from the myoneural junction and the action potential from the adjacent electrically excitable muscle fiber membrane (Chap. 10). We hope these examples come readily to mind.
2. This is just what you would expect as the amount of depolarization (EPSP) increased. If this is not clear, perhaps you should review the section on latency and stimulus strength in Chap. 6.

267

268

While it is easy to see how hyperpolarization could cause inhibition in the motor neuron (see below), inhibitory synapses are known that inhibit by depolarizing! In such cases, the inhibitory change is primarily an increase in I_{Cl} (where E_{Cl} is less negative than the resting potential). As a consequence, during the inhibition the membrane potential becomes **less negative.** It **still inhibits** by reducing the effects of an increase in g_{Na}, however! You can see how this might be if you play with the steady-state equation of Chap. 4, assuming $g_{Cl} \gg g_{Na}$. In such a case, a g_{Na} increase will depolarize the cell only slightly, in spite of the large driving force on Na^+. [1]

Sometimes this process is described as inhibition due to a large chloride "shunt." That is, a very low resistance resulting from chloride movement provides an alternative return path for some of the Na^+ current that otherwise would have depolarized the membrane capacitance. When the membrane potential is held at various levels by injection of current through one side of a dual micropipette, the reversal potentials seen are 0 to −10 mV for the EPSP and typically about −80 mV for the IPSP (resting level at about −70 mV) [12, p. 49; 37, pp. 1246–1249; 40, p. 322; or 47, pp. 175–178]. [2]

INTERACTION AT INITIAL SEGMENT

We have been describing the EPSP and IPSP as recorded by a microelectrode inserted in the cell body, or *soma.* **The postsynaptic membrane in which the conductance changes occur is the subsynaptic membrane, i.e., that just underneath the synaptic buttons** (Figs. 11-5 and 11-3). Thus, both EPSPs and IPSPs could be occurring at the same time at different synapses on the soma. **The cell response is the net effect of all the potentials generated by the simultaneously occurring EPSPs and IPSPs. The point of the cell at which this algebraic summation occurs is the initial segment** (Fig. 11-14). [3]

If the net effect of these potentials is to depolarize the initial segment to threshold, an action potential will propagate down the axon. The initial segment is the point of lowest threshold in the cell. Let us look at it this way: in order to depolarize the initial segment, there must be an outward I_{cap} in that region. [4]

This outward current is driven by the summed *inward* currents derived from the activated excitatory synapses (EPSPs). [5]

These summed inward currents drive not only the I_{cap} needed to depolarize the initial segment, but also the capacitative regions of the soma and dendrite membranes. And a further component of outward current occurs through activated inhibitory synapses, thus reducing the current available to depolarize the initial segment. Thus, by implication, the effects of inhibitory synapses are related to their proximity not only to the initial segment, but also to activated excitatory synapses. Thus, spatial arrangement, as well as temporal pattern, will affect the response pattern of a given neuron. [6]

Fig. 11-14. Diagram of parts of motor neuron of physiological importance. It is thought that a microelectrode in the soma cannot detect depolarization of point on distant dendrites, e.g., point *A*.

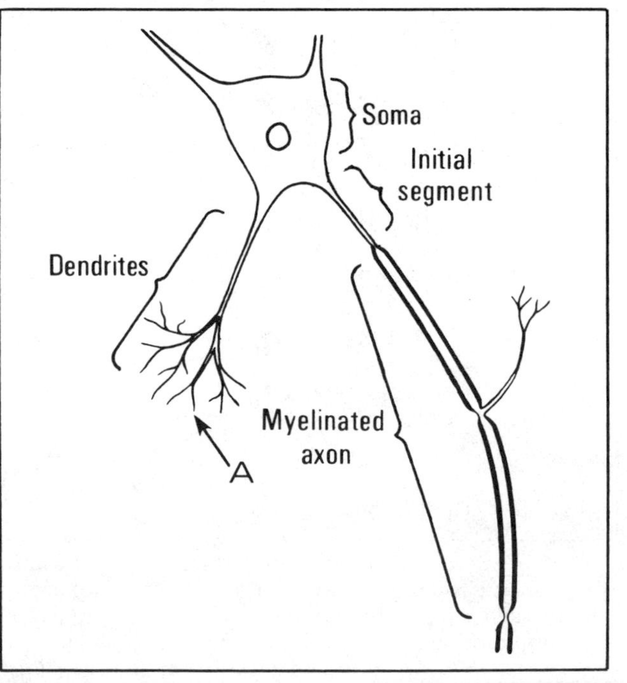

(Modified from J. C. Eccles. The central action of antidromic impulses in motor nerve fibres, *Pflugers Arch.* 260:385, 1955.)

268

269

QUESTION: Assuming that you obtained the results shown in Fig. 11-9, i.e., that the action potential starts off at 10-mV depolarization, is the threshold of the initial segment greater than 10 mV, 10 mV, or less than 10 mV? (Hint 3↓)

QUESTION: If there is a 10-mV hyperpolarization at the soma, will the membrane at the initial segment be hyperpolarized, at resting level, or depolarized? (Hint 4↓)

While the initial segment is the region with the lowest threshold for electrical excitation, **the soma itself can be electrically excited.** When the initial segment generates an action potential, the potential is great enough (after electrotonic spread back into the soma) to bring the soma to threshold. This leads to an electrical depolarization of the soma and main dendrites. (It is not clear how far up the dendritic arborization the process continues.) The sequence of events in the motor neuron is summarized in Table 11-1.

In antidromic activation of the cell, the sequence is obviously different, but the initial segment also depolarizes the soma. Because of the different pattern of activation, the waveshape with orthodromically induced depolarization is somewhat different from that with antidromically induced depolarization [1, pp. 137–140; 12, pp. 101–113].

This is the basis for the identification of motor neurons, since it is possible to verify by waveshape in antidromic activation that the electrode is in the soma [40, pp. 328–329]. You *might* think that any cell that can be activated at all by stimulation of the ventral roots must be a motor cell, but it's a little more complicated than that! Some cells can be activated *synaptically* by ventral root stimulation (see the section on inhibitory systems later in this chapter for discussion of Renshaw cells).

The preceding description requires that **part of the cell** be **electrically excitable, and part of it not.** Since exactly the same thing occurs in the muscle cell (where the neuromuscular junction is chemically excitable and the muscle membrane electrically excitable), it seems plausible that similar conditions might prevail in the motor neuron.

What goes on in the dendrites is far from clear. If there were a gradation of threshold such that the initial segment were the lowest, the soma higher, and the dendrites higher still, perhaps decremental conduction (see page 247) would occur in the dendrites, since the only requirement for such an effect would be raised threshold.

There are other theories of dendritic function: passive electrotonic spread might be the only mechanism. Computations have been attempted, but the problem is made especially difficult by not only the complex geometry of dendrites, but also uncertainty as to whether the values for membrane resistance, membrane capacitance, and intracellular and extracellular resistance are the same for such small structures as they are in the larger structures in which direct

Table 11-1. Sequence of Events in Synaptic Activation in the Motor Neuron

1. Depolarization of one or more afferent presynaptic endings.
2. Release of transmitter substances (chemical structure unknown).
3. Changes in conductances of the postsynaptic (subsynaptic) membranes due to action of the transmitter.
4. Changes in ionic currents, due to the changed conductances (a net inward subsynaptic current in the EPSP, a net outward subsynaptic current in the IPSP).
5. Changes in the charge on the soma membrane capacitance, due to the net effect of all the currents acting (including electrotonic spread). If there is a net depolarization, there has been a net outward current through the soma capacitance; the opposite occurs in a net hyperpolarization.
6. Generation of an action potential at the lowest threshold region of the electrically excitable membrane: i.e., the initial segment, if it is depolarized to threshold.
7. Propagation of the action potential down the axon, and propagation back into the soma and main dendrites as the higher-threshold, electrically excitable portions of the soma are brought above threshold by electrotonic spread from the initial segment (plus any residual depolarization remaining from the long-lasting EPSPs). Note that the initial segment action potential is a much larger depolarization than is the EPSP.

269

measurements have been possible. (We now have enough types of membranes and membrane responses to make the number of theoretically possible mechanisms in the dendrites very large!) 1

At this point, let us reiterate that as far as soma synapses (and probably the action of dendrites as well) are concerned, the initial segment acts as the final "judge" of the net effects of all excitatory and inhibitory inputs acting at the same time. Thus, the result is that **the initial segment responds to the net result of the spatial and temporal summations of both excitatory and inhibitory inputs.** 2

Both spatial and temporal summation are shown in a motor neuron in Fig. 11-15. **Spatial summation** occurs because **as stimulus strength increases, the number of peripheral afferent axons stimulated increases and hence the number of excitatory synapses activated is increased. Temporal summation** (shown in Fig. 11-15) **occurs because both the EPSP and IPSP are potentials that outlast the presynaptic events which cause them.** 3

QUESTION: Consider two excitatory synapses, call them 1 and 2, to a cell. In each of the following, will it be spatial summation, temporal summation, or both?

1. Synapses 1 and 2 are activated at the same time.
2. Synapse 2 is activated; then 2 is activated again immediately.
3. Synapse 1 is activated, then 2 is activated immediately after.

(Hint 5↓) 4

The amount of depolarization of the initial segment determines the firing rate of the axon. 5

The mechanism is presumably the same as that seen in the repetitive firing of sensory neurons (page 226). 6

In addition, another mechanism occurs in motor neurons: a prolonged after-hyperpolarization, lasting 70 to 170 ms. Such a long hyperpolarization could affect the firing of the initial segment, since the amount of depolarization needed to bring the potential of the initial segment to threshold would be greater soon after a firing. Thus, the greater the synaptically induced depolarization, the sooner after a firing it would be possible to reactivate the initial segment. This hypothesis gains additional credence from the finding that the motor neurons serving a slow muscle (soleus) have both a slow firing rate (10 to 20/s) and a long after-hyperpolarization (greater than 130 ms), while motor neurons to a fast muscle (gastrocnemius) showed a faster firing rate (30 to 60/s) and a shorter after-hyperpolarization (50 to 110 ms). 7

The importance of the after-hyperpolarization to the normal functioning of these cells is indicated by the following observation. When the membrane po- 8

Fig. 11-15. Diagram to show spatial summation (upper recordings) and temporal summation (lower recordings). Traces show responses in postsynaptic cell to stimulation (at arrows) of presynaptic fibers. The continuous, horizontal lines indicate zero potential, and broken lines show threshold for action-potential production in postsynaptic cell. In upper recordings, stimulation of more and more fibers (indicated by thickness of arrows) produces progressively larger EPSPs, the final one being large enough to produce an action potential. In lower recordings, EPSPs of same size (produced by activity in same group of presynaptic fibers) sum if time interval between one EPSP and the next is short enough.

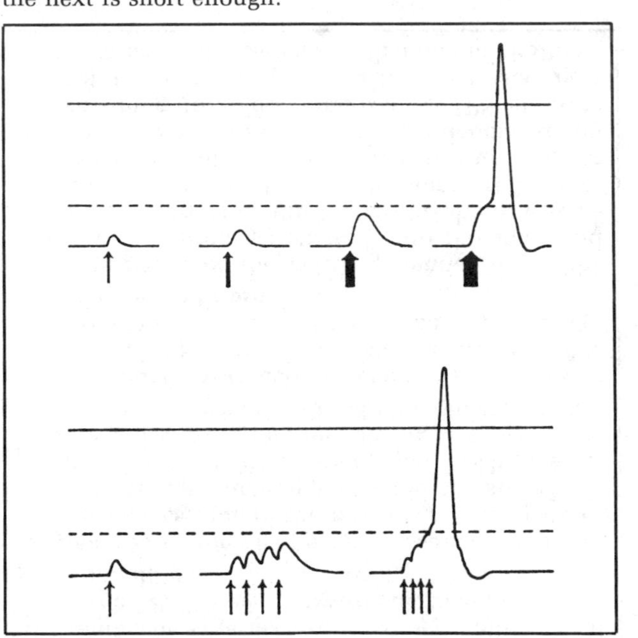

(From D. J. Aidley, *The Physiology of Excitable Cells*. Cambridge: Cambridge University Press, 1971.)

271

tential is permitted to return to rest after prolonged electrical hyperpolarization, firing occurs in the initial segment [40, pp. 335–338]. It follows that the initial segment, like the squid axon, must be partially accommodated even at resting potential. If the excitability of the initial segment were not "reset" by the normal after-potential, it seems likely that the initial segment would become steadily more refractory with each impulse. But this after-hyperpolarization would be vastly less effective if there were no soma-dendrite spike. How could the small initial segment generate a sufficient current to repolarize itself effectively if the large soma were not also repolarizing at the same time? Thus one of the most curious features of this mechanism, the apparently antidromic invasion of the soma by the initial segment spike, might be a vital component of the normal functioning of the motor neuron!

LONGER-LASTING CHANGES IN SYNAPTIC TRANSMISSION

Up to now, we considered mainly the effects of presynaptic endings activated simultaneously by a single stimulus to afferent fibers. However, interesting and important effects can be produced by double stimuli, trains of stimuli, and continuous, repetitive firing of afferent fibers.

In a study of motor neurons, by using two stimuli, one after the other to the same afferent nerve fibers, **both facilitation and depression are observed.**

The meanings of these terms are the same as previously described for the EPP. That is, the second stimulus can give an EPSP that is up to 120 percent of that obtained from the first stimulus. This **facilitation occurs in the first 15 to 20 ms after the first stimulus, during which time temporal summation also occurs.** However, **for longer time séparations between the two stimuli, there is a decrease in the second response,** i.e., a *depression* that lasts many seconds! The evidence indicates that in the EPSP, as in the EPP, **these changes** are presynaptic in origin and probably **are due to varying amounts of transmitter released** during the action potential at the presynaptic terminals.

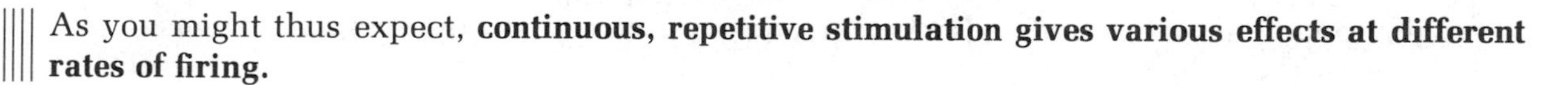

As you might thus expect, **continuous, repetitive stimulation gives various effects at different rates of firing.**

HINTS

3. It must be *less* than 10 mV since the recording is made in the soma and there must be decrement as a result of electrotonic spread to the soma from the initial segment.

4. It will be hyperpolarized, but not as much as at the soma. The principle is this: Electrotonic spread decrements hyperpolarizations, just as it does depolarizations. *The tendency is for the membrane at the initial segment to be changed less than any change at the cell body.*

271

272

At low rates, there is a long-lasting depression. At intermediate rates (60 to 100/s), there is a relative facilitation. At higher rates, there is again depression of the heights of the individual EPSPs, but they are so close in time that there is sufficient temporal summation to give a maintained postsynaptic depolarization. 1

A final method of studying longer-term changes in synaptic transmission is by trains of stimuli, *a tetanic stimulation*. Here a stimulus frequency is maintained for varying periods (conditioning stimuli). Then a single (test) stimulus is applied at various times after the end of the repetitive stimulation (stimulus train). **A common finding** in this sort of experiment **is posttetanic potentiation.** 2

Posttetanic potentiation can be used to show that **synaptic pathways can be facilitated for minutes after rather brief stimulation.** For example, in some experiments [37, pp. 1262–1263], a 9-s train of stimuli gave potentiation for over 1 min, and a 120-s train gave over 5 min of potentiation. The term *potentiation* is a carryover from the time before intracellular recording from these neurons became possible. Measurements of soma potentials in motor neurons indicate that **potentiation is due to presynaptic facilitation,** as previously described. 3

Such facilitation is limited to presynaptic terminals that are activated by incoming action potentials. This has been shown by experiments in which there are two different nerves that can activate a given cell. Tetanic stimulation of one nerve can show posttetanic potentiation, but does not potentiate the effect of stimulation of the second nerve. 4

Motor neurons also exhibit accommodation and "reverse" accommodation, i.e., changes in threshold during prolonged changes in membrane potential. These effects can combine with the presynaptic effects of facilitation and depression to make the responses of synaptic systems quite variable (and hence not easily predictable). 5

PUMP POTENTIALS AND SYNAPTIC ACTION

As long-lasting as the effect of tetanic stimulation may be, it is still hardly long enough to explain such obvious qualities of the nervous system as short-term memory. However, very recent findings suggest that **some cellular effects of neural activation can last for tens of minutes.** These effects are **caused by release of synaptic transmitter.** However, the mechanism of the prolonged change in the membrane potential is due not to a conductance change, but to change in the rate of pumping of an electrogenic pump. 6

Such long-acting effects have been observed in mammalian sympathetic ganglia. It was shown that the potential changes were caused by a chemical transmitter, but were *not* caused by a change in membrane conductance. 7

272

273

Apparently, the transmitter increases the pump rate, which affects the membrane potential, which in turn influences cell excitability, thus producing a long-term *modulating action* on the effectiveness of normal synaptic input.

As an exercise for the student, this mechanism can be added to those in the preceding section, so as to increase markedly the number of possible methods the CNS can use in producing unpredictable results, all of which are synaptically mediated by chemical transmitters!

Potentials that are due to electrogenic pumps should not surprise you. Recall (see pages 85 to 86) that the change in membrane potential resulting from the effects of a 3:2 electrogenic sodium-potassium pump is extremely small (about 2 to 3 mV) when the cell is in steady state. However, if the pump rate is increased by transmitter action, substantial potentials can be achieved by electrogenic pumps.

When cat papillary muscles are rewarmed after a period of cold storage, potentials of the order of −300 mV have been observed while the muscle cells pump out the excess $[Na^+]_i$ that accumulated when the pump was slowed by cold storage.

The economy of producing long-term changes in cell potentials by an electrogenic pump should be appreciated. If the potential change were due to conductance changes, many more ions would have to be pumped! Recall that all moving ions will have to be pumped, so an increased conductance implies a lot of extra work since such potential changes are sustained for long periods. Thus, we see that **conductance changes are utilized** in the nervous system **when rapid, short changes occur, while changes in pumping may be used when slow, long changes are needed.**

ELECTRICAL SYNAPSES

While we are breaking down cherished notions, we had better remind you that **not all synapses are chemical. Some synapses are electrical.** That is, a membrane depolarization on the presynaptic side will depolarize the postsynaptic side enough to cause either an action potential or increased excitability on the postsynaptic side.

For the most part, such synapses have been found and studied electrically in invertebrates, but **synapses with similar histologic features** (as shown by electron microscopy) **have been**

HINT

5. (1) spatial; (2) temporal; (3) both.

273

274

Table 11-2. Properties of Chemical and Electrical Synapses

I. Properties of chemical synapses exclusively:
1. Liberation of a transmitter substance
2. Ability to block or prolong activity with drugs
3. Miniature postsynaptic potentials
4. Reversal potentials
5. Synaptic vesicles

II. Properties that may be shared by both chemical and electrical synapses:
1. One-way transmission
2. Similar time course of postsynaptic potential
3. Increased membrane conductance in postsynaptic element (during depolarizing current or transmitter action)
4. Electrical inexcitability of the postsynaptic membrane in chemical synapses and of the presynaptic and postsynaptic membrane in electrical synapses (!)

III. Properties that distinguish chemical and electrical synapses:

Property	*Chemical Synapse*	*Electrical Synapse*
1. Synaptic delay	Present	Negligible
2. Electric coupling (presynaptic to postsynaptic)	Negligible	Large
3. Structure as shown by electron microscopy	Clear separation	Close coupling

Adapted from J. C. Eccles, *The Physiology of Synapses.* New York: Springer-Verlag, 1964. Pp. 262–264.

found in the CNS of mammals. The functional significance of such synapses in the CNS is unknown at present. 1

However, **electrical synapses are present and important in cardiac and smooth muscle** (see pages 157 to 158). 2

One might expect that electrical synapses would conduct in both directions, in contrast to the one-way conduction in chemical synapses, but some electrical synapses are unidirectional as well! The distinctions between chemical and electrical synapses are summarized in Table 11-2. 3

There is one synapse in the chick ciliary ganglion that is both electrical and chemical! What a mind-boggler! See Eccles [12, pp. 145–147]. 4

EPHAPTIC TRANSMISSION

Last, but certainly not least, **under a few conditions, transmission can occur from cell to cell in the absence of a synapse.** This is called **ephaptic transmission.** 5

274

275

Ephaptic transmission can occur when a nerve that has been cut or otherwise damaged is stimulated, so as to cause a massive synchronized depolarization. Also, it has been postulated that in some types of "phantom limb" pain, the ends of the motor and sensory nerve fibers become so enmeshed in the stump that when the patient "moves" the phantom limb, the efferent motor impulses to the stump may activate afferent pain fibers by ephaptic transmission.

Furthermore, it can be shown that activity in one axon can change the excitability of other axons in the same nerve trunk. But it also has been shown that the change in excitability is not large enough to do more than produce small changes in conduction velocity. It is not clear, however, whether such changes have any physiological function in normal, undamaged nerves.

Table 11-3. Summary of Properties of Excitable Membranes

	Action Potential	Generator Potential	Postsynaptic Potential
1. Cause of change in membrane potential	Current through membrane capacitance	Same	Same
2. Cause of change of current	Conductance increase	Same	Same
3. Mechanism of return of potential to resting level (in addition to electrotonic spread)	g_K; g_{Na}	Passive (resting-level g's)	Passive (resting-level g's)
4. Cause of change in conductance	Electrical depolarization	Specific energy (transduction)	Chemical transmitter
5. Conductances that change	g_{Na} fast response, short duration g_K slower response, long duration	g_{Na}, g_K, g_{Cl} (various combinations at different receptor membranes)	EPP: g_{Na}, g_K EPSP: g_{Na}, g_K IPSP: g_K, g_{Cl}
6. Response to input (stimulus)	Graded below threshold, maximal above threshold	Graded only	Graded only
7. Refractory period	Yes (above threshold only)	No	No
8. Temporal summation	Only below threshold	Yes	Yes
9. Spatial summation	Yes	Yes	Yes
10. Membrane generating potential change is electrically excitable	Yes, but not during absolute refractory period	No	No
11. Acts to depolarize electrically excitable membrane, at a distance from the site of conductance change	Yes	Yes	Yes (except IPSP)
12. Facilitation	No	Not usually	Yes
13. Duration of longest aftereffects	Milliseconds	Tens of milliseconds	Minutes

275

276

SUMMARY

At this point, we might as well pull all the major ideas about electrical potentials in nervous system tissues together in one grand table, Table 11-3. Both similarities and differences among the different types of membrane potentials are presented. All the points have been covered previously in this book, so if you do not understand parts of the table, this indicates parts of the material that you should review.

You are now in a position to understand the following generality. **Membrane potentials are due to movements of ions across the membrane.** And, with a few minor exceptions, **the movement of the ions is due to conductance changes, and the differences between membranes are determined by the ions that move and the various stimuli required to bring about the conductance changes.**

This brings us full circle, for this idea was one of the first in Chap. 4 (page 42)! You might find it interesting to look back at Table 2-7 and to the ideas of page 17.2 and Table 2-7 (page 19) to see how much you **now** understand of what you read there.

INHIBITORY SYSTEMS, IN PARTICULAR

In describing synaptic activity, we mentioned inhibition, but have concentrated on excitation. Now it is important to emphasize inhibitory systems because **inhibition plays an important role in the functioning of the CNS.**

Historically, **excitatory systems must be discovered before inhibitory ones,** since, unless one is recording intracellularly (a recent development), one can detect inhibition only by the reduction of an excited reflex response. Thus, our ideas of the nervous system have progressed from theories based exclusively on excitation to those in which inhibition plays an increasingly dominant role. It is easy to show that any system based exclusively or extensively on excitation soon "turns on" maximally with predictable results: the system becomes incapacitated! Grand mal seizures and strychnine convulsions are examples of the results of excess excitation (or the absence of inhibition). Thus, inhibitory neurons may be crucial in the functioning of the entire system. They may also play an important role in directing reflex responses through the CNS (see the section "Overall View").

We now take up four different inhibitory systems: the Golgi bottle neuron, the Renshaw cell, presynaptic inhibition, and lateral inhibition.

The Golgi bottle neuron. The neural system shown in Fig. 11-16 has been extensively studied and used to obtain many of the results described previously in this chapter.

Fig. 11-16. Probable anatomic connections responsible for inhibiting antagonists to muscle contracting in response to stretch. Activity is initiated in spindle in agonist muscle. Impulses excite motor neurons supplying the same muscle and, via collaterals, the Golgi bottle neurons that end on motor neurons of antagonist muscle.

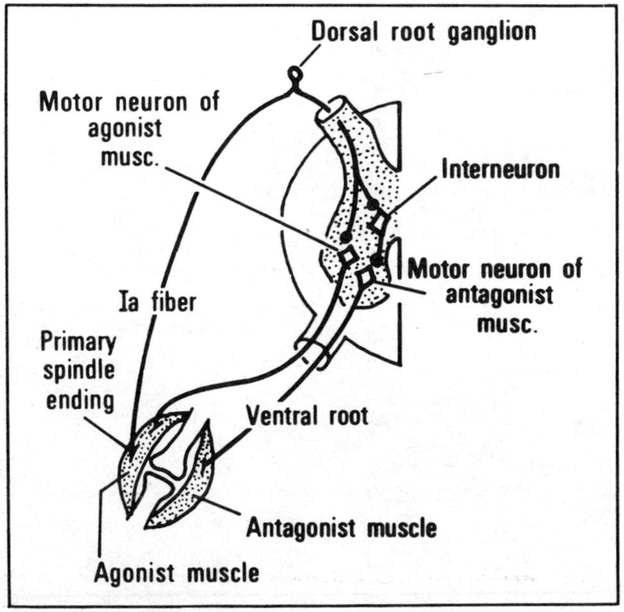

(Modified from W. F. Ganong, *Review of Medical Physiology* [9th Ed.]. Los Altos, Calif.: Lange, 1979.)

276

277

Sensory impulses from muscle stretch receptors (from a given *agonist* muscle) pass through the dorsal roots and then excite at least two types of cells: motor neurons of the same (agonist) muscle and inhibitory interneurons (Golgi bottle neurons). The inhibitory interneurons, in turn, inhibit the motor neurons of the opposing (antagonist) muscle (Fig. 11-16). Both pathways are important in spinal cord reflexes (see Chap. 13).

1

As you can see, **by recording from the appropriate cell and stimulating the appropriate nerve, either EPSPs or IPSPs can be experimentally recorded.** The reason for the extra latency of the IPSP is now obvious. Even though the distances traveled are similar, the reflex must pass through an *extra* synapse (with its 0.5-ms "synaptic delay") and along the short, slower-conducting axon of the inhibitory interneuron.

2

Note that there seems to be a "necessity" to interpose an inhibitory neuron in a reflex path when an excitation is to be changed to an inhibition. This may be the result of the principle that a neuron secretes only a single transmitter at its various axon terminals, even if they are far removed from one another. This idea is called **Dale's principle** and it seems to hold for all systems studied so far.

3

However, since we know so little about synaptic transmission and the elaborate interconnections of the nervous system, the "principle" is more of a hypothesis that has not yet been refuted than a fully established "law" (the Jewett-Rayner law strikes again!).

4

In one invertebrate system, direct excitation and direct inhibition were traced to the action of a single neuron. At first, this was thought to be a violation of Dale's principle, but further investigation revealed that ACh (acetylcholine) was released at both terminals. One postsynaptic membrane was depolarized by ACh, while the other was hyperpolarized! Thus, the differences lay not in the transmitter, but in the response of the postsynaptic membrane.

5

The Renshaw cell. The anatomy of this cell is diagrammed in Fig. 11-17. This cell, like the Golgi bottle cell, turns an excitation into an inhibition.

6

QUESTION: The release of acetylcholine (ACh) at the synapse exciting the Renshaw cell has been postulated. On what basis, do you think? (Hint 6↓)

7

Since the **Renshaw cell is activated** by action potentials in the motor neuron axon, **it is excited by either orthodromic or antidromic impulses** and so at first was difficult to distinguish from motor neurons in microelectrode experiments. The function of the circuit (other than annoying neurophysiologists) is uncertain.

8

Fig. 11-17. Negative feedback inhibition of spinal motor neuron via inhibitory interneuron (Renshaw cell).

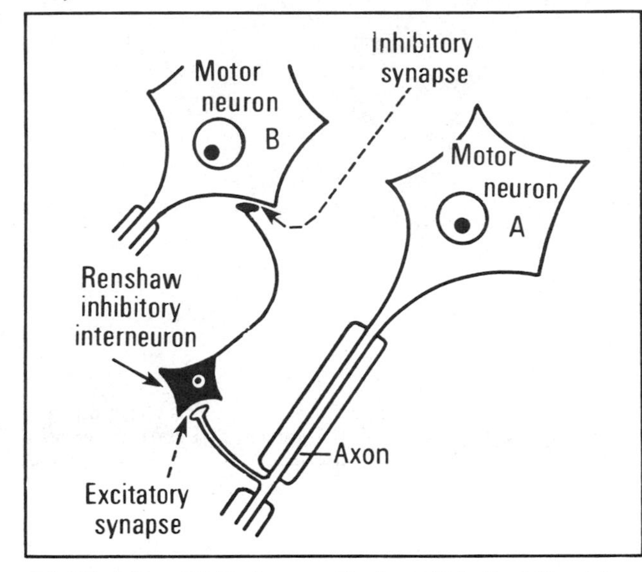

(Modified from W. F. Ganong, *Review of Medical Physiology* [9th Ed.]. Los Altos, Calif.: Lange, 1979.)

277

This is the neuronal circuit that was mentioned previously (page 269) as causing problems in identifying motor neurons by intracellular recordings. However, as previously mentioned, the response of the motor neuron can be identified on the basis of waveshape [1, pp. 137–140; 40, pp. 327–329].

It may be that the Renshaw cell circuit acts in much the same way as *lateral inhibition* in sensory systems (see the next section). Note that there is mutual inhibition; i.e., motor neuron *A* (Fig. 11-17) inhibits *B* (and many others), while *B* also inhibits *A* and many others (not shown). This tends to make the first and fastest-firing neuron "dominate" the others. Perhaps (according to the "Jewett theory") this is the mechanism by which the firings of motor units to a muscle are interrelated, to fire some muscle fibers and let others rest during sustained, moderate contractions. (There is so little evidence for this that we were tempted to call it "Jewett's law"!)

If there were a mutual cross-connection between cells *A* and *B*, as described in the preceding paragraph, then this neural network would form a positive feedback system (see Chap. 12). Note that the Renshaw cell in Fig. 11-17 is not shown sending a collateral back to neuron *A* (as is the case in many textbooks). Such a collateral would inhibit the firing of a cell that had just fired. Renshaw's data seem to indicate that the inhibition occurs on adjacent motor neurons, but not on the motor neuron activating a given Renshaw cell. This would make a considerable difference in the effects of the network (see the next section, on lateral inhibition).

Presynaptic inhibition. An entirely different mechanism for inhibition from that already described occurs with presynaptic inhibition, shown in Fig. 11-18. In this case, the inhibition occurs *before* the synapse, hence is termed *pre*synaptic inhibition.

This inhibition rests on the special anatomy of a synaptic ending on an axon terminal (Fig. 11-18). **Both synapses are excitatory. The depolarization of the axon terminal of axon *A*** (Fig. 11-18) **by the action of axon *B* reduces the amount of transmitter released by the terminal of axon *A*.**

This can be better understood by realizing that **one of the factors** that **determines the amount of transmitter released by a terminal is the spike height in the terminal** (Fig. 11-19), where the spike height is **the difference between the resting membrane potential and the peak of the action potential.** Thus, depolarization of the presynaptic membrane reduces spike height. As you can see from Fig. 11-19, as the presynaptic spike size is reduced as little as 10 mV (80 to 70 mV), the postsynaptic EPSP size is reduced to less than half its maximum size. **Since the amount of transmitter released is so sensitive to presynaptic spike size, small, prolonged depolarizations of the axon terminal of the neuron ending on the motor neuron** (Fig. 11-18) **can greatly reduce the amount of transmitter released, and hence the amount of depolarization of the motor neuron.** The neuron that ter-

Fig. 11-18. Arrangement of neurons producing presynaptic and postsynaptic inhibition on a motor neuron. The neuron producing presynaptic inhibition is shown ending on excitatory synaptic knob. Many of these neurons actually end higher up along axon of excitatory cell.

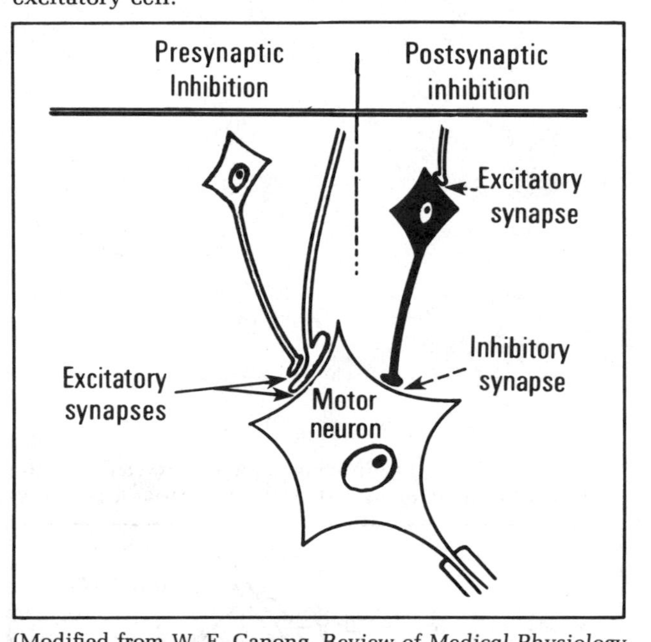

(Modified from W. F. Ganong, *Review of Medical Physiology* [9th Ed.]. Los Altos, Calif.: Lange, 1979.)

279

minates on the presynaptic terminal controls the spike height, and hence the influence of the presynaptic neuron on the motor neuron. This mechanism of inhibition is apparently widespread in the CNS.

Note that **presynaptic inhibition is accomplished without an inhibitory synapse!** Contrast presynaptic inhibition with postsynaptic inhibition in this regard (see Fig. 11-18). Perhaps more importantly, **presynaptic inhibition is accomplished without change in the excitability of the postsynaptic cell to other incoming stimuli.** Whereas "normal" postsynaptic inhibition tends to affect the excitability of the cell to all sources of stimulation, presynaptic inhibition affects specific inputs in a selective manner. This must surely be an important attribute to add to our growing stock of mechanisms by which subtle and complex CNS function can be obtained.

You can imagine the problems that such inhibition caused researchers, since it could be shown that a response was inhibited, but no IPSP could be demonstrated [12, pp. 220–234]!

The depolarizations of axon terminals by presynaptic endings can be recorded at the surface of the cord, where the potentials are called *primary afferent depolarizations* (see Eccles [12, pp. 226–231]).

One of the functions of presynaptic inhibition in the spinal cord will be discussed in Chapter 13.

Fig. 11-19. Relationship of presynaptic spike magnitude on transmitter release as it affects EPSP size. Data from squid giant synapse.

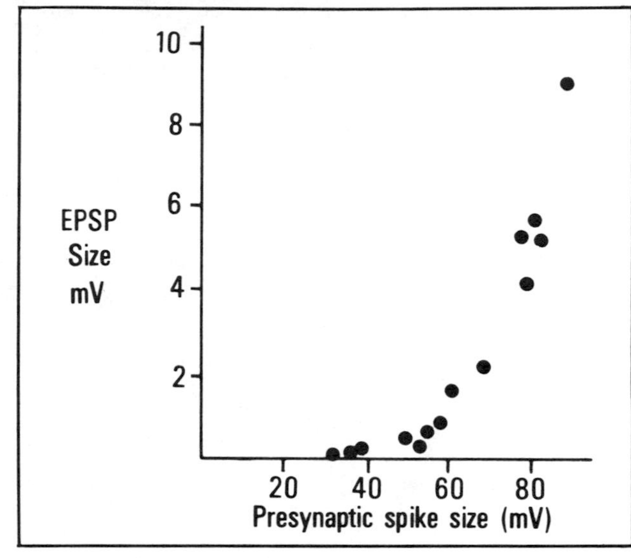

(Modified from A. Takeuchi and N. Takeuchi. Electrical changes in pre- and postsynaptic axons of the giant synapse of *Loligo*, *J. Gen. Physiol.* 45:1181, 1962.)

LATERAL INHIBITION

Our final example of an inhibitory system is important enough to have a whole section devoted to it. **Lateral inhibition refers to a neural network in which elements inhibit their neighbors with a magnitude that decreases with distance.**

A classic example of lateral inhibition is provided by the interactions in the lateral eye of the horseshoe crab, *Limulus*. This invertebrate eye is similar to an insect eye, having many separate lenses, each over a group of cells that together act as a receptor to transduce light and generate action potentials in the efferent fibers of the optic nerve (Fig. 11-20). These same fibers also make up a *lateral plexus* interconnecting the receptors (Fig. 11-20).

Lateral inhibition can be shown in experiments in which the firing rates of two single receptors are studied (Fig. 11-20). **The receptors** studied are close together, so that **each can inhibit the other.** The light to both is kept steady, so that a uniform firing rate is ob-

HINT

6. Because of Dale's principle. If you don't understand this answer, go to Hint 7.↓

279

280

Fig. 11-20. Schematic drawing of *Limulus* eye experiment arranged to permit steady light on receptor B while varying light to receptor A while recording from two single optic nerve fibers, coming from receptors A and B.

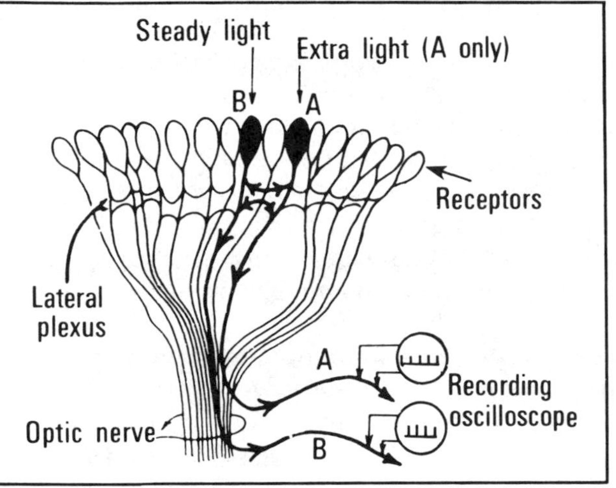

(Modified from F. Ratliff. Inhibitory Interaction and the Detection and Enhancement of Contours, in W. A. Rosenblith [Ed.], *Sensory Communication*. Cambridge: M.I.T. Press, 1961. Chap. 11.)

Fig. 11-21. Firing rates of two single receptors (A and B of Fig. 11-20) when light to A is suddenly increased for 2 s. Before time zero, steady light on A and B was adjusted to give firing rate of about 30 Hz for each receptor.

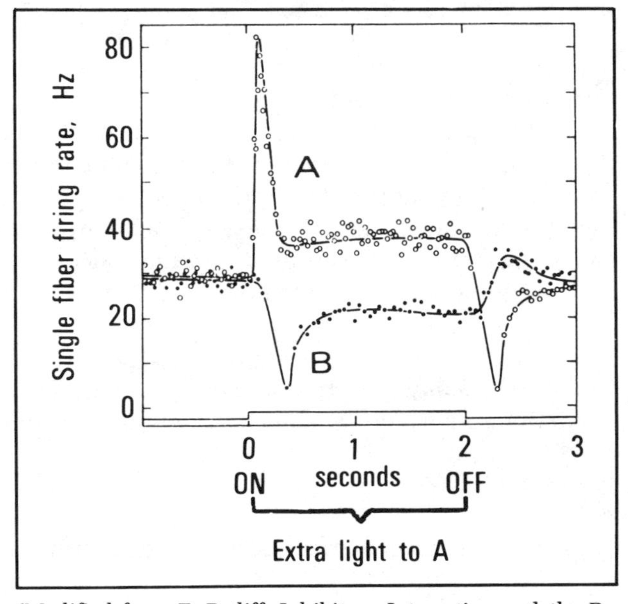

(Modified from F. Ratliff. Inhibitory Interaction and the Detection and Enhancement of Contours, in W. A. Rosenblith [Ed.], *Sensory Communication*. Cambridge: M.I.T. Press, 1961. Chap. 11.)

Fig. 11-22. Two possible types of mutual inhibition: 1, recurrent; 2, nonrecurrent. Points X indicate where action potentials are generated by light-sensitive receptors.

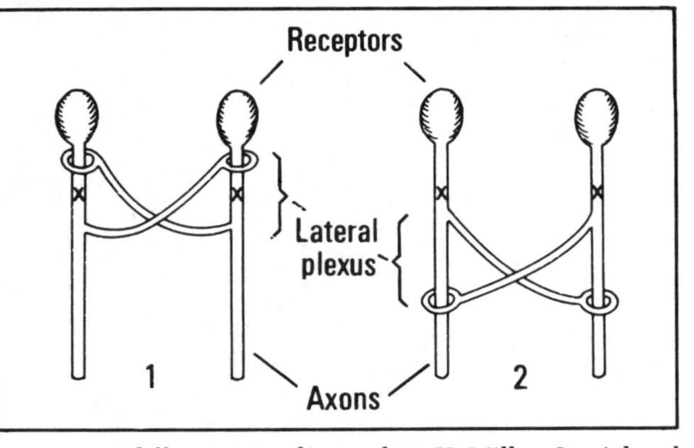

(From F. Ratliff, H. K. Hartline and W. H. Miller. Spatial and temporal aspects of retinal inhibitory interaction, *J. Opt. Soc. Am.* 53:110, 1963.)

280

281

tained (Fig. 11-21). When the light to receptor A is suddenly increased, and all other conditions are kept the same, the firing rate of the efferent axon from receptor A suddenly increases, but then rapidly adapts (Fig. 11-21). The inhibitory effect on receptor B of the increase in firing of receptor A is easily seen (Fig. 11-21).

This inhibitory effect can be shown to be mutual. That is, if the light is increased on B, then its firing rate increases and it inhibits A. Furthermore, **the inhibitory effect diminishes with distance.** The effect is greatest with adjacent cells and least with cells that are far removed from one another.

The exact connections of the lateral plexus have not been described in the preceding. Two possible sorts of functional interconnections are shown in Fig. 11-22. Both could show the decrease in firing of cell B of Fig. 11-21. However, network 1 has been shown to be the form of the connections. Functionally, network 1 acts in the same way as the Renshaw cell network described above. Indeed, Renshaw cell activation has been termed *recurrent inhibition*. Again, a network such as 1 in Fig. 11-22 forms a positive feedback loop (see Chap. 12), but the gain is low—less than 0.01 [51, p. 109].

The evidence for network 1 of Fig. 11-22 is provided by experiments in which a third receptor (call it C) was stimulated by light. Now C was close enough to A to inhibit it, but far enough from B to have little effect. If, when A was inhibiting B, C was illuminated, then A was inhibited and B fired faster, since it was now released from the inhibition of A. Such "disinhibition" could occur only in a network such as 1 in Fig. 11-22.

As shown in Fig. 11-21, the response to an increased light rapidly adapts, so obviously **this system is suited to detect temporal changes in light intensity.** It is less obvious, but also true, that this system detects **also spatial changes in light intensity.**

A system showing lateral inhibition accentuates differences in the firing rates of nearby receptors. In turn, **this accentuates the detection of "contrast,"** i.e., differences in light intensity, especially at "edges." This can be illustrated by Fig. 11-23, where it is assumed that a lateral inhibition "array" of receptors are illuminated with an edge (a sudden increase in light between receptors 9 and 10). The firing rates of the various receptors are

Fig. 11-23. Firing rate of group of receptors that have lateral inhibitory interconnections when illuminated with spatial pattern of intensity shown.

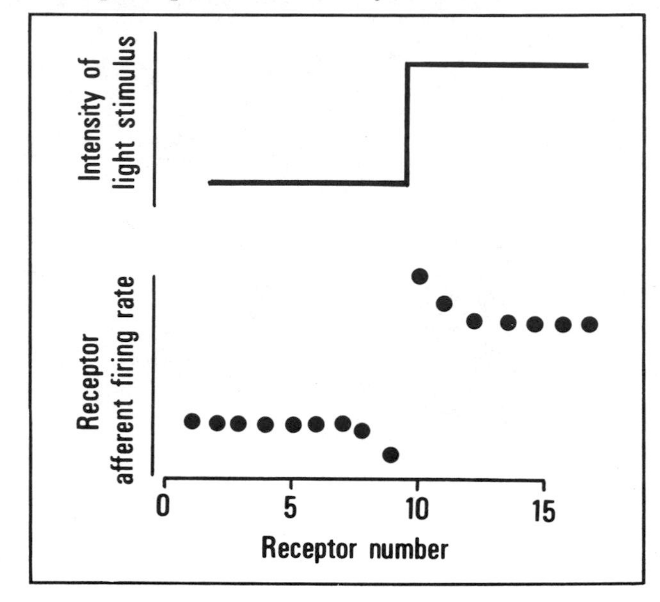

HINT

7. Since the same cell innervates both the muscle and the Renshaw cell, it follows immediately from Dale's principle that both terminals should produce the same transmitter. What is the transmitter at the myoneural junction? (Hint 8↓)

281

282

shown. Note that **where the illumination is uniform, adjacent receptors fire at the same rate,** since they have the same amount of excitation (light) and mutually inhibit one another by the same amount (e.g., receptors 5 and 6 or receptors 14 and 15). However, the cells at or near the edge do not have the same amount of inhibition (or the same firing rate) as other receptors with the same illumination. For example, receptor 9 is inhibited more than 5, though both have the same amount of illumination, because 9 is adjacent to receptor 10, which is on the bright side of the edge. However, receptor 10 fires more than 15 because 10 receives less inhibition from its neighbors than does 15, since 10 is close to 9, which is firing much less than those on the bright side. Thus, the **difference in firing** between receptors 9 and 10 **is more than would be expected on the basis of the difference in light intensity alone.** In this way, a group of receptors showing lateral inhibition is especially sensitive to edges. 1

Such lateral inhibition occurs in vertebrate eyes as well and is the probable basis for some visual illusions, such as Mach bands [51, 59]. 2

Teleologically, it isn't hard to see why visual edges should be important to detect. After all, prey, predators, and possible mates all have edges! If you don't think edges are important, then look at a copy of a halftone photograph made on an older type of xerographic copying machine. You will be surprised to find that most of what is important can still be seen. The xerographic copying process, though it has no biological lateral inhibition, does accentuate differences in shading (contrast). High-contrast photography can do the same thing. 3

It seems that **lateral inhibition is widespread in the nervous system.** Groups of cells are interconnected with inhibitory pathways, with the inhibition being greater for nearby cells than for those widely separated. Evidence for lateral inhibition is found in a very large number of nervous systems, as delineated below. 4

First, in vertebrate and mammalian retinas, there are known lateral connections. These connections are probably the mechanism by which the illumination surrounding a receptor can influence whether that receptor will activate its efferent retinal fibers. Thus, the stimulus "surround" influence the stimulus response as transmitted in the optic nerve. This is true in both rod vision and cone (color) vision. In color vision, the color of the surround can influence the color perceived in the center. Furthermore, "off" responses (seen in retinal recordings) in which there is greater activity at the *end* of a stimulus than at the beginning are also the result of an inhibition occurring in laterally connected networks. Note that cell *B* in Fig. 11-21 increases its firing rate at the end of the stimulus to receptor *A*. This rebound represents the inverse adaptation due to a *prior inhibition*. 5

Second, in psychophysical measurements, the influence of the sensation from one group of receptors by activation of an adjacent group of receptors can be found in the following 6

282

283

1 sensations: vision, sound, touch, vibration, heat, cold, taste, smell, and electrical stimulation [59].

2 Third, single unit recordings in the CNS have suggested that the following areas of the CNS have lateral inhibitory interconnections: the spinal cord (Renshaw cells described above), cerebellum, hippocampus, and cerebral cortex.

3 Of course, in so many different systems, the idea of "edge" becomes a bit difficult to visualize. So you must understand that **lateral inhibitory connections accentuate differences in firing rates in a group of cells,** whatever "message" is occurring in the group. However, if it is important to the brain to detect small differences in firing rate, in order to determine, for example, the cells that are firing the fastest, then the mechanism of lateral inhibition is well suited to such a task.

INTERNEURONS, IN GENERAL

4 **Interneurons make up the bulk of the neurons in the CNS,** and they are undoubtedly a varied lot.

5 Because often they have short axons and multiple connections, interneurons are difficult to study by either anatomic or physiological methods. By a corollary of Murphy's law (Hint 9↓), interneurons probably are crucial in many of the higher functions as well as the more mundane operations of the spinal cord.

6 What *is* known about interneurons indicates that they may differ in several significant ways from motor neurons. They may have **high firing rates, spontaneous activity,** and **prolonged discharge to a brief input.**

7 **The firing rate of interneurons can be very high.** As shown in Fig. 11-24, the firing rate is up to 300 per second, noticeably higher than the firing rates of motor neurons to fast muscles of 30 to 60 per second. Some interneurons have been recorded with rates up to 1000 per second.

8 Such high rates are possible because there is little or no postspike afterhyperpolarization, so that there is no period of prolonged lowered excitability (in contrast with motor neurons).

Fig. 11-24. Firing rate of interneuron of cat spinal cord, showing "spontaneous" firing rate, and increase in rate due to single volley in dorsal root at time zero.

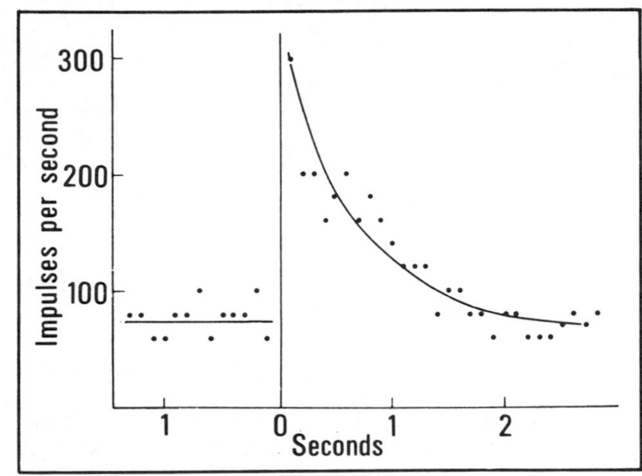

(Modified from K. Frank and M. G. F. Fuortes. Unitary activity of spinal interneurons of cats, *J. Physiol.* [*Lond.*] 131:424, 1956.)

HINTS

8. If you can't remember, you can guess that it is ACh, since that is the transmitter we hypothesized to excite the Renshaw cell! Why do you think that is? (Hint 6, p 279)

10. Convergence: $A + B \rightarrow X$ $B + C \rightarrow Y$
Divergence: $B \rightarrow X + Y$ $C \rightarrow Y + Z$

283

284

The high firing rate of spontaneous activity shown in Fig. 11-24 could be due to various continuous sensory inputs, as from slowly adapting sensory fibers, such as proprioceptors and thermal receptors. It could also be due to "reverberating circuits" (see next paragraph) or to spontaneous rhythmic activity, such as is found in some pacemaker cells in invertebrates or in the axon under unusual conditions. 1

The prolonged discharge of an interneuron resulting from a brief input is shown in Fig. 11-24. In some cases, a single sensory volley may cause reflex polysynaptic activation of motor neurons for 100 ms or more. **Such prolonged discharge may be due to a prolonged presence of synaptic transmitter, but also could be due to reverberating (recirculating) neuronal circuits,** such as diagrammed in Fig. 11-25. 2

Fig. 11-25. Hypothetical recirculating neuronal chain that might provide sustained output with brief input.

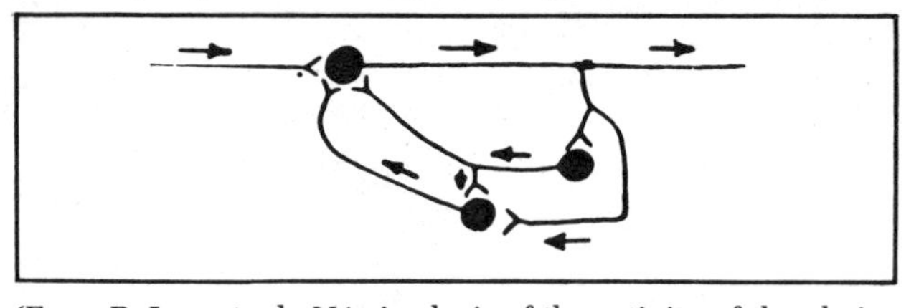

(From R. Lorente de Nó, Analysis of the activity of the chains of internuncial neurons, *J. Neurophysiol.* 1:207, 1938.)

Mountcastle and Baldessarini [37, pp. 1265–1266] point out that although the physiological data and anatomic findings support the idea of such circuits, there is no *direct proof* that the same *single cell* in such a circuit is repeatedly activated and participates in such a recirculation of impulses. The multiplicity of interconnections and the large numbers of synaptic inputs to cells (many times more than 1000) make such evidence difficult to obtain. Furthermore, such recyclic activity may be a property of a neuronal pool, rather than of a specific neuronal circuit. Thus there could be generalized increased activity in the pool, as a result of a high interconnectedness, without there being specific delineated circuits that are the same from moment to moment. 3

Fig. 11-26. Hypothetical multiple chain of internuncial neurons providing extended activation of neuronal pool.

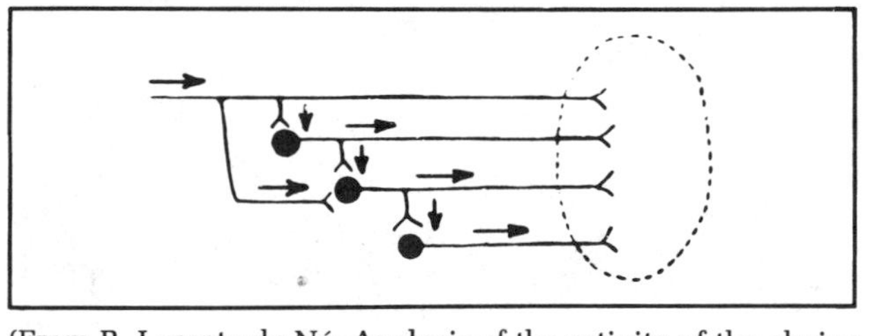

(From R. Lorente de Nó, Analysis of the activity of the chains of internuncial neurons, *J. Neurophysiol.* 1:207, 1938.)

Even without a reverberating circuit, an input to interneurons could provide a prolonged barrage of impulses into a neuron or group of neurons, as in the parallel, multiple-chain model shown in Fig. 11-26. 4

CONSEQUENCES OF CONVERGENCE AND DIVERGENCE

The CNS is characterized by both convergence (the termination of more than one afferent ending on a cell) **and divergence** (the efferent connection of one cell to two or more cells). **As one ascends the neuraxis, the number of cells increases. Presumably, the amount of both divergence and convergence also is increased.** We mention this to encourage you to view the CNS interconnectiveness as almost always much more complex than indicated in textbooks of neuroanatomy and neurophysiology! 5

You can figure out for yourself that given either divergence or convergence, the other must occur in the mathematical sense. 6

The consequence of convergence and divergence in even simple systems is important to understanding the workings of the nervous system. As you will see shortly, **spatial facilitation, occlusion, and subliminal fringe are all consequences of convergence and divergence.** 7

284

285

1 The effects of convergence and divergence in a nerve net illustrate how one neuronal pathway can influence other pathways.

2 Consider the neuronal network diagrammed in Fig. 11-27, which shows both convergence and divergence.

3 QUESTION: Which interconnections show convergence, which divergence? (Hint 10↑)

4 If we assume all the synapses to be excitatory, what are the consequences? Let us further assume a high threshold for firing of a postsynaptic neuron; e.g., both inputs (say, *A* and *B*) must occur at the same time. Thus there must be spatial summation; i.e., both inputs must depolarize the postsynaptic membrane at the same time. Finally, assume that we detect the effects of *X* and *Y* together (as in a muscular contraction).

5 If we now excite *A* and *B* synchronously, *X* will be activated, but not *Y*. Thus *Y* will be said to be in the **subliminal fringe.** This means that a further input (from *C*) will cause *Y* to fire, but not if *C* alone is activated. Thus, the term *subliminal fringe* refers to excitation of groups of cells whose responsiveness to other inputs has been changed. But this excitation does not, in itself, cause a detectable change in the response.

6 If we activate cell *C* at the same time as cells *A* and *B*, the response is greater than with *A* and *B* alone (since *Y* was in the subliminal fringe and now both *X* and *Y* fire). Thus, input *C* is said to "facilitate the response" to inputs *A* and *B*. [Note that this use of the term *facilitation* is different from its utilization earlier with regard to temporal facilitation in the myoneural junction (Chap. 10). The use here is reasonable in that the overall neural net response is larger with spatial facilitation than without.] Thus, **subliminal fringe and facilitation are consequences of high thresholds in a convergent neural network.** Another way to express this is to say that **the activity of *B* can "gate"** (i.e., control the passage of) **the pathways *A* → *X* and *C* → *Y*** (see also the next section).

7 If we now change our assumptions, so that only a single depolarization is sufficient to fire *X* or *Y* (low threshold), then occlusion can be demonstrated. If the overall response to stimulation of *A* and *C* together is compared with stimulating *A*, *B*, and *C* together, the response is found to be no greater, yet the response to *B* alone is also just as great. Thus, the combined effect of stimulating all three cells (*A*, *B*, and *C*) is not as great as would be expected from simple addition of the results of stimulating *B* alone and *A* and *C* together. Thus, the response is said to be *occluded* (meaning less than that expected). **In this case, the fastest-firing cell can "command" the output and "block" other pathways by activity.**

8 If you start imagining more complex networks, with inhibitory as well as excitatory synapses, the possibilities quickly multiply! Try it!

Fig. 11-27. Simple nerve net. Neurons *A*, *B*, and *C* are presynaptic; neurons *X*, *Y*, and *Z* are postsynaptic.

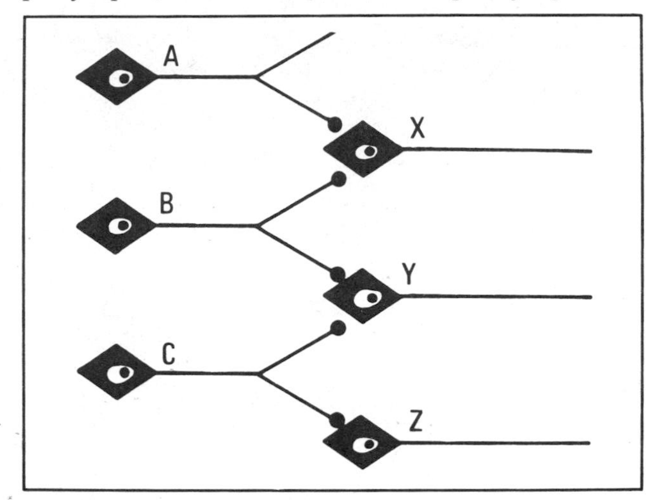

(From W. F. Ganong, *Review of Medical Physiology* [9th Ed.]. Los Altos, Calif.: Lange, 1979.)

HINT

9. The harder something is to discover, the more important it is.

285

286

OVERALL VIEW

In studying the CNS, it is common practice to study specific neural systems with specific neural connections. Most of the mammalian central nervous systems are more complex (and less well understood) than those described. However, there is no reason to believe that the basic principles that you have learned up to this point suddenly become inapplicable! In fact, it would be a surprise to find a system that did not operate along the lines already presented—such is the egotism of neurophysiologists. 1

But in the host of details to be encountered within the CNS, it is sometimes difficult to see the woods for the trees, i.e., the overall working of the system in the welter of single cell recordings! The remainder of this chapter is one form of (highly personal) overview. 2

There is no need for a synapse unless you wish to (1) transform or modify the output of other cells, (2) control information flow, or (3) amplify a biological result (as in the EPP). 3

The largest numbers of interneurons seem to occur at locations in the CNS where neural firing patterns are modified by the "integrative action" of convergence. This will be most apparent to you if later you study the details of the transformations in the CNS parts of sensory systems. 4

The computer-minded might like to contemplate the number of Boolean functions that the excitatory and inhibitory mechanisms already described can generate. Surprising, isn't it? Can you design a flip-flop out of neural elements? 5

Remember that *all* Boolean functions can be generated out of just NANDs or just NORs, although obviously many more functions are already present in the CNS. Given all Boolean functions and a memory system, you've got a Turing machine, and hence any computable function—given enough time and brain power! All this does not contradict the idea that the brain is *not* operating on digital principles, as previously argued. 6

With such a rich entanglement of converging and diverging fibers, **it is likely that parts of the nervous system form subsystems that tend to operate somewhat independently of one another.** As you study the CNS, you will become able to name many: muscular reflexes, postural mechanisms, brainstem autonomic reflexes, eye-hand and ear-eye coordinations, neuroendocrine systems, etc. *However, at times it is important for systems that are usually independent to become closely interrelated* (coupled). For example, respiration is often independent of hand movement, but when you are playing a musical wind instrument, the two must be closely coordinated. **Threshold is a mechanism that allows systems to be coupled and uncoupled readily.** In Fig. 11-28, neural system I and neural system II can operate independently if neurons X and Z are not activated. However, I and II can interact in two ways. 7

Fig. 11-28. Hypothetical neuronal connections to show how synapses can act to couple or uncouple one set of neurons from another. Dots represent axonal branch points.

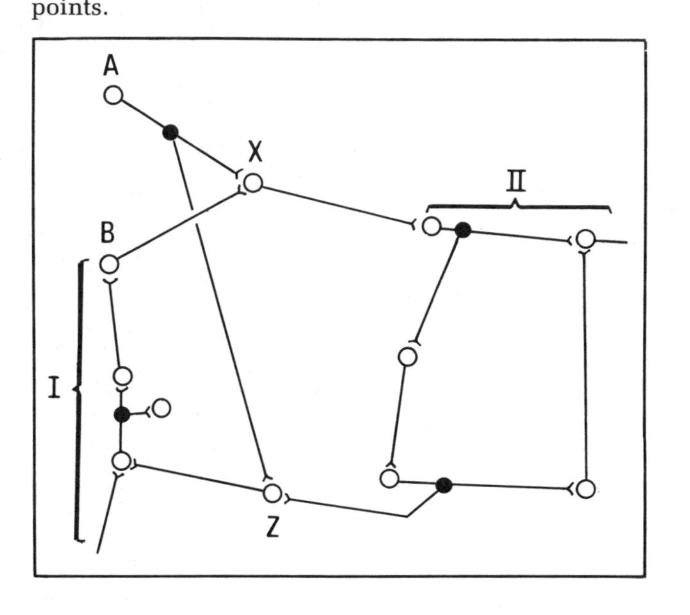

286

287

First (we assume *A* to be quiescent), the activity of system I may become great enough for increased activity of *B* to rise above threshold of neuron *X* and thus cause the activity of system I to influence the activity of system II.

Second, neuron *A* might increase its firing rate so that both *X* and *Z* are brought to threshold. Of course, *A* could also influence the connectivity between the two loops if it were inhibitory on *X* and *Z* or if it made presynaptic connections.

Figure 11-28 conveys the idea that there are **"loops" in the interconnections in the nervous system,** as is most likely. Now **loops have strange and useful properties that are crucial to the operation of the body.** For this reason, Chap. 12 is devoted to them. As you will see, the behavior of some loops can be predicted on the basis of the interconnections, if known. Conversely, some interconnections can be predicted if the behavior is known.

We reserve to the last the marvelous and penetrating overview of synaptic interconnections of Rushton [53, pp. 180–181]. He is concerned with the basic idea, expressed in Fig. 11-29, that one consequence of axonal conduction is that it transmits information concerning chemical *concentrations* (transmitter concentrations) from one place to another, rapidly.

"The simplest animals have no nerves; yet they react with purpose, seeking and avoiding. Life for them presumably lies in their sensitivity to the chemical environment and the flow of protoplasm or thrash of flagella by which they can move in it, in the change of permeability with the mixing of cell ingredients, and in the secretion and removal of the hormones by which activity is controlled.

"At the dawn of life, urges and efforts must have been chemical—have they ever been otherwise?

"*Nerves do not replace chemicals: they secrete them—instantly, exactly,* and at a distant location. Nerves are the biological response to the needs of an animal who thinks with its hormones but has grown so large that diffusion can no longer distribute them fast enough nor precisely enough. We do not need to wait for adrenalin to be delivered even by an efficient blood flow, for we may secrete it immediately and intimately from a thousand nerve endings, and we possess other nerves that are faster and discharge their hormones more particularly.

"In the long and faltering journey of evolution one great stride permitted animals to increase in size and specialization without loss of a unified chemical control—the appearance of a system of nerves whose essential function was to respond to a hormone here and at once to secrete a linear replica of it exactly there. . . .

"Slight as are these hints they lend a faint color to the view that the material correlate of cold thought or hot passion may be the play of chemicals in a chemical playground. . . . Upon this view we think and feel with our hormones. The suggestion is almost as groundless and implausible as that we do it by trains of nerve impulses."

Fig. 11-29. How axonal conduction can "transmit" the information of changing concentrations of a chemical over some distance. The middle neuron in the chain has as an input the chemical concentration changes that led to the EPSP of that neuron. Action potentials then act to release chemical transmitters at a distance, in some cases minimizing the input concentrations. Note that exact time courses of concentrations at either end are not known.

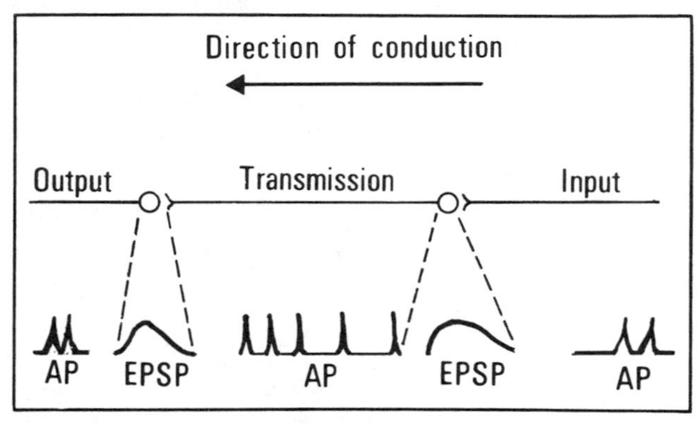

(Modified from H. Grundfest. Electrical inexcitability of synapses and some consequences in the central nervous system, *Physiol. Rev.* 37:337, 1957.)

287

289

12 Servo-Control Mechanisms

290

1 **The most powerful generalizations can be applied to many different situations or conditions.**

Biology has relatively few generalizations (compared with the physical sciences); variation and complexity are rampant in living systems! Yet the material of this chapter concerns some very important generalizations that have considerable applicability to *all* living systems. To be this powerful, the generalizations must be *nonspecific*. Although many examples are given, it will be up to you to find additional examples of these ideas in the rest of the course, in the rest of physiology or biology, in medicine, and even in society! (Note: This is *not* an idle suggestion!) 2

Servo-control theory deals with the means by which a system can control or regulate a variable in the face of factors that tend to disturb it. 3

Relative constancy is characteristic of many bodily functions: your body temperature is relatively constant over a wide range of ambient temperatures; your blood sugar level stays relatively constant in spite of all the candy that you eat or the variable times between your meals; your equilibrium is maintained even when you walk on uneven ground; your blood CO_2 level remains quite uniform in spite of wide variations in the amount of CO_2 you produce; you don't go insane in spite of the pressures of school; and so on. **The purpose of this chapter is to describe some of the properties such systems must have to be so remarkably resistant to disturbances.** 4

Some of these ideas have been expressed in biology as the "constancy of the internal environment" (Claude Bernard) and the "homeostasis" of living systems (Cannon). However, nonquantitative biology has done little with such important ideas as compared with the quantitative approach of engineering! 5

HISTORY

An early servo-control device was the speed governor on stationary steam engines. The mathematical basis of the stability of such mechanisms was analyzed in the 19th century. Modern servo-control theory dates to developments in engineering and applied mathematics at the time of World War II. 6

At that time, a group of engineers and mathematicians (centered at the Massachusetts Institute of Technology) was devising electronic and electromechanical war equipment, such as the automatic gun sight, and radar. The principles that they used to accurately control equipment such as guns, radar antennas, and so forth under a wide variety of environmental conditions were developed and codified into a body of knowledge called **servo-control theory.** 7

290

291

This field is now developed to the extent that it is a specialty in engineering, with its own textbooks, courses, etc.

In the last 20 years, the applicability of these ideas to biological and social systems has become widely recognized. Now several books are available on the applications of servocontrol theory to biological systems. (See the reading list at the end of this chapter.)

Although the mathematics used in the engineer's approach can be rather high-powered, this chapter uses nothing more complex than elementary algebra.

Students wishing a more quantitative approach should consult more advanced engineering texts.

ANALOGOUS SYSTEMS

In this chapter, simple electrical circuits are used to explain the principles involved. Then such electrical circuits can be utilized as models of a biological or mechanical circuit. That is, the electrical system can be analogous to a biological system. You are already familiar with such models; just recall chaps. 4, 6, and 7.

It is important to realize just how powerful and useful such analogous systems can be. Of course, "mental models" are employed in *all* scientific fields to simplify a complex body of facts.

Table 12-1 lists several analogous physical systems. You can see that in each system there is a something that corresponds to the abstract idea of "effort," another to the idea of "flow," etc.

Such parallels do not imply exact equivalence: the hydraulic resistance of a moving fluid is complexly related to the flow rate, as compared with the electrical system,

Table 12-1. Analogous Systems

General	Mechanical	Electrical	Fluid	Thermal
Effort	Force	Voltage	Pressure	Temperature
Flow	Velocity	Current	Flow	Entropy flow
Resistance	Friction	Resistance	Resistance	Insulation
Capacitance	Spring	Capacitance	Tank storage	Specific heat
Inertness	Mass	Inductance	Momentum	

291

292

where $E = IR$. However, in the fluid system, a head of pressure acting on a resistance generates a flow, just as in an electrical system, a voltage acting on a resistance generates a current.
1

For those who are mathematically fastidious in a strict sense, two systems can be said to be fully analogous if they have governing equations of the same type. Values of the coefficients are not crucial in the comparison.
2

The main idea is that since an electrical system can model a mechanical or fluid one, **it may be much easier to determine the important characteristics of the analogous electrical system,** rather than to attempt to analyze the interactions in a mechanical or hydraulic system.
3

This is particularly true nowadays, since relatively inexpensive electrical parts are available in a wide variety of accurate values. For example, it is easy to get electrical resistors accurate to 1 percent. But you cannot find a variety of mechanical friction devices calibrated to such accuracy, if you can find accurate frictional resistances at all! Thus, the electrical analog computer found widespread use in modeling the behavior of complex systems.
4

Recently, digital computer simulations have come into widespread use because they can be more accurate and less costly for some simulations. In the case of either digital or analog computer simulation, the solution can predict the behavior of a system when a solution by traditional mathematical techniques is not feasible or even possible!
5

Since the terminology of electrical engineers dominates the field of servo-control theory, it is very helpful to describe biological systems in the words and concepts of electrical engineering, as we do in this chapter.
6

LINEARITY

The ideas developed in this chapter strictly apply to linear systems only. What is a linear system?
7

Linear systems have linear relationships between the variables. In Fig. 12-1, line A shows a linear relationship between X and Y. This can be expressed as

$$X = kY \quad \text{Eq. 12-1}$$

where k is a constant of proportionality between X and Y.
8

Fig. 12-1. (A) Linear and (B) nonlinear relationships between X and Y.

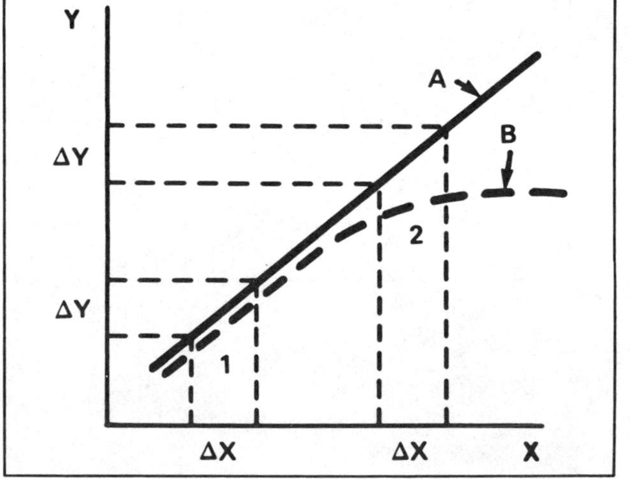

292

293

Another way of defining a linear relationship is to note that the slope of the line is *constant*. That is, for all values of X, a given change in X (that is, ΔX) gives the same change in Y (that is, ΔY). In a nonlinear relationship (curve B of Fig. 12-1), ΔX gives a different ΔY depending on the absolute value of X (the slope is not constant).

2 The mathematics presented here is strictly applicable only to linear systems.

"Well," you say, "I guess I can stop reading right here, because if there is anything that is nonlinear, it's a biological system." Hold on! Actually, *everything is nonlinear,* and yet the physical sciences haven't done so badly in advancing our understanding of the world! Consider a spring. It is a linear device (1 g of force: 1 cm of stretch; 2 g of force: 2 cm of stretch, etc.), until you reach a force where it won't stretch farther, and then you find a nonlinear relationship (as in curve B).

Another example of an apparently linear system is the physical characteristics that are proportional to temperature (e.g., expansion). These characteristics are linear at reasonable temperatures, but at 500,000°C they are definitely nonlinear! Thus we can use the tricks of the physical sciences and examine our nonlinear biological systems over small ranges in which they are linear. For example, we can study system B in areas 1 and 2 of Fig. 12-1 separately, since over these small ranges they are linear! In this way, we can apply the principles of this chapter to parts of nonlinear relationships, even if we can't explain the total range of behavior with a single linear model. **So a strict mathematical limitation need not prevent us from using these ideas to understand biological systems better.**

5 And now, on to some engineering terminology and concepts.

OPEN-LOOP SYSTEMS

An **amplifier** is diagrammed in Fig. 12-2. **The output is merely a constant multiple of the voltage at the input.**

The G stands for gain, which is

$$\text{Gain} = \frac{\text{amplifier output in volts}}{\text{amplifier input in volts}} \qquad \text{Eq. 12-2}$$

Thus G is merely the amount by which the input is multiplied to get the output:

$$\text{Amplifier output} = G \times \text{amplifier input} \qquad \text{Eq. 12-3}$$

Fig. 12-2. Open-loop system with amplifier.

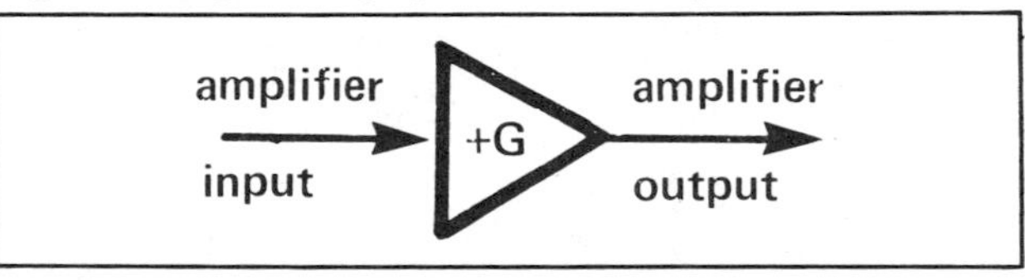

293

1 Note that G is shown in Fig. 12-2 as positive, which means that the amplifier output and input have the same algebraic sign. (It does *not* mean that the amplifier "adds.") If the output and input had opposite signs (polarity), there would be a −G in Fig. 12-2, as you can see by looking at Eq. 12-2.

2 Amplifiers usually can be arranged to have either a plus or a minus G. *For the purposes of this exposition, it is easier to follow if G is positive.*

3 The amplifier system of Fig. 12-2 is "open-loop" because the input is not influenced by the output. (This sentence will become a lot clearer when you see what a closed loop is.)

4 Now, this electrical diagram can symbolize other systems as well. For example, it can represent the power amplification of a public address system, such as that of a hospital paging system where the voice of the page operator is amplified to the sounds emitted by the loudspeakers. In this sort of system, note that the paging operator may have no immediate information as to how (or if!) the system is working, since the loudspeakers may be out of the operator's hearing.

5 Technically, the "amplifier" is any multiplicative element in which the input and output are expressed in the same physical units.

6 Open-loop systems are very common and often work quite satisfactorily as long as no significant disturbance affects the output.

7 The addition of a disturbance to the system is shown in Fig. 12-3. There the amplifier output is added algebraically (at the circle) to the disturbance ($V_{disturbance}$) to give V_{out}. **It is output V_{out} that we are interested in** and wish to keep relatively constant. Thus, the paging system might work well until a new additional noise (e.g., street repairs) became too large a proportion of the total sounds. A similar result is seen when the efficacy of the paging system in an empty, quiet cafeteria is compared with the same system when the cafeteria is full at mealtime.

CLOSED-LOOP POSITIVE FEEDBACK

8 One of the goals of this chapter is to exhibit some of the unusual properties of closed loops. **A closed loop is one in which the output "feeds back" to influence the amplifier input.** Such closed-loop feedback comes in two forms, positive and negative. Let us see the consequences of a closed loop with **positive feedback**.

9 Such a system is shown in Fig. 12-4. The quality of interest, V_{out}, has been connected to the amplifier input. The **amplifier input *e* is the algebraic sum** (at the circle) **of a steady input voltage V_{in} together with V_{out}.**

Fig. 12-3. Open-loop system with disturbance affecting output.

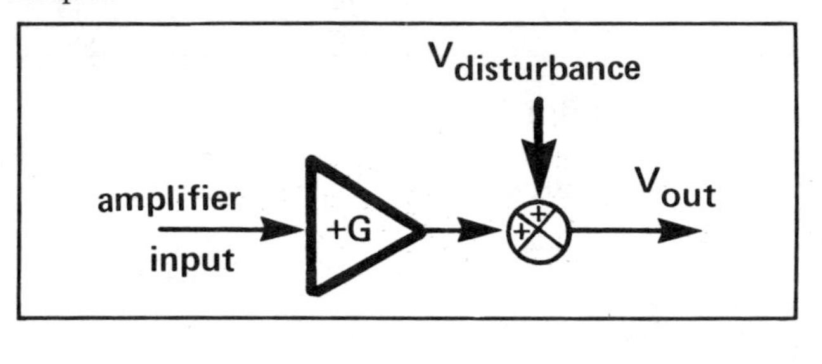

Fig. 12-4. Positive feedback closed loop with disturbance.

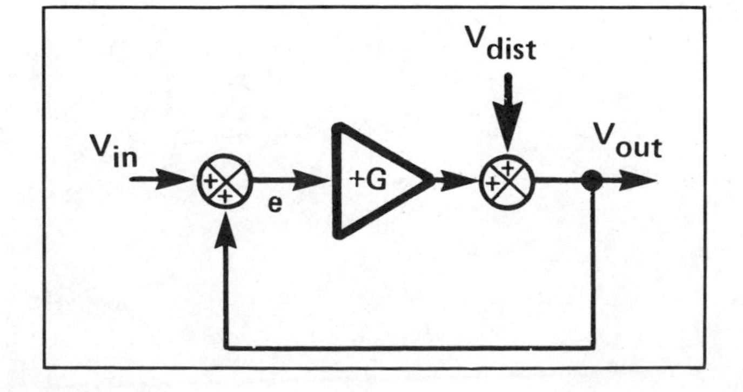

295

A positive-feedback closed loop shows a very characteristic behavior (when $G > 1$): V_{out} always goes to the system's maximum.

Example 1. You can see how this happens if you imagine that $G = 10$, $V_{dist} = 0$, and $V_{in} = 1$. Initially, the amplifier input e is 1 since $V_{out} = 0$. A 1-V input to the amplifier gives a 10-V amplifier output. Since $V_{dist} = 0$, V_{out} is now 10 V. By following the loop, you can see that now V_{out} and V_{in} sum, so that the amplifier input must be 11 V. This gives an amplifier output of 110 V. Coming around the loop a second time, a V_{out} of 110 V with another 1 V from V_{in} gives an amplifier input of 111 V and an output of 1110 V! It isn't hard to see that the voltage will rise rapidly toward infinity!

QUESTION: What is the result if $G = 10$, $V_{dist} = 0$, and $V_{in} = 10$ V at the start? (Hint 1↓)

QUESTION: What is the result if $G = 100$, $V_{dist} = 0$, and $V_{in} = 1$ V? (Hint 2↓)

Of course, the amplifier output cannot *reach* infinity. Instead, the output rises asymptotically to the power supply voltage, i.e., the highest value that the amplifier can generate. Thus, the amplifier, although linear over its working range, can be driven out of the linear range. Then it shows the nonlinear characteristics of curve B in Fig. 12-5/12-1 (past region 2, where a further increase in input voltage X gives no further increase in output voltage Y).

If V_{in} had been negative, then V_{out} would have gone toward minus infinity.

QUESTION: Do you believe the preceding sentence? If not, try Hint 3.↓

In addition, note that such extreme behavior (going toward plus or minus infinity) would also have occurred if V_{in} had been zero and there were a plus or minus value for V_{dist}.

QUESTION: If you don't believe the preceding sentence, what should you do?

In fact, **with any (nonzero) value at all, anywhere in the loop, V_{out} would be driven toward either plus or minus infinity.** The random noise resulting from thermal agitation of the electrons in the wire would be enough to do it!

Therefore, remember this: **a positive feedback loop** (with a gain greater than 1) **drives the system to its maximum** (either positive or negative).

Perhaps a "biological" example will fix the idea in your mind. Consider the situation in a two-bed dormitory bedroom, where each student has an electric blanket. Imagine that a practical joker (having mastered this chapter) sneaks into the room after A and B are asleep and quietly switches the electric blanket controls so that *the control from A's blanket is next to B and vice versa* (see Fig. 12-6). The joker quietly leaves the room. What will hap-

Fig. 12-5/12-1. Linear (A) and nonlinear (B) relationships between variables X and Y.

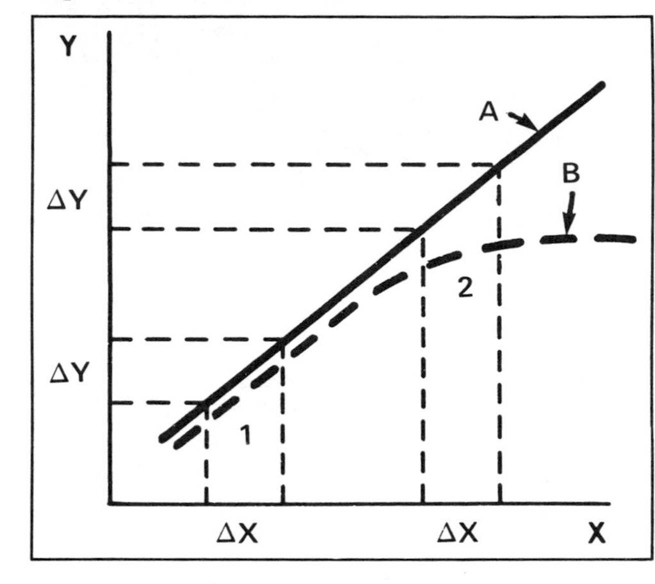

Fig. 12-6. Diagram of practical joke involving positive feedback.

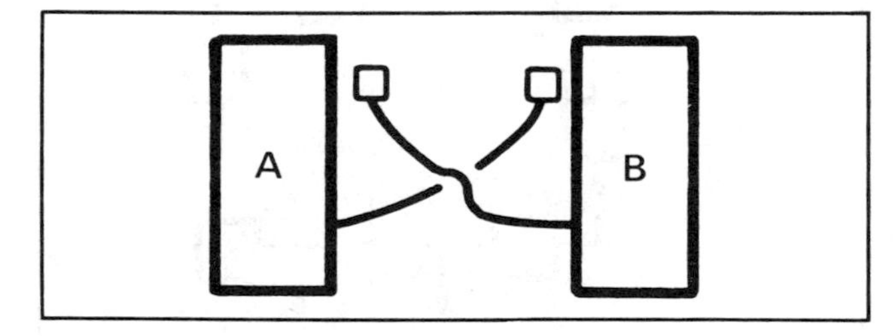

295

pen? Nothing—until *A* turns over and, feeling just a little chilly, turns *up* the blanket control by the bed. Some time later, *B* awakens enough to turn the control near the bed *down*, for *B*'s bed has gotten warmer. Sooner or later, *A* will turn the control *up still further*, with the thought, "Maybe it's going to snow," and *B* will turn it down *still further*, thinking, "Why is it so hot tonight?" Finally, *A* will turn the control near the bed *all the way up*, and *B* will turn the other control *all the way down*. Positive feedback in action! Note that the result of the positive feedback loop is to drive the parts of the system to a maximum, plus or minus. (P.S. Don't think that this is a hypothetical example—a dual-control electric blanket for a double bed, when put on wrong-side-up, will give exactly the same connections and the same results! Don't say we didn't warn you—check the connections when you get into an unfamiliar double bed!) 1

2 Now the electric blanket was an *obvious* example of a positive feedback loop.

How can we say this? After all, the whole system is rather complex. Consider the electric blanket controls, the physical properties of electric flow through resistances that generate the heat, the complex heat transfers in a cool room from blanket to bed, the processes involved in the skin thermoreceptors that lead to generator potentials. Then, of course, the generator potentials, in turn, give rise to trains of axonal action potentials that ascend over complex neural pathways. And the conductances change here and there until the recticular system is activated to desynchronize the cortical waves and bring about semiconsciousness. Also memory mechanisms are brought into action with a complex analysis involving many cells and God knows what transmitters. Next impulses are sent down the descending tracts to activate motor neurons and muscles with Ca^{2+}-filled cisterns, which, in turn, finally twist the control knob. *You* know how complicated it all is. How can we be sure that this complex system forms a positive feedback loop? Easy. 3

The general rule is this: **To find out whether a loop is a positive feedback loop, break the loop at any point and then assume a change on the far side of the break** (see Fig. 12-7). Follow the results of the assumed change all the way around the loop. **If,** when you get back to the "near" side of the break, **the result of the far-side change is to increase the value on the near side, the loop is a positive feedback loop.** 4

Since increasing the value at *A* in Fig. 12-7 causes the value at *B* to increase, the effect of the loop, when *B* is connected to *A*, will be for *A* to be increased again, and the process will continue. If you reread the electric blanket saga, you will see that what was done was to start out by *assuming a change* ("*A* turns up the blanket control") and then to follow all the way around the loop until a *change* in the *same* direction occurred at that point ("and *A* turns the control up *some more*"). 5

6 **This rule applies no matter where in the loop the break is made!**

Fig. 12-7. Complex positive feedback loop, which is shown operationally to be positive by breaking the loop and applying or assuming a change past the break and then detecting the resulting effect entering the break.

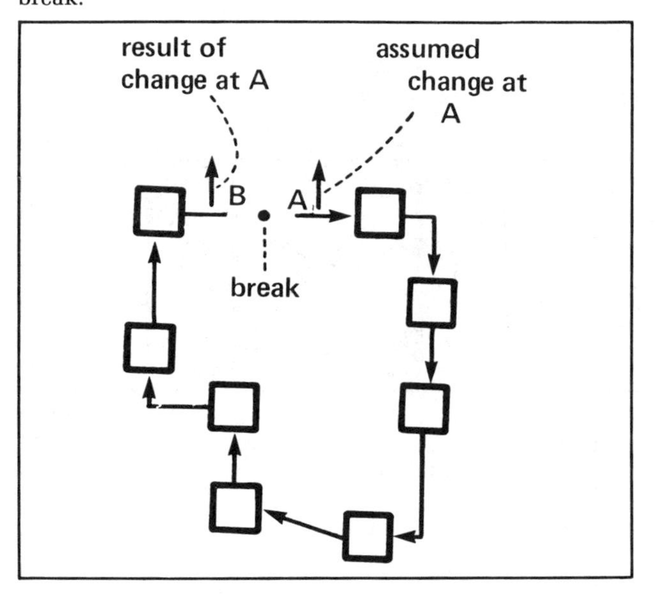

297

This is very useful, since you can test the loop in any location that is convenient. (Try it yourself on the electric blanket model, choosing different parts of the loop to increase or decrease at the start.)

QUESTION: Is the loop described in Fig. 12-8 positive? (Hint 6↓)

Fig. 12-8.

The more I know,
the more I worry.
The more I worry,
the less I sleep.
The less I sleep,
the more I read.
The more I read,
the more I know.
The more I know . . .

QUESTION: Assuming that a person has some hesitancy about taking a first drink, alcohol reduces the inhibitions against taking a second drink. Is this loop positive? (Hint 5↓)

With this idea of a positive feedback loop, you can deal with very complex systems; **you need only understand the system enough to predict the behavior of the variable you are observing.** The actual values of intermediate variables need not be known as long as the overall behavior is predictable. For example, you can determine the positive feedback loop in the electric blanket example without knowing any details of the nervous system, just as long as you know that a drop in temperature will lead to a turning up of the blanket control.

Another "biological" example can be diagrammed even more simply (Fig. 12-9). The figure can symbolize two grammar school children on a playground. Child A calls B a nasty name. Child B calls A an even nastier name. Child A shoves B. Child B pushes A harder. Child A hits B with his open hand. Child B hits A harder. Child A closes his fist, and . . . you get the idea. *It is easy to predict the outcome.* Now if, instead, Fig. 12-9 symbolizes nation A and nation B, you can easily predict the outcome of what is politely called "escalation." (One of the instabilities that makes the present and future so frightening is the positive feedback inherent in so many international situations!) **Note that a very simple diagram can be used to analyze the consequences of complex systems.**

Fig. 12-9. Another positive feedback system.

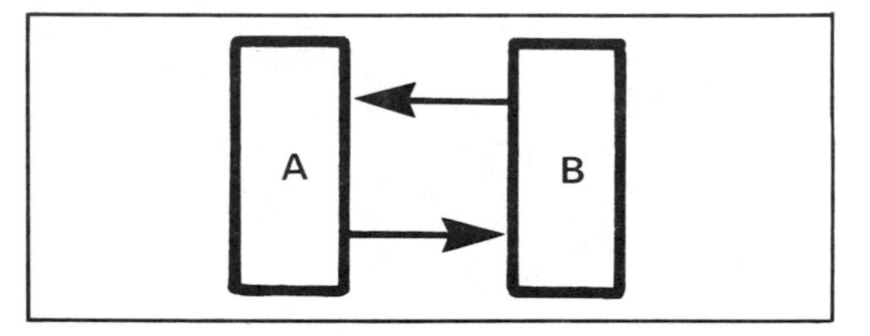

If you know the working of an atomic bomb, you know another form of positive feedback: the chain reaction. The energy in one moving neutron is sufficient to release two more neutrons, which in turn each release two more neutrons, etc. This is but a specific case of exponential growth, which can be viewed as the result of positive feedback, since in exponential growth the size of the variable determines the magnitude on the next "cycle." Thus, if the birthrate is greater than the death rate, the population size will follow an exponential rate of growth, with the positive feedback loop being: parents ⟶ more children ⟶ more parents (when children mature) ⟶ more children ⟶ more parents, etc. The positive-feedback elements of ex-

HINTS

1. The same as in Example 1: the output goes off toward infinity, with just a slight headstart!
2. Off it goes, but faster!
3. What if $G = 10$, $V_{dist} = 0$, and $V_{in} = -1$? (Hint 4↓)

297

ponential population growth are well described in *The Limits to Growth* (see the reading list at end of this chapter). When you consider the implications of unlimited population growth, you can see why such a positive feedback loop is often described as a "vicious circle," since the result is so inevitable and escape so difficult! 1

The effects of positive feedback described so far occur only when the **gain** of the amplifier **is 1 (unity) or greater.** Since **the magnitude of gain in a closed loop can influence the behavior of the loop,** let's see the result of a positive feedback loop with a gain less than unity. 2

Example 2: Consider what would occur in the system of Fig. 12-4 if $G = 0.1$ while $V_{dist} = 0$ and $V_{in} = 1$ V (as before). Now, going around the loop, the input to the amplifier is 1 V, but the output is only 0.1 V! This sums with V_{in} to give 1.1 V at the amplifier input and 0.11 V at the output, on the second time around the loop. Subsequent cycles yield values of V_{out} of 0.111, 0.1111, 0.11111, 0.111111 V, and so on. Clearly, V_{out} is approaching a value that is less than 0.12 V! Thus **a positive feedback loop with a gain less than unity** will not continue toward infinity, but **will approach asymptotically a value determined by the input and the gain.** However, **if the gain is 1 or greater** (even by the slightest amount), then the consequences are as previously described, and **the system is rapidly driven to its maximum.** 3

Even in a complex system such as in Fig. 12-7, the **loop gain** can be readily determined by seeing whether the change at B is equal to or greater than the change at A, which initiated it. While the system remains linear, the loop will go toward infinity (as in Example 1) if B changes more than A. But, if the change at B is small (say, $1/10$) compared with that at A, then the next time around the loop, the change will be even smaller (as in Example 2). 4

ACTION POTENTIAL AS POSITIVE FEEDBACK LOOP

One positive feedback loop that you know of **is the action potential!** The loop is shown in Fig. 12-10, just to remind you. 5

You can see how the loop involves positive feedback—break the loop, assume a change, and see what happens! 6

Now, when an initial depolarization is above threshold, the gain is greater than 1, and the system soon drives to a maximum—the peak of the action potential! You can see why **the absolute refractory period is absolutely necessary.** If g_{Na} were *always* sensitive to membrane depolarization, you'd have the membrane potential stuck at the maximum, unable to come down! By opening the loop (g_{Na} is no longer increased by further membrane depolarization) it is possible to return to the resting level (which requires an additional factor—K^+—part of a simultaneously acting negative feedback loop, described later). 7

Fig. 12-10. Positive feedback loop of rising phase of action potential.

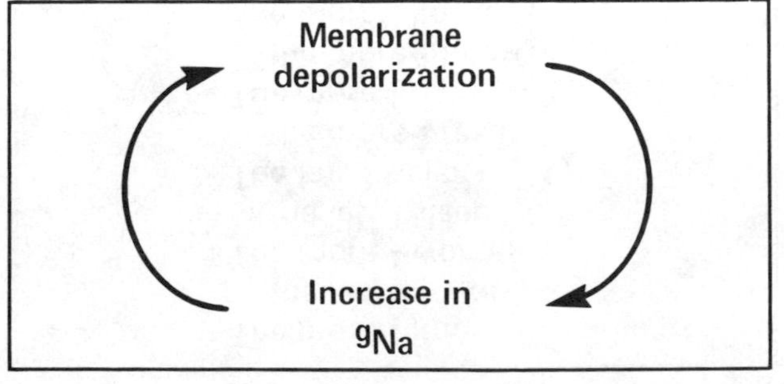

299

What makes the action potential more complex than the simple positive feedback loops you studied is not only the opposing action of K^+, but also that the *gain* is not constant, but varies, depending on the amount of depolarization (see levels 3 and 4 in Chaps. 6 and 8). Consequently, the membrane potential goes to a maximum, but the mathematics is more complicated than a simple exponential growth. (After all, the rising part of the action potential is more rapid than a fixed-gain exponential growth.) 1

OTHER EXAMPLES OF POSITIVE FEEDBACK IN THE BODY

At first, you might think that positive feedback would have little usefulness to the organism. But, it *is a great way to make something go to a maximum, fast.* Here are some additional examples. 2

1. In Head's paradoxical reflex, an inflation of the lungs causes an inspiratory movement on the part of the animal. This reflex has been implicated in the first breath of the newborn, when the lungs, empty of air, must be inflated. The reflex also has been thought to play a role in the inspiratory phase of sighing, where the function of the sigh may be to inflate portions of lung that have been collapsed by prolonged sitting (as in lectures). Ever try to suppress (inhibit) a spontaneous deep breath before a sigh? 3
2. The presence of free fatty acids in the duodenum leads to the release of pancreozymin, which leads to release into the duodenum, by the pancreas, of lipase, which in turn liberates more free fatty acids. If you draw out this description, you find a positive feedback loop. Not a bad idea, if the body is to digest a meal rapidly! 4
3. Biochemical systems that have enzymes activated by the product(s) of the reaction are equivalent to the preceding example. The system, once activated, will go toward maximal activity with all enzymes activated until all substrate is consumed. 5
4. "Bearing down" in the labor of childbirth is a reflex in which internal pelvic pressure causes the mother to forcibly increase intra-abdominal pressure. This is a useful mechanism if the baby is to be expelled from the uterus! 6
5. Viral infection, viral multiplication, and then further infection of other areas of the body (like a chain reaction) can be considered a positive feedback loop—which goes to a maximum. 7

HINTS

4. If you don't get the output growing more negative, probably you have forgotten that with a positive value of G, the input and output of the amplifier have the same sign.
5. It is certainly positive! Toward what maximum is the loop driving? (Hint 7↓)
6. It certainly is positive feedback, no matter how depressing or true! Note that you can assume the change anywhere in the sequence (loop).
8. The drinker passes out and opens the loop before the fatal dose of 80 proof can be consumed. Higher concentrations (which reduce volume) or rapid, large ingestions (e.g., "chugalugging" a pint of 80 proof) can be fatal.

299

Fig. 12-11. Feedback loops involved in circulatory shock.

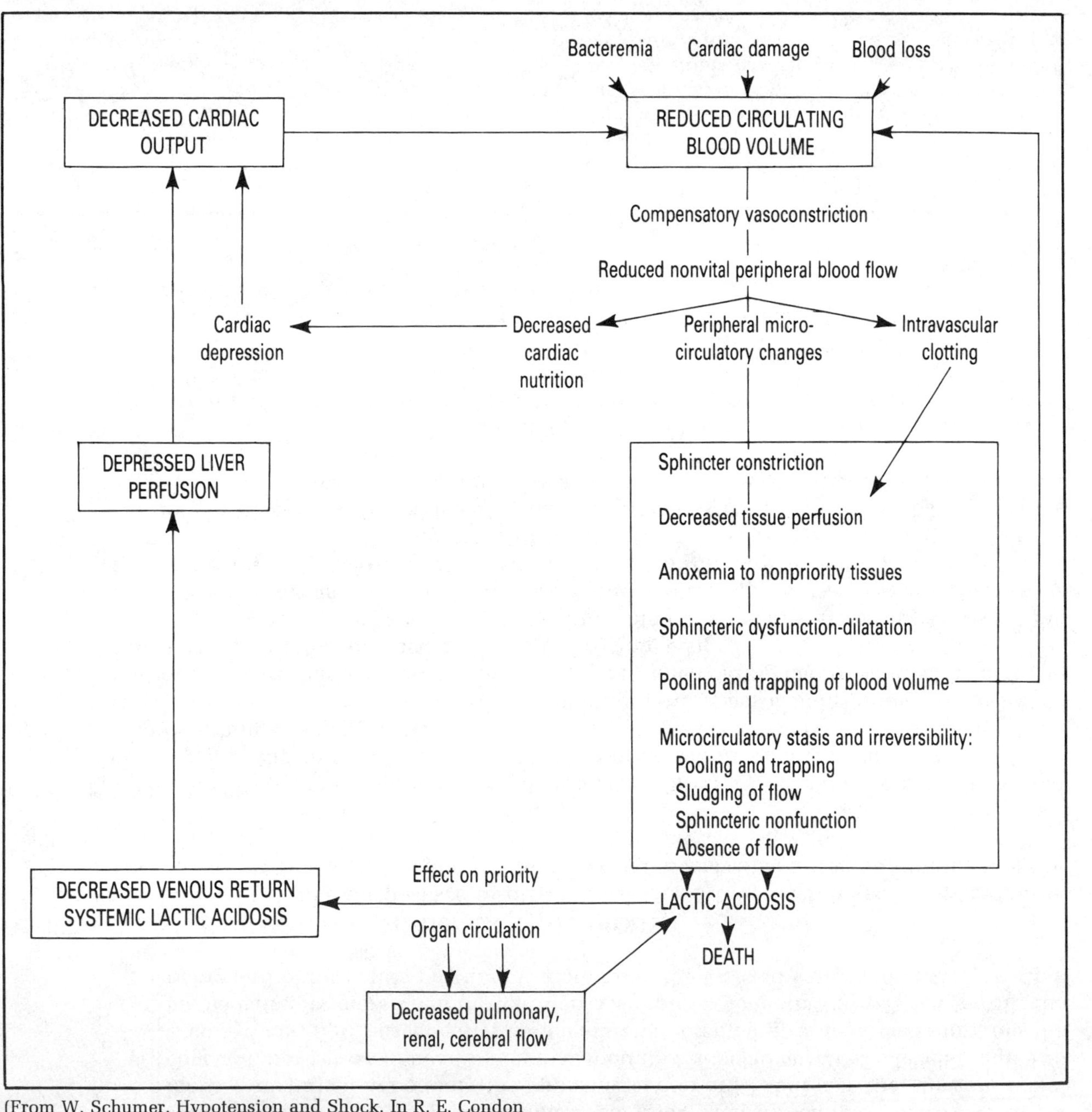

(From W. Schumer, Hypotension and Shock. In R. E. Condon and L. M. Nyhus (Eds.), *Manual of Surgical Therapeutics* (2nd Ed.). Boston: Little, Brown, 1972.)

301

6. Sudden death may have elements of positive feedback. If one of the coronary arteries is blocked, this may damage the heart so that it cannot pump enough blood to maintain the oxygen supply to the heart muscle, so the heart pumps less blood, so there is even less oxygen, so. . . . (See the positive feedback aspect?) Figure 12-11 shows three positive feedback loops involved in circulatory shock (reduced circulatory blood volume)! Note that this illustration does not label the loops as positive. But it is clear that if the gain is greater than unity, such loops will lead to the death of the patient.

7. As described in Chap. 11, lateral inhibition (and probably the Renshaw cell system as well) forms a positive feedback system with a gain less than 1. Even though there is inhibition in the loop, an even number of inhibitions in a loop give a positive feedback for the loop as a whole!

8. Repeated shear forces to skin cause thickening of the skin (calluses). On the sole of the foot, the presence of calluses combined with shoes leads to an increase in the localized shear forces, which increases the thickness of the calluses, etc. Thus, corns and calluses have their basis in a positive feedback loop.

9. When the thigh bone (femur) is broken, the pain can cause a reflex contraction of the thigh muscles, which increases the pain. Such involuntary muscle spasms can be most distressing and require strong traction to reduce the gain in the loop. A given amount of muscle contraction causes less shortening and hence less pain once traction is applied.

10. In certain urinary reflexes, an increased bladder pressure causes the bladder to contract, increasing the pressure. And in some gut reflexes, an increase in internal pressure causes the gut to contract, increasing the pressure (presumably to move the contents along). These are obvious examples of positive feedback, where it is useful for the system to be driven to a maximum, temporarily.

11. We leave it to you to spend a few interesting moments contemplating all the positive feedback loops involved in courting and mating, from the first shy glance (returned) to the time where the gain is obviously greater than 1!

Positive feedback loops also occur in medical treatment. Epilepsy may be characterized by involuntary twitchings of muscles. Treatment with the drugs diphenylhydantoin and phenobarbital over long periods can lead to disturbances in calcium blood levels. The resulting hypocalcemia can lead to muscle twitching (remember hypocalcemic tetany?), and so the physician, interpreting this as a recurrence of the epilepsy, *increases* the amount of drugs taken by the patient. By now you should realize the possible consequences of this loop!

HINT

7. A fatal overdose of alcohol. What normally prevents this outcome? (Hint 8↑)

301

302

QUESTION: What are the servo-control implications of a societal situation in which accumulation of power and wealth enhances accumulation of power and wealth? (Hint 9↓) [1]

At this point, take a few moments to let your imagination work. You probably can think of several more examples! [2]

CLOSED-LOOP NEGATIVE FEEDBACK

It is surprising how the behavior of the loop is changed by making the loop negative, instead of positive. [3]

Figure 12-12 shows a negative feedback loop. The only difference, as compared with Fig. 12-4, is that as V_{out} returns to the left side, it is *subtracted* from V_{in}, rather than added to it. This small change creates a large difference in the behavior of the loop! [4]

Now *obviously* this is a negative feedback loop. How can we say that? Well, the same idea holds as before: Break the loop, assume a change, and come around the loop to see what happens. [5]

In a negative feedback loop, the action of the loop is to oppose the assumed change (Fig. 12-13). [6]

Try the idea yourself, on the loop of Fig. 12-12. If you assume that V_{dist} decreases, then V_{out} will decrease. This will *subtract less* from V_{in}, so the amplifier input (labeled *error signal*) will become larger, opposing the decrease in V_{dist}! The result is that V_{out} is hardly changed, i.e., is stable in the face of a disturbance! [7]

NOTE: at this point, do not try to fill in actual numbers (as we did with the positive feedback loop) and then go around and around the loop trying to find out what will happen! The unusual behavior of the loop under such conditions is explained later (see "Stability and Instability"). [8]

The action of the system in a quantitative way can be easily understood by means of simple algebra. [9]

From Fig. 12-12, you can easily see that the following is true:

$$G = \frac{\text{amplifier output}}{\text{amplifier input}} = \frac{V_{out} - V_{dist}}{V_{in} - V_{out}} \qquad \text{Eq. 12-4}$$

(If you have trouble understanding this, go to Hint 10.↓) [10]

Fig. 12-12. Negative feedback loop with external disturbance.

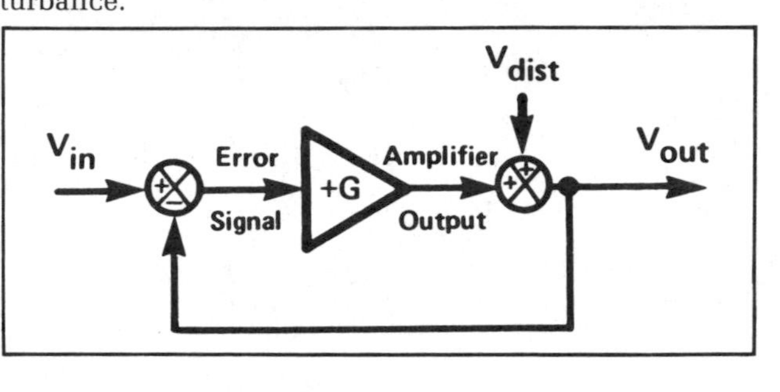

Fig. 12-13. Complex feedback loop, shown operationally to be negative.

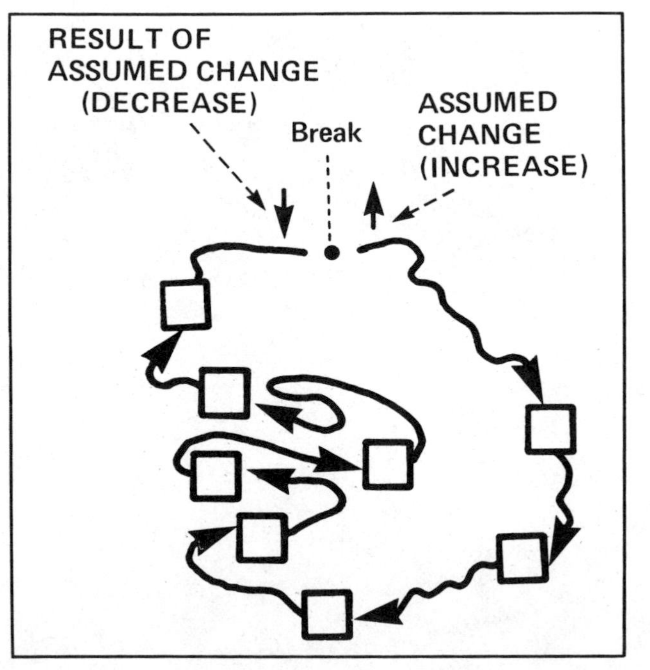

302

303

Now simply solving for V_{out} algebraically gives

$$V_{out} = V_{in}\frac{G}{1+G} + V_{dist}\frac{1}{1+G} \qquad \text{Eq. 12-5}$$

This equation is worth studying, since it can tell you a great deal about how a negative feedback loop operates! You can use Eq. 12-5 to tell you what happens to V_{out} (the variable we are interested in) under a variety of conditions.

We now ask: What factors, if changed, will affect V_{out} (that is, change V_{out})? Assume for a moment that V_{in} is constant. So we can ignore the product $V_{in}G/1 + G$, since it is constant (let's call it K) and won't affect changes in V_{out}. Thus Eq. 12-5 simplifies to

$$V_{out} = K + V_{dist}\frac{1}{1+G} \qquad \text{Eq. 12-6}$$

And you can see that any change in V_{dist} will cause *very* little change in V_{out}, as long as G is fairly large. For example, if $G = 1000$, then a change in V_{dist} of 1 V changes V_{out} by only 1 mV! This is an example of the negative feedback loop as a regulator. This sort of loop is present in thermostatically controlled temperature baths, where the water temperature (V_{out}) remains almost constant, in spite of temperature changes in the room (V_{dist}). If G is made sufficiently high, the change in V_{out} can be made negligibly small.

If G is made infinitely great, then V_{dist} has no effect on V_{out} (!), because the loop fully compensates for the disturbance. However, if G becomes very small (say, zero), then the loop has little effect and V_{dist} is unopposed.

Now, for another example, assume that $V_{dist} = 0$; that is, assume there is no disturbance. In this case (see Eq. 12-5), V_{dist} $[1/(1 + G)]$ becomes zero, and Eq. 12-5 becomes

$$V_{out} = V_{in}\frac{G}{1+G} \qquad \text{Eq. 12-7}$$

From this equation you can see that V_{out} *almost* equals V_{in}, since if $G = 1000$, then $G/(1 + G)$ is *very* close to 1. In this way, **a negative feedback loop acts as a controller.** A specified value is given as input, and the output will come very close to that value! This is why the amplifier input is labeled error signal. **The error is the difference between what you want, and what you've got, namely, $V_{in} - V_{out}$.** The action of the negative feedback loop is to reduce the error almost (but not quite) to zero. Thus, if $G = 1000$ and $V_{in} = 1.00$ V, then V_{out} will equal 0.999000999000. . . (from Eq. 12-6) while the error signal will be

303

304

0.000999000999. . . . This sort of mechanism is seen in temperature-controlled rooms (heater and air conditioning), where the system will follow the temperature set on the control. 1

The mathematically inclined may want to go back and, using the method of Eq. 12-4, derive the equation for a *positive* feedback loop. Ignoring V_{dist} (as zero), you can easily find that for the positive feedback case, $V_{out} = V_{in}G/(1 - G)$, which should be compared with Eq. 12-6. In this case, the equation predicts V_{out} for $G < 1$. For $G < 0.5$, V_{out} is less than V_{in}! For $G > 0.5$, V_{out} is greater than V_{in}. If $G = 1$, then $V_{out} = \infty$ no matter what the value of V_{in}. For $G > 1$, you have passed a discontinuity, which makes the equation invalid. 2

Now you should be able to see that the negative feedback loop corrects many problems of open loops:

1. There is no problem in knowing what input will give you what output since V_{out} will be very close to V_{in}. All you need to do is specify V_{in}.

2. The loop still gives good results, even if the components change. You can see that V_{out} will be close to V_{in} even if G drops from 1000 to 100 (a whole order of magnitude). Again, V_{dist} will affect V_{out} only a little, even if the gain drops from 1000 to 100. Instead of 0.1 percent control, it is now 1 percent, which still isn't so bad!

3. As you can see, **the loop automatically compensates for disturbances,** so as to oppose them, and in this way prevents disturbances from affecting V_{out}. 3

At some point, you might find it of interest to reread the parts on the voltage clamp in Chap. 8, since this experimental technique is based on a negative feedback loop that keeps membrane potential constant, in spite of changes in membrane conductance that otherwise would change the membrane potential. 4

Any system that comes into equilibrium after a displacement ("disturbance") has a negative feedback loop. You do not understand the system sufficiently until you can describe the loop. For example, the charging of a capacitor (as described in Chap. 3) may be viewed as a negative feedback loop, where the "back-EMF" acts to oppose the current imposed on the capacitor. Similarly, the Nernst equation represents an equilibrium condition that can be thought of as a negative feedback system in which a movement of ions due to a concentration gradient is opposed by the movement of ions due to an electrical potential. In such a case, the system will return to the equilibrium position regardless of the direction in which it is displaced by an outside force. 5

Thus, many physical systems can be analyzed in servo-control terms. Obviously, as described at the beginning of this chapter, two systems are analogous if they can be described by the same governing equations, no matter how different 6

304

305

their physical embodiments. So, again you are encouraged to look about you in the world to find examples of negative feedback loops.

BIOLOGICAL EXAMPLES OF NEGATIVE FEEDBACK

There are scads of examples of negative feedback in the body. The general rule is this: **If a variable is maintained relatively constant, in spite of significant disturbances, there is a negative feedback loop somewhere!**

In the body there are so many "variables" that fit this description that you can easily make up your own list: temperature, blood pressure, heart rate, blood glucose, CO_2, O_2, pH, eye position when the head is moved, CSF pressure, muscle length when a limb is held in a steady position while the load varies, and so forth. **The study of biological systems is the study of multiple, interconnected negative feedback loops!** The more you study, the more you encounter them.

One negative feedback system that you studied in considerable detail (Chaps. 4 and 6) is the action of K^+ in bringing the membrane potential back to resting level, if it has been displaced (by a "disturbance" called the action potential). In this way, K^+ is part of a negative feedback loop that keeps the membrane potential at a steady level! Of course, the negative feedback loop involving K^+ also opposes a disturbance that hyperpolarizes the membrane.

QUESTION: How? (Hint 11↓)

In a biochemical system, if the product of a reaction acts to *inhibit* the enzymes involved, then the amount of product will tend to stay constant even when the amount of substrate varies. In some genetic diseases, the inherited defect prevents this negative feedback, so that products reach higher-than-normal levels [34, pp. 20–21].

You should realize that **feedback loops may be completed outside the body.**

HINTS

9. Accumulation of increasing amounts of power and wealth by a minority of the population. Note that the question implied that $G > 1$. The answer assumes that power and wealth accumulate faster than the proportion of the wealthy increases. Hereditary succession to only the eldest child (as in countries with hereditary nobility) is a societal mechanism to keep the number of wealthy from increasing too rapidly.

10. The numerator ($V_{out} - V_{dist}$) follows immediately because we know from Fig. 12-12 that amplifier output + $V_{dist} = V_{out}$. Just solve this for amplifier output to get the numerator.

    The denominator is a direct statement of Fig. 12-12 at the left, where the circle indicates that $V_{in} - V_{out}$ = error signal.

305

306

You already considered such positive feedback loops when thinking about courting and mating behavior. (You did think about it, didn't you?) The patient with tabes dorsalis (the condition of late syphilis in which limb proprioception is lost) provides another example of a negative feedback loop closed outside the body. When walking, patients must look at their feet to determine their position, or they may fall. With eyes closed (with proprioception absent) their movements are open-loop. They close the loop visually, through the external world.

1

The control of eye movement in following a moving visual object is another example of a negative feedback loop closed outside the body (as you will see later).

2

EXAM QUESTION: Which of the following is a negative feedback loop?

1. Increased Na^+ conductance in response to depolarization of the membrane
2. Increased K^+ conductance in response to depolarization of the membrane
3. Pupillary dilatation in darkness
4. The muscle stretch reflex, in which increased stretch causes increased muscular contraction, shortening the muscle
5. The Hering-Breuer reflex, in which inflation of the lungs inhibits inspiratory efforts of the animal
6. Constriction of the middle ear muscles in response to a sound heard by bone conduction only
7. The consensual light reflex (light flashed in one eye causes constriction of the pupil of the other eye), assuming that the stimulated eye has been treated with atropine so that the pupil is unresponsive

Mark your answers, then on to Hint 12.↓

3

Of course, social systems have abundant negative feedback loops (if they are stable in the face of disturbances!). You might spend a little time considering the negative feedback aspects of elections in a democracy. The famous system of checks and balances of the U.S. Constitution can be well described in terms of negative feedback loops designed to oppose the positive feedback loop that occurs when the acquisition of power in a society permits actions resulting in the accumulation of more power, etc. [48, pp. 108–131].

4

QUESTION: Are antitrust laws and government regulatory agencies negative feedback loops, in theory? In practice?

5

QUESTION: When an industry can make sufficient "campaign contributions" to influence legislators, what happens to the societal feedback loop if the actions of the legislators can increase the wealth of the industry? Does history support the predictions your answer suggests?

6

306

307

QUESTION: In theory, is the maxim "Do unto others as you would have them do unto you" an open or a closed loop with positive or negative feedback? In practice?

QUESTION: If parents do not have to pay for educating their children or providing space and facilities for them once they are grown, is this part of the population-control feedback positive, negative, or open-loop?

QUESTION: Compare, in terms of loops, an industry in which all costs are passed on to consumers and an industry in which some costs (pollution control, employee illness) are paid for by non-consumers or the government, but not by the consumers of the product. What are the differences in the behaviors of these two loops?

STABILITY AND INSTABILITY

If that were all there is to feedback loops, it really would be pretty simple. And we wouldn't need to do much more than wave our hands at a little algebra and you would understand the whole thing. Unfortunately, it isn't quite that simple.

The preceding description is completely valid for negative feedback loops that are stable. But in order to understand this statement, you need to have some more experience with both stable and unstable negative feedback loops. (And who are we to shirk such a situation?)

First let's consider what happens when a step input occurs at V_{in}. Further, let's consider what happens *when there is a finite time delay in the feedback loop.*

Example 3. Assume a negative feedback loop, such as in Fig. 12-14, where $G = 0.1$ and the delay in the feedback loop prevents V_{out} from reaching the input summing junction for a time $= T$. (Because V_{dist} is not important in this example it is left out.) At the start T_0, let $V_{in} = 0$; but at T_1 and all following times, let $V_{in} = 100$. What happens? Well, you can easily figure this one out yourself. Just fill in Table 12-2.

QUESTION: If you are completely at a loss to fill in Table 12-2, then see Hint 13.↓

QUESTION: If you want to check your answers for the blank spots in Table 12-2, see Hint 15.↓

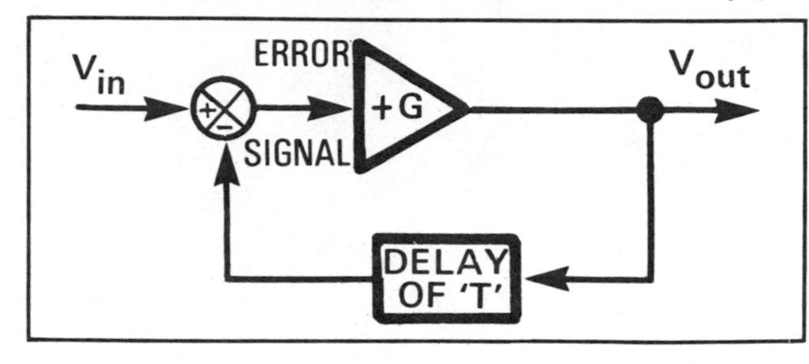

Fig. 12-14. Negative feedback loop with time delay (T).

Table 12-2. Values for Loop Parameters of Fig. 12-14 and Example 3, with $G = 0.1$ and Step Change in V_{in} of 100. Blank spaces are to be filled in by the reader.

Time	V_{in}	Error ($V_{in} - V_{out}$ of Previous T)	V_{out} (Error × G)
0	0	0	0
1	100	100	10
2	100		
3	100		
4	100	90.9	9.09
5	100	90.91	9.091
6	100	90.909	9.0909

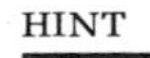

HINT

11. By reducing I_K (both by reducing g_K and by reducing the driving force, since V is closer to E_K), which is also acting in a direction to make the membrane potential more negative.

307

308

Fig. 12-15. V_{in} and V_{out} for Example 3.

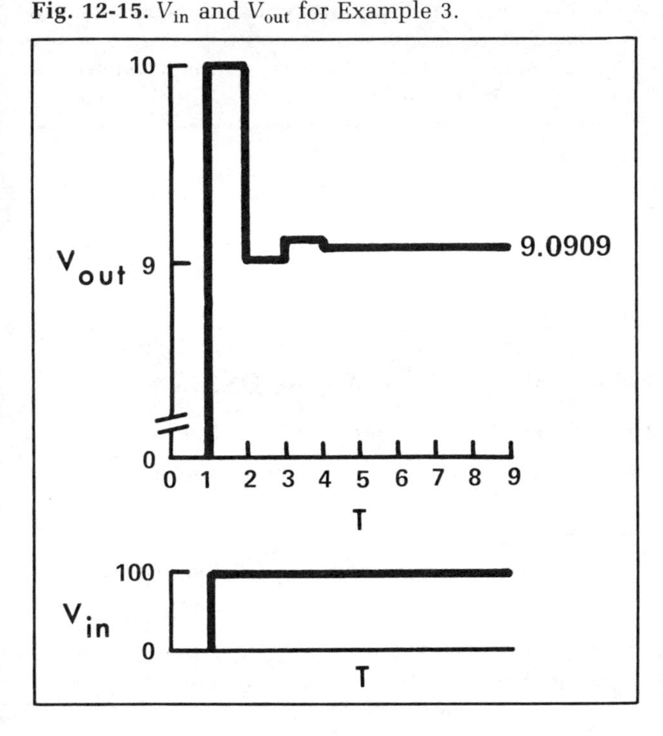

Table 12-3. Values of Parameters in Negative Feedback Loop of Fig. 12-14 and Example 4 with $G = 0.7$ and Step Change in V_{in} of 100. Blank spaces are to be filled in by the reader.

Time	V_{in}	Error ($V_{in} - V_{out}$ of Previous I)	V_{out}	Difference ($V_{out} - V_{out_{final}}$)
0	0	0	0	
1	100			28.8
2	100			−20.2
3	100	79	55.3	14.1
4	100	44.7	31.3	−9.89
5	100	68.7	48.1	6.92
6	100	51.9	36.3	−4.84
7	100	63.7	44.6	3.39
8	100	55.4	38.8	−2.37
9	100	61.2	42.8	1.66
10	100	57.2	40.0	−1.16
11	100	60.0	42.0	0.814
12	100	58.0	40.6	−0.570
13	100	59.4	41.6	0.399
14	100	58.4	40.9	−0.279
15	100	59.1	41.4	0.195
16	100	58.6	41.0	−0.137
17	100	59.0	41.3	0.096

Values of parameters are rounded to three significant figures.

Fig. 12-16. Time course of V_{out} and V_{in} with $G = 0.7$, of Example 4.

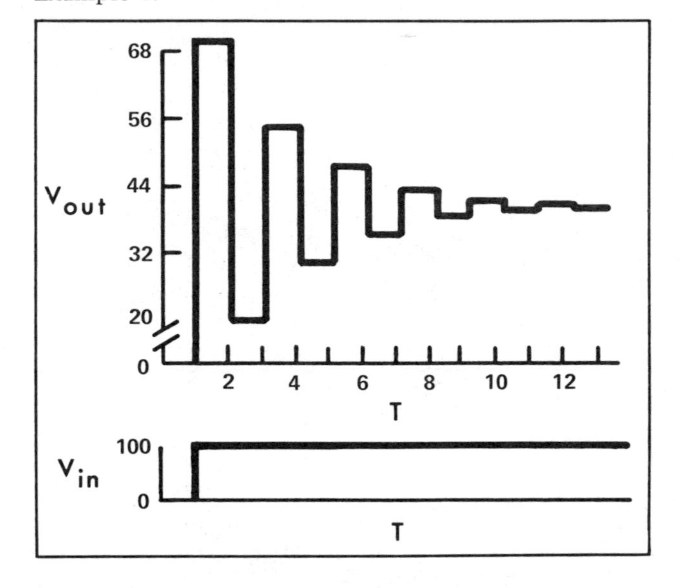

308

309

Now study the completed Table 12-2. *Note how the value of V_{out} rather quickly reaches just the value you would expect from Eq. 12-7,* shown again here:

$$V_{out} = V_{in} \frac{G}{1 + G}$$

For those of you who don't have your calculator at hand, 100 (0.1/1.1) = 9.0909 (to five significant figures). You can see in Fig. 12-15 how V_{out} responds to the step change of V_{in}. After a single overshoot, it rapidly settles down to the final value.

Example 4. *The overshoot becomes more obvious as a "damped" oscillation when the gain is higher (but still less than 1).* Let's try a gain of 0.7, with the same conditions as in Example 3. From Eq. 12-7 you can easily calculate that V_{out} will equal 41.2 when things settle down to a steady state. But V_{out} jumps around a bit before reaching this value, as seen in Table 12-3. We now add a new column, the difference between V_{out} and its final value, so you have some idea of how far off it is.

QUESTION: What are the blank values in Table 12-3? (Hint 18↓)

Now, from Table 12-3 you see how V_{out} takes a number of oscillations (of a gradually decreasing size) to reach the final steady-state value predicted by Eq. 12-7. This damped oscillation can be seen by looking at the values in the "Difference" column of Table 12-3 as well as by looking at Fig. 12-16. Strange behavior, isn't it? Why this is important will become clear shortly. The important thing to realize is that **the oscillations become larger and last longer as the gain gets closer to that magic number—unity.**

QUESTION: What was the difference in behavior of a positive feedback loop with a gain less than 1 versus one with gain greater than 1? (Hint 16↓)

You may have encountered a graph such as that of Fig. 12-16 previously in your physics course—it is a graph of a spring-mass-damper system (or a pendulum with friction) that has been displaced from equilibrium and then released. The governing equations for the negative feedback system and for such a mechanical system are similar.

HINTS

12. 2, 3, 4, 5.
13. At time 2 (that is, T_2), the error is V_{in} minus the V_{out} that occurred at time 1 (T_1) since the delay in the feedback prevented the value of V_{out} from reaching the summing junction until T_2. So what is the size of the error signal at T_2? (Hint 14↓)
15.

T	Error	V_{out}
2	90	9
3	91	9.1

309

310

Example 5. Just to show you how extreme the oscillations can get, let's try the same thing with a gain of 0.99. The results (which you needn't bother to calculate) are shown in Table 12-4. Note that even after 15 oscillations, V_{out} is still a long way from the steady-state value of 49.7 (predicted from Eq. 12-7). 1

Note, however, that the difference *is* becoming smaller (by small stages), so that you can predict that **the system will finally settle down to the predicted steady-state value,** after a considerable time. 2

Example 6. Now, **if the gain is greater than 1, the behavior is markedly different.** Let's try $G = 2$, with $V_{in} = 100$ again. 3

QUESTION: What final (steady-state) value would you expect? (Hint 17↓) 4

Why not fill in the blank parts of Table 12-5 to see what will happen? **With the gain greater than 1, the negative feedback loop (with a pure time delay) becomes unstable.** That is, **the output will grow without bound,** until it oscillates between plus infinity and minus infinity. 5

In practice, infinity is never reached; instead, the nonlinear part of the amplifier curve (Fig. 12-1) is soon reached, and then the oscillations go from one maximum to another. You can visualize this by imagining that in Example 6 the amplifier is linear up to a value of V_{out} of 1000 and then will produce no further increase. Thus V_{out} will be +1000 at T_5, −1000 again at T_6, etc. 6

At this point, you can see why, when negative feedback loops were introduced in this chapter, you were warned (just before Eq. 12-4) not to compute the values in a negative loop in the same way as you did in the positive feedback loops. In such a calculation, if $G > 1$, the values of V_{out} will oscillate, and students always pick a $G > 1$!! Such a calculation method (as used in positive feedback, Examples 1 and 2) implies a built-in time delay, although the student does not realize it. In the positive feedback loop, it makes no difference in the result; but in the case of the negative feedback loop, wow! 7

With higher gains, the system just takes off at a faster rate. With gains that are close to, but still greater than, 1, V_{out} still increases in size with each cycle around the loop, but not at such a fast rate. To illustrate this, see Table 12-6, which gives the result when the gain = 1.01 (that's pretty close to unity). Note how the V_{out} oscillations slowly grow in magnitude. You shouldn't find it hard to guess where it will all end! 8

Finally, you should realize that **with a gain greater than 1, this loop (with fixed delay) will go into oscillations even if the input is only transient** (brief), as shown in Table 12-7. 9

Thus, a loop with the properties described is always unstable. That is, **it will always go into increasing oscillations,** because all systems contain some "noise" (small random variations) which will "set it off." 10

Table 12-4. Loop Parameters for $G = 0.99$ When V_{in} Is Step Function of 100 in Example 5

Time	V_{in}	Error	V_{out}	Difference ($V_{out} - V_{out_{final}}$)
0	0	0	0	—
1	100	100	99	49.3
2	100	1	0.99	−48.8
3	100	99	98	43.3
4	100	1.98	1.96	−47.8
5	100	98	97.1	47.3
6	100	2.94	2.91	−46.8
7	100	97.1	96.1	46.4
8	100	3.88	3.84	−45.9
9	100	96.2	95.2	45.4
10	100	4.80	4.76	−45.0
11	100	95.2	94.3	44.5
12	100	5.71	5.65	−44.1
13	100	94.3	93.4	43.7
14	100	6.6	6.53	−43.2
15	100	93.4	92.5	42.8

Values of parameters are rounded to three figures.

Table 12-5. Values for Gain of 2 for Feedback Loop of Fig. 12-14 and Example 6 with Step Function Input. Blank spaces are to be filled in by the reader.

Time	V_{in}	Error	V_{out}
0	0	0	0
1	100	100	200
2	100		
3	100		
4	100		
5	100		

If you want to check your calculations, see Hint 19.↓

310

311

In the conductance and resistors of electrical systems, the random movements of electrons resulting from thermal agitation, called *Johnson noise,* are equivalent to the thermal agitation of molecules in a liquid medium which results in Brownian movement of microscopic particles. Such random electron movements in electrical systems are sufficient to start the oscillatory behavior. Of course, biological systems are replete with noise. Some cynics say that is all they have!

For simplicity, V_{dist} was left out of the preceding. You can go back, if you wish, and verify that the loop will behave, in all cases, in a manner similar to that described, if V_{dist} varies as V_{in} did. So leaving out V_{dist} does not change any of the ideas or conclusions. Just to be sure that you are convinced of this, you might take a look at Fig. 12-17, which shows a negative feedback loop with a delay and a disturbance. If we assume that $V_{in} = 0$ and that at time T_1 the disturbance suddenly becomes 100 and stays that way, then we obtain the values seen in Table 12-8, for the simple case when $G = 0.1$. Remember that when $V_{in} = 0$, the value that V_{out} will approach can be calculated from $V_{out} = V_{dist}\,[1/(1 + G)]$, as described (Eq. 12-6). Note how V_{out} approaches the final value by the sixth cycle.

Tables similar to all those shown for the negative feedback loops can be constructed for changes in V_{dist} instead of changes in V_{in}. As shown in Table 12-8, compared with Table 12-2, the **qualitative behavior of the loop is the same, being either stable or unstable, depending on the gain, and showing more or less amounts of damped oscillations, depending on the gain.** Furthermore, the behavior of positive feedback loops with gains greater or less than 1 is the same whether the loop is perturbed by V_{in} or V_{dist}.

The whole point of the preceding is to show that, first, **negative feedback loops can show different forms of behavior, some stable and some unstable,** and, second, **the behavior can depend on the gain of the amplifier in the loop, in both positive and negative loops.**

Let's sum up these ideas in Table 12-9 in order to clarify the points we have been trying to get across.

Table 12-6. Loop Parameters with Time Delay of Fig. 12-14 and $G = 1.01$

Time	V_{in}	Error	V_{out}
0	0	0	0
1	100	100	101
2	100	−1	−1.01
3	100	101	102
4	100	−2.02	−2.04
5	100	102	103
6	100	−3.06	−3.09
7	100	103	104
8	100	−4.12	−4.16
9	100	104	105
10	100	−5.21	−5.26

Table 12-7. Loop Parameters with Time Delay of Fig. 12-14 and $G = 2$, but with Only Transient Input at V_{in}

Time	V_{in}	Error	V_{out}
0	0	0	0
1	100	100	200
2	0	−200	−400
3	0	400	800
4	0	−800	−1600
5	0	1600	3200
6	0	−3200	−6400

HINTS

14. Error signal = 100 − 10 = 90. And then V_{out} is just the error times the gain, or $90 \times 0.1 = 9$. Now just carry on from here!

16. With a gain less than 1, the positive feedback loop approaches a value determined by the input and the gain. But with a gain greater than 1, the loop goes toward infinity (or to the nonlinearity of the amplifier). If you don't remember this, go back and review quickly the sections on positive feedback.

18.

T	Error	V_{out}
1	100	70
2	30	21

If you didn't get this, then back you go to Example 3.

311

312

Fig. 12-17. Negative feedback loop with disturbance as well as delay.

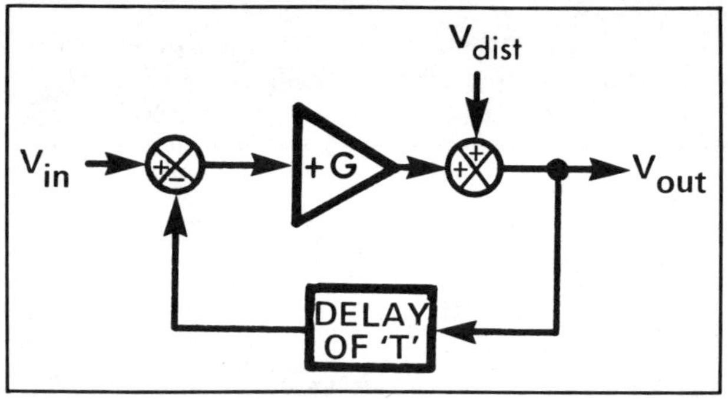

Table 12-8. Values of Parameters of Loop of Fig. 12-17 When $V_{in} = 0$, V_{dist} Becomes 100 at T_1, and $G = 0.1$

Time	V_{dist}	Error	V_{out}
0	0	0	0
1	100	0	100
2	100	−100	90
3	100	−90	91
4	100	−91	90.9
5	100	−90.9	90.91
6	100	−90.91	90.909

The asymptotic value of V_{out} is 90.909 (to five places).

Table 12-9. Feedback and Gain Characteristics Influencing the Behavior of a System

Gain	Positive Feedback	Negative Feedback
< 1	Approaches asymptotically a value dependent on input and gain. Example 2.	Rather inaccurate control. Can show damped oscillations. Examples 3, 4.
> 1	Goes to infinity (or to amplifier maximum). Example 1.	Stable: good, accurate control. Examples in preceding sections. Unstable: uncontrollable oscillations (± infinity, or ± amplifier maximum). Examples 5, 6.

312

313

STRANGE SACCADES

By now, you may feel the need for a "biological" example. Here is one, which you wouldn't have been able to understand if you hadn't worked your way through the preceding ideas. 1

In the visual system, **when a target in the center of vision suddenly moves, an automatic reflex moves the eye so as to bring the target back to the center of vision.** 2

The movement of the eye under these circumstances is called a **saccade** (pronounced "sack-caid"). The size of the saccade obviously varies with how far the target is off from the center of vision. The whole process of detecting a target shift, determining its magnitude, computing the appropriate action, and sending the signals to the eye muscles to bring about the change in eye position requires 150 to 250 ms. During this time, no additional information concerning the target position is processed. When the eye stops moving, the whole cycle is repeated. Thus, **this system has an inherent delay in correction** (of about 200 ms) that is directly analogous to the delays in the feedback loop in the preceding examples. 3

This negative feedback loop for eye position is stable in the normal person, whenever the eye can move faster than the target moves. 4

However, by means of electronic gadgetry (some might call it electronic trickery!), it is possible to *add* an additional loop to this system. Thus the system can be made to have either positive or negative feedback, and so the gain can be varied by the experimenter. What will happen under the various conditions? 5

The experiments involved having a subject look at the spot of light on an oscilloscope [57, pp. 231–296]. The subject's eye position was recorded and used to determine the spot position. Thus, if the spot moved exactly with the eye, then the position on the retina would remain unchanged, even though the extraocular muscles were moving the eye. In such a case, the visual system would be open-loop since then the position of the eye does not determine the position of the spot on the retina. 6

This is possible because **the biological loop is closed through the environment.** If the environment acts so that "corrective action" gives no result, the loop is operationally "open." 7

HINTS

17. 66.7. If you didn't get this, then back to Eq. 12-7, and do it on a piece of paper this time!

19.

T	Error	V_{out}
2	−100	−200
3	300	600
4	−500	−1000
5	1100	2200

313

The resulting overall feedback loop of the experimental situation is shown in Fig. 12-18. The eye control system of the subject has the built-in time delay of about 200 ms mentioned previously. So this system is directly analogous to that shown in Fig. 12-17—and you know how that one works! (Note that the summing junction circle at the left shows a plus and a minus—the loop can be made positive or negative by a switching arrangement.) In all cases, V_{in} is left equal to zero. And this means that the eye should remain on the target, with the target being displaced by V_{dist}.

The behavior of the normal eye can be demonstrated by setting the gain of the amplifier to zero (Fig. 12-19). In this way, V_{dist} can affect V_{out} (the target position), but the feedback loop has no effect (since the amplifier is not amplifying). In Fig. 12-19 and in the subsequent figures, V_{out} (the target position) as well as the eye position (as detected by the experimental apparatus—Fig. 12-18) is shown. There is about a 200-ms lag before the eye position follows the suddenly moved target. With a small overshoot, the eye quickly reaches the target position.

If the gain is set at 0.75 with the feedback loop positive, then the behavior of Fig. 12-20 is observed. Each time the eye reaches the previous position of the spot, the spot has moved on. But since the gain is less than 1, the amount that the target moves is less and less. So the system approaches the asymptote as the eye finally reaches the target, when the difference is less than 0.5° (the resolving power of the eye detection system). You should recognize this behavior as that of a positive feedback loop with a gain less than 1.

If the loop is kept positive and the *gain* is set at 1, then **each eye movement causes an equal target movement,** and the target and eye move off to infinity (i.e., the maximum) in equal steps (Fig. 12-21). Furthermore, *if the gain* of the positive feedback loop is greater than 1, for example, 2 in Fig. 12-22, then **each eye movement results in an even greater target movement,** and the system takes off even faster (Fig. 12-22).

However, if the system is made into a negative feedback loop, then **for a given eye movement, the target moves in the opposite direction!** If the gain is less than 1 (for example, G = 0.75 in Fig. 12-23), then the system approaches the expected position asymptotically, with damped oscillations, just as in Fig. 12-16 (sound familiar?).

QUESTION: What will the behavior of this system be if the gain is greater than 1 (negative feedback)? (Hint 20↓)

Thus, this experimental (partially biological) system behaves just as we would expect when the sign of the feedback loop is changed and when the gain is varied.

Furthermore, just as you should predict, **if the gain is greater than 1 with either type of feedback, without any changes in either V_{in} or V_{out}, then the random movements of the eye are enough to trigger either an increasing movement to a maximum** (positive feedback, $G > 1$), as shown in Fig. 12-25, **or a sustained oscillation** (negative feedback, $G > 1$), as shown in Fig. 12-26.

Fig. 12-18. Experimental feedback loop for study of eye saccades. Note that eye control system involves delay of about 200 ms, so this system is directly analogous to Fig. 12-17.

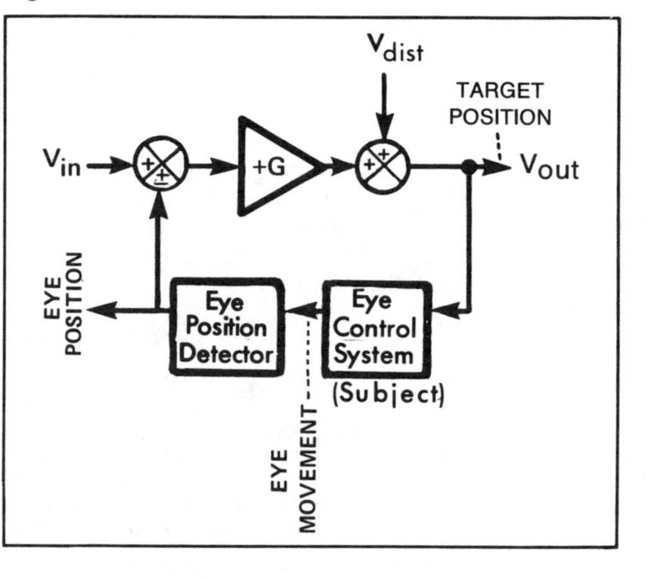

Fig. 12-19. Saccadic movements of eye to displaced visual target.

(OPEN LOOP)
G = O
TARGET
POSITION
EYE
T
Vdist
3°
0
T
200 ms

(Modified from L. Stark, *Neurological Control Systems: Studies in Bioengineering.* New York: Plenum, 1968.)

315

Fig. 12-20. Target and eye movements after step change in target position with gain shown. Experimental arrangement as in Fig. 12-18.

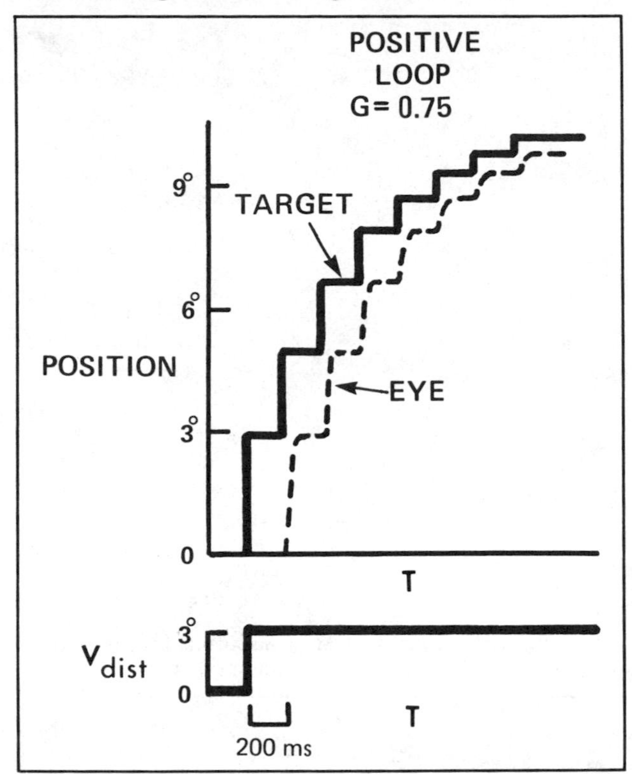

(Modified from L. Stark, *Neurological Control Systems: Studies in Bioengineering*. New York: Plenum, 1968.)

Fig. 12-21. Same experimental arrangement as in Fig. 12-20.

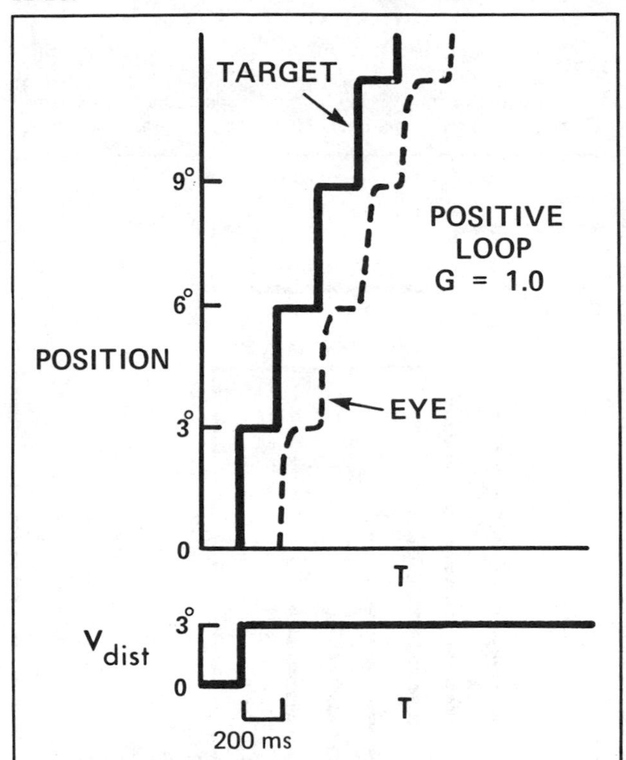

(Modified from L. Stark, *Neurological Control Systems: Studies in Bioengineering*. New York: Plenum, 1968.)

Fig. 12-22. Same experimental arrangement as in Fig. 12-20.

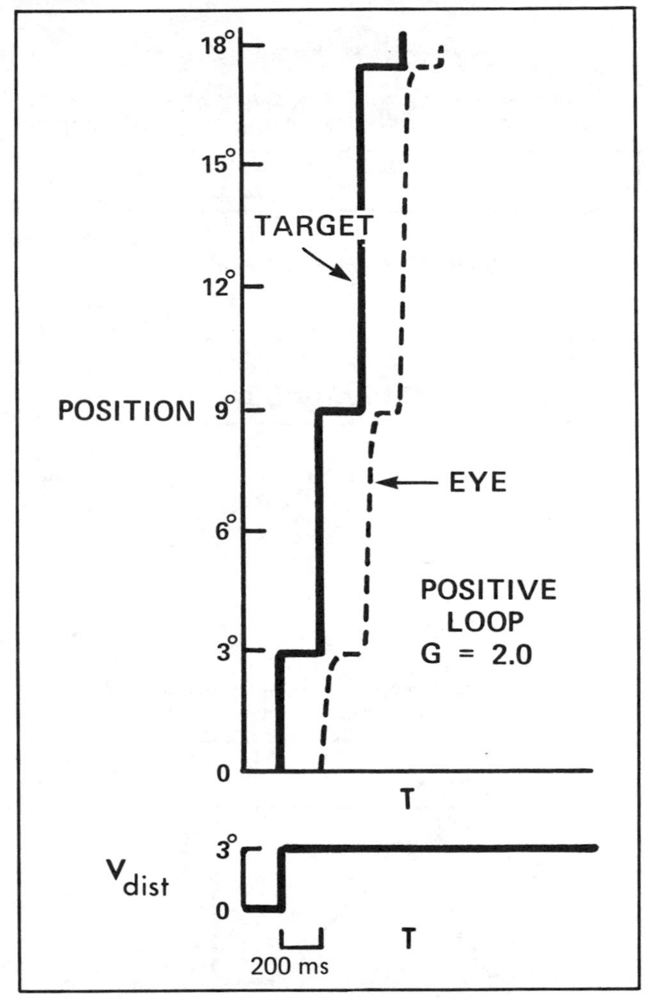

(Modified from L. Stark, *Neurological Control Systems: Studies in Bioengineering*. New York: Plenum, 1968.)

315

316

Fig. 12-23. Same experimental arrangement as in Fig. 12-20, except for negative loop.

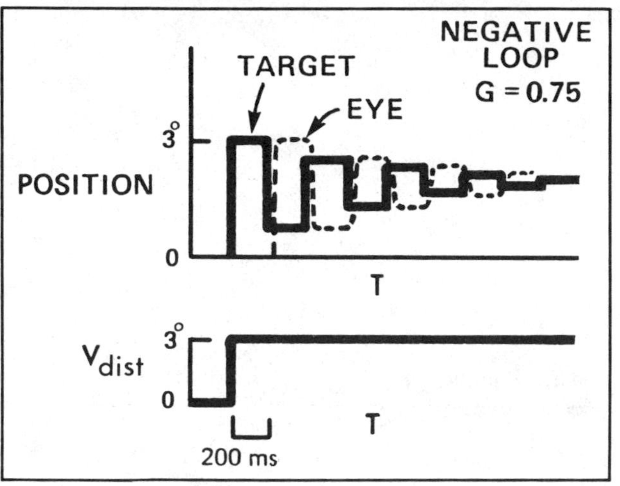

(Modified from L. Stark, *Neurological Control Systems: Studies in Bioengineering*. New York: Plenum, 1968.)

Fig. 12-24. Same experimental arrangement as in Fig. 12-23.

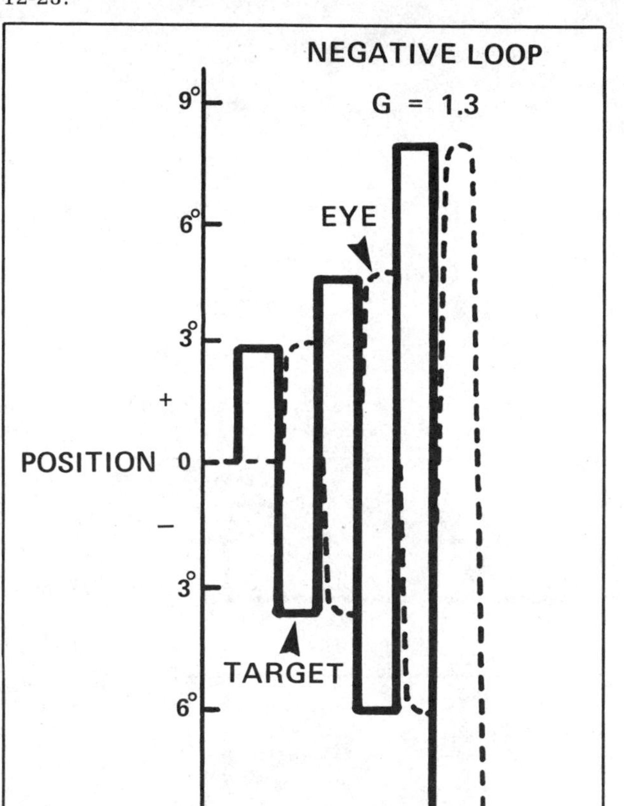

(Modified from L. Stark, *Neurological Control Systems: Studies in Bioengineering*. New York: Plenum, 1968.)

Fig. 12-25. Saccadic movements with $V_{dist} = 0$ showing positive feedback going toward maximum once set off by random eye movements. At *X* some unknown factor (perhaps a blink?) reversed the direction, so that curve headed toward opposite maximum.

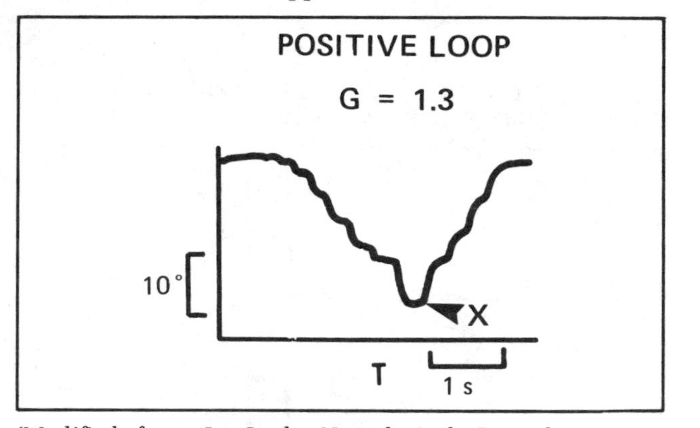

(Modified from L. Stark, *Neurological Control Systems: Studies in Bioengineering*. New York: Plenum, 1968.)

Fig. 12-26. Saccadic movements showing instability of negative feedback loop. Oscillatory behavior was set off by random eye movements.

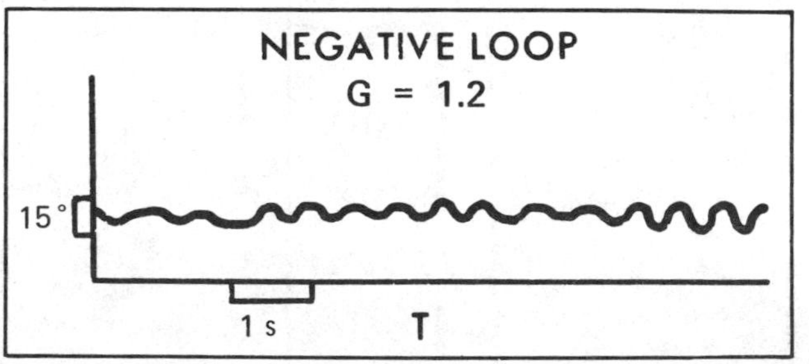

(Modified from L. Stark, *Neurological Control Systems: Studies in Bioengineering*. New York: Plenum, 1968.)

316

317

1 Thus, we hope that you are fully convinced that the principles described here will work in practice, assuming feedback loops meet the criteria described.

STABLE NEGATIVE FEEDBACK LOOPS

2 By this point, you should be convinced that negative feedback loops can be either stable or unstable. Furthermore, **the loop gain can determine whether the loop will be stable.** It remains only to explain how a negative feedback loop with a gain greater than 1 can also be stable!

3 In all the examples presented, the amplifier is "ideal" (completely linear) and the delay in the loop "pure" (delaying all signals the same amount). **Under these circumstances, the loop cannot be stable when the gain is 1 or greater.** However, **if the amplifier gain is reduced when the rate of change in input dV/dt is high, then it is possible to construct a system with high amplifier gain that is still stable.** Basically, such an amplifier will have a high gain for slowly changing inputs, but will have a low gain (less than 1) for rapidly changing inputs. This attenuates the loop oscillations since the amplifier has a gain less than 1 at the frequency at which the loop would otherwise oscillate. However, since the loop is stable, a high amplifier gain is present to control a slowly changing disturbance or input.

4 Electronically, a low-pass filter has just the attenuating characteristics described. So if the filter is inserted into the system, it will prevent the oscillations, even when the amplifier gain is greater than 1 at low frequencies.

5 From here on, the discussion of the stability requirements of negative feedback loops involves greater detail, leading to complex engineering concepts beyond the scope of this book. Therefore, if you are willing to take on faith the idea that electrical engineers who study and build negative feedback systems have techniques by which high-gain negative feedback loops can be made stable, then you need not get into the next section, which is only for those with an insatiable curiosity! Do remember that **the good features of negative feedback loops with high gain** (small change with disturbances, V_{out} follows V_{in}, small variations with changes in the system) **can be obtained in a stable system.** But under some circumstances, **when loop conditions are changed in various ways, the system may become unstable, begin oscillating, and hence become uncontrollable.**

OPEN-LOOP ANALYSIS AND STABILITY CRITERIA

6 The following section is at the third and fourth levels, to supply background information for those who wish to know more about engineering approaches to biological systems.

HINT

20. See Fig. 12-24.

317

318

In the two preceding sections, loops are presented that are "time-sampling," or discontinuous. This is reasonable for a system such as eye saccades, but does not apply directly to "continuous" systems, dealt with in this section. 1

In reality, all amplifiers have a gain that is *frequency-dependent,* i.e., which varies depending on the dV/dt of the input. This is shown by the arrangement of Fig. 12-27. There a sine wave generator that produces sine waves of constant amplitude, but of varying frequency, is hooked up to an amplifier input. Then we can see what happens to the amplifier output as the frequency is increased at the amplifier input while the input amplitude remains constant. Figure 12-28 shows the sort of result that might be obtained from an amplifier in which $G = 2$. The input amplitude is the same at different frequencies. At low and medium frequencies, the amplifier output is twice that of the input, as one would expect. However, at high frequencies, the output of the amplifier is less than you would expect. 2

The actual result from one system (discussed later) is shown in the upper part of Fig. 12-29, where the peak amplitude of the amplifier output is plotted. Note that the **amplitude drops off noticeably at higher frequencies.** This is **characteristic of all amplifiers.** 3

Note that the ordinate of the upper part of Fig. 12-29 is labeled *amplitude* and *gain.* Since the maximum (peak-to-peak) amplitude is measured, that is what is plotted. However, since gain is output divided by input and since input (peak to peak) is constant, the curve also plots *gain* (differing only in having a different scale). Thus, **at high frequencies the gain of the amplifier is said to decrease.** 4

In electronic circuits, such decreased gain is due to capacitances that form, with their associated resistances, a *low-pass* filter. 5

At higher frequencies, in addition to the decreased amplifier gain, **there is also a phase shift between the input and output.** 6

What is a phase shift? **A phase shift is the measure of how much one sinusoidal wave is displaced from another sinusoidal wave in time.** This shift can be measured in *degrees.* Remember way, way back in trigonometry that a sine wave can be generated in a circle by projecting the radius of a changing angle that is moving from 0° to 360° (Fig. 12-30)? If the hypotenuse is equal to 1, then the side opposite the angle θ (theta) will be equal to the value of sine θ. Thus as θ varies from 0° to 360°, the length of the line will vary as sine θ. Now, if we plot this against time, we get a sine wave (solid line, Fig. 12-31). Along this sine wave, we can specify a point by reference to the number of degrees through which θ has moved (Fig. 12-31). Now, if you want to compare two waves, e.g., the solid line versus the dotted line in Fig. 12-31, you can say that the dotted line is 180° *out of phase* with the solid line; meaning that the start of the second (dotted line) wave occurs when the first (solid line) wave is at 180°. (We know you are beginning to wonder what this is leading to, but hang on!) 7

318

319

Fig. 12-27. Open-loop test of frequency response of amplifier.

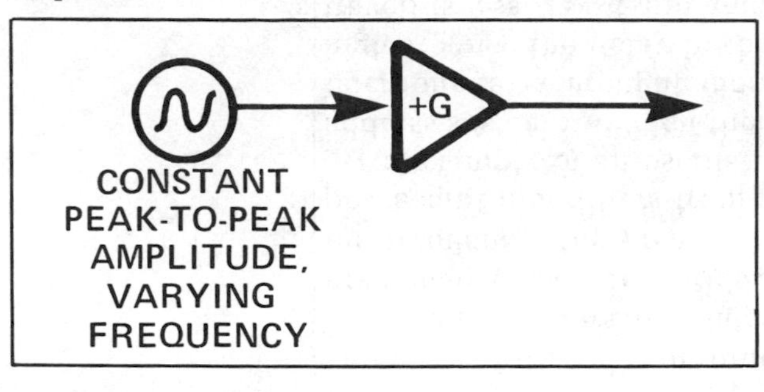

Fig. 12-28. Variations in output of circuit of Fig. 12-27, with differing frequencies.

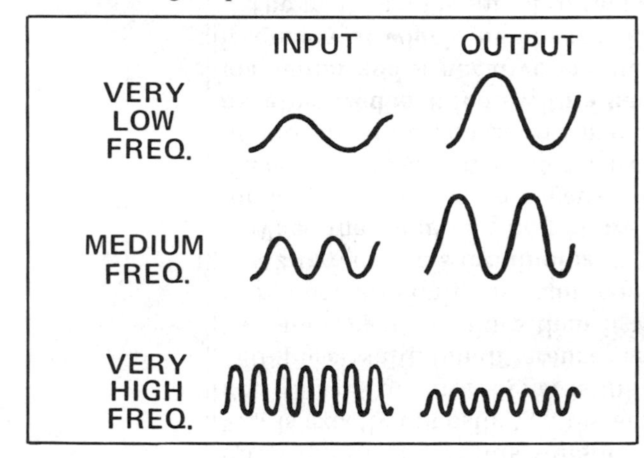

Fig. 12-29. Amplitude and phase shift at differing frequencies for open-loop biological system.

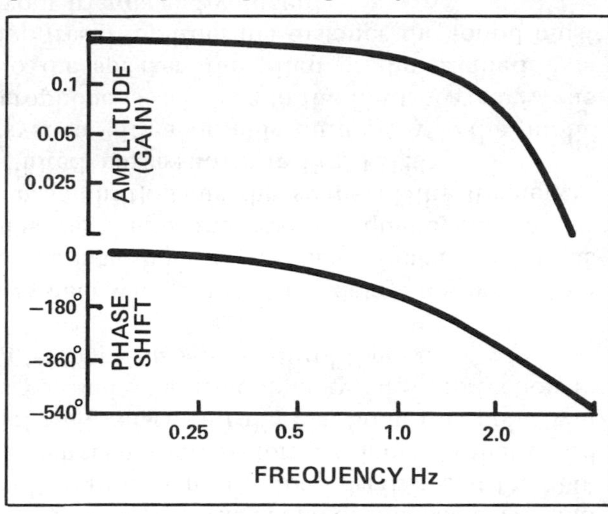

(Modified from L. Stark, *Neurological Control Systems: Studies in Bioengineering*. New York: Plenum, 1968.)

Fig. 12-30. Generation of sine wave changing an angle θ from 0° to 360°, and measuring the length of side opposite the angle.

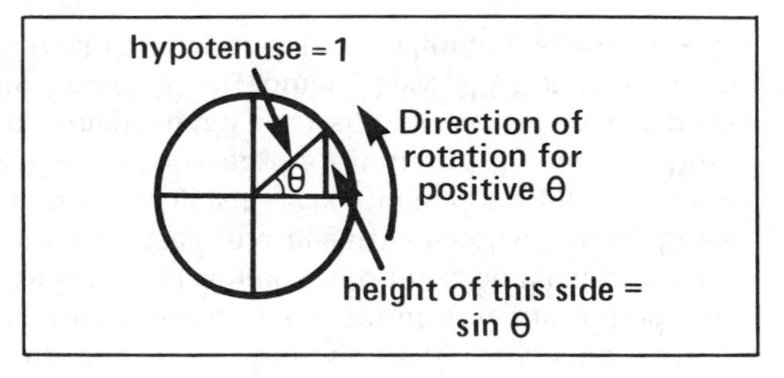

Fig. 12-31. Input (solid line) and output (dotted line) with 180° phase shift.

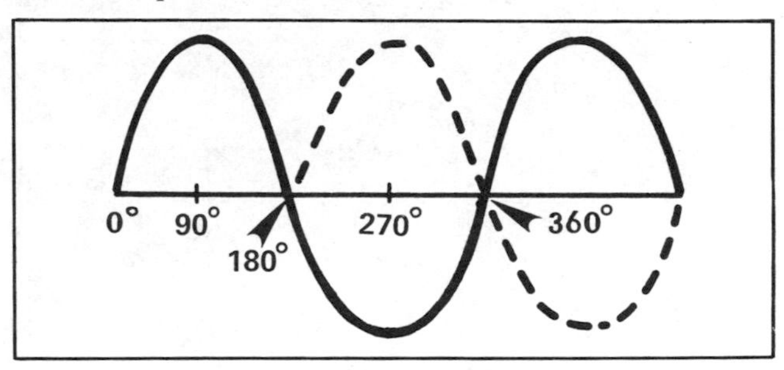

319

320

An increasing phase shift (or lag) in the amplifier output, compared with the input, is shown at the bottom of Fig. 12-29 (where a lag is plotted in minus degrees, hence downward). Thus, in the amplifier, as input frequency is increased, output phase lag increases (as gain decreases). Phase lag is crucial since **at a phase lag of 180°, a negative feedback loop has the reinforcing properties of a positive feedback loop!** The meaning of a 180° phase lag is that when the amplifier input is cycling up and down, the amplifier *output* is also cycling up and down, but is *exactly opposite!* This is shown in Fig. 12-31, where the solid line represents the input and the dotted line indicates the output (with a 180° phase lag). You can begin to imagine what problems this would create if this situation were occurring in a negative feedback loop. Since V_{out} is exactly opposite V_{in}, the action of the loop is to accentuate the input. Consider Fig. 12-31. The error signal is the *difference* between the input and the output. With 180° phase shift, the differences are maximal, so the error signal reaches greater positive and negative values than it would reach at any other phase shift or if the loop were open and only the input could affect the amplifier (error signal) (see Fig. 12-32/12-12 to recall this terminology). 1

When the input and output are 180° out of phase, it adds an additional reversal in the loop, which turns a negative feedback loop into a "*quasi*" positive feedback. It is "quasi" because it does not act completely as the positive feedback loop described previously: it doesn't go to a maximum and stay there. Instead, this loop, **an unstable negative feedback loop with a gain greater than unity, oscillates between a positive maximum and a negative maximum** (as you already know), and it oscillates **at the frequency at which the phase shift is 180°!** 2

The mathematical proof of this idea is well beyond the scope of this book. Without the necessary mathematics, it is difficult to intuit why the loop oscillates at the 180° phase shift frequency. And we have never seen a satisfactory explanation that could be understood by students who are not well versed in calculus. So at this point you will have to take this on faith unless you have the time, energy, and motivation to seek a more fundamental, engineering textbook on servo-control theory. 3

Of course, in the discontinuous system with a pure time delay, the oscillating frequency is inversely related to the magnitude of the delay. Such a delay (as previously considered) adds a phase lag that is linear with frequency (the higher the frequency, the greater the phase shift). Thus the shape of the frequency-phase shift plot *differs* from the "filter" curve shown in Fig. 12-29. 4

An example of an oscillating system with a variable pure delay is the public address system in which the microphone is too close to the loudspeaker. As the microphone is brought closer to the speaker, the pitch of the oscillation is higher, since the delay in the loop (resulting from the distance the sound must travel) is less. Try it and amaze your friends and yourself! 5

Fig. 12-32/12-12. Negative feedback loop with external disturbance.

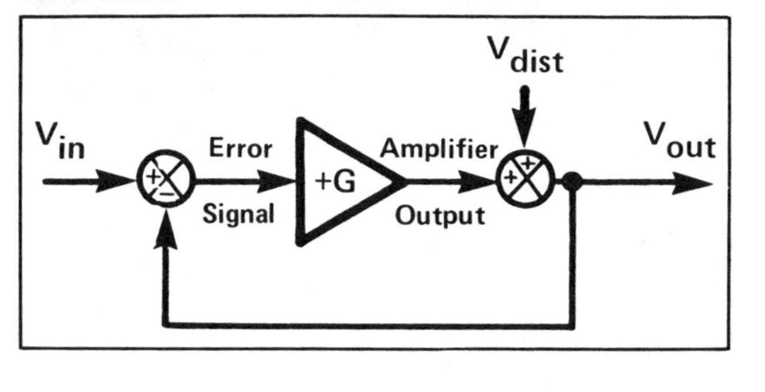

320

321

QUESTION: How do you stop the oscillations of such a public address system once begun? (Hint 21↓)

1

You can probably see quite easily that the driving of the system into increasing oscillations can occur only if the gain is greater than 1. That is, each time around the loop, the effect must be greater than the cause if the system is to be driven to the maximum. But, if the gain is less than 1, the buildup is self-limiting and dies out, just as when there is a gain of less than 1 in a positive feedback loop. That is, the effect changes asymptotically toward an intermediate value. Thus it is useful to have a plotting of gain at various phase angles.

2

The plots of the frequency characteristics of an open-loop amplifier, i.e., the plots of gain versus frequency and phase shift versus frequency previously shown (Fig. 12-29), are together called the *Bode* ("bo-dee") *plot*.

3

Bode, an engineer at the Bell Telephone Laboratories, pioneered analyses of networks and feedback amplifiers.

4

Now what we really want to know is the gain versus phase shift. The best way to plot this is by the Nyquist diagram.

5

Nyquist, also at the Bell Telephone Laboratories, is famous for classic studies in systems dynamics (among other things).

6

The **Nyquist diagram is plotted on polar coordinates** rather than the rectilinear or logarithmic coordinates of the Bode plot. Polar coordinates are those in which the angle θ is plotted as an *angle* and (at a given frequency) the **gain is plotted as the distance from the point of origin** (Fig. 12-33). When the results from the Bode plot of Fig. 12-29 are replotted on polar coordinates (the Nyquist diagram), an inward-spiraling curve is produced. (Note that phase lag is plotted as a negative angle, hence clockwise.)

7

Now this plot is especially useful because it shows the **critical point!** The critical point on the Nyquist diagram is the point at $G = 1$ and $\theta = -180°$. **It is critical because of the Nyquist stability criterion:** a negative feedback loop will be stable (i.e., will not oscillate) only if the plot of the open-loop frequency characteristics does not encircle the critical point. This is a more precise definition than the intuitive ideas previously presented. You can see, however, that if the Nyquist plot does encircle the critical point, the loop *will* oscillate, since the gain *will* have to be greater than 1 at $\theta = -180°$ (the requirement presented previously as the condition for increasing oscillations). Thus, the Nyquist plot of the amplifier open-loop characteristics shown in Fig. 12-33 will give a stable system when a closed negative feedback loop is formed. However, if an amplifier showed the open-loop characteristic such as that of Fig. 12-34, in which the critical point is encircled, then such a system would be unstable when the loop was closed.

8

Fig. 12-33. Open-loop Nyquist diagram of system that will be stable if loop is closed.

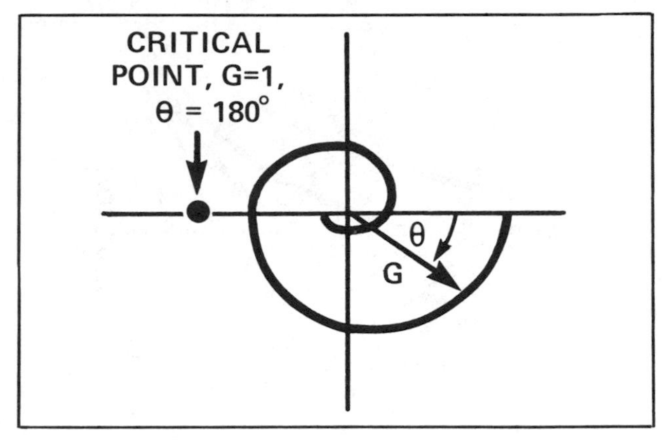

Fig. 12-34. Nyquist diagram of open-loop characteristics of system that will be unstable when loop is closed.

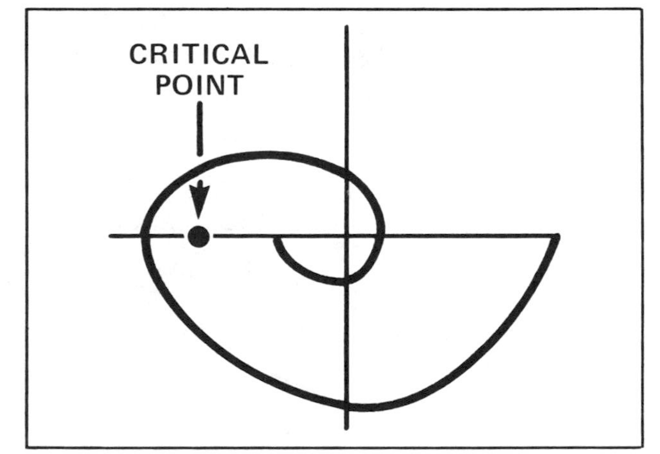

321

The Nyquist diagram is useful since a quick glance will tell you whether a negative feedback system will be stable (given the Nyquist diagram for a particular amplifier that might be put in a negative feedback loop). Some systems can meet the Nyquist criteria with a rather large gain! Actually, either of the Nyquist plots shown in Fig. 12-35 will be stable when connected in a negative feedback loop (!) since in neither is the critical point encircled. So, *by clever design* it is possible to build a stable negative feedback loop containing an amplifier with a high gain.

EXAM QUESTION: The Nyquist diagram encircles the critical point. This implies that (choose one or more):

1. The gain is less than 1.
2. The system will oscillate when the loop is closed as a negative feedback only if a sine wave generator is applied to the input.
3. The system has significant time lags (phase shifts).
4. Bode is plotting.
5. None of the above.

(See Hint 22.↓)

Here are some mathematical ideas. The *minimal* requirement for a system to oscillate is one described by a second-order differential equation. (You can get a second-order equation from two first-order differential equations that are interrelated; i.e., the variables of one appear in the other, and vice versa.) These equations (in a minimum system) describe the elements of the negative feedback loop.

Second-order systems can show a variety of behaviors to a step function input (a sudden shift of input to a new, steady value). Figure 12-36 shows a family of curves obtained when the values of the parameters in the system are changed. (If you are familiar with it, you can consider this the spring-mass-damper model.) The responses of the system can be overdamped (*A*), critically damped (*B*), or underdamped (*C* and others). At the extreme, the system goes into continuous oscillations. The interesting point is that **as the parameters are changed to give these effects, the corresponding Nyquist diagram** of the open-loop amplifier characteristics **comes closer and closer to the critical point** (as you go from *A* to *B* to *C*, etc.) **until the critical point is encircled at the point where the system makes sustained oscillations.**

Another mathematical idea of some interest is the following. In an oscillating system that does not go from plus infinity to minus infinity, there must be, in the system, a nonlinearity (that makes it stop at the maximum and minimum voltages). Thus, in the design of an electronic oscillator for laboratory use, a nonlinearity must be built in, or else the system will find one by pushing the amplifiers out of their linear range (Fig. 12-1).

Fig. 12-35. Nyquist diagrams of open-loop characteristics of two stable systems. Note: the critical point is at G = 1. The scales of the two diagrams differ.

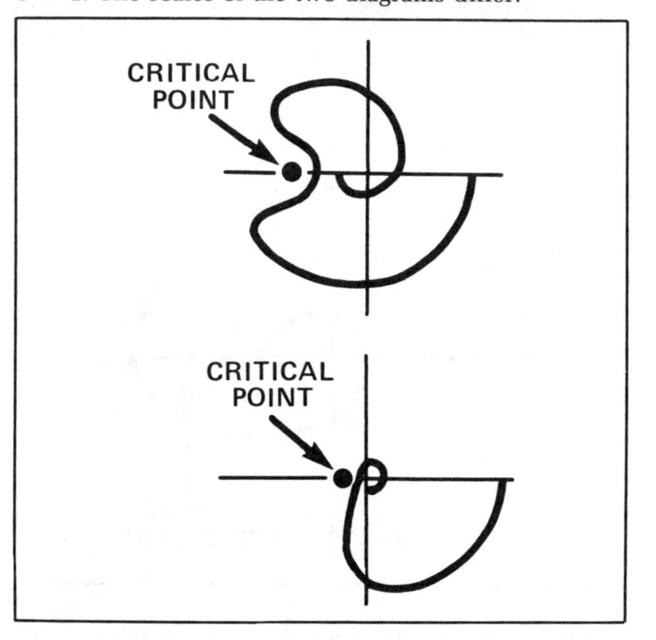

Fig. 12-36. Negative feedback system (with different degrees of damping), returning to steady value after disturbance.

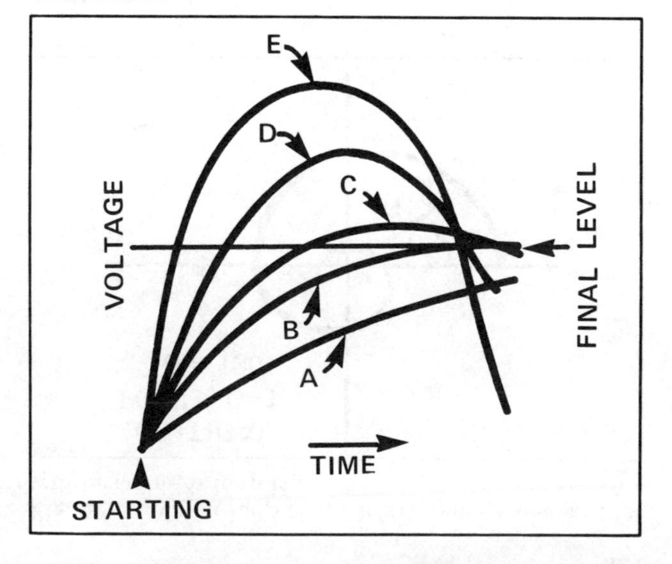

323

The ideas expressed here can be demonstrated in a biological system—the pupillary reflex to light. The pupil's reflex to light has been studied beautifully as a negative feedback loop [57, pp. 73–184]. The reflex is *obviously* a negative feedback loop.

QUESTION: How can we say that "obviously" part? (Hint 23↓)

To study this negative feedback loop, it was necessary to open the loop and study the open-loop characteristics of the system. This was done in a very clever way (Fig. 12-37). In the normal case (see Fig. 12-37), light strikes the entire pupil, and the loop is closed. However, if the light is made so small that no matter how small the pupil gets, it cannot change the light reaching the retina, then the system becomes open-loop. That is, the light intensity may vary, but the movement of the pupil cannot affect the light intensity reaching the retina. Now, the light intensity (input) can be varied sinusoidally, and the pupil diameter (the output) can be measured simultaneously, so the open-loop characteristics of the system can be determined.

This, in itself, was no mean task, since the pupil diameter was measured in the dark! It was done by measuring the area of the pupil by means of infrared reflection off the iris.

Under these experimental conditions, the data shown in the Bode plot of Fig. 12-29 were obtained! The same data are shown on the Nyquist diagram (Fig. 12-38).

QUESTION: Is the pupil stable as a negative feedback (closed) loop? (Hint 24↓)

From these data we can conclude that **the pupil is a negative feedback system that is stable, in spite of the many lags in the reflex, resulting from a very low gain.** That is, it doesn't really regulate the amount of light reaching the retina very much! We could be even more secure in our analysis if we could make the system oscillate. In other words, if we could somehow increase the gain, we might make the Nyquist plot encircle the critical point, and then the system should oscillate at the frequency where the phase shift was −180°. The system was made high-gain by the maneuver shown in Fig. 12-37 (bottom), i.e., by shining the small light on the edge of the pupil. Under these conditions, a small change in the pupil diameter that under normal conditions might have changed the light reaching the retina by, say, 10 percent now re-

Fig. 12-37. A spot of light of different sizes and at various positions was used to change pupillary negative feedback loop.

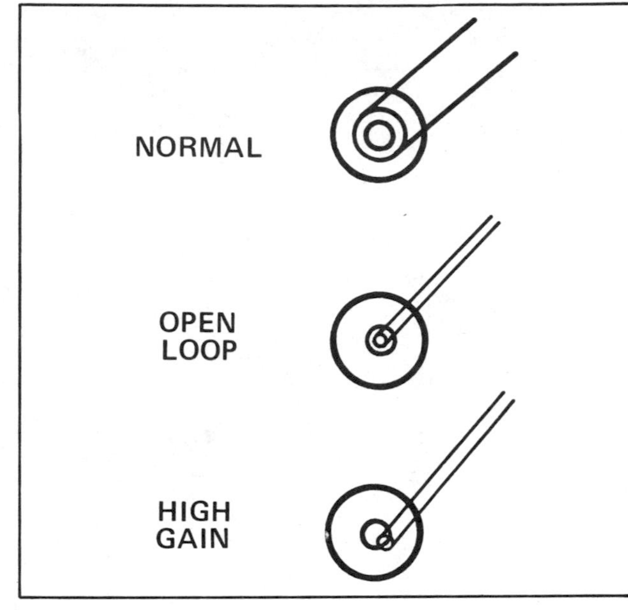

(Modified from L. Stark, *Neurological Control Systems: Studies in Bioengineering.* New York: Plenum, 1968.)

Fig. 12-38. Nyquist diagram for pupil (open loop).

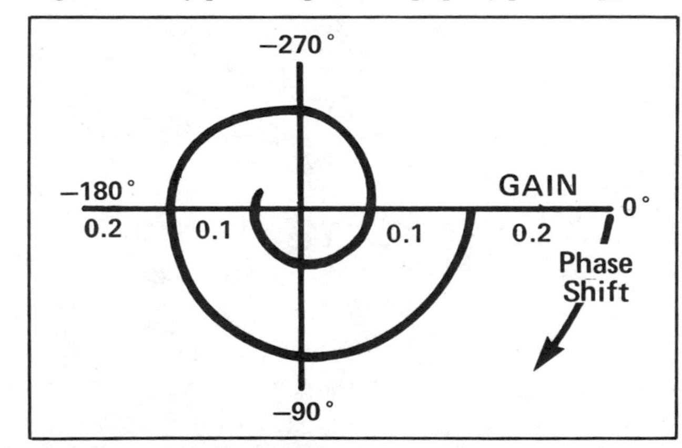

(Modified from L. Stark, *Neurological Control Systems: Studies in Bioengineering.* New York: Plenum, 1968.)

HINT

21. Either open the loop (shut off microphone) or reduce gain (turn down the volume control; put hand over microphone, move microphone to location where speaker sound is less).

323

duces it much more, say 70 percent. At the extreme, if the pupil response is enough, the light will be shut off entirely, so that no light reaches the retina. Then the pupil will dilate (since the retina is in the dark), and suddenly the full intensity of the light will again strike the retina, causing the pupil to close until there is no light, over and over (i.e., it's oscillating!). 1

Gratifyingly, when the experiments were done, the pupil *did* oscillate, and just at the frequency expected from the 180° phase shift frequency (which you can determine for yourself on Fig. 12-29). 2

QUESTION: O.K., what was the frequency? (Hint 27↓) 3

Not clinically relevant? How many times have ophthalmologists looked at a patient's eyes by a slit lamp and accidentally directed the light on the edge of the pupil so that it oscillated? We don't know, but it must be pretty common. How many realized what the basis of the oscillation was? Moreover, how many realized that they might be able to establish useful clinical standards for pupil response? All they would have to do would be to get it oscillating and then time 5 or 10 oscillations, with a stopwatch or nurse, in order to determine the frequency—*a frequency that is highly characteristic of the lags of the system.* When you see patients in your clinical training and flash a light in and out of their eyes to see whether the pupil is "sluggish," why not remember that such a subjective measure of performance ("sluggishness") might be *more accurately measured and recorded.* Well, we might not have convinced you. But as a parting comment, we should tell you of the time a patient came to the ophthalmology department with only one complaint: "My pupil oscillates." (Unfortunately, they didn't know servo-control theory, so they didn't know how to analyze further!) 4

If you encounter the following disease in your studies, you might consider its servo-control aspects. 5

Parkinsonism. How is it that a stable system (hand position) becomes unstable and oscillating (pill-rolling tremor)? It might occur if the gain is increased, the lag is increased, or both. Gain can be increased by *less inhibition* (presumably less "descending influences" on the spinal cord from above). Lags can be increased if short, fast pathways become inoperative and longer, slower pathways control the reflexes. 6

Oscillations can occur in the feedback loops inherent in medical practice. 7

QUESTION: Assume that a physician forms part of a "therapeutic" feedback loop that is oscillating. For instance, the physician finds that on successive visits of the patient, the doctor is prescribing high doses of a drug, then low doses, then high doses again, etc. Briefly 8 describe in servo-control terms (and also illustrate in practical terms) two basic approaches that the physician can take to prevent such oscillations *and* still maintain therapeutic control. (Hint 28↓)

1

While we are considering the feedback loops in medical practice, you should note that when a patient returns to see the M.D., the physician receives feedback as to how well the therapy is working. Thus the doctor can take appropriate corrective action if the therapy is inappropriate. But, if the patient becomes so dissatisfied as to not return, then the physician is operating "open-loop" without the self-correcting features of a feedback system necessary for the practice of good medicine. It should be obvious that advances in medicine have occurred only where feedback is available. Conversely, where feedback is not available, medical practice is most likely to go awry (and astray!).

2

An article you might find interesting describes the nervous system as a hierarchical organization of control loops and takes the position that a servo-control viewpoint explains the organization of behavior better than does a stimulus-response approach [49].

3

An engineering approach to nervous system feedback loops is found in Partridge [44].

4

An interesting analysis of the ambiguities in diagrams of biological loops has been made by Allweis [2]. The importance of closed-loop control of lecturing has been described by Jewett [26].

5

It has been found that the simplest system that will show the rapid transitions of "avalanche" in a chemical system involves three elements, and the interaction dynamics must include one positive feedback loop. By now this should seem intuitively correct, since the rapid transitions of avalanches or other mathematical "catastrophes" could be driven by positive feedback loops with high gain.

6

HINTS

22. Answer: 3. Item 1 is obviously wrong. Choice 2 is not correct, since the system will oscillate under *any* conditions, as long as the loop is intact. The sine wave generator was used *open-loop* to study the characteristics of the amplifier.

23. Consider what happens when a light is suddenly increased on an open pupil, thus increasing the light on the retina. Many neural mechanisms come into play, but the net result is this: The pupil constricts so as to reduce the amount of light on the retina, for the response of the loop is to *reduce* the amount of light on the retina. Hence the loop *must* be negative.

24. Well, probably *you can guess* at this by looking around you and seeing whether anyone's pupils are oscillating! But the real question is, Can you predict the stability from the Nyquist diagram? (Hint 25↓)

26. It certainly won't be on the graph! Not since the gain calibrations show only 0.1 and 0.2! So there isn't the slightest chance that the curve encircles the critical point, which is way off to the left at $-180°$.

326

An excellent description of the interlocking positive and negative feedback loops in the ecology of the world can be found in *The Limits to Growth* [35, pp. 149–151]. Here, from that book, is a description of the impact of natural delays (lags) that can bring a technological society to disaster. You will understand the concepts presented very easily now. You also can see that the concept of feedback applies to many important systems, even social ones!

1

"The basic behavior mode of the world system is exponential growth of population and capital, followed by collapse. . . .

"It is not really difficult to understand how the collapse mode comes about. Everywhere in the web of interlocking feedback loops that constitutes the world system we have found it necessary to represent the real-world situation by introducing time delays between causes and their ultimate effects. These are natural delays that cannot be controlled by technological means. They include, for example, the delay of about fifteen years between the birth of a baby and the time that baby can first reproduce itself. The time delay inherent in the aging of a population introduces a certain unavoidable lag in the ability of the population to respond through the birth rate to changing conditions. Another delay occurs between the time a pollutant is released into the environment and the time it has a measurable influence on human health. This delay includes the passage of the pollutant through air or rivers or soil and into the food chain, and also the time from human ingestion or absorption of the pollutant until clinical symptoms appear. This second delay may be as long as 20 years in the case of some carcinogens. Other delays occur because capital cannot be transferred instantly from one sector to another to meet changing demands, because new capital and land can only be produced or developed gradually, and because pollution can only slowly be dispersed or metabolized into harmless forms.

"Delays in a dynamic system have serious effects only if the system itself is undergoing rapid changes. . . . the delays in the feedback loops of the world system would be no problem if the system were growing very slowly or not at all. Under those conditions any new action or policy could be instituted gradually, and the changes could work their way through the delays to feed back on every part of the system before some other action or policy would have to be introduced. Under conditions of rapid growth, however, the system is forced into new policies and actions long before the results of old policies and actions can be properly assessed. The situation is even worse when the growth is exponential and the system is changing ever more rapidly.

"Thus population and capital, driven by exponential growth, not only reach their limits, but temporarily shoot beyond them before the rest of the system, with its inherent delays, reacts to stop growth. Pollution generated in exponentially increasing amounts can rise past the danger point, because the danger point is first perceived years after the offending pollution was released. A rapidly growing industrial system can build up a capital base dependent on a given resource and then discover that the

2

326

327

exponentially shrinking resource reserves cannot support it. Because of delays in the age structure, a population will continue to grow for as long as 70 years, even after average fertility has dropped below the replacement level (an average of two children for each married couple)."

SUMMARY

Of course, you won't find the electrical circuits described in this chapter in the clinic or basic science courses. You will have to "see" them in the flesh and blood examples that you will encounter from here on, whatever you do (even if all you do is read the next chapter, where the feedback loops involved in muscle control are described).

A negative feedback loop diagram, modified with "biological" terms, is shown in Fig. 12-39.

The effector might be a motor nerve, a muscle, or a gland. The sensor may be a sensory nerve or a part of an endocrine gland, etc. The point is that **loops are so common in biological systems, you need to have some idea as to what strange behaviors such loops can generate.**

Note: what has been presented in this chapter certainly does not exhaust the types of behavior that a loop can exhibit. Recall that the descriptions have mainly dealt with

Fig. 12-39. Negative feedback loop in biological system.

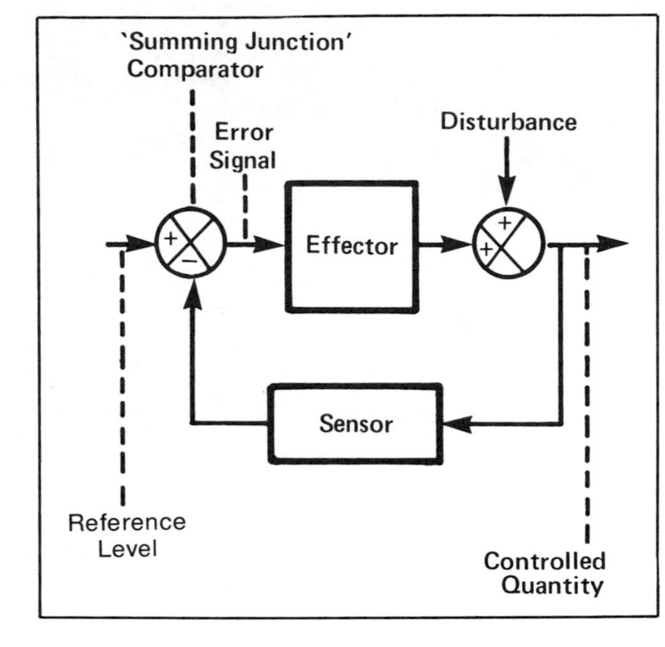

HINTS

25. That's right, you need to know whether the critical point is encircled. Where the devil is the critical point in Fig. 12-38? (Hint 26↑)
27. A little over 1 per second.
28. In the Nyquist diagram of the open-loop data of an oscillating system, the critical point is encircled. There are basically two simple ways of preventing this: reduce the gain or reduce the lag (phase shift). First, the gain can be reduced if the drug dosage prescribed at a given time is decreased. (For example, a gain of more than 1 occurs in this way: First visit: "Sick? I'll give you lots of a powerful drug." Next visit: "Too many side effects, let's cut way back on the dose," etc. You may think that this is far-fetched—you may be surprised. Second, the lags might be reduced. (For example, the physician might not be seeing the patient often enough, or the drug dosage may depend on some lab test that could be obtained faster by phoning the lab, rather than waiting for the mail, etc.) Both types of problems occur when common, powerful drugs (e.g., digitalis and coumarin anticoagulant) are administered. Don't say we didn't tell you!
29. Refer to Table 12-9 in the section "Stability and Instability."

327

328

simple loops. The behavior can become much more complex when multiple, interconnected and parameter (gain) changes are also considered! Let's leave that for other authors!

1

Now **go back and review Table 12-9.** If you understand what it summarizes, then you understand this chapter! Most interesting systems, whether they are physical, biological, or social, will have many interconnecting loops. Even if you can identify one of the behaviors shown in Table 12-9 in a given system, you will not understand what underlies it until you can find the predominant loop that gives that result! (Motto: Cherchez la loop!)

2

EXAM QUESTION: What combinations of gain and type of feedback (in a system with a lag) lead to

1. Unstable oscillations
2. Going rapidly to a maximum
3. Control of a variable in spite of disturbances
4. Sluggish, somewhat inaccurate control
5. Progression to an intermediate value (not amplifier maximum)

(If you have difficulty, see Hint 29.↑)

3

RECOMMENDED READING

Freeman, W. J. *Mass Action in the Nervous System. Examination of the Neurophysiological Basis of Adaptive Behavior through the EEG.* New York: Academic Press, 1975. Chaps. 5 and 6.

Grodins, F. S. *Control Theory and Biological Systems.* New York: Columbia University Press, 1963.

Hardin, G. Will Xerox kill Gutenberg? (Editorial) *Science* 198:883, 1977.

Mackey, M. C., and Glass, L. Oscillation and chaos in physiological control systems. *Science* 197:287, 1977.

McKusick, V. A. *Heritable Disorders of Connective Tissue,* 4th Ed. St. Louis: Mosby, 1972.

Meadows, D. H., Meadows, D. L., Randers, J., and Behrens, W. M., III. *The Limits to Growth: A Report for the Club of Rome's Project in the Predicament of Mankind.* New York: Universe Books, 1972.

Milhorn, H. T., Jr. *The Application of Control Theory to Physiological Systems.* Philadelphia: Saunders, 1966.

Platt, J. R. *The Step to Man.* Huntington, N.Y.: Krieger, 1966.

Randall, J. E. *Elements of Biophysics,* 2d Ed. Chicago: Year Book Medical Publishers, 1962. Chap. 4.

Stark, L. *Neurological Control Systems: Studies in Bioengineering.* New York: Plenum Press, 1968.

Stolwijk, J., and Hardy, J. Regulation and Control in Physiology. In V. B. Mountcastle (Ed.), *Medical Physiology,* 13th Ed. St. Louis: Mosby, 1974. Chap. 57.

328

329

Sensory and Motor Components of Posture and Movement

330

The **goal of this chapter is** to bring together the ideas and principles of the previous chapters to deal with one of the functions of the nervous system: **control of posture and movement.** Thus, we ask, what mechanisms are available to the nervous system to control the action of muscles needed in posture and movement? An answer will involve sensory endings, axonal transmission of information, integrative functions of the CNS and their underlying mechanisms, and a study of the properties of the muscles themselves. In this way, this chapter attempts to bring together your hard-earned knowledge of the separate parts of the "reflex arc" into an overall perspective of a functional unit of the nervous system (Fig. 13-1/2-14). 1

We can consider muscles as a system of bringing about movement and the nervous system as the means by which the movement is initiated and then controlled. **Such a system must have servo-control loops,** since the components in the system can vary so easily from moment to moment. An obvious example is muscle strength, where the discharge in the motor neuron innervating the muscle will result in different amounts of movement, depending on a variety of factors: frequency of neuronal firing, length of the muscle, type of load, and fatigue. The muscular system as a whole is certainly quite complicated. So you should expect that **the system by which the muscular system is controlled is complicated as well.** 2

You will see that our knowledge has advanced to the point where we can see a large number of the elements that function together to make the muscular system an integrated, correlated functioning unit. However, our ignorance often will frustrate us when we look for solid evidence on how the elements interact to achieve the successful final result. 3

To avoid entering into a discussion of the entire brain, **this presentation deals primarily with the spinal cord and the peripheral nervous system.** 4

WHAT CAN MUSCLES DO, IN TERMS OF FUNCTION?

At this point, you should fix firmly in your mind *the large range of different ways in which skeletal muscles can act and interact.* 5

Perhaps you should think about some graceful ballet dancers you have seen, some fast-moving athletes, or fix your gaze on any person in the vicinity who happens to be moving and fun to look at! You should walk around a bit and move your arms and legs. Thus you can combine a little needed stretching with a little observation—consider it sort of brief lab exercise. (But don't be gone too long, there's a lot to learn in this chapter!) 6

The first thing to notice is that muscles can act for both posture and movement. That is, **when no movement is occurring, a muscle may have a static function,** that of maintaining posture (i.e., *position*). **During movement the muscle has a dynamic function** (i.e., *change* of position). 7

This rather simple-minded distinction is important because **many of the physiological mechanisms involved in the motor system divide easily into two categories: static and dynamic.** 8

Fig. 13-1/2-14. Sequence of topics in this book, showing chapter numbers.

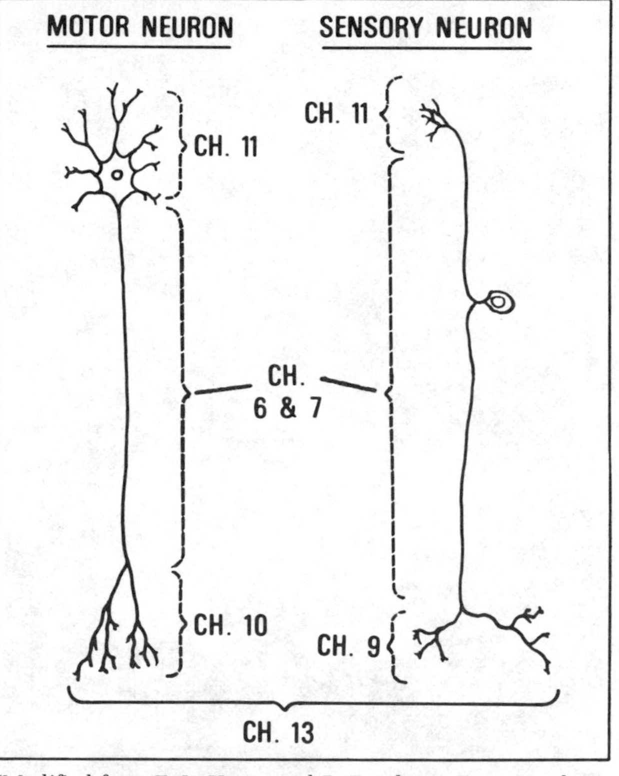

(Modified from E. L. House and B. Pansky, *A Functional Approach to Neuroanatomy* [2d Ed.]. New York: McGraw-Hill, 1967.)

330

331

It may also occur to you that this distinction approximates the two "abstract" conditions of muscle contraction studied by physiologists, namely, *isometric contraction* (change of tension without change of muscle length) and *isotonic contraction* (change of muscle length without change of tension). However, although these terms can be applied readily to the small class of muscles that provide direct force of movement without interposed mechanical linkages (e.g., the diaphragm, ocular muscles, muscles of the hyoid bone, and most smooth muscles), **skeletal muscles normally are arranged to provide relative movement between different elements of the skeletal system.** In the case of a muscle causing movement about a joint (or joints), **the nature of the mechanical linkage can be crucial to the analysis of muscle function,** and it is rare to find either static or dynamic situations that correspond exactly to the physiologist's abstractions (alas!).

1

Following the convention that a muscle causing an action is called the *agonist* while an opposing muscle is called an *antagonist,* we can list the following examples of muscle action:

2

1. **Contraction of the agonist without antagonist contraction.**
   a. *The muscle can shorten against a constant load,* as when a weight is lifted. (Notice that a constant weight at the end of a limb does *not* provide a constant load on the muscle as the joints move through an arc because of changes in the effective lever arm.)
   b. *Tension can change while muscle length remains constant,* as when one attempts to lift an immovable object.
   c. *A constant velocity of movement may be maintained.* But remember that a constant rate of movement at the end of a limb may not be at all constant at the muscle, since the angle (and hence the effective lever arm) is constantly changing.
   d. *The agonist may bring about an acceleration,* as in throwing or striking.
   e. There may be brief *periods of activity* at an appropriate point in the *cyclic movement of a limb.* For example, during walking, the extensors of the hip and knee fire only a brief burst at their resting length (length of maximum strength) and then are silent as the inertia of the leg carries the movement forward.
   f. *Muscle activity may vary according to the requirements of the movement and the conditions under which it is being performed* (e.g., variation in external load, etc.). Here none of the abstractions such as tension, length, or velocity need remain constant.
   g. *Muscles often contract while being lengthened by the load,* as when a heavy weight is set down. Indeed, lengthening during activity is probably as frequent a muscle function as is shortening under load.

3

2. **Alternation of agonist and antagonist activity.** Many movements are *cyclic,* so that first the agonist contracts and then the antagonist, as the action reverses. Furthermore, *the antagonist may come into action just before the completion of a movement, so as to*

4

331

slow down or stop the movement. (If you rapidly extend your elbow while palpating your biceps, you will find that the biceps suddenly becomes tense just as you stop the movement.)

3. **Simultaneous contraction of the agonist and antagonist.**
   a. *If both contract* vigorously, the limb will become rigid and inflexible.
   b. *If one contracts more than the other,* a smooth, graded, and controlled movement results (e.g., slow movements in modern dance).
   c. When a muscle acts across two joints, it may have to act differently depending on the *position* of the two joints, contracting at times to oppose the agonist of one joint while contracting at other times as the antagonist of the second joint.
4. **Coordinated movements of a limb.** For these to occur, the activity of numerous muscles must be complexly interrelated in the complete act. For instance, the movement of the leg during a single step requires precise timing of muscle length and tension of varying amounts during the cycle.
5. **Coordination of muscles in two limbs.** This is necessary during many movements, e.g., those of walking.
6. **Coordination of many muscles.** For some actions, a very large percentage of muscles in the body may be called into action, e.g., during throwing of a weight or fast running. In a four-legged animal, the hindlimbs are coordinated with the motions of the forelimbs.

The examples above, if you think about them a bit, suggest that the nervous system that controls such actions is not likely to be simple! So although we can delineate some general principles, the details can be pretty messy. In this way, we are trying to forewarn you against hoping that *everything* will be clear by the end of the chapter! Why not take it as a sign of your advancing knowledge if you end up with more questions than you started with (we hope they'll be more sophisticated questions, though)? K. G. Johnson has defined *education* as "going forward from cocksure ignorance to thoughtful uncertainty!" Have a nice trip.

The general plan of the chapter is first to describe the *muscles and their efferent nerves.* Thus you can see what the nervous system has to work with in trying to get something done. Second, we discuss the *sensory system and its actions,* so you can see what kind of information the nervous system receives. Finally, we describe (though do not necessarily explain) what is known of the workings of the *spinal cord.*

FAST AND SLOW MUSCLES

You learned about muscle in Chap. 10. Now is the time to reveal that it is a *bit* more complicated than you might have gathered from that chapter. **Muscle fibers of posture and movement can be divided into** static and dynamic groups, i.e., **the slow and fast muscle fibers.**

Figure 13-2 compares the *tension* developed by two different motor units in the same muscle, one fast, the other slow. At a stimulus frequency of 7 stimuli per second (lowest part of

Fig. 13-2. Tension in same lateral segmental muscle in rat tail when single slow motor unit (*S*) or single fast motor unit (*F*) is stimulated at different frequencies.

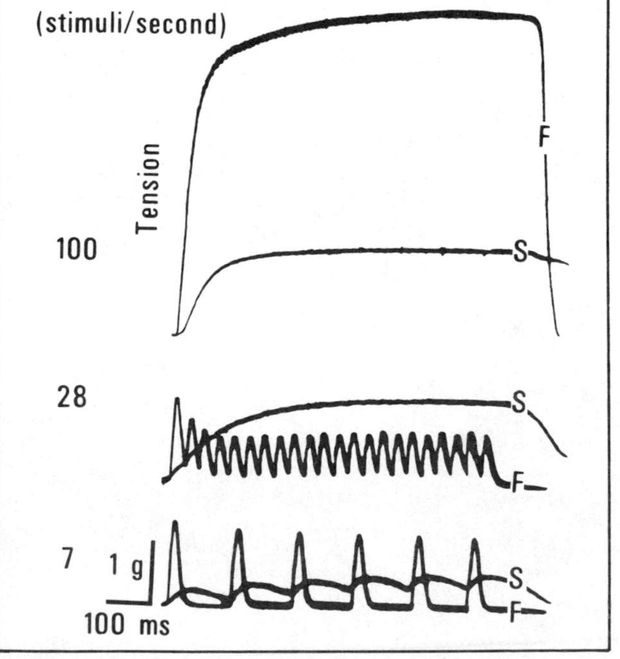

(Modified from G. Steg, Efferent muscle innervation and rigidity, *Acta Physiol. Scand.* 61 [Suppl. 225]:1964.)

333

Fig. 13-2), **the fast unit** (labeled *F*) **has a faster rise of tension, although the duration is shorter** than that of the more slowly active slow unit (labeled S). (At 28 stimuli per second, the slow unit shows greater tension than the fast unit. But this is because the slow unit has fused into a tetanus at a frequency at which the fast unit is still producing twitches.) **When both fast and slow units are tetanized** (at 100 stimuli per second), **the greater tension developed by the fast unit** (as well as its faster rise in tension) **is readily apparent.**
1

As clearly shown in Figs. 13-2 and 13-3, the fast muscle fiber also returns to the resting tension more rapidly than the slow muscle fiber after stimulation ceases. Thus, the **slow fiber acts** as if it had greater viscosity than the fast fiber, being **slower in reaching a final value** of tension in either the contracted or the relaxed state.
2

As you might expect, the fast unit pays a price for being faster and stronger—it can't keep it up as long! Figure 13-3 shows that **the slow unit can maintain its tension after repeated tetanizations, while the fast unit fatigues quickly.** Such an effect is related to the metabolic machinery of the muscle cell.
3

On one hand, the **slow muscle cell** is admirably **suited to maintain sustained contractions,** having a *high myoglobin content,* prominent pathways for *anaerobic glycolysis,* and *high lipid and oxidative metabolism.* On the other hand, the **fast fiber is more oxygen-dependent,** having a *low myoglobin content and metabolic pathways that are very glycogen-dependent* (see Table 13-1).
4

The metabolic differences have been studied by means of enzyme-specific stains (e.g., for NAD) examined histologically [15, pp. 1–6].
5

While it is useful to separate fibers into categories, we must emphasize that actually **the fibers form a continuous spectrum with regard to the properties of Table 13-1.** That is, **many intermediate forms have been found.** This is true not only of the metabolic aspects and speeds of contraction of muscle fibers, but also of the size of the nerve fibers that innervate the muscle fibers. An example of the correlation between contraction time and conduction velocity of the innervating nerve fibers is shown in Fig. 13-4.
6

Not only is there a spectrum of fiber types, but also the **different types are distributed throughout various muscles.** Some muscles are predominantly "slow" (e.g., soleus), while **other muscles have mixtures of all types** (e.g., gastrocnemius).
7

QUESTION: Have you ever had experience in distinguishing between fast and slow muscles? (Hint 1↓)
8

A rather surprising finding is that **the type** (fast or slow) **that a given muscle fiber becomes is determined by the nerve fiber that innervates it!**
9

Fig. 13-3. Effects of fatigue in slow and fast muscle. Tension recordings of single slow and fast motor units, from the lateral segmental muscle in tetanization at 100 stimuli per second for 1 s repeated every 2 s. Response of slow unit remains the same, while fast unit "fatigues."

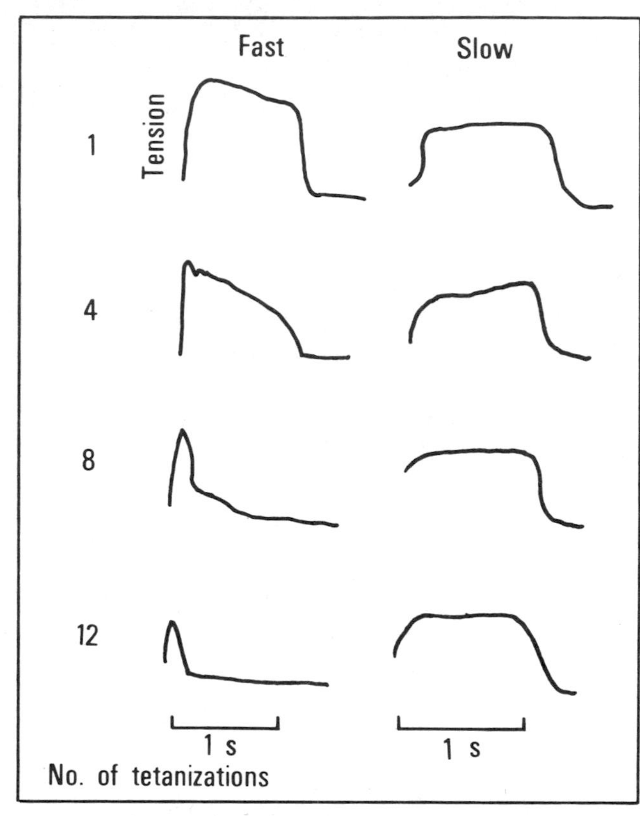

(Modified from G. Steg, Efferent muscle innervation and rigidity, *Acta Physiol. Scand.* 61 [Suppl. 225]:1964.)

333

334

Table 13-1. Differences between Fast and Slow Muscle Fibers

Parameter	Fast Fibers	Slow Fibers
Rate of tension increase or decrease	Fast	Slower
Force during twitch or tetanus	High	Less
Twitch-fusion frequency	High	Low
Rate of fatigue	Fast	Slow
Myoglobin content	Low	High
Color	Light	Dark
Metabolism		
Glycogen metabolism	High	Low
Oxygen dependence	High	Less (anaerobic glycolysis prominent)
Lipid metabolism	Low	High
Mitochondria	Less	More
Diameter	Larger	Smaller
Size of motor nerve fiber	Large alpha	Smaller alpha

Fig. 13-4. Conduction velocity of nerve fibers innervating muscle units, as related to speed of contraction of unit. Data from six experiments in cat superficial lumbrical muscle.

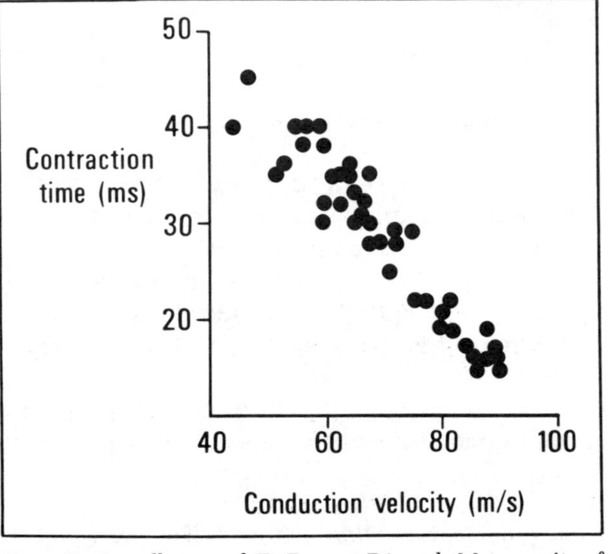

(From B. Appelberg and F. Emonet-Dénand, Motor units of the first superficial lumbrical muscle of the cat, *J. Neurophysiol.* 30:154, 1967.)

334

335

This was discovered in experiments on kittens where a nerve innervating a predominantly slow muscle was cut and interchanged with a nerve from a predominantly fast muscle. **Upon reinnervation, the muscles changed to take on the characteristics appropriate for the nerve fiber** [15, p. 14]. Recent experiments with chronic stimulation of motor nerves suggest that the firing pattern may be crucial in determining the muscle cell's biochemical mechanisms. That is, the muscle fibers seem to adapt to different patterns of depolarization, developing the characteristics of slow fibers if persistently bombarded with action potentials at a slow rate, compared with intermittent, high-frequency bursts.

1

LENGTH-TENSION DIAGRAMS WITH DIFFERENT MUSCLES AND VARIOUS LOADS

Well, as you are beginning to gather, things are always a bit more complicated than we prefer. Take, for example, the length-tension relationship of muscle, based on a static measurement made under isometric conditions. Do you expect the length-tension relationship to be the same for slow and fast muscles? Well, they are different, as Fig. 13-5 shows.

2

The slow soleus muscle shows an almost flat length-tension relationship over its working range in the body (Fig. 13-5). In contrast, the fast muscle shows a sharper peak, the "classic" length-tension diagram contour. Thus, we can expect muscles with various intermixtures of fast and slow fibers to show a variety of different shapes in length-tension diagrams. Of course, muscles with markedly different length-tension relationships behave quite differently under load (as you will see shortly).

3

QUESTION: Why is the slow muscle in Fig. 13-5 almost as strong as the fast muscle, when the previous figures indicate that the fast muscle shows a much greater tetanus tension? (Hint 3↓)

4

The steady tension developed by a muscle is determined by not only the length, but also the number of muscle fibers activated and the frequency of activation. That is, the total steady tension of the muscle is the sum of the tensions of all the individual muscle fibers. The tension of each fiber is determined by both its length and the frequency of its discharge (if the action potentials are close enough together in time for the tensions to sum).

5

Fig. 13-5. Length-tension diagrams for two muscles in cat, one slow, the other fast, having similar passive curves. Length shown at top indicates range of motion of slow soleus muscle in vivo.

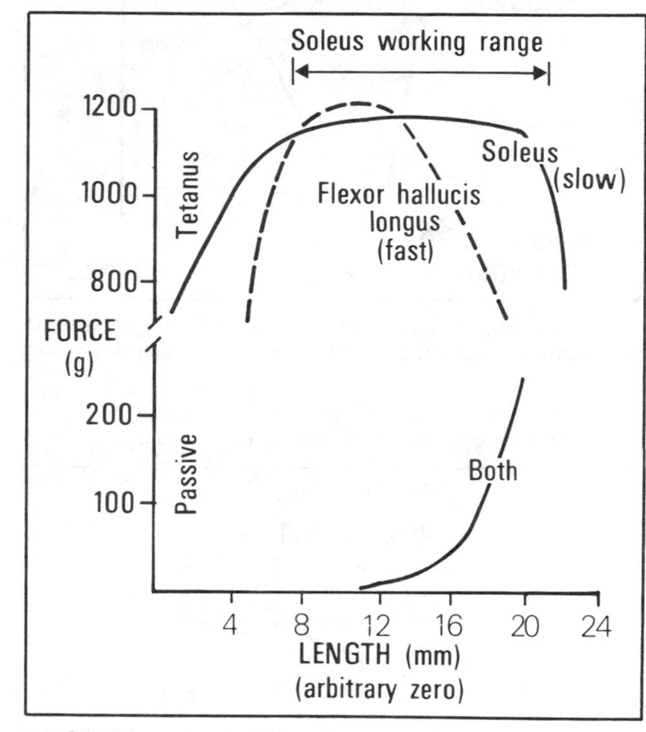

(Modified from A. J. Buller and D. M. Lewis, Factors Affecting the Differentiation of Mammalian Fast and Slow Muscle Fibres, in E. Gutmann and P. Hník [Eds.], *The Effect of Use and Disuse on Neuromuscular Functions*. Amsterdam: Elsevier, 1963.)

HINT

1. Have you ever tried to decide between the light or dark meat from a chicken or turkey? The color of the meat is due to the myoglobin content. Of course, the leg muscles are postural, while the breast (flight) muscles in these fowl are not used for sustained flight.

   Is the breast muscle light in a flying bird such as a pigeon? (Make a guess if you don't know.) (Hint 2↓)

335

336

Fig. 13-6. Length-tension diagram at various degrees of muscle activity; 0 percent = passive stretch, 100 percent = maximum during tetanus. Intermediate curves are hypothetical. Data from experiments on human triceps. (Units of ordinate are pounds.)

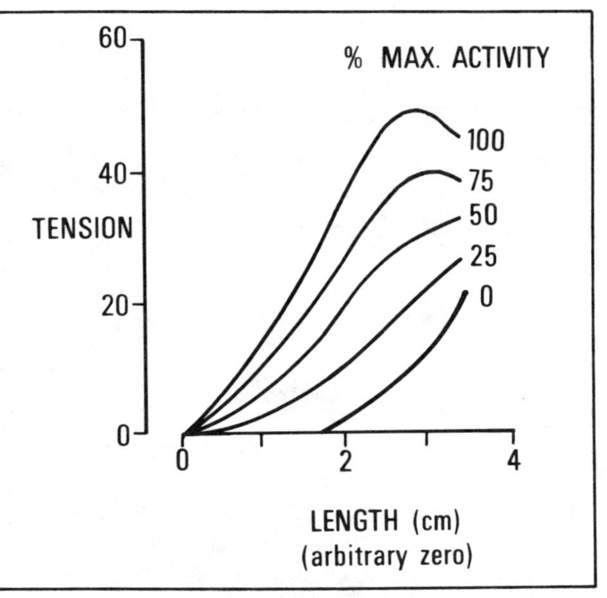

(After Prosthetic Devices Research Project, *Fundamental Studies of Human Locomotion and Other Information Relating to Design of Artificial Limbs*, Berkeley, 1947; modified from W. F. Ganong, *Review of Medical Physiology* [9th Ed.]. Los Altos, Calif.: Lange, 1979.)

Fig. 13-7. Length-tension diagrams for cat soleus muscle, at same intensity of stimulation, at various frequencies of stimulation. Bottom curve is passive length-tension curve.

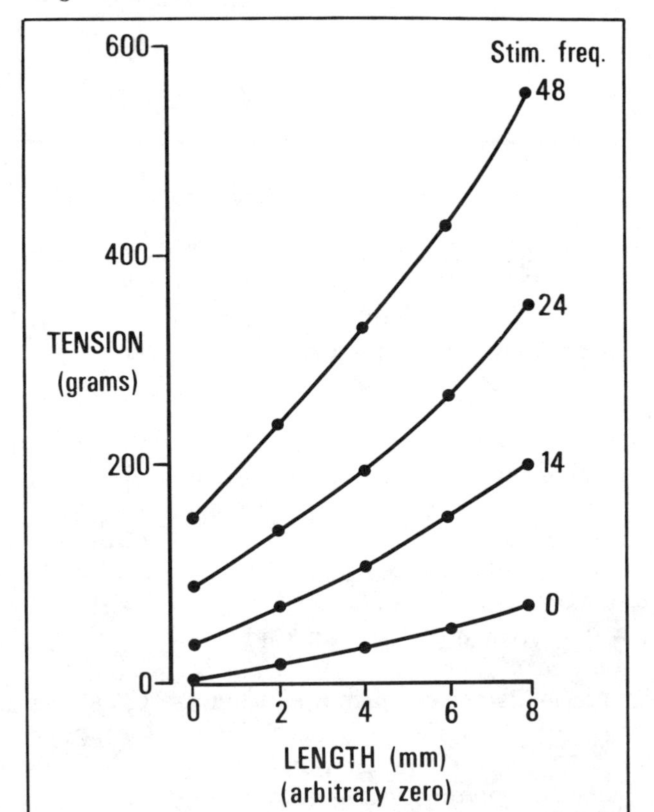

(Modified from R. Granit, Neuromuscular interaction in postural tone of the cat's isometric soleus muscle, *J. Physiol.* [*Lond.*] 143:387, 1958.)

Fig. 13-8. Length-tension diagrams for cat soleus muscle, at same frequency of stimulation, but at different stimulus intensities (i.e., with different numbers of axons activated). Bottom curve is passive length-tension curve at zero stimulus intensity.

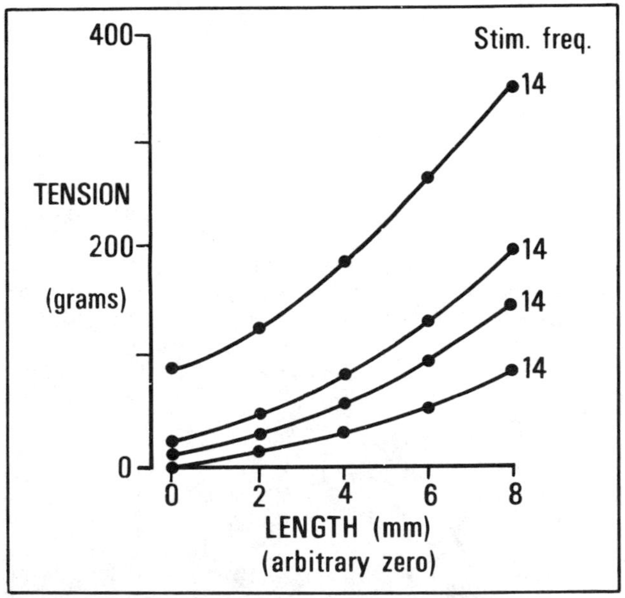

400
TENSION
200
(grams)
0
Stim. freq.
14
14
14
14
0 2 4 6 8
LENGTH (mm)
(arbitrary zero)

(Modified from R. Granit, Neuromuscular interaction in postural tone of the cat's isometric soleus muscle, *J. Physiol.* [*Lond.*] 143:387, 1958.)

Fig. 13-9. Length-tension plots for muscles contracting against different types of load shown in Fig. 13-10.

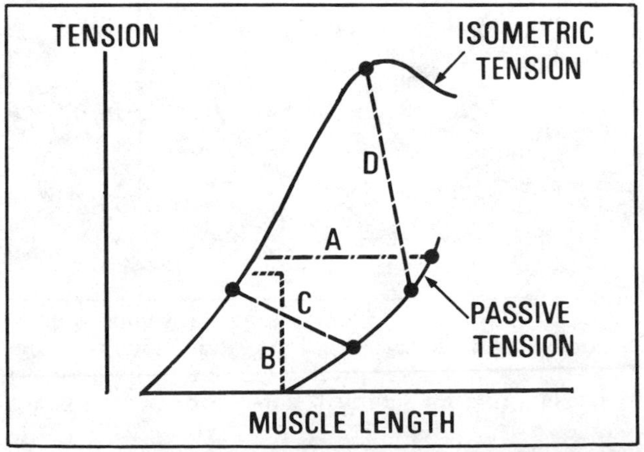

TENSION
ISOMETRIC TENSION
D
A
C
B
PASSIVE TENSION
MUSCLE LENGTH

336

337

Thus, on any length-tension graph, there should be **a family of curves between the passive tension and the tetanus tension,** which **represent intermediate degrees of activation,** as shown in Fig. 13-6, where activation is a function of both the number and frequency of motor units firing.

The experimental evidence for such a family of curves is shown in Figs. 13-7 and 13-8. In Fig. 13-7, the number of nerve fibers stimulated is kept constant, but the stimulus frequency is increased, thus increasing the amount of activation by increasing the amount of summation. In Fig. 13-8, the stimulus frequency is held constant, but various numbers of axons are activated, which changes the amount of activation. In either case, a family of curves can be generated in between the full-off (passive) and full-on (tetanus) curves of the length-tension plots.

Question: Why aren't the tensions reached in Figs. 13-7 and 13-8 attained in Fig. 13-5? (Hint 5↓)

Question: Since Figs. 13-7 and 13-8 are from cat soleus, why do they not show the flat portion seen in Fig. 13-5? (Hint 4↓)

It should be noted in passing that the **length-tension diagrams presented are obtained from steady-state measurements,** and so do not take into account the complex results that occur during dynamic changes [1, pp. 237–265; 65, pp. 1137–1142].

Now, by using Fig. 13-6 **it is possible to see how a muscle will perform against different types of load at various levels of activation.** But to do this, we must first examine the way in which **the trajectory of a contraction on the length-tension plot is determined by the nature of the load.** Different types of load produce different trajectories.

We consider first some simple types of load that do not involve the trigonometric complexities associated with lever systems (see Figs. 13-9 and 13-10). Lines A and B in Fig. 13-9 refer to *isotonic* contractions in which the load remains constant. Notice that in line A, before the stimulus was applied, the free-hanging weight stretched the muscle to a point

Fig. 13-10. Different types of load against which muscle may work. (A) Isotonic pulley; (B) "after-loaded" isotonic system; (C) soft spring (high compliance); (D) stiff spring (low compliance).

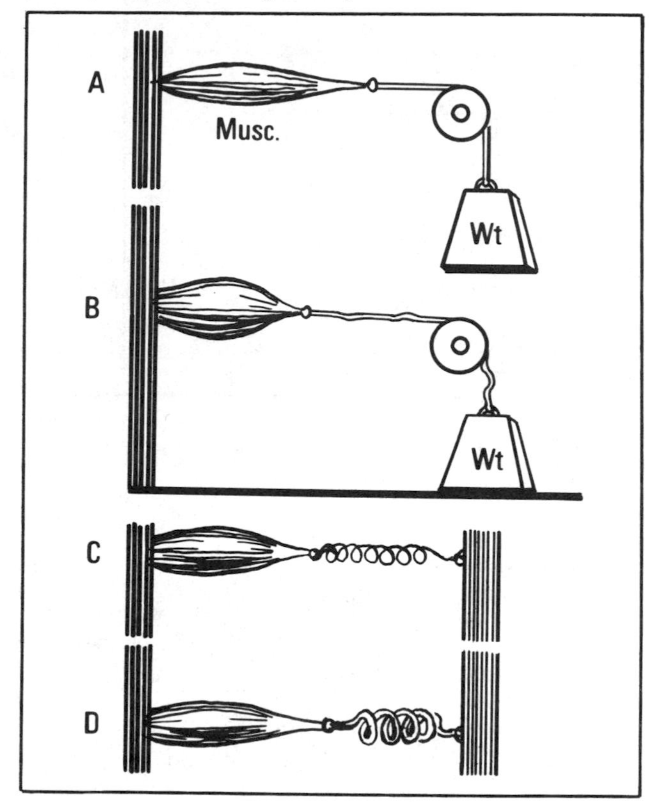

HINTS

2. No. In the pigeon, the breast muscle is dark, but the legs are relatively light in color.

3. The soleus is a large muscle acting at the ankle, while the flexor hallucis longus is a small muscle acting on the great toe. If the data were normalized (e.g., tension per unit of cross-sectional area), then the fast fibers would be much higher on the graph than the slow fibers.

337

338

determined by the passive properties of the muscle. During stimulation, the muscle will contract until it reaches a length at which the load is equal to the isometric tension produced under equivalent conditions of stimulation.

1

QUESTION: If the weight in A were increased, would the length-tension trajectory change? (Hint 6↓)

2

In line B, the weight is supported while the muscle is unstimulated (technically this is called an *after-loaded isotonic contraction*). Notice that in B the tension rises without change in length until the developed tension is equal to the applied load. Afterward, the length changes without further change in tension.

3

Lines C and D in Fig. 13-9 show contraction against a soft and stiff spring, respectively. Line D shows the length-tension diagram of a nearly *isometric load,* where the length changes very little while the tension varies markedly. (Note that the abscissa is a measure of *muscle,* not spring, length. Soon you will see why we want the length-tension load expressed in terms of the muscle length.)

4

Question: In Fig. 13-9, if you drew a line parallel to C, to the right of C, what would this represent in terms of the mechanical system? (Hint 7↓)

5

Now let's see how different loads affect the action of muscle. **If the muscle** whose family of length-tension curves is shown in Fig. 13-11 **were acting against load A, then the muscle-load system would operate only along line A, as the amount of neural activity changed.** If the amount of muscle activity were varied from 0 to 100 percent, the length would go from N to P with little or no change in tension.

6

Note that when we consider the *contracting* muscle, tension in the length-tension plot becomes the independent variable, while length becomes the dependent variable. That is, the depolarization of the muscle leads to increased tension, which in turn determines muscle length, depending on the "length-tension" characteristics of the load. (While acknowledging the purist, the graph "as is" is probably the lesser sin than switching axes in mid-explanation.)

7

But, **if the same muscle were working against load D, it could operate only along line D.** And if the amount of muscle activity were changed from 0 to 100 percent, the muscle length would change from W to V when the tension was changing from S to R. Thus, against load A there are **some tensions the muscle would never reach** (i.e., greater than the level of line A). Against load D there are **some lengths the muscle would never reach** (i.e., greater than W or shorter than V). Load C in Fig. 13-11 represents an intermediate between A and D, and yet *the muscle still would not reach all lengths or tensions.*

8

Fig. 13-11. Combined muscle and load length-tension diagram. Load lines are as in Fig. 13-9; muscle lines, as in Fig. 13-6.

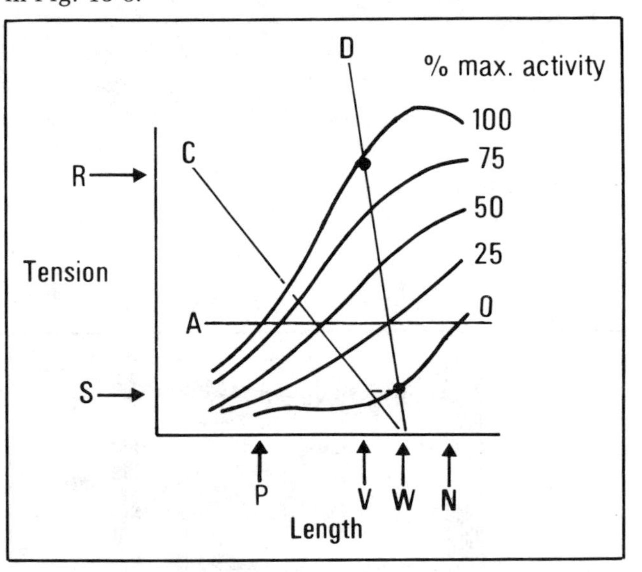

Fig. 13-12. Length-tension diagram of change in *isotonic* load A.

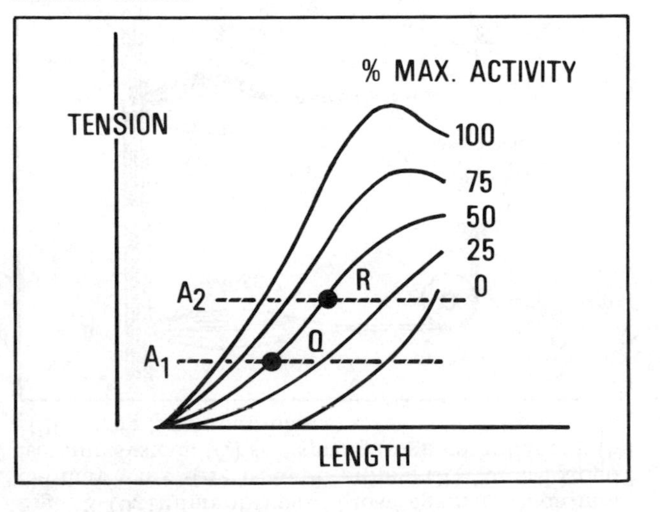

338

339

In the preceding description, the load is established, and then the consequences (in terms of length and tension) of changes in the activity of the muscle are deduced. Of course, the inverse also can be considered. What happens when the activity of the muscle is held constant and the magnitude of the load varies? Under such circumstances, **the length-tension curve of the muscle** (at a fixed level of activation) **determines how a change in load will affect length and tension.**

For example, *if the muscle activity* in Fig. 13-12 *were held constant at 50 percent of the maximum*, then as the load changed from A_1 to A_2, the length and tension would move from Q to R. Similarly, in Figs. 13-13 and 13-14, if a fixed muscle activity is assumed at 50 percent, then the muscle moves from Q to R as the load changes. The differences among Figs. 13-12, 13-13, and 13-14 are in the types of load.

Now, the nervous system basically controls the activation of the muscles, but not the characteristics of the load. So let's see what range of possible actions is open to the nervous system when faced with different loads.

When the *isotonic* load changes from A_1 to A_2 in Fig. 13-12, it should be clear that if **the muscle** were at point Q initially (at 50 percent of maximum activity), it **would have to increase its activity to keep its length constant. Conversely,** from Fig. 13-12 you can see that as the load changes from A_1 to A_2, **there is no way that changing the activity of the muscle can keep the tension constant.** Again, if we are dealing with an *isometric* load changing from D_1 to D_2 in Fig. 13-13, we can see that if the muscle were initially at point Q, then length would be kept constant by decreasing activity, or both length and tension would vary if the activity did not change much. **In no such case is it possible for muscle activity to keep both length and tension constant in the face of a changing load.**

The important point to realize is this: **The characteristics of the load determine in part what the muscle can or cannot do.** Yet the "plain" length-tension diagram (without a "load line") shows only those conditions of load that are commonly encountered in physiological experiments on isolated muscle.

The interaction between a muscle and its load becomes even more complex when a mechanical linkage is added to the system. For example, the same load may have markedly different effects on a system, depending on which way "up" is (i.e., the orientation of the system with respect to gravity).

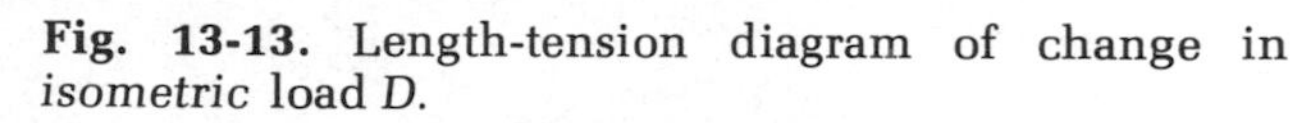

Fig. 13-13. Length-tension diagram of change in *isometric* load *D*.

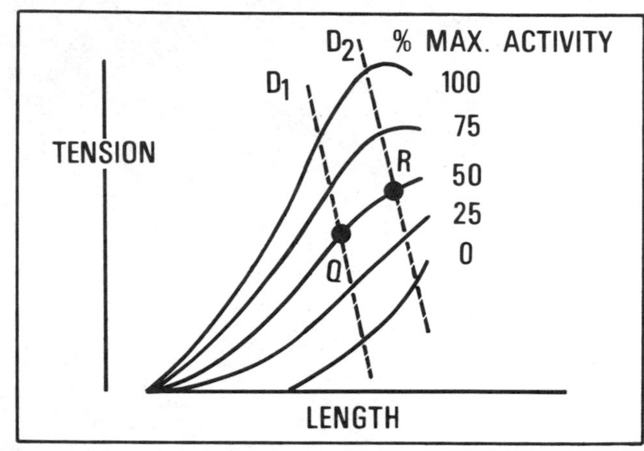

Fig. 13-14. Length-tension diagram of change in load *C*.

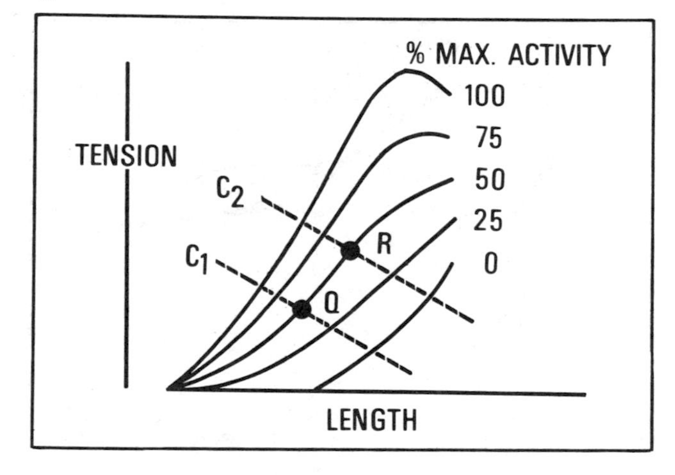

HINTS

4. Take a look at the length in all the figures.

5. In Figs. 13-7 and 13-8, not all the nerve is stimulated at maximum frequency, as in Fig. 13-5.

339

You can demonstrate this quite easily. Stand up and, keeping both legs straight, raise one leg in front of you until it is nearly horizontal. Notice that this becomes more and more strenuous as the leg approaches the horizontal position. Now lie down and raise your leg until it forms the same angle with the rest of your body as you achieved when standing up. This time the greatest effort is at the beginning of the movement. 1

The effects of gravity are investigated in greater detail in Figs. 13-15 and 13-16. In orientation A of Fig. 13-15, the load tends to zero as the joint angle reaches 90° and becomes negative when the angle is greater than 90°. By contrast, in orientation C, the load is negative at angles greater than 90°, is zero at 90°, and reaches a maximum as the angle tends to 0°. In orientation B, the load is positive at all angles between 0° and 180°. 2

In Fig. 13-16, the length-tension trajectories are plotted for the three orientations of Fig. 13-15. Numerical values were calculated by using the approximate dimensions of the biceps system and assuming a hand-held load of 25 kg. Joint angles refer to the elbow joint, and the range of movement studied is from 170° to 20°. 3

The angle of maximum flexion is about 20°. While most people can extend the elbow joint to at least 180°, the range beyond 170° requires a more sophisticated treatment than was used here, since the effective lever arm does *not* become zero at a joint angle of 180° as our analysis predicts. 4

One might expect in orientation B that muscle tension would be maximal at 90°. Trajectory B of Fig. 13-16 shows that this is not the case, since the change in length of the effective lever arm (distance a in Fig. 13-16) entirely compensates for the increased effective load that occurs at this angle. 5

Muscle tension T can be calculated readily since the moment Ta must equal the moment Wb. Hence $T = Wb/a$. 6

The change in length permitted by the mechanical linkage can be quite small compared with the full range of length change that can be investigated in experiments on isolated muscles. The change in length in the biceps model in Fig. 13-16 was about 30 percent of the resting muscle length. In the case of the thigh muscles, the entire range of normal movement can be as little as 5 percent of muscle length. Where the "working range" is very small, it may correspond to only the "flat top" of the isometric length-tension diagram seen for the isolated muscle. 7

The preceding description dealt with length-tension diagrams of fast muscles only. **Similar diagrams could be drawn of the effects of different loads on slow muscles** by using their characteristic length-tension diagrams. Furthermore, remember that in the body, loads can vary in 8

Fig. 13-15. Same musculoskeletal system shown in three different orientations in respect to gravity. In fourth position (not shown), load would be negative at all joint angles between 0° and 180°.

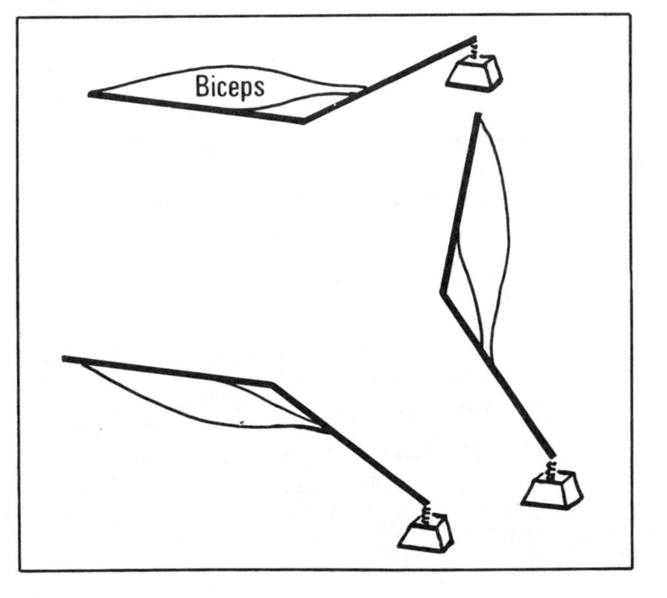

341

more complicated ways than have been described. Thus the compensations necessary to achieve a given movement can be complex indeed. **Real loads can be frictional and inertial as well as gravitational.**

1

Now, to make matters a bit worse, try to imagine what happens when the slow and fast components of a muscle are activated independently, for different purposes (e.g., posture or rapid movement)! The next section describes some of the evidence for such independent action.

2

Finally, recall that the preceding discussion of length-tension diagrams dealt with statics, not with the dynamic aspects of muscles and loads (i.e., how results differ when muscles and loads are moving comparatively rapidly).

3

For example, the velocity of shortening can depend on both the length of the muscle and the magnitude of the load [1, p. 249; 65, pp. 1137–1139].

4

Since this analysis is complicated enough, let us set these considerations aside with respect to length-tension diagrams and deal with only the static cases. (We know the dynamic situation is even harder to understand.)

5

LARGE AND SMALL ALPHA MOTOR NEURONS

Now that you have some idea of what muscles are like and the loads they must cope with, we can talk about the part of the body that gives the orders: the central nervous system. The muscle cells are innervated by motor neurons whose cell bodies are in the ventral (anterior) horn of the spinal cord. The myelinated axons of the neurons course outward to form the efferent portions of the peripheral nerves. (For review, see Figs. 2-5 and 2-9.) First, let's look at the situation in the periphery. **A single axon, as it reaches its muscle of destination, branches and rebranches so as to innervate many muscle fibers.** However, a muscle cell is usually not innervated by more than one motor neuron. The axon and its associated muscle fibers are called the **motor unit.**

6

The size of the motor unit (i.e., the number of muscle fibers innervated by a single axon) **varies from muscle to muscle.** For the extraocular muscles, the ratio of axons to muscle fibers can be as much as 1:3, whereas some muscles have ratios as small as 1:1750 (medial gastrocnemius) [15, pp. 10–11]. In the cat, the soleus muscle has a ratio of 1:170 [32, pp. 94–95]. It seems that the **muscles that produce finely graded, accurate movements have a larger ratio,** whereas coarser muscles have smaller ratios.

7

The muscle cells of a motor unit are not closely packed, but **are spread out, intermingled with cells from other motor units.** In the biceps, the muscle cells of a single axon may

8

Fig. 13-16. Upper drawing demonstrates method of graphic analysis in orientation *A* (see Fig. 13-15). Lower drawing shows calculated length-tension trajectories for movement from 170° to 20° with 25-kg load in orientations *A*, *B*, and *C*.

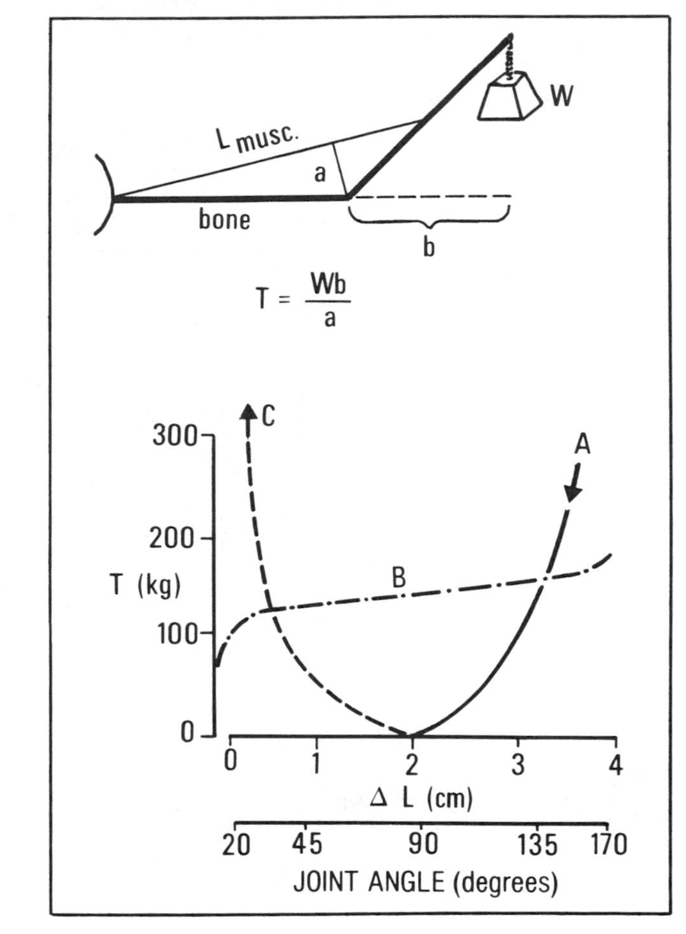

HINTS

6. Yes, it would be parallel to line *A*, but above it.

7. Moving the "wall" at the right farther away from the muscle relative to a given length of spring.

341

342

be distributed in 1 cm^2, and they are even farther separated in the diagram. **The spatial intermixing of motor units ensures a rather even distribution of tension in the tendinous attachments of the muscle,** even when (during weak contractions) only a few motor units are active.

QUESTION: Are the muscle fibers of one motor unit all the same type (i.e., slow or fast)? (Hint 8↓)

The diameters of the motor axons innervating a muscle have a characteristic distribution (Fig. 13-17). The distribution has **two main humps:** that which gives rise to the *alpha* (α) wave in the compound action potential and that which gives the *gamma* (γ) wave. (The significance of the gamma fibers is described later in connection with the muscle spindle.) **The alpha-fiber hump shows a rather wide distribution of diameters** (and hence conduction velocities), e.g., from about 9 to about 15 to 16 μm in Fig. 13-17.

Since you know that the conduction velocity of the motor axon matches the slowness or fastness of the muscle fiber (Fig. 13-4), you might suspect that **the distribution of alpha-fiber conduction velocities gives some idea of the spectrum of slow and fast fibers in the muscle.** This idea is further reinforced in Fig. 13-17 by the differences in the distributions of alpha-fiber diameters to the soleus (a predominantly slow muscle) and to the gastrocnemius (a predominantly fast muscle).

Cell bodies of motor neurons that have large axons (identified by their fast conduction velocities) **are also large,** as can be determined by intracellular microelectrode recordings. This makes it possible to study, to some extent, the firing patterns of slow and fast muscles.

The size of a cell body can be determined by measuring the electrical resistance of the cell by passing current when the microelectrode is intracellular. The higher the resistance, the smaller the cell, since the membrane area through which current can pass is smaller.

QUESTION: Which cells in the spinal cord, would you guess, are found to be tonically active: the small cells or the large cells? (Hint 10↓)

The distinction is important since the small cells are those most readily excited by the tonic stretch reflex (described later).

Selective activation of the small motor neurons can be accomplished by lateral inhibition in the spinal cord by means of Renshaw cells. Recall (from the section "Inhibitory Systems, in Particular" in Chap. 11) that collaterals from motor neuron axons excite Renshaw cell interneurons, which in turn inhibit other motor neurons. **Such a circuit is a low-gain positive feedback loop that accentuates small differences in ex-**

Fig. 13-17. Diameters of motor axons to soleus and gastrocnemius of cat. Equivalent conduction velocities are shown on abscissa. Alpha (α) and gamma (γ) distributions are indicated.

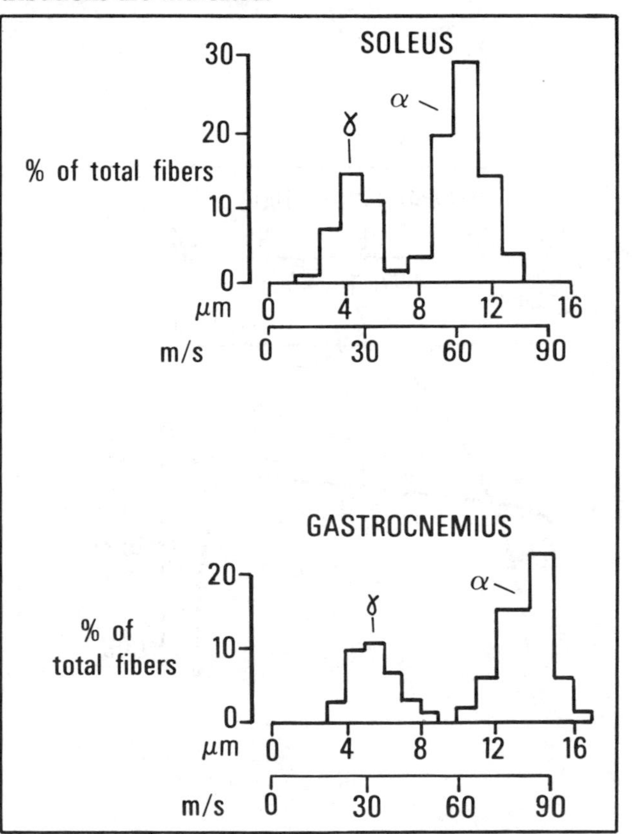

(Modified from J. C. Eccles and C. S. Sherrington, Numbers and contraction-values of individual motor-units examined in some muscles of the limb, *Proc. R. Soc. Lond.* [*Biol.*] 106:326, 1930.)

342

343

citability, such that the more excited cell inhibits less excited neighboring cells. This, in turn, leads to disinhibition of the more excitable cell. Hence, such an arrangement could give rise to what might be called *motor contrast* [15, pp. 155–156], since this system is analogous to lateral inhibition in sensory systems. Lateral inhibition enhances differences in firing rates of adjacent receptors (see the section on lateral inhibition in Chap. 11 for further description of this mechanism). 1

The changes in response of a pool of motor neurons seem to occur by recruiting new units from the subliminal fringe of cells already active (see the section "Consequences of Convergence and Divergence" in Chap. 11; see also 15, pp. 178–179). Such action is likely in those muscle movements in which there is only a brief burst of activity during a phase of the cycle (as in the swing-through phase of human walking). Obviously, in other cases with a different type of load, change in firing rate may be the only way that the muscle can control the movement (e.g., when most of the motor units are already active and then the load increases). 2

Small motor neurons tend to have larger EPSPs, and this may account for their greater excitability. 3

Question: What might be the mechanism of the larger EPSP? (Hint 16↓) 4

Thus, there is some evidence that **the static and dynamic aspects of motor control are evident not only in the muscles and their nerves, but also in the distinctive ways by which the central nervous system activates neurons connected with muscles of different characteristics** (fast, slow). This theme recurs frequently throughout this chapter. 5

An interesting finding in clinical electromyography **confirms the existence of two types of motor units in humans.** If the firings of a single motor unit are recorded for some time, during different amounts of contraction, in weak contractions not only is the mean time between firings large, but so is the variability (i.e., standard deviation about the mean). In a single muscle, various motor units are found to have different types of mean-variability curves. For example, in Fig. 13-18 one unit is from a fast fiber (dynamic), the other from a slow fiber (static). 6

Thus, at low levels of excitation, discharge of static, slow fibers is more regular than that of dynamic, fast fibers. When many muscles were studied, it was found that **while each muscle has motor units separable in this way, between muscles there is considerable overlap,** so that a dynamic unit in one muscle may be very similar to a static unit in another (Fig. 13-19). Thus, **the generalization concerning static and dynamic properties should be considered in each muscle, in terms of its own function rather than on any absolute basis.** 7

Fig. 13-18. Upper graph: Two single motor units in soleus muscle in human leg; relationship of mean interspike interval (abscissa) to standard deviation of firing rate (ordinate) under various degrees of contraction. Motor unit marked *K* is presumed to be a dynamic, fast motor unit; *T*, a static, slow motor unit. Lower graph: Example of raw data, showing increased variability when discharge rate is low (larger discharge interval).

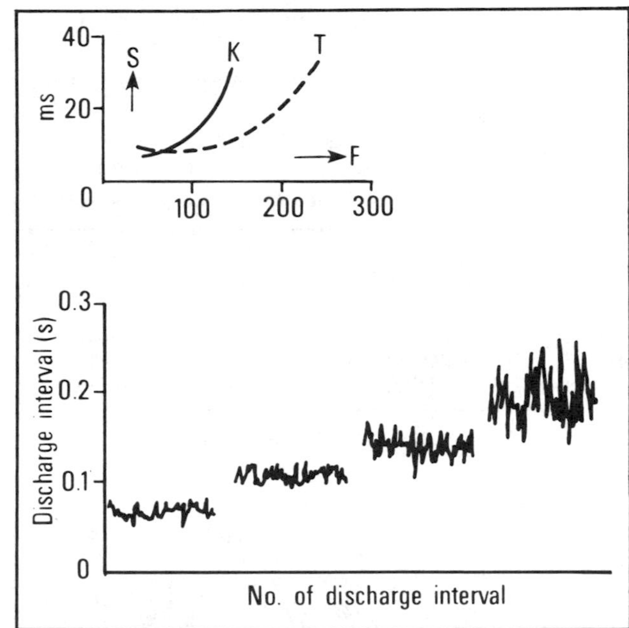

(Modified from T. Tokizane and H. Shimazu, *Functional Differentiation of Human Skeletal Muscle.* Tokyo: University of Tokyo Press, 1964.)

343

The increased variability of interspike time at low contraction strengths may represent a statistical shifting of excitation from motor unit to motor unit so as to maintain (by an unknown mechanism) an even distribution of activity across units of the same type, thus avoiding fatigue of the most excitable unit by repeated activation. As muscular contraction increases, variability decreases as more and more units are activated (brought above threshold). Finally, **when all units are active, additional tension is developed by an increase in the firing rate of all units, with the variability of firing continuing to be low.** 1

Question: Imagine a disease "X," something like poliomyelitis, which attacks motor neurons, selectively killing the larger motor neurons but sparing the smaller. What might be the consequence as seen clinically? (Hint 17↓) 2

SENSORY SYSTEM COMPONENTS OF MOTOR CONTROL

3 Now let's look at the sensory aspects of motor control.

The importance of the sensory aspects of muscle control is apparent from the amount of neural machinery involved with sensing muscle parameters. For example, **there are more myelinated fibers involved with muscle sense organs than there are alpha motor fibers!** The calculation with respect to the cat soleus is shown in Table 13-2. 4

Note that Table 13-2 ignores the contributions of joint receptors to the sensory control of muscle movement. Further, it does not list unmyelinated axons and their functions. Thus, Table 13-2 is presented more for curiosity than as a serious calculation of the percentage of the nervous system devoted to sensory processes. 5

In this chapter we cover the sensory endings in *the following order:* muscle spindles, tendon organs, joint position receptors, and free nerve endings. These various endings subserve a variety of functions. But among the most important, as you might expect from the preceding description of the intricacies of the length-tension diagrams, are the detection of length and tension! **Only by detecting length and tension can these variables be controlled to meet the needs of posture and movement in the face of changing loads.** In addition, some of the sensory responses can be classified into static and dynamic components. Thus, **afferent fibers signal not only changes in length and tension, but also the rate of change in length (velocity) and in tension.** 6

Tension is measured by the tendon organ, which is mechanically "in series" with the muscle fibers, **while length is measured by muscle spindles,** which are mechanically "in parallel" with the muscle fibers (Fig. 13-20). These mechanical connections explain immediately the observations that **both spindles and tendon organs are excited by any force** 7

Fig. 13-19. Same plots as in Fig. 13-18 for various human muscles. Dotted lines are for static units; solid lines, for dynamic units. Note that a dynamic unit of leg is similar to a static unit of eye! (Units of axes in milliseconds.) The abscissa is time, counted as number of discharge intervals.

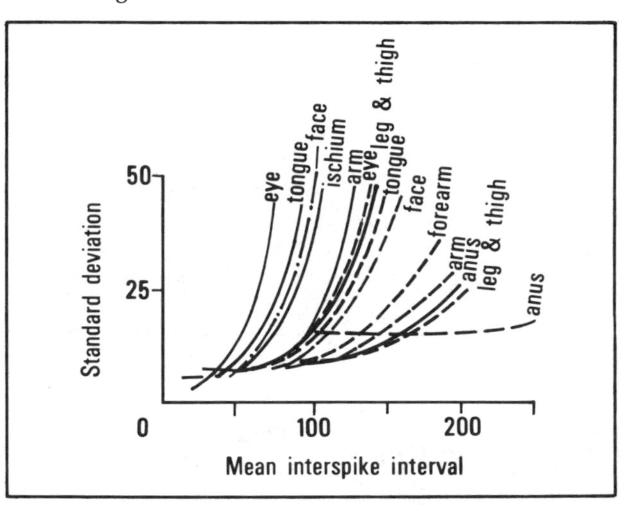

(Modified from T. Tokizane and H. Shimazu, *Functional Differentiation of Human Skeletal Muscle.* Tokyo: University of Tokyo Press, 1964.)

Table 13-2. Myelinated Nerve Fibers to Cat Soleus Muscle

Extrafusal Muscle Control	
Alpha motor neurons (to 25,000 muscle fibers)	150
Total	150
Muscle Sensory System	
Afferents	
Spindle primary endings	50
Spindle secondary endings	50
Tendon organs	40
Efferents	
Gamma motor neurons to spindles	100
Total	240

Adapted from P. B. C. Matthews, *Mammalian Muscle Receptors and Their Central Actions.* London: Edward Arnold, 1972. P. 92.

345

that elongates (stretches) the muscle** (Fig. 13-21), while **the responses are opposite from each other when the muscle actively contracts** (Fig. 13-22). The passive stretch (Fig. 13-21) increases both tension and length, whereas the contraction (Fig. 13-22) *increases* tension but *decreases* length. We deal with these responses in greater detail, but first some anatomic details and terminology must be introduced.

1

The nerve classification scheme that uses Greek letters (alpha, beta, etc.) **is applied only to motor fibers. The classification using Roman numerals distinguishes different diameters of sensory fibers.**

2

There is really no good reason for this convention except historical accident [32, pp. 78–85], but it does make communication somewhat easier, since any fibers referred to by Roman numerals are sensory, whereas motor fibers are referred to by letters, both Roman and Greek!

3

Like the motor fiber classification, the sensory fiber classification is based on axonal diameter and function to some extent (Table 13-3).

4

Fig. 13-20. Highly schematic diagram of relationships of muscle spindle and Golgi tendon organ to muscle fiber, including efferent and afferent innervations.

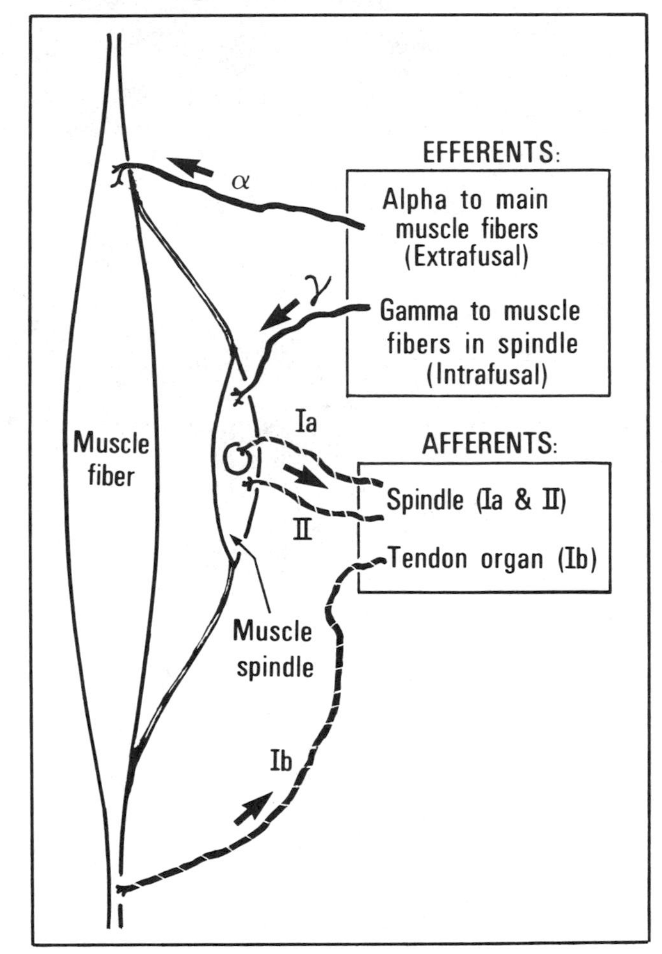

HINTS

8. Yes. On what basis can this be asserted, considering the technical difficulties? (Hint 9↓)
10. The small cells. If you missed this, carry on to Hint 11.↓
12. It is a small alpha fiber. What type of muscle fiber does the small alpha fiber innervate? (Hint 13↓)
14. Posture. What type of discharge would be expected from postural (static) muscles? (Hint 15↓)
16. (1) The smaller cells might receive more synchronously discharging afferents. (2) The afferents might innervate more synaptic knobs on the smaller cells compared with the larger ones. (3) The same synapses might be farther out the dendritic tree in the larger cells, so that the amount of depolarization detected at the soma by the microelectrode (as well as the initial segment) is greater. (4) The synapses of the smaller cell may release more transmitter. Or (5) if all the previous factors were equal between the large and small cells, the small cells would still have a larger EPSP, as a result of the smaller area of membrane available for the outward-directed depolarizing currents entering at the activated synapses. This is equivalent to saying that the constant current flowing through synaptic regions has a greater effect on the soma membrane potential when the resistance (as load) of the soma is higher. At this point, you should be able to guess that the question of why the EPSP is larger has not been settled yet!

345

346

Fig. 13-21. Responses to passive stretch of muscle in single sensory receptors of Golgi tendon organ and muscle spindle (primary) in cat.

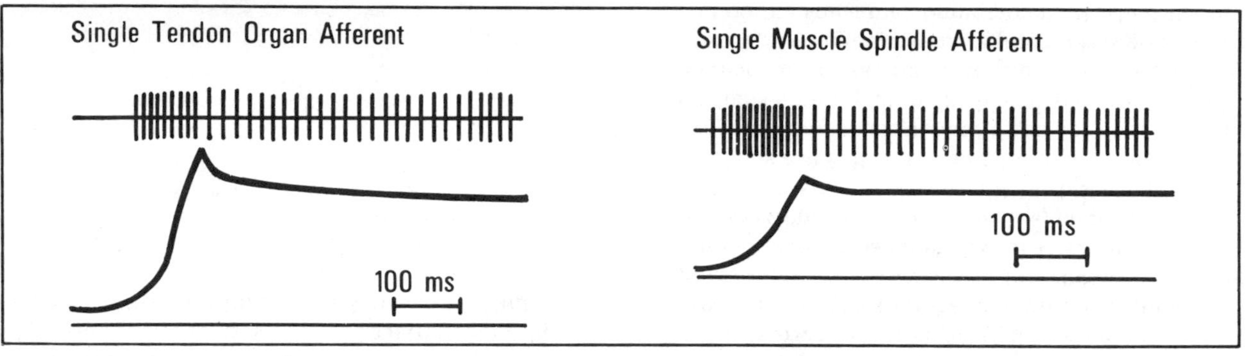

(Modified from B. H. C. Matthews, Nerve endings in mammalian muscle, *J. Physiol.* [*Lond.*] 78:1, 1933.)

Table 13-3. Classification of Afferent Nerves in Hindlimb of Cat

Group	Name	Diameter (μm)	Velocity (m/s)	Muscle Receptors	Joint Receptors
Ia	Thick myelinated	12–20	72–120	Spindle primary; ? some free endings	Some Golgi endings
Ib	Thick myelinated	12–20	72–120	Tendon organs	
II	Medium myelinated	4–12	24–72	Spindle secondaries	More Golgi endings
				Paciniform corpuscles	Ruffini endings, some free endings, and Paciniform corpuscles
				Some free endings	
III	Fine myelinated	1–4	6–24	Free endings	Free endings
IV	Unmyelinated	<1	<2	Free endings	Free endings

Adapted from P. B. C. Matthews, *Mammalian Muscle Receptors and Their Central Actions.* London: Edward Arnold, 1972. P. 90.

Fig. 13-22. Contrasting responses of Golgi tendon organ and muscle spindle (without gamma efferent activation, i.e., passive) to muscle twitch caused by single shock to efferent nerve. Note the "silent period" when spindle was shortened during contraction.

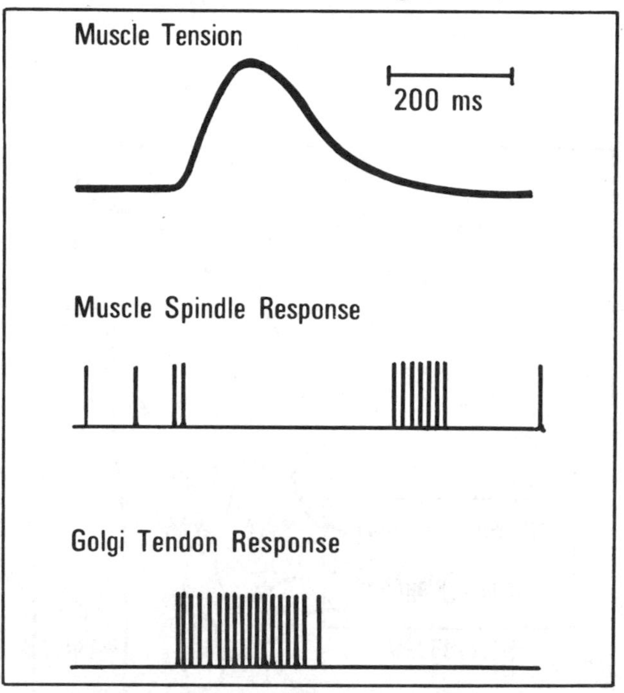

(Modified from P. B. C. Matthews, *Mammalian Muscle Receptors and Their Central Actions.* London: Edward Arnold, 1972.)

346

347

As you can see from Table 13-3, **Roman numerals indicate separations based on anatomic differences** (diameters), while **letters** (as between the Ia and Ib fibers) **indicate functional differences.** Note also that the afferent fibers range from the smallest, unmyelinated fibers (0.2 μm) to the largest, myelinated fibers in the body (20 μm). The Roman numeral classification scheme is referred to repeatedly in this chapter, so be sure you are familiar with Table 13-3 before going on. Table 13-3 applies to the cat, in terms of the absolute diameters and conduction velocities of the various types of fibers. **A different set of values applies for different species.** For example, in the cat the fastest afferent fibers are 100 to 120 m/s, whereas in humans and monkeys the fastest are about 90 m/s.
1

In the rat the fastest are 60 to 80 m/s, so there seems to be no clear relationship to body size.
2

The diameters given in Table 13-3 are those in the main nerve trunk. **Near the axon terminals the diameters can be less,** especially if the axon branches.
3

The classification scheme, although useful, is somewhat arbitrary. So you should realize that **the range of fiber sizes is more or less a continuum with clusters** without any absolute separation points. (In the same way, the bell-shaped curve of student test scores is usually a continuum without clear separations.)
4

As shown in Fig. 13-23, the various groups form a continuum with respect to both conduction velocity and electrical threshold, even though electrical threshold usually is used to distinguish the reflex effects of different groups! Only by picking an afferent nerve that contains some of the groups or disproportionate numbers of one group or the other is it possible to make any rational use of this classification method. But it still is useful, as you will see!
5

Fig. 13-23. Conduction velocity plotted against electrical threshold for 68 single fibers dissected from dorsal roots; stimulation in nerve to semitendinosus muscle.

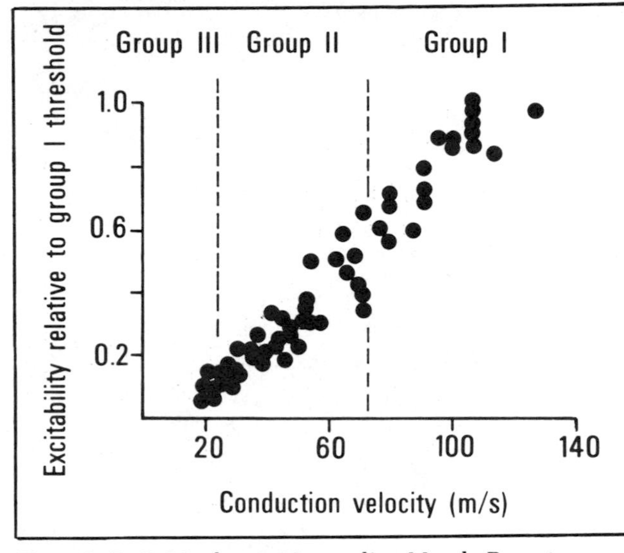

(From P. B. C. Matthews, *Mammalian Muscle Receptors and Their Central Actions.* London: Edward Arnold, 1972.)

HINTS

9. The cross-innervation experiments described earlier in this chapter showed that the type of muscle fiber was determined by the innervating nerve. Since all the muscle cells are innervated by only one (and the same) axon, they must all be the same type.
11. What is the relative size of the axon of a small motor neuron? (Hint 12↑)
13. The slow muscle fibers. Which type of activity does a slow muscle fiber tend to be useful in, posture (static) or movement (dynamic)? (Hint 14↑)
15. Continuous (tonic) activity.
17. The patient's phasic movements would become very weak as the larger cells died, but posture might not be affected *as much.* As the axons of the motor neurons degenerated, the denervated muscle cells would show an absence of MEPPs and denervation hypersensitivity. Then the remaining axons would reinnervate the postsynaptic membranes abandoned by the dead axons, so that muscle power would begin to increase again. How could this last event be distinguished from recovery from a weakness that was due to a temporary, reversible inactivation of the larger alpha motor neurons? (Hint 18↓)

347

348

The separation of Ia and Ib fibers in experiments involving electrical stimulation often depends on small differences in threshold. As shown in Fig. 13-24, in the nerves from some muscles it is possible to choose an intensity of stimulation so as to stimulate over 50 percent of the Ia fibers but only a few of the Ib fibers (e.g., in the semitendinosus). However, in other muscles almost as many Ib fibers are activated as Ia fibers at all levels of stimulation. Thus, experiments based on differential stimulation may be difficult to interpret in some cases. In the past this led to some confusion concerning reflex activity. 1

Another technical difficulty involves *"contamination."* For example, an experiment designed to investigate reflex effects of electrical stimulation of spindle (Ia) afferents may have inadvertently stimulated axons (Ia) from joint receptors (Table 13-3). 2

Fig. 13-24. From several different muscle nerves, comparison of percentage of Ia and Ib fibers that are activated at same stimulus strength over wide range of stimulus strengths. Curves, left to right, from the following muscles: 1: peroneus longus; 2: medial gastrocnemius; 3: soleus; 4: semitendinosus.

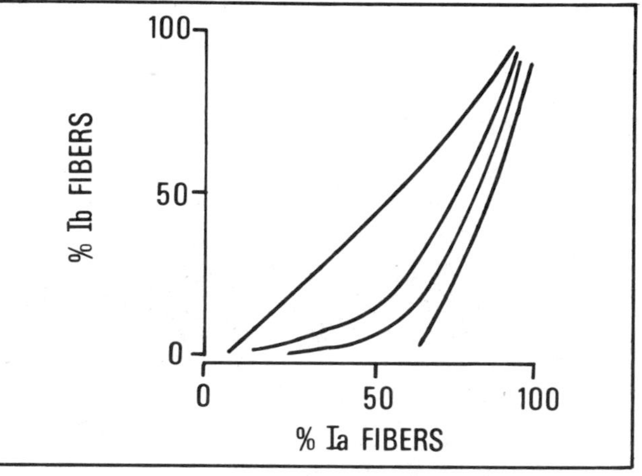

(Modified from P. B. C. Matthews, *Mammalian Muscle Receptors and Their Central Actions.* London: Edward Arnold, 1972.)

ANATOMY AND DISTRIBUTION OF MUSCLE SPINDLE

3 **The muscle spindle is a complex sensory organ found in almost all striated muscles.**

Muscle spindles are not found in the extraocular muscles of the cat and dog, but are found in these same muscles in humans and goats! (You can draw your own conclusions from this observation. The more we think about it, the wilder are our speculations, ending just in cheap humor.) 4

The spindle is relatively large, sometimes measuring as much as 7 to 10 mm in overall length. **The spindle consists of a group of specialized muscle cells** (variable in number—up to 20) **whose central region is surrounded by a capsule.** A bulge in the central region gives it the spindle shape. 5

Within the capsule are two types of specialized muscle fibers (called *intrafusal* fibers since they lie within the *fusiform* spindle): **nuclear bag fibers** and **nuclear chain fibers,** distinguished by the arrangement within them of the multiple nuclei in the central region (Fig. 13-25). The ordinary muscle fibers are called *extrafusal* (in this context). 6

The nuclear bag fibers are somewhat longer (by 1 to 2 mm) than the nuclear chain fibers. And nuclear bag fibers extend past the limits of the spindle capsule to insert on the perimysium of the extrafusal (ordinary) muscle fibers. The shorter, nuclear chain intrafusal muscle fibers usually end at about the limit of the spindle capsule and insert either into the capsule or onto nuclear bag fibers (both insertions are diagrammed in Fig. 13-25) [15, pp. 31–32]. 7

Across a large range of muscles, **the number of muscle spindles is relatively constant, with about 15 to 30 spindles per 100 alpha motor fibers** (hence per 100 motor units) [32, p. 49]. Thus, muscles that have relatively small motor units (fine, fast movement) appear to have more spindles per muscle mass than muscles with large motor units (larger, slower muscles). 8

348

349

In many muscles, the spindles are distributed rather uniformly throughout both the length and the mass of the muscle. However, in some muscles (such as the extraocular muscles of humans) the spindles tend to be located toward one end of the muscle.

The complex innervation of the spindle is shown in Fig. 13-26. **There are two types of afferent fibers and two types of efferent fibers.**

The Ia fiber is called the *primary receptor*—you can remember this since I is primus. It innervates the central regions of all the muscles in a spindle, i.e., the one to four nuclear bag and two to eight nuclear chain fibers per spindle. Thus, there is **only one primary ending per spindle.** There can be from zero to five secondary endings per spindle, innervated by a group II axon. (You can remember that group II axons are *second*ary). The *secondary fiber* makes connections primarily (and sometimes exclusively) with the nuclear chain fibers in the spindle, always off of the center of the spindle. The nuclear bag fibers receive their efferent supply from a dynamic gamma fiber (γ-D), while the nuclear chain fibers receive a static gamma fiber (γ-S). Each gamma efferent fiber goes to 15 to 30 intrafusal fibers in four or more spindles. (That is, there is divergence just as in alpha fibers.) Each spindle receives 6 to 12 separate gamma fibers, since there are several fibers in each spindle.

Fig. 13-25. Anatomic and mechanical connections of intrafusal fibers (not drawn to scale, nor is neural innervation shown). The two reported types of attachment of nuclear chain fiber are shown. Spindles normally contain more than one of each type of fiber.

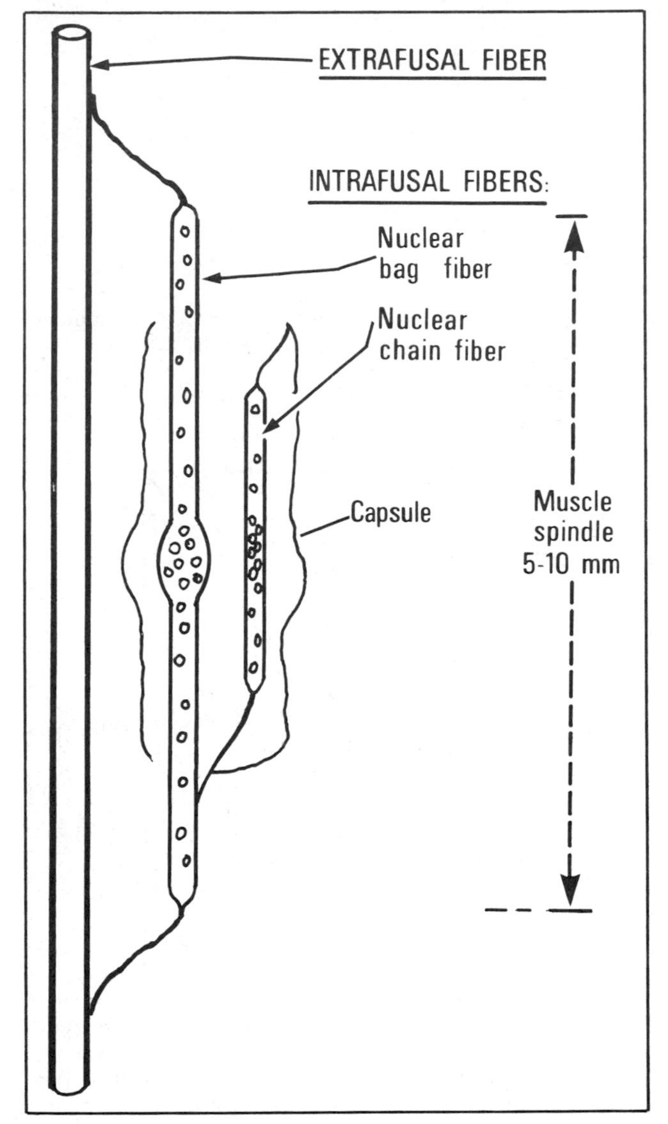

HINT

18. In the dead: by histology at autopsy. In the living, by the following: (1) measurement of the conduction velocity of peripheral motor nerves—the alpha peak would be slower than normal in the case of disease X; (2) measurement of the speed of contraction—since in the case of reinnervation, all muscle cells will be slower than normal; (3) possibly seeing whether the gradients of fine control were coarser than normal, since the average size of motor unit would be significantly increased; (4) comparing the peak power and endurance characteristics of the muscles with normals (a person with predominantly slow muscles should be less able to exert maximum effort, but better able to maintain prolonged static position); (5) electromyographic recordings of single units to see whether the relationship between mean firing interval and its standard deviation is shifted toward the static type (Fig. 13-18).

Now consider disease "Y," which selectively attacks *small* motor neurons. What might be the consequences of this disease? (Hint 19↓)

349

350

Fig. 13-26. Diagram of innervation of intrafusal muscle fibers within spindle.

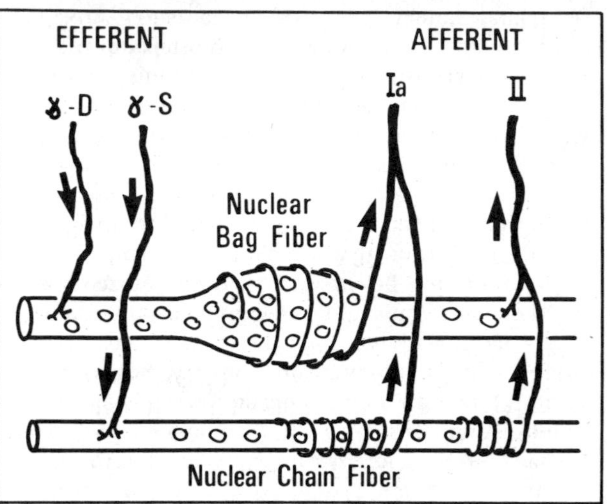

(Modified from P. B. C. Matthews, Muscle spindles and their motor control, *Physiol. Rev.* 44:219, 1964.)

Fig. 13-27. Firing rate, at different degrees of very slow stretch, of primary and secondary ending from cat soleus muscle before and after cutting ventral roots (V.R.). Note that the effect of removing gamma efferent control is similar for both primary and secondary, for slow stretching.

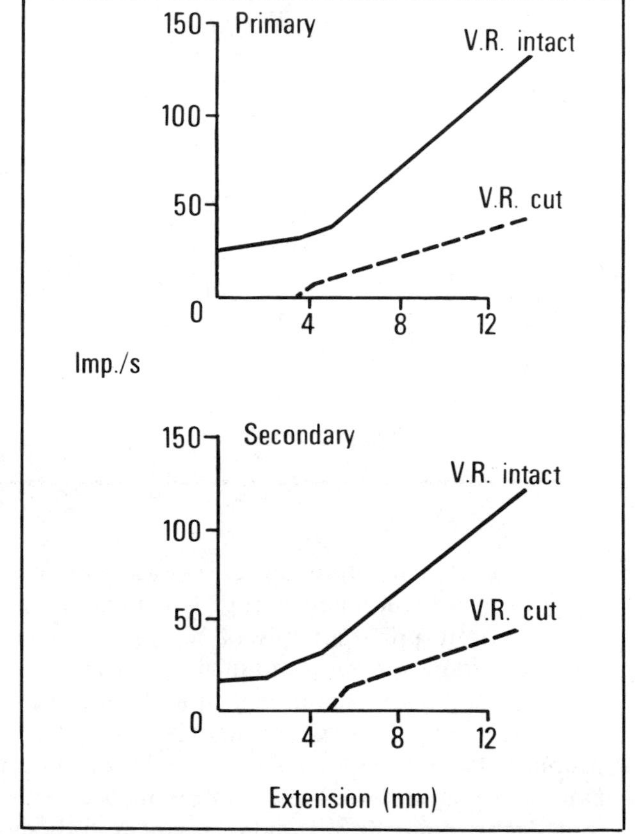

(Modified from J. K. S. Jansen and P. B. C. Matthews, The effects of fusimotor activity on the static responsiveness of primary and secondary endings of muscle spindles in the decerebrate cat, *Acta Physiol. Scand.* 55:376, 1962.)

Fig. 13-28. Firing frequency of single primary muscle stretch receptor during differing rates of extension (change in length). Note that there is greater response with faster changes in length, but firing rate is about the same in all cases when muscle has reached its constant length.

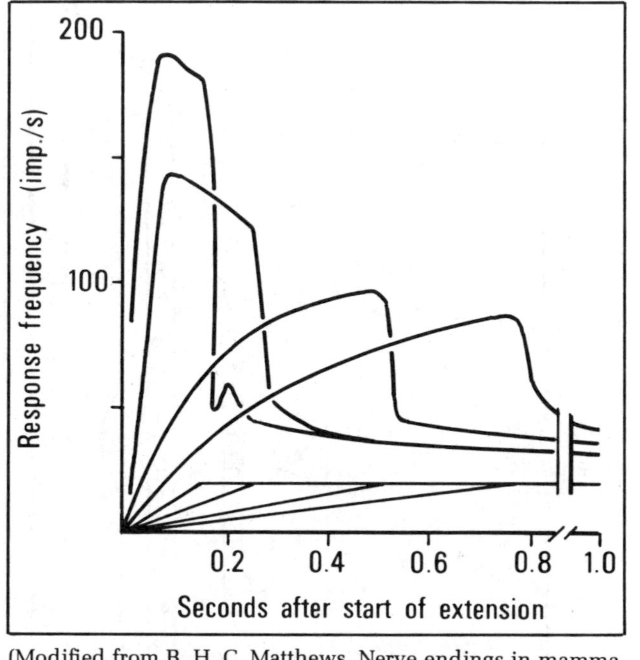

(Modified from B. H. C. Matthews, Nerve endings in mammalian muscle, *J. Physiol.* [*Lond.*] 78:1, 1933.)

350

351

The gamma (efferent) fibers have a range of diameters of 2 to 8 μm (as previously shown in Fig. 13-17), which correspond to a conduction velocity of 10 to 50 m/s. Of course, the diameters and velocities just mentioned apply to the main nerve trunk. **Near the spindle, the axons become smaller, especially as they branch to innervate several structures** (not shown in Fig. 13-26 with respect to gamma fibers).

1

The γ-D and γ-S fibers can be distinguished on the basis of their anatomy. (The γ-S fibers end in rambling "trail" terminals, in contrast to the more localized "plate" endings of the γ-D fiber.) In addition, γ-D and γ-S fibers also can be distinguished physiologically (described in a later section, "Static and Dynamic Gamma Efferent Fibers").

2

LENGTH RECEPTION—PLAIN AND FANCY

The complex anatomy of the muscle spindle serves physiological purposes—some complex, others simple. Let's deal with the simple actions first. The simple actions are just the **static responses** (when there is little or no movement) **in which the spindle receptors,** both primary and secondary, **signal muscle fiber length by means of their firing rates** (Fig. 13-27).

3

Figure 13-27 shows that **the firing rates of Ia** (primary) **and II fibers** (secondary) **in the same muscle are determined by the length of the muscle and by the amount of contraction of the intrafusal fibers caused by gamma efferent activity.** The effects of gamma efferent activity are taken up in the next section in greater detail. It is sufficient here to note that the firing rate of the spindle afferents can signal length quite nicely, given otherwise constant gamma efferent activity and slow changes (if any) in length.

4

The situation is more involved when it comes to the **dynamic responses** of the muscle spindles. The **spindle primary** (Ia fiber) **responds markedly to both length and rate of change in length (velocity)** (Fig. 13-28).

5

The following generalizations can be verified from Fig. 13-28. The spindle primary, being highly sensitive to the rate of change in length (velocity) of the muscle, shows a marked *dynamic sensitivity*. In addition, the **primary also detects length during static** (nonchanging length) **conditions** (see also Fig. 13-27). Recall from Chap. 9 that many types of sensory receptors (e.g., joint receptors, photoreceptors, etc.) signal both absolute values and their rate of change, often by means of adaptation (see pages 216 to 217).

6

HINT

19. Well, if you can't answer this one by now, you'd better go back and reread the last few pages—after a cup of coffee!

351

352

The spindle secondary is also *slightly* sensitive to rate of change in length, but nowhere near as much as the spindle primary is. **The secondary endings signal primarily the length of the muscle,** whereas the primary ending sends a signal dependent on both length and rate of change in length of the muscle. The secondary and the primary are about equally sensitive to length under static conditions (shown in Fig. 13-27).

Figure 13-29 demonstrates how the spindles primary and secondary differ in their responses to two types of lengthening. **The response of the primary is** best understood as **a combination of both the length and velocity components of movement, while the secondary ending signals, for the most part, only length.**

The combined length and velocity sensitivity of the primary ending is probably due to the fact that the primary ending innervates both the nuclear bag and nuclear chain fibers (Fig. 13-26), **which have different mechanical properties.** The two or more branches of the spindle primary may interact in the following way: The most active branch (i.e., that with the highest firing rate), which we call branch *A*, will control the firing rate of the afferent axon, since impulses arriving at the branch point, from the most active branch, will travel in both directions—toward the CNS unimpeded and antidromically out the other branch (*B*) to collide with any action potential moving orthodromically on *B*. As long as the firing rate of *A* is greater than that of *B*, all the action potentials of *B* will be blocked by the antidromic action potentials of *A*. However, when the firing rate of *A* falls below that of *B* (for instance, because *A* adapts), then *B* will control the firing in the afferent axon.

By such a mechanism, a single axon may carry information first about velocity and then about length. Alternatively, if the action potential were first generated at a point *proximal* to the branch point, then there would be summation of the generator potentials (from the two types of ending) by electrotonic spread from the two branches. In this way, the greater depolarization would still dominate the axonal firing pattern.

Evidence for the differing mechanical properties of nuclear bag and nuclear chain fibers is based on high-speed microcinephotography of single nuclear bag and nuclear chain fibers immediately after a sudden, small, sustained stretch, as shown in Fig. 13-32. There is little movement of a point on a nuclear chain fiber during a sustained stretch of the muscle, whereas a point on the nuclear bag fiber returns rapidly to its initial position. **Such "mechanical adaptation" on the part of the nuclear bag fiber might well be the basis for the adaptation of the firing rate seen in the primary ending response,** which is, of course, the velocity sensitivity that characterizes the primary ending. Since the primary axon innervates both the nuclear bag and the nuclear chain fibers (Fig. 13-26), it is not surprising that the static responses of primaries and secondaries should be very similar (Fig. 13-27). **The length sensitivity of the spindle primary probably**

Fig. 13-29. Responses of spindle primary and spindle secondary to two different types of muscle lengthening (bottom). Note that the primary's response is a combination of both the length and the velocity of movement in both cases, while secondary responds almost entirely to length only.

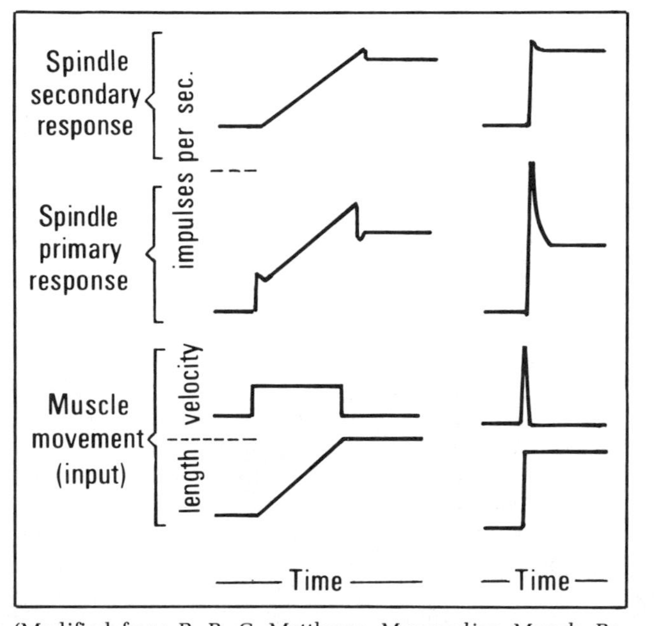

(Modified from P. B. C. Matthews, *Mammalian Muscle Receptors and Their Central Actions.* London: Edward Arnold, 1972.)

352

353

comes from the nuclear chain fibers after the nuclear bag fiber adapts mechanically.

The adaptation of the primary ending (velocity sensitivity) may have multiple causes, but most probably is due to the mechanical properties just mentioned, rather than to some adaptation of the portion of the axon that generates the action potential, since sustained electrical depolarization does not show adaptation, while mechanical movement does (Fig. 13-30). This conclusion is further supported by data from experiments on isolated stretch receptors whose length or tension was maintained constant automatically by a servo-controlled "puller," i.e., a "length clamp" and "tension clamp" (Fig. 13-31). In these experiments, the change in generator potential (which would be reflected in the firing rate of the unit) seems determined by slow, mechanical changes in the receptor when external length is held constant, as in Fig. 13-32.

The threshold sensitivity of the primary ending is much greater than that of the secondary ending.

At threshold, the primary can be excited with a movement of as little as 10 to 100 μm, while the secondaries require a movement of 160 to 500 μm at threshold [15, p. 65].

Thus the spindle primary is almost specifically stimulated by small, rapid lengthening of the muscle.

QUESTION: In Fig. 13-33, which is the primary ending, *A* or *B*? (Hint 21↓)

QUESTION: Which ending, primary or secondary, is more sensitive to vibration of a muscle tendon? (Hint 20↓)

At this point, we can summarize the spindle responses by describing the **primary ending** as **more sensitive to the dynamic components of a stimulus** (change in muscle length), while **both the primary and secondary endings are approximately equally sensitive to the static components of the stimulus.**

For very small stretches (e.g., less than 0.1 mm), over which the response of the primary ending is linear, and at vibrations up to 30 stimuli per second, the situation may be entirely different with regard to the differences between the primary and secondary endings. For further details, see Matthews [32, p. 181].

Muscle afferents do not appear to excite pathways that reach consciousness, as shown by human experiments in which tendons exposed by an anesthetized skin incision have been pulled. Similarly, traction on eye muscles cannot be perceived [32, pp. 498–499]. Furthermore, selective stimulation of Ia fibers in

Fig. 13-30. Adaptation to stretch and lack of adaptation to applied current in single muscle stretch receptor.

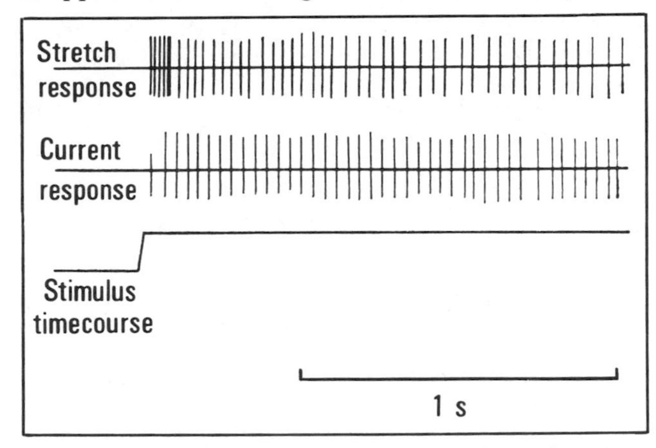

(Modified from O. C. J. Lippold, J. G. Nicholls and J. W. T. Redfearn, Electrical and mechanical factors in the adaptation of a mammalian muscle spindle, *J. Physiol.* [Lond.] 153:209, 1960.)

353

Fig. 13-31. Adaptation of generator potential of isolated crustacean stretch receptor under two servo-controlled stretches: where length is suddenly changed and then held constant and where tension is suddenly changed and then held constant. Note that there is "adaptation" of generator potential in constant-length case, but not in constant-tension situation, suggesting that there is internal movement in receptor at a fixed length.

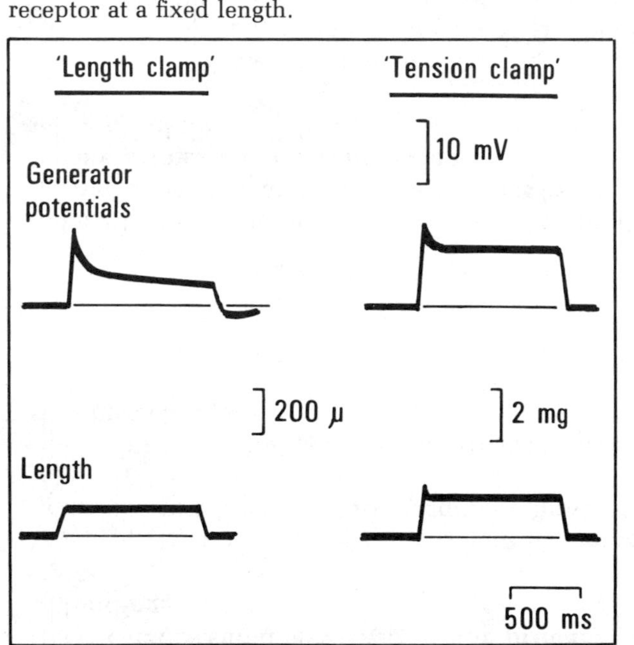

(Modified from S. Nakajima and K. Onodera, Adaptation of the generator potential in the crayfish stretch receptors under constant length and constant tension, *J. Physiol.* [*Lond.*] 200:187, 1969.)

Fig. 13-32. Mechanical property (plastic deformation) of two types of intrafusal muscle fiber as shown by movement of marked point after sudden, sustained change of 40 μm in muscle length, recorded by microcinephotography.

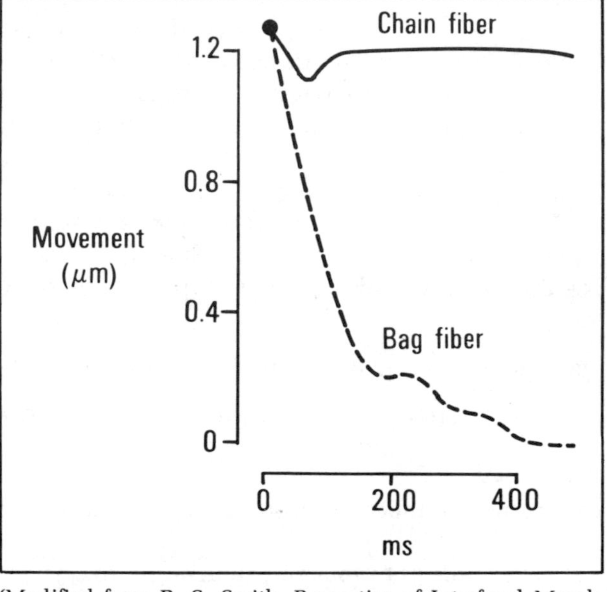

(Modified from R. S. Smith, Properties of Intrafusal Muscle Fibres, in R. Granit [Ed.], *Muscular Afferents and Motor Control: Nobel Symposium I.* Stockholm: Almqvist and Wiksell, 1966.)

Fig. 13-33. Different responses (*A* and *B*) of primary and secondary ending to tendon tap. Which response is from the primary, and which the secondary? (See Hint 21.↓) (There is a continuous background activation of gamma efferents in this experiment.)

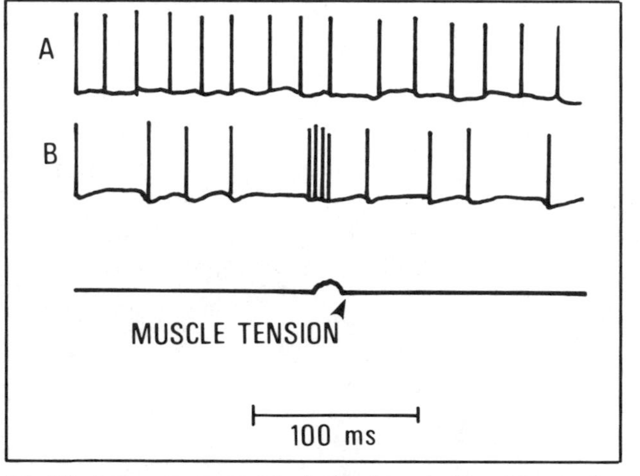

(Modified from P. B. C. Matthews, *Mammalian Muscle Receptors and Their Central Actions.* London: Edward Arnold, 1972.)

355

awake animals causes no change in behavior or in "arousal" as measured by the EEG, nor can stimulation of group I fibers be used to establish a conditioned response [32, pp. 498–499].

However, investigators have shown that vibration can influence subjective sensation of joint angle. Therefore, though there is no direct perception, it is still possible that Ia afferents do influence some sense modalities. (A wild hypothesis is that since Ia fibers don't excite pathways that reach consciousness, the fibers are part of the mechanism in some of the nonconscious effects of acupuncture when the needle is vibrated, since vibration is a highly selective means of stimulating Ia fibers.)

STATIC AND DYNAMIC GAMMA EFFERENT FIBERS

Up to this point, we have not described the actions of the motor fibers to the spindle, the γ-D and γ-S fibers (Fig. 13-26). The importance of these fibers should not be underestimated.

They not only influence markedly the behavior of the spindle afferents, but also make up a significant proportion of the neural machinery concerned with motor control. Gamma efferent fibers make up about 30 percent of the total efferent fibers in the ventral roots [15, p. 41]! The changes in afferent response wrought by stimulation of the two types of gamma fibers are summarized in Table 13-4.

The γ-D efferent fiber affects both the dynamic and the static responsiveness of the spindle primary, as shown in Fig. 13-34. However, as you might expect, **the γ-D fiber does not affect the spindle secondary at all,** since the group II fiber usually does not innervate the nuclear bag fiber (Fig. 13-26). **The γ-S fiber affects the secondary's responsiveness to stretch** (as shown in Fig. 13-27), as you might expect since the nuclear chain fiber is innervated by both the γ-S and group II nerve fibers (Fig. 13-26). In addition, the γ-S fiber also influences the spindle primary's responses, as shown in Figs. 13-34 and 13-35. **Increased γ-S activity increases the static responses of the spindle primary to muscle length** (Figs. 13-34 and 13-35).

This is certainly easy to understand, if the static response of the primary is due to the branch of the Ia fiber that innervates the nuclear *chain fiber* (Fig. 13-26).

Finally, **increased activity of the γ-S fiber decreases somewhat the dynamic responsiveness of the spindle primary,** as can be seen by careful, close scrutiny of

Table 13-4. Changes in Afferent Response of Primary and Secondary Endings according to Change in Activity of Gamma Efferents

Ending	Response	↑ γ-D	↑ γ-S
Primary	Dynamic	↑↑	↓
	Static	↑↑	↑↑↑
Secondary	Static		↑↑↑

Fig. 13-34. Response of single spindle primary ending to muscle extension with and without stimulation of γ-S and γ-D fibers to that spindle.

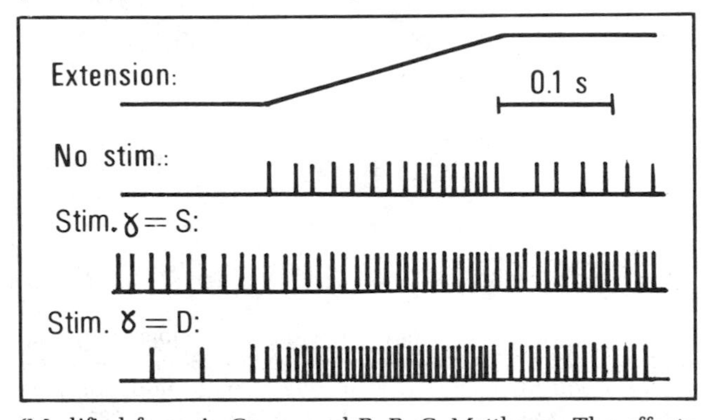

(Modified from A. Crowe and P. B. C. Matthews, The effects of stimulation of static and dynamic fusimotor fibres on the response to stretching of the primary endings of muscle spindles, *J. Physiol.* [*Lond.*] 174:109, 1964.)

HINTS

20. The primary ending. Actually, experiments indicate that vibratory stimuli are very specific stimuli for just the primary endings [32, p. 169].

21. *B* is the primary ending. If you missed this, go to Hint 22.↓

355

356

Figs. 13-34 and 13-35. There seems no simple explanation for this effect, which could be due to mechanical interaction or, possibly, to some electrotonic interactions in the Ia fiber branches (Fig. 13-26).

QUESTION: In Fig. 13-27, the static responses of a primary and secondary from a spindle were affected by gamma efferent discharge. Although it is not clear what affected the Ia fiber, why is it clear that the firing of the secondary was influenced by a γ-S fiber and not a γ-D fiber? (Think simply.) (Hint 23↓)

The significance of the gamma efferent fibers lies in the fact that the **CNS can control the responsiveness of the peripheral sensory receptors in this system,** just as it does in other sensory systems, e.g., the pupil, the middle ear muscles, etc. **Such control is an essential element in the servo-control loops** needed in posture and movement. The functions of the gamma efferent fibers may be several, and are still not really understood. But it is not hard to **hypothesize** that **at least one of the functions of the fusimotor fibers is to allow a receptor sensitive to length (and rate of change in length) to operate over a much larger range of muscle lengths than otherwise would be possible.**

For example, if there is no gamma efferent activity, a spindle easily will become sufficiently shortened by an extrafusal muscular contraction that it will stop firing (e.g., Fig. 13-22). Under these circumstances, **contraction of the intrafusal fibers can be sufficient to again make the ending sensitive to changes in muscle length, at lengths that previously did not excite the ending. In some cases (described later), it is common for a muscle spindle to increase its firing during a muscular contraction! This could occur only if the gamma fibers were also firing, shortening the spindle somewhat faster than the extrafusal fibers were contracting.**

The speed of contraction of the nuclear bag fibers and nuclear chain fibers has been well studied. **Paradoxically, the nuclear bag fibers contract more slowly than the nuclear chain fibers** (Fig. 13-36). In keeping with the differences in speed, the twitch response (not shown) of the nuclear chain fiber is faster and larger than the twitch response of the nuclear bag fiber.

The distinction between nuclear bag and nuclear chain intrafusal fibers with respect to the speed of contraction is also reinforced by the finding that **nuclear chain fibers have propagated action potentials** (at least down part of their length), whereas **nuclear bag fibers contract only by means of local (nonpropagated) responses** (i.e., the nuclear bag fiber is not electrically excitable) [32, p. 233].

In the frog, slow and fast muscle fibers are distinguished in the same way; i.e., the slower "postural" fibers are not electrically excitable, yet can hold a contraction for minutes without fatigue [15, pp. 7–9]. Thus, the nuclear bag fiber in

Fig. 13-35. Effect of stimulation of *static gamma fiber* (at different rates) and of muscle lengthening on firing rate of spindle *primary*. Note that at increasing frequency of stimulation of static gamma fiber, static response S increases, while dynamic response to lengthening D actually decreases slightly.

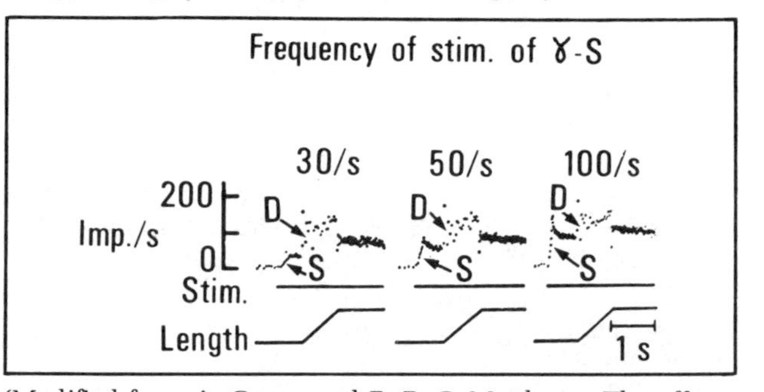

(Modified from A. Crowe and P. B. C. Matthews, The effects of stimulation of static and dynamic fusimotor fibres on the response to stretching of the primary endings of muscle spindles, *J. Physiol.* [*Lond.*] 174:109, 1964.)

356

357

the mammal can be considered to be the last evolutionary remnant of this more primitive system (since both the slow and fast extrafusal fibers in the mammal are electrically excitable). Calculations show that if the nuclear bag fiber has the same length constants as frog muscle, then local potentials from a single neuromuscular junction can still cover a significant portion of the length of the intrafusal contractile mechanism [32, p. 288]. 1

The significance of the differing speeds of contraction is unclear. But the higher speed of contraction of the nuclear chain fiber suggests that it may play an important role in movement as well as posture. So it is of interest that there are about twice as many γ-S as γ-D fibers. 2

Another fact of unclear significance is that gamma motor neurons do not show much recurrent inhibition, in contrast to alpha motor neurons [15, p. 156]. Such differences are not surprising since the functions of alpha and gamma motor neurons are so different. 3

Clearly the differences between γ-S and γ-D fibers are utilized by the body, since it has been shown that stimulation in various locations (such as anterior cerebellum, reticular formation pyramidal tract, pons) can selectively depress or activate the two fiber types [32, pp. 525–529]. 4

EXAM QUESTION: What motor cells are functionally part of a sensory system? (Hint 24↓) 5

MONOSYNAPTIC ACTION OF SPINDLE PRIMARY—THE TENDON JERK

Now we can consider some of the physiological actions of the muscle spindle. The simplest and best studied of the spindle effects is the **tendon jerk,** which is **due to a monosynaptic reflex brought about by the Ia afferent fiber** from the spindle primary. 6

The Ia fiber makes monosynaptic excitatory connections with the alpha motor neurons of the same muscle, causing *autogenic excitation* (Fig. 13-37). 7

Fig. 13-36. Movement of point on nuclear chain and nuclear bag intrafusal fibers during tetanic stimulation of gamma efferent fibers (starting at time zero). Data obtained by microcinephotography. Time courses of relaxations at end of contraction differ in the same way.

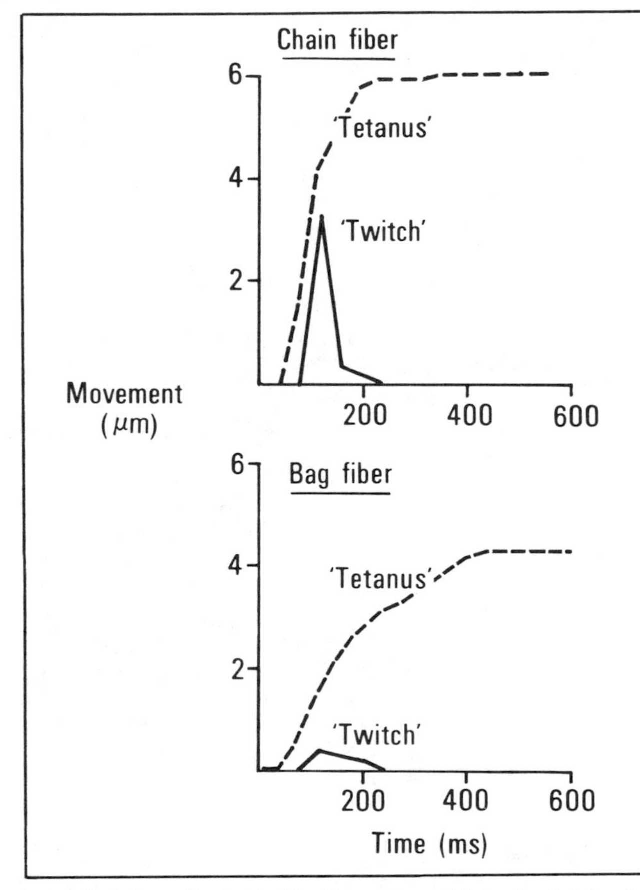

(Modified from R. S. Smith, Properties of Intrafusal Muscle Fibres, in R. Granit [Ed.], *Muscular Afferents and Motor Control: Nobel Symposium I.* Stockholm: Almqvist and Wiksell, 1966.)

HINTS

22. The tendon tap is a very *fast* lengthening of the muscle, but it does not lengthen the muscle very much. So the combination of these two ideas must lead to the idea that the tendon tap is more likely to stimulate the primary than the secondary ending.

30. The primary ending responds to fast changes, i.e., velocity, much more than the secondary ending does.

357

358

There is no autogenic excitation of *gamma* motor neurons [32, pp. 448–449], one example of the significant differences between alpha and gamma motor neurons [32, p. 451].

In addition, **the Ia fiber excites an interneuron, which in turn inhibits the alpha motor neuron of the antagonist muscle** (Fig. 13-37). This is the basis of *reciprocal inhibition* (discussed later).

The connections shown in Fig. 13-37 form the major basis for the study of the EPSP and IPSP described in Chap. 11.

In the next section, the *poly*synaptic actions of the Ia and II fibers are discussed. Here we are concerned with only the *mono*synaptic tendon jerk reflex.

Figure 13-38 shows the earliest satisfactory evidence that the reflex from Ia stimulation is monosynaptic (although this had been suspected from earlier indirect evidence). This work also established the approximate value of the synaptic delay as 0.5 ms.

Question: The inhibitory connections to the motor neuron in the experiment of Fig. 13-38 also should have been activated by the spinal cord stimulation. Is such activation evident in Fig. 13-38? If not, why not? (Hint 25↓)

Question: Why do you suppose the monosynaptic tendon reflex is one of the most studied, best understood actions of the muscle spindle primary? (Hint 26↓)

A given Ia fiber will innervate many motor neurons, including those of synergistic muscles [32, p. 354]. Furthermore, **every alpha motor neuron is probably innervated by almost every Ia fiber from its muscle.** Thus, there is **considerable divergence and convergence in the system,** including connections between muscles having similar functions with regard to movement.

Between motor neurons of the same muscle, there can be considerable variation in the number of monosynaptic contacts, hence in the size of the EPSPs generated by a single Ia afferent in different alpha motor neurons. The variation can be as great as 10-fold (EPSPs ranging between 20 and 200 μV) [32, p. 355].

However, it would be unwise to assume that Ia fibers provide the major input to the alpha motor neuron. In fact, **Ia fibers provide only about 0.5 percent of the synaptic terminals on the soma and proximal dendrites of the motor neuron** [32, p. 355]!

Fig. 13-37. Connections of spindle primary (Ia) fibers: monosynaptically to agonist motor neuron (i.e., autogenic excitation) and disynaptically to antagonist motor neuron (i.e., reciprocal inhibition).

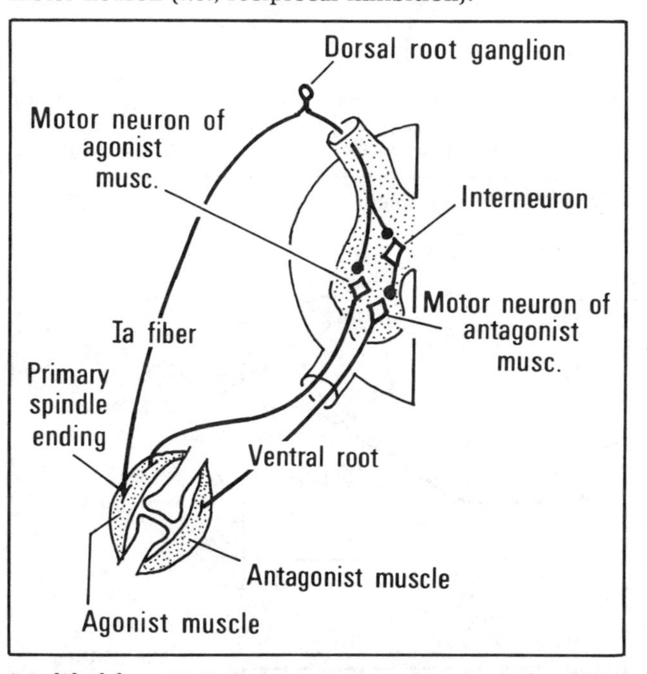

(Modified from W. F. Ganong, *Review of Medical Physiology* [9th Ed.]. Los Altos, Calif.: Lange, 1979.)

358

359

1 The clinically important tendon jerk (known clinically as a *tendon reflex*) is initiated by a rapid stretch of the muscle that occurs when the tendon is tapped with a reflex hammer. **The resulting reflex can be detected by the motion of the muscle** under isotonic conditions (the usual clinical measure), **by increased muscle tension** under isometric conditions (Fig. 13-39), **or by increased activity shown in the electromyogram** (EMG), indicating synchronous depolarization of the muscle (Fig. 13-40).

2 Much physiological investigation has gone into the now well-established basis for this reflex response. **The spindle primary is particularly sensitive to brief, rapid stretch** (Fig. 13-33). **The Ia fibers from the spindle primaries make monosynaptic terminations on the alpha motor neurons** of the same muscle (Fig. 13-37) **and cause depolarizing EPSPs. The synchronous efferent volley,** which can be **detected** in the motor neuron axons (Fig. 13-38) and **in the depolarization of the muscle fibers** (Fig. 13-40), **causes contraction of the muscle.** In turn, this increases muscle tension (Fig. 13-39) sufficiently to effect movement. **This is the simplest of the CNS reflexes, involving only two neurons** (one sensory, one motor) **without interneurons.**

3 Of course, conceivably the tendon tap might also stimulate Golgi tendon organs, which are sensitive to muscle tension. However, as described later, the action of the tendon organ's Ib fiber is to inhibit (via an interneuron) the motor neuron, so the firing of the tendon organ cannot explain the reflex contraction of the tendon jerk.

4 Question: The tendon jerk reflex is only a brief contraction of the muscle. What factors can you list that could contribute to its brevity? (Hint 27↓)

5 As you might expect, **the reflex response to vibration is closely related to the tendon jerk.**

6 QUESTION: Why should you expect this? (Hint 30↑)

7 Vibration is almost a specific stimulus for the spindle primary. The reflex response of a muscle to vibration of its tendon is shown in Fig. 13-41. As with the tendon jerk, the **reflex response to vibration** has been shown to be **due to the monosynaptic reflex from primary spindle receptors.**

Fig. 13-38. Demonstration (by Renshaw) of monosynaptic nature of dorsal root stimulation (later called Ia fibers). Recording from ventral root while stimulating (*A*) in spinal cord with needle electrode or (*B*) on dorsal root of same segment. Spinal cord stimulation elicits an initial wave from direct motor neuron stimulation and *s* wave after minimum (approximately 0.5 ms) delay from monosynaptic activation of motor neuron. The *s* wave is due to depolarization of motor neuron afferents that make monosynaptic excitatory connection with the motor neuron. Dorsal root stimulation activates motor neurons with delay comparable with monosynaptic activation (*s* wave).

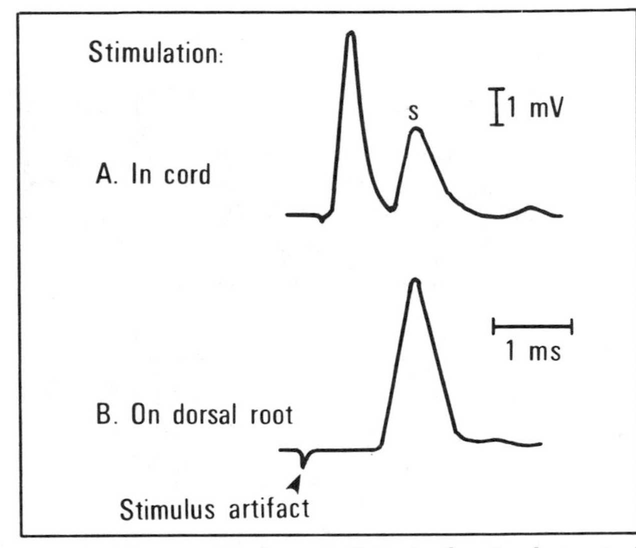

(Modified from B. Renshaw, Activity in the simplest spinal reflex pathways, *J. Neurophysiol.* 3:373, 1940.)

HINTS

23. The γ-D fiber does not affect the firing of the secondary (see Table 13-4, based on studies of stimulation of single efferent fibers).

24. The gamma motor neurons. This seems paradoxical at first, but is not unreasonable when you think about it. If you missed this question, you may be trying too hard, or not hard enough. So reread the last few first-level paragraphs!

359

360

Fig. 13-39. Tendon jerk recorded by means of muscle tension. Small tension wave is due to tendon tap, while larger wave is reflex contraction of muscle. Also shown in this record is long (about 1 s) inhibition of reflex caused by single afferent stimulation of cutaneous afferent fibers.

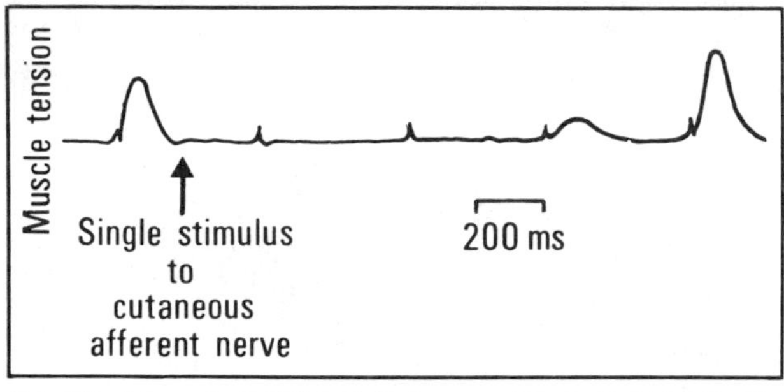

(Modified from L. Ballif, J. F. Fulton, and E. G. T. Liddell, Observations on spinal and decerebrate knee-jerks, with special reference to their inhibition by single break-shocks, *Proc. R. Soc. Lond.* [*Biol.*] 98:589, 1925.)

Fig. 13-40. Achilles tendon jerk recorded electromyographically (electrical activity from surface electrodes) in humans. The tap causes electrical artifact, followed in about 30 ms by large, synchronous depolarization of muscle.

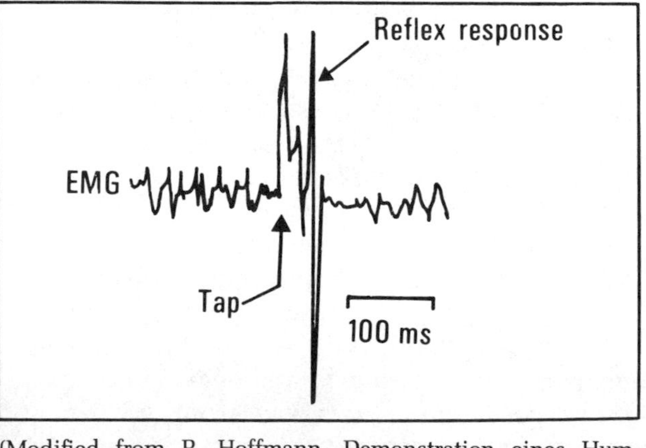

(Modified from P. Hoffmann, Demonstration eines Hummungsreflexes in menschlechen Rückenmark, *Z. Biol.* 70:515, 1920.)

Fig. 13-41. Increased muscular tension reflexly induced by high-frequency vibration (200 stimuli per second), during the intervals indicated by horizontal bars labeled with magnitude of vibration. Recorded from a decerebrate cat. Similar, though less spectacular, results have been obtained in conscious humans. Note that increased amplitude of vibration gives greater response.

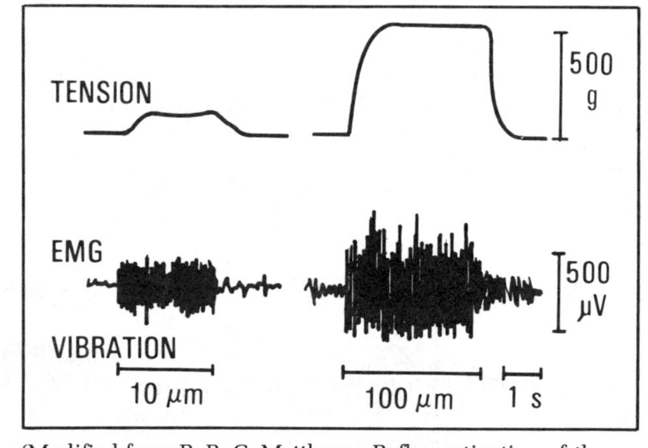

(Modified from P. B. C. Matthews, Reflex activation of the soleus muscle of the decerebrate cat by vibration, *Nature* 209:204, 1966.)

Fig. 13-42. Vibration-induced reflex tension in decerebrate cat as related to magnitude and frequency of vibration. Data from similar experiments as Fig. 13-41.

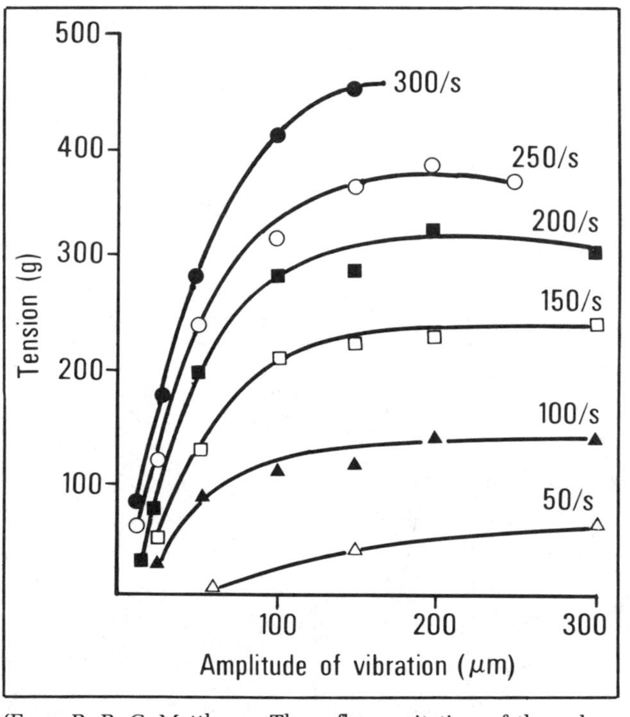

(From P. B. C. Matthews, The reflex excitation of the soleus muscle of the decerebrate cat caused by vibration applied to its tendon, *J. Physiol.* [*Lond.*] 184:450, 1966.)

360

361

1 The strength of the reflex, as a function of both vibration rate and magnitude, is shown in Fig. 13-42. The marked sensitivity to amplitudes of stretch less than 100 μm is evident.

2 Question: What explanation can you offer as to why, in Fig. 13-42, the response at a given frequency levels off and does not increase when the magnitude of the vibration is increased? (Hint 29↓)

3 QUESTION: When vibration is applied before and during tendon tap, the response to the tap is reduced or abolished [15, p. 174]. Why? (Hint 31↓)

4 In humans, the reflex increase in tension induced by vibration is involuntary, just like the tendon jerk. However, both reflex responses can be influenced by "higher centers," such as by the direction of attention [15, p. 172]. (The mechanisms of "attention" and its method of affecting these spinal cord functions are entirely unknown.)

HINTS

25. Remember, first, that these are ventral root recordings; hence any IPSP that is present cannot be seen. Inhibition must be detected by a diminution in an excitatory response, as *can* be seen in this record. The magnitude of the ventral root response (which in a compound action potential such as this measures the *number* of motor neurons activated) caused by *dorsal root* stimulation is roughly the same as that elicited by direct stimulation of the motor neurons by the electrode in the spinal cord (Fig. 13-38). However, the synaptic excitation of the motor neurons by spinal cord stimulation (wave *s* in Fig. 13-38) is less. While such diminution might be due to refractoriness, it is also explainable on the basis that some motor neurons were sufficiently inhibited so as not to fire. (If you are not convinced, then you see why intracellular recordings, as described in Chap. 11, were such a step forward in understanding excitatory and inhibitory mechanisms in the spinal cord!)

26. (1) It is activated by a brief (transient) input, the tendon tap, which can be reasonably mimicked by a synchronous volley resulting from electrical stimulation. (2) The Ia fibers are the largest afferents, so they are most readily stimulated without contamination from other fiber types (but see Fig. 13-24). (3) The reflex is monosynaptic, hence is rapid, with no chance for complications arising from inhibition from competing pathways. The variability characteristic of ıelatively long multisynaptic pathways is also absent. (4) The reflex involves excitation of all elements—such effects are much easier to detect than inhibition (which usually is measured as a reduction of an established excitation). (5) Although of less importance than the preceding reasons, there is considerable interest in the tendon jerk since it is such a common clinical tool in the evaluation of neurological disorders in humans.

27. All the following could (and probably do) contribute to the shortness of the muscular contraction. (1) The stimulus itself is short and not sustained. (2) The synchronous volley may involve most of the muscle cells. Hence there will follow a brief period in which most of the cells are refractory and thus silent. (Note that you can see such a "silent period" immediately following the reflex response shown in Fig. 13-40.) (3) The recurrent inhibition (by Renshaw interneurons) of alpha motor neurons caused by their own activity can immediately follow a burst of synchronous activity. (4) The contraction will shorten the muscle and briefly "unload" tonically active spindles (as in Fig. 13-22), thus reducing excitatory input to the alpha motor neurons. (5) The *reflex contraction* will stimulate Golgi tendon organs, which in turn inhibit the motor neuron (described later, although this idea was just mentioned). Did you get all items in the list? This was a tough question! Note that a silent period can be observed also in the synergists of a muscle when the muscle's tendon jerk is elicited.

361

Fig. 13-43. Polysynaptic response shown by compound action potentials recorded from ventral roots, with single stimuli of increasing strengths delivered to dorsal roots of same segment. Note that first wave, which is from monosynaptic reflex, has reached its maximum at lowest stimulus strength shown; at higher strengths, monosynaptic response remains the same, but polysynaptic reflex firing increases.

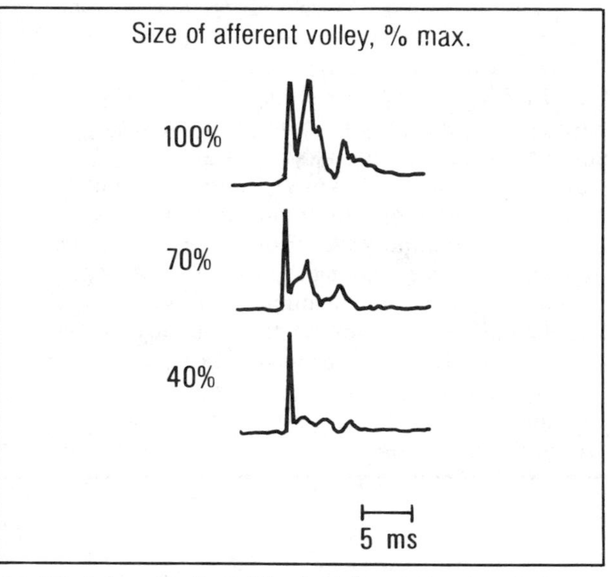

(Modified from D. P. C. Lloyd, Reflex action in relation to pattern and peripheral source of afferent stimulation, *J. Neurophysiol.* 6:111, 1943.)

Fig. 13-44. Stretch reflex, in which increase in tension on ordinate is due to increase in length of muscle during stretch, over time shown in tracing *M*. Passive curve *P* was obtained after denervation of quadriceps muscle of decerebrate cat. Extra tension ($M - P$) developed by reflex is apparent. Change in length of stretched muscle is shown in tracing labeled *L*, with its own "upside-down" ordinate.

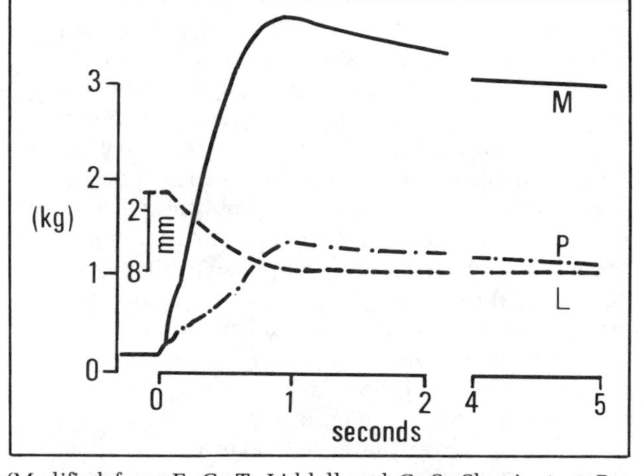

(Modified from E. G. T. Liddell and C. S. Sherrington, Reflexes in response to stretch [myotatic reflexes], *Proc. R. Soc. Lond.* [*Biol.*] 96:212, 1924.)

Fig. 13-45. Combined effects of stretch and vibration on muscle tension during *static* portion of stretch reflex. Note that the 200-Hz vibration, which can be expected to stimulate majority of Ia fibers (e.g., see Fig. 13-42), is not occluded by larger tension reflexly induced by 9-mm extension. Thus, the parts of reflex activated by stretch are probably independent of those activated by vibration (which is specific for spindle primary).

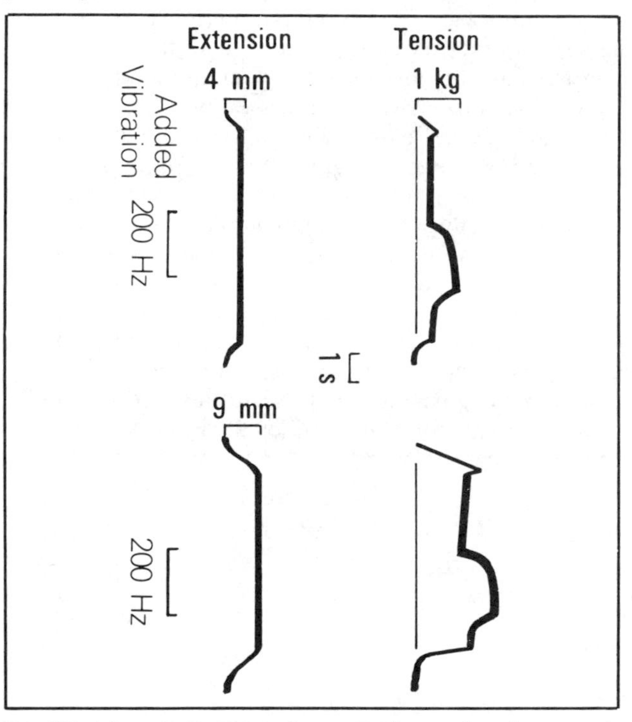

(Modified from P. B. C. Matthews. Evidence that the secondary as well as the primary endings of the muscle spindle may be responsible for the tonic stretch reflex of the decerebrate cat, *J. Physiol.* [*Lond.*] 204:365, 1969.)

POLYSYNAPTIC REFLEXES—THE STRETCH REFLEX

Although the preceding descriptions emphasized the monosynaptic reflexes from the spindle primary, there is some indirect evidence that activation of the **Ia afferent fibers can also activate polysynaptic excitatory reflex pathways** [32, pp. 360–361]. 1

A classic example of the electrical recording of a polysynaptic reflex is shown in Fig. 13-43. 2

Now let us take up the **stretch reflex,** which **has both dynamic and static components** (sound familiar?), as shown in Fig. 13-44. 3

In Fig. 13-44, the muscle of a decerebrate cat is stretched, as shown by the change in length recording. The change in length causes an increase in tension because of reflexly increased alpha motor neuron firing—the *stretch reflex.* The *static* portion of the reflex is evident by the active tension present even 5 s after the start of the lengthening (Fig. 13-44). The *dynamic* part of the reflex is indicated by the overshoot of the peak at about 1 s. It is of interest that **stretch reflexes are especially prominent in postural (red) extensor muscles,** even though stretch reflexes can also be seen in flexors [32, pp. 417–418]. 4

What receptors are active in the stretch reflex? Most probably there are at least three: *the spindle primary, the Golgi tendon organ,* and the *spindle secondary.* 5

The spindle primary obviously plays a part during the active stretching. It can contribute also to the static phase [32, p. 422], since the spindle primary transmits impulses in proportion to steady stretch, after a lengthening (Figs. 13-28, 13-29, and 13-34). 6

The Golgi tendon organ is stimulated by the increased tension (described later) and reduces the tension of the reflex (by inhibition) by about 50 percent [32, p. 426]. 7

Finally, as their length increases, the spindle secondaries probably also polysynaptically activate the alpha motor neurons. 8

There is controversy over the action of the spindle secondaries, and the evidence for their activation of the alpha motor neurons (as described) is indirect. As shown in Fig. 13-45, the reflex responses to stretch and to vibration are independent. This im- 9

HINTS

29. Above an amplitude of about 100 μm, all the Ia fibers are being activated. "Now wait a minute," you say. "Then how is it that increasing the frequency from 100 to 200 stimuli per second can increase the tension if all Ia fibers were activated at 100 stimuli per second?" (Hint 28↓)

31. The same sensory and motor elements are active in both responses. If the vibration is large enough and fast enough, a large proportion of the fibers already will be active and hence cannot be further activated by the added tap. This is a good example of *occlusion,* described in Chap. 11, on page 285.

 364

plies that the response to stretch is *not* due to the spindle primary. Hence, the only likely candidate is the spindle secondary. This view is not yet fully established, since traditional teaching states that the group II afferents are excitatory to flexors and inhibitory to extensors [32, p. 370]. 1

The traditional view is based on studies that compared electrical stimulation above and below the threshold for group II fibers. However, such experiments also involve changes in the number of group I fibers firing (see Fig. 13-24) and certainly test the effects of group II activity only in the face of simultaneous massive group I input. **It may be** that both views are correct and **that the nervous system switches between one mode of action and another under different conditions and inputs** [32, p. 373]. Perhaps, at this point, you can be somewhat sympathetic to the evaluation by Matthews of another similar problem: "It is all becoming unpleasantly complicated and it seems unlikely that the normal mode of operation of even this simple reflex pathway can be sorted out simply by the continued use of the technique of activating it by single synchronous afferent inputs, but no better method is yet to hand" [32, p. 367]. 2

Additional evidence that activation of spindle secondaries can cause increased muscle tension in the stretch reflex comes from measurements of the stretch reflex tension developed from *slow* (static) stretch in the decerebrate cat. The relationship of amount of tension to amount of stretch is often linear over a large range [32, p. 419], while independent measurements show that the linear range of the spindle secondary is large, compared with the primary [32, p. 177]. Of course, the primaries are probably involved in those situations in which secondaries are not likely to be activated, such as the dynamic phase of the stretch reflex and where high sensitivity to small stretches (that are less than threshold for many spindle secondaries) can be observed [32, p. 421]. 3

And, as the sun sets in the golden West, let us take our leave of the motor system neurophysiologists, still at work in their laboratories, trying in vain to sort it all out, and return to . . . 4

The **lengthening reaction** often is observed in the same experiments as the stretch reflex. It consists of an **increase in tension during lengthening,** but the **tension is not sustained even though the muscle is kept in the lengthened state.** 5

This response, which at times can be observed clinically as *increased tone* when stretching a patient's muscle rapidly, may be a reflection of the dynamic component of a stretch reflex that shows little static activity [32, p. 446]. Thus, it is logical to presume that the lengthening reaction is reflexly mediated by spindle primary activity. 6

 364

365

SERVO-CONTROL OF MUSCLE LENGTH

1 What you presently know about muscle and its control can be cast in servo-control terms. You will see that there are a variety of mechanisms for control of *muscle* length.

2 The first negative feedback loop occurs within the muscle itself. As delineated in the intramuscular loop (Fig. 13-46), **an increase in the load** on a muscle **causes an increase in length and hence in tension** (at a constant motor neuron discharge rate), **which tends to oppose further increase in length.**

3 At a fixed motor neuron discharge, the muscle can be considered to be acting as a spring being stretched by the change in load. The description of the system as a negative feedback loop is perfectly appropriate under these circumstances. Actually, many purely physical processes in which a change induces an opposing force can be described in negative feedback terms, e.g., the back electromotive force that occurs as a capacitor is charged (Chap. 4).

4 A second, more powerful negative feedback loop involves neural control of muscle activity.

5 As shown in the neural loop of Fig. 13-47, **an increase in the load** on the muscle **leads to a series of events in the spindle afferent, spinal cord, and alpha motor neuron,** which together **also oppose the change.** A more formal diagram of the loops is shown in Fig. 13-48. It is not necessary to understand all the details of Fig. 13-48 to grasp the major points of the next pages.

6 The details of Fig. 13-48 are described here for those whose passion is explicit analysis. The muscle is the amplifier in the loop, but it is a nonlinear amplifier that transforms the alpha motor neuron firing rate into a tension whose magnitude depends on the length of the muscle. This transformation is merely a regraphing of Fig. 13-6 with frequency as the independent variable.

7 The muscle tension, in turn, is transformed to a change in length according to the characteristics of the "load." The axes of the load *T-L* (tension-length) diagram are reversed from the previous length-tension diagrams (e.g., Fig. 13-9) because tension is the independent variable and the loop diagram is easier to follow if the abscissa is consistently the independent variable. (This means that on this plot an isometric load would be horizontal and an isotonic one vertical. So be sure to look at the labels of the axes carefully!)

8 The characteristics of the load can change, as indicated by the ΔT-L input to the load transformation (discussed later). Note that the linear relation between T and L

Fig. 13-46. Intramuscular loop, which opposes increase in muscle length.

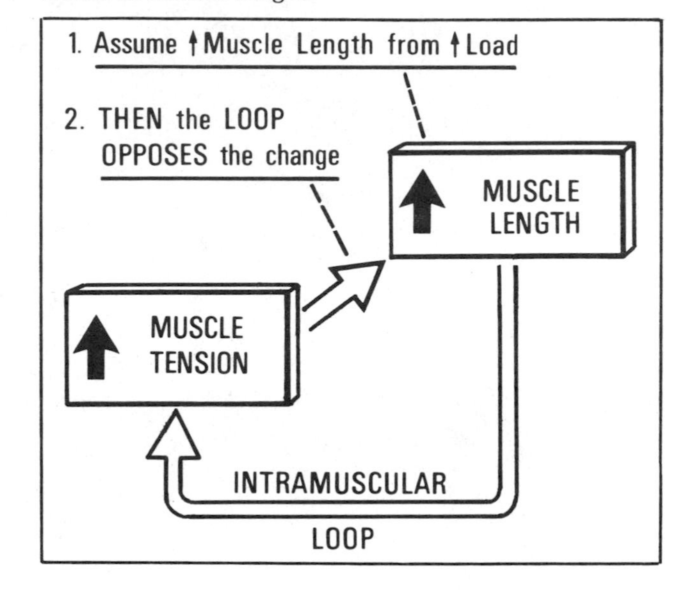

HINT

28. Temporal summation strikes again!

365

366

Fig. 13-47. Neural (reflex) loop, which opposes increase in muscle length.

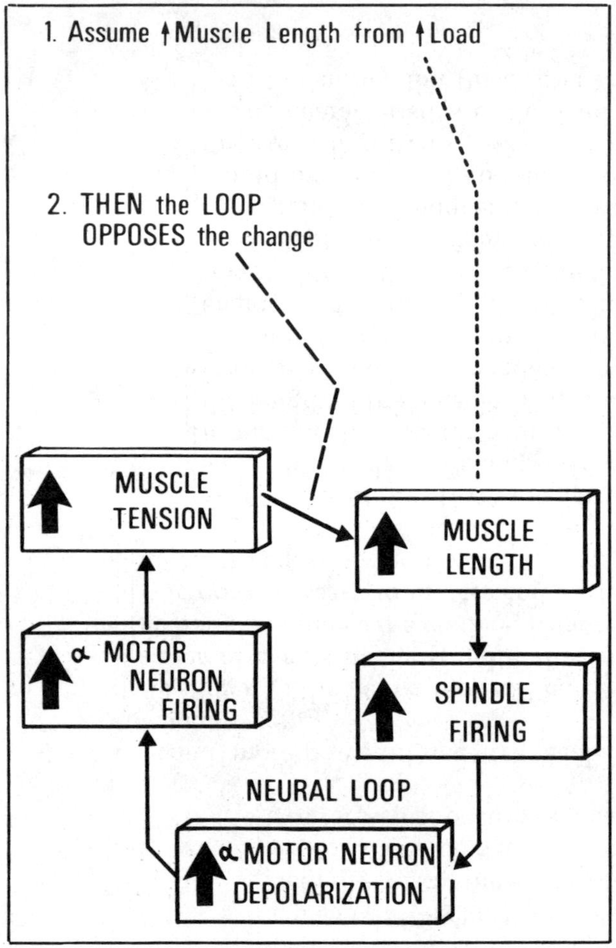

1. Assume ↑Muscle Length from ↑Load

2. THEN the LOOP OPPOSES the change

↑ MUSCLE TENSION

↑ MUSCLE LENGTH

↑ α MOTOR NEURON FIRING

↑ SPINDLE FIRING

NEURAL LOOP

↑ α MOTOR NEURON DEPOLARIZATION

Fig. 13-48. Negative feedback loops for length control in single muscle, diagramming only Ia static response. Note that inserted graphs shown are only diagrammatic (i.e., approximate). Symbols: L = length, T = tension, F = frequency of impulses.

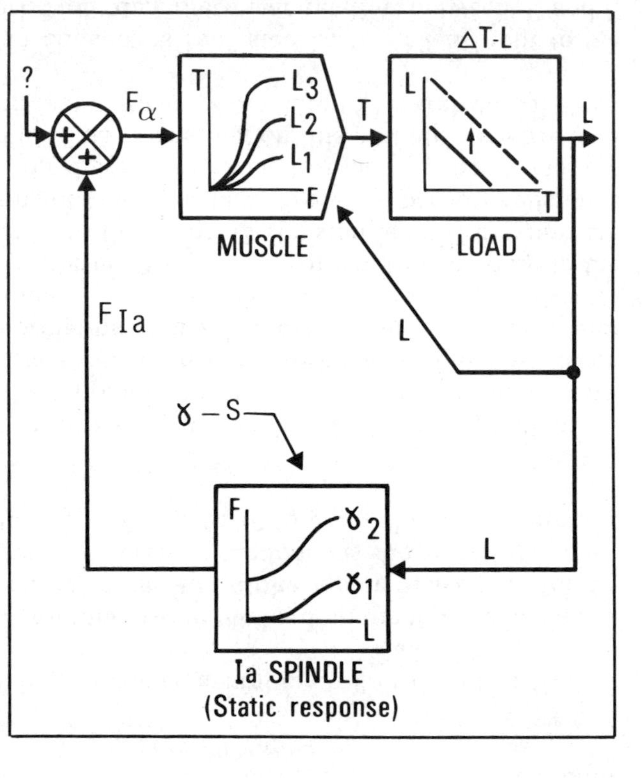

366

367

shown here is a didactic oversimplification, as will readily be apparent if you check Fig. 13-16. **Note** that the load and load changes may be external to the body, so **that the loop may go through the environment.**

1

The length (the controlled quantity in this diagram) in turn affects both the muscle and the spindle. First, the change in length of the muscle alters the relationship between motor neuron firing frequency and developed muscle tension, as indicated by the family of curves. This loop forms a negative feedback since a length increase changes the family of curves so as to increase the tension, which in turn will reduce (oppose) the increased length. Second, the spindle transforms a change in length to a change in frequency of firing, which in turn feeds back on the alpha motor neuron "comparator." The comparator generates the firing frequency of the error-signal input to the muscle. At this point, we can just consider the γ-S input to the spindle to be constant.

2

The comparator's *set-point input* is shown as a question mark since for this analysis the source is not critical, as long as it is constant.

3

Since the action of the Ia fiber is to excite the alpha motor neuron, at first glance the plus sign in the comparator circle seems to indicate a positive feedback loop. However, **the loop is negative because of the inverse tension-length relationship of the load.**

4

The importance of Fig. 13-48 is to show that **the characteristics of the load are part of the negative feedback loop's attempt to keep the length of the muscle relatively constant.** Thus, **a change in the load,** and more especially a change in the type of load (remember Figs. 13-11, 13-12, 13-13, and 13-14?), **will have marked effects on the loop.** Thus the loop, as it is diagrammed here with fixed relationships at the spindle and in the spinal cord, could not have great accuracy (i.e., high gain) under all conditions and under all loads. Conversely, for highly accurate control, we must assume that this analysis is deficient and represents a great simplification of the actual situation. Figure 13-48 serves only as a good starting point for discussing servo-control of muscle.

5

Later in this chapter, we consider the contribution of the γ-S fibers (and interneurons in the spinal cord, not shown in Fig. 13-48) in changing the properties of the neural position of the loop under higher control and in response to changing load characteristics.

6

You might note that only the static Ia spindle response is diagrammed in Fig. 13-48 for two reasons. First, in the whole preceding analysis of length-tension relationships between muscle and load, at no point did we consider changes with respect to velocity. Second, the effect of the spindle secondary (group II) fibers both is controversial

7

367

368

(as previously mentioned) and requires an interneuron (since the group II reflexes are definitely polysynaptic), which would make the diagram more complex without adding any new characteristic to this loop. 1

Of course, if we diagrammed the spindle secondary as traditionally taught, the extensor inhibition of group II fibers on the alpha motor neuron would make the loop positive (with respect to the *extensors*), since it would contain *two* inversions. (It is interesting to try to imagine what would be the physiological usefulness of such a positive feedback loop.) The traditional idea of group II excitation of the *flexor* motor neuron would be a negative feedback loop, against which the assumed extensor positive feedback loop would battle. This again exemplifies why this complication is best left to further research. 2

Several parts of the loop are (or can be) nonlinear: (1) the frequency-tension transformation of muscle, (2) the tension-length transformation of the load, (3) the length-frequency transformation of the spindle (see Fig. 13-27), and (4) the frequency-to-frequency transformation between the Ia fiber and the alpha fiber (especially if interneurons were added as well). **With all these nonlinearities, obviously one cannot apply the linear analysis for stability** (previously described in "Open-Loop Analysis and Stability Criteria" in Chap. 12). But the usefulness of the approach at least indicates that a neural negative feedback loop for control of muscle length is present, even though, because of its complexity, not all the consequences of the loop may come readily to mind. 3

At this point, you can test your understanding of this diagram by assuming the change in load ΔT-L, shown as the shift from the solid to the dotted tension-length line in the load box, and following the consequences of the change around both loops. Prove in both cases that the loops oppose the assumed change. (Note that ΔT-L is diagrammed as a disturbance that the loop opposes.) 4

The combined effects of the two negative feedback loops stabilizing muscle length shown in Fig. 13-48 are not large enough to describe the whole system as one of high-gain control. 5

The evidence for this is indirect, but comes from several observations [32, pp. 591–594]. (1) A load of 120 g on the outstretched finger can displace it as much as 4 cm in a person who is not trying to maintain the finger in a steady position. (2) When one tries to lift an object that turns out to be unexpectedly heavy (e.g., a dark bottle filled with mercury), one may fail to raise it, even though it is within the muscle's capacity. (3) The flexible walk of the cat would not be possible if there were a strong negative feedback loop controlling length. (4) Stretch reflexes (where the action of the loop is most clearly seen) are most easily found only under experimental conditions that change markedly the "normal" balances of excitation and inhibition in the spinal 6

368

369

cord, e.g., in decerebrate preparations. (5) Vibration, which stimulates Ia fibers rather specifically, causes only modest increases in tension and does not prevent (although may somewhat impair) voluntary movements, as in workers handling vibrating sanders and pneumatic hammers.

(Although vibration gives only modest increase in tension by reflex action, repeated bursts of vibration can potentiate the reflex sufficiently that the tension can be as great as a subject can produce by voluntary effort [32, p. 598]. This is another complication that makes simple generalizations difficult!)

The preceding first-level generalization concerning the two loops is correct with respect to large muscular excursions. But note that the spindle primary response may have a much higher *gain* for small movements than for large ones. Hence **the system may be nonlinear in such a way that it controls well for small disturbances, but not for large ones.**

So much for static considerations. What about the dynamic aspects of motor control? Because even less is known about this subject, we offer only some preliminary generalizations, which may serve until an adequate analysis is available.

Dynamic feedback control implies that velocity must be sensed, as it is by the spindle primary. Such an addition to the feedback system must be essential for adequate control of a system that contains significant masses (and hence inertia), as the musculoskeletal system does.

The larger the mass set in motion, the larger is the need for damping to prevent overshooting of a target and the subsequent oscillation. In engineering terms, a *velocity sensor* can be used in a negative feedback system to counteract this tendency toward overshoot and oscillation in a dynamic system.

In the mathematical description of a servo-loop, a velocity-sensitive element can be used to "damp" the system without the energy losses associated with true viscous damping [32, p. 576]. (In muscle control, some viscous damping is provided by the physical characteristics of the muscle.)

Another function of a velocity sensor in a feedback loop is to overcome the inherent delays of the loop, which otherwise might lead to oscillations under some conditions (Chap. 12). The large-diameter, rapidly conducting fibers that make up both the sensory (Ia) and motor (alpha) sides of the feedback loop (combined with the minimum synaptic delay of a monosynaptic reflex) suggest that, for adequate muscular control, it has been necessary to minimize the loop delay in the reflexes. Thus, in descriptive terms, determination of velocity of movement may allow the CNS to predict the length of the muscle after the delay time of the reflex and to initiate sufficient increase or decrease of response to compensate for the predicted delay [15, p. 271; 32, p. 576].

369

In engineering terms, the primary ending gives a signal that is phase-advanced on the mechanical input, in order to oppose the phase lags of the other parts of the system [32, p. 576].

Clearly, the stability of these negative feedback loops is important, since oscillations in the system can be observed so easily.

There are several examples of such muscular oscillations. In normal humans, there is a *tremor* when the muscle is fatigued and still exerting considerable effort to maintain a position against a load (as just before "losing control"). In normal humans, there is the usual unsteadiness, called *physiological tremor* (best seen at the tips of the fingers when the effect is exaggerated by nicotine and caffeine), which is reflexly mediated [32, pp. 578–579]. In normal humans, shivering (which is partly controlled by the stretch reflex arc) exhibits muscular oscillations [32, p. 580]. In diseases causing clonus, a series of sustained contractions is initiated by a rapid extension of the muscle—probably an exaggeration of the stretch reflex. In disease causing parkinsonian tremor, there is a sustained, involuntary oscillation of motion at rest (e.g., the "pill-rolling tremor").

FUNCTION OF GAMMA EFFERENTS

As you may have noticed, the gamma efferents have hardly been mentioned as part of motor control. Not only do the gamma efferents create additional complications, but also, unfortunately, **the role of gamma efferents in the control of posture and movement is still far from settled.** There is no question, however, that **gamma efferents are active in many movements** and cannot be ignored in any theory attempting to explain motor control.

A clear example of the effect of firing of gamma fibers is shown in Fig. 13-49. If the muscle is shortened passively (Fig. 13-49), then the spindle is "unloaded," i.e., is shortened, and so firing ceases. However, if the CNS controls both alpha and gamma fibers during a coordinated movement (Fig. 13-49), the spindle fires *during* the contraction, which implies that **the gamma efferent fiber must have been active** before or concurrent with the alpha fibers to that muscle. That is, the intrafusal fibers must have shortened *before* the extrafusal fibers. One must reach a similar conclusion from the experiments in which spindle afferent discharge occurs during an isometric contraction (when there is certainly no lengthening!), as shown in Fig. 13-50.

The simplest hypothesis about the function of the gamma efferent fiber is based on the idea of a spindle sensitivity range that is less than the total range of movement of the muscle.

In this view, the range of movement over which the spindle has to operate is too large for the spindle to adequately span (given its range of firing rates—say, 1 to 300 per second). **In order to maintain the accuracy of the detection of changes in length, the spindle itself is**

Fig. 13-49. Spindle primary from jaw muscle during active and passive movement. Active movement was initiated by causing lightly anesthetized animal to swallow fluid placed in mouth. Passive movement was obtained by servo-controlled device that duplicated active movement recorded on magnetic tape. Note marked difference in spindle activity in two cases, which must mean that in the active case, gamma fibers were firing.

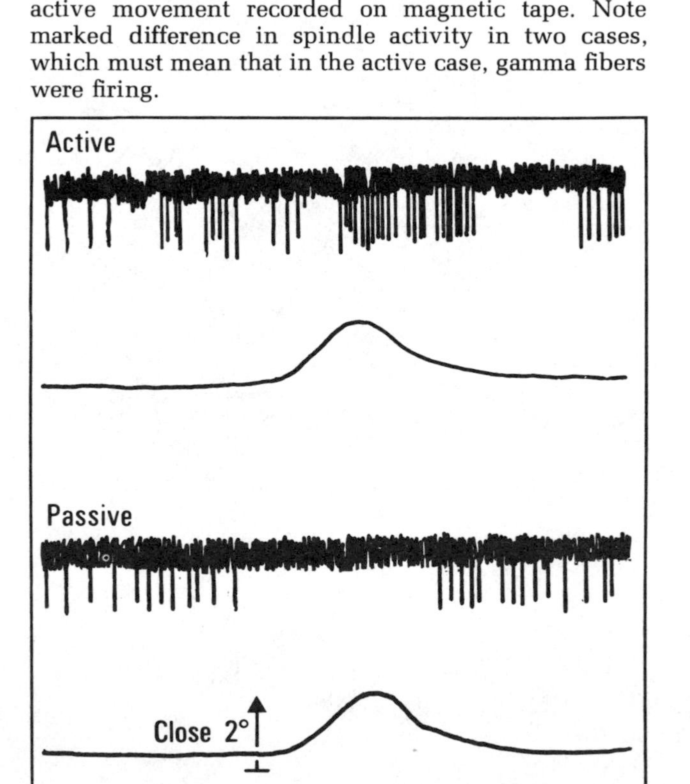

(Modified from A. Taylor and M. R. Davey, Behaviour of jaw muscle stretch receptors during active and passive movements in the cat, *Nature* 220:301, 1968.)

371

changed to match the muscle length at any given moment. So if there are changes in muscle length as a result of load variations, the negative feedback loops involving the spindle will operate immediately. 1

Such is very likely to be the case with respect to the spindle primary, since it is sensitive to very small movements (about 100 μm) and can be driven to high rates of firing by rapid movement. Thus, **the spindle primary can easily be "saturated"** (i.e., made to fire maximally), at which point it can no longer send additional information concerning movement. 2

This hypothesis would gain support if measurements were available concerning the percentage of the range of movement of a muscle that is "covered" by the spindle endings. Unfortunately, obtaining this information is experimentally difficult and has not been accomplished yet [33]. 3

This hypothesis would predict that alpha and gamma fibers have very similar firing patterns, as long as the muscle is actually shortening. 4

In some experiments, similar patterns of firing of alpha and gamma fibers have been clearly observed (e.g., Figs. 13-49 and 13-50). 5

Such observations have been used to promote the terms *alpha-gamma activation* and *alpha-gamma linkage*. Both mean operationally nothing more than the preceding description [15, p. 167; 32, p. 522]. Such similarity between the firing patterns of alpha and gamma fibers indicates that the mammalian system (in which the intrafusal and extrafusal muscle fibers have separate innervation) can operate in the manner of the more primitive systems in amphibians and reptiles, in which the intrafusal and extrafusal fibers are innervated by branches from the same alpha fiber [32, pp. 42–43]. Such simultaneous innervation also has been clearly shown in *some* mammalian experiments [32, p. 235], although it is not common. 6

(These mixed skeletofusimotor fibers have been labeled, on a functional basis, *beta* fibers, even though anatomically their size has been found to overlap with both alpha and gamma groups [32, p. 235]!) 7

With fibers such as these, the intrafusal fibers must receive the same pattern of input as the extrafusal fibers, which tends to support the hypothesis advanced. But . . . 8

None of the above suggests why there should be separate intrafusal and extrafusal innervation in the vast majority of spindles in mammals. Furthermore, **independent activity of alpha and gamma efferents has been clearly observed in a number of cases** [15, p. 170; 32, p. 521]. 9

Fig. 13-50. Activity of single spindle afferent and electromyogram during voluntary movement of finger in human. Electromyogram indicates extrafusal activation (alpha fibers). Firing of spindle during isometric contraction implies activation of gamma efferents to spindle. Note that electromyogram activity slightly precedes spindle firing.

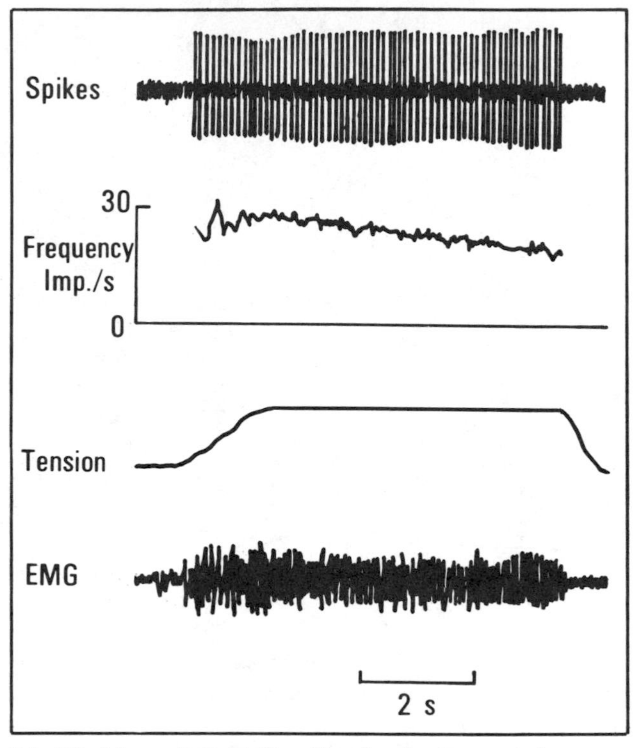

(Modified from A. B. Vallbo. Slowly adapting muscle receptors in man, *Acta Physiol. Scand.* 78:315, 1970.)

371

372

For example, decerebrate rigidity caused by cutting the brain stem between the superior and inferior colliculi seems to be brought about by increased gamma efferent activity (so-called gamma rigidity). Decerebrate rigidity brought about by *anemic decerebration* (caused by tying off the basilar and internal carotid arteries) seems to be predominantly due to direct excitation of alpha motor neurons (so-called alpha rigidity). 1

Question: How might one experimentally determine whether a given decerebrate rigidity was due to excessive alpha or excessive gamma activity? (Hint 32↓) 2

Although decerebrate rigidity has been much studied, the exact spinal cord mechanisms responsible are still a matter of dispute (for example, see Matthews [32, pp. 451–463]). 3

In other experiments, the gamma efferents can be made to fire when the alpha motor neurons are silent [32, p. 558]. Conversely, it seems likely that very fast (ballistic) movements may be carried out by direct activation of alpha motor neurons [15, p. 185] without any feedback control! Even in slower movements, recordings in humans have shown that alpha motor neurons can fire somewhat before stretch receptors (Fig. 13-50) (and thus are independent). 4

The whole motor control system is so complicated that it is not yet possible to document the static and dynamic aspects of posture and movement that utilize the differences between slow and fast muscle fibers, between the effects of spindle primaries and secondaries, or between static and dynamic gamma fibers. Thus, generalizations at this point are premature. 5

Independent activation of gamma efferents supported a hypothesis that is now generally rejected [15, p. 271; 32, p. 572]: the "follow-up servo." This is described briefly here because the idea is easily reinvented by those interested in servo-control (and even those who have diligently read Chap. 12!). The idea is that the muscle-length *set-point* input in the servo-loop of Fig. 13-48 is the gamma efferent firing rate, *rather* than an input in the spinal cord (the input in the figure represented as a question mark). This is a reasonable conjecture, for if the transformation between the Ia firing and the alpha motor neuron were simple (say, one-to-one), the small circle (comparator) could be removed and the muscle spindle "box" put in its place. Then the "error signal" would be the Ia firing, transformed to alpha activity. According to this hypothesis, alpha activity is controlled not directly by higher centers, but only indirectly via the loop. Thus the sequence to cause a movement would be as follows: gamma efferents fire, establishing a new set-point length; then the spindle afferents fire an error signal, and the alpha motor neurons and muscle act to reduce the error. 6

372

373

The arguments against this hypothesis are dealt with in detail by Matthews [32, p. 551]. However, some of the problems are not hard to understand. (1) For accurate control, the loop gain would have to be high—which it is not. (2) Different loads can affect the loop characteristic markedly (e.g., producing large changes in loop gain), which would negate CNS control based solely on the gamma system. (3) Gamma efferent firing changes the relationship between length and the spindle afferent discharge (Fig. 13-27)—the CNS would have to allow for this variation. (4) Only the spindle secondary could be involved, since the nuclear bag fiber of the spindle primary does not maintain its length with a fixed γ-D firing rate (Fig. 13-32). But the actions of the group II axons are complicated and as yet uncertain. (5) Under some circumstances, the alpha fibers fire before the gamma fibers (Fig. 13-50). Conclusion: it seems likely that the loop can act to aid in muscle length control, but most likely the gamma efferents are not the *sole* means of CNS control, as postulated by the "follow-up servo" hypothesis.

With facts rare and speculation abundant, we suggest a combination of the simplest hypothesis presented earlier and some of the ideas of the follow-up servo. Let's hypothesize that the gamma efferent input is programmed to cause the intrafusal fibers to follow the *predicted* change in length of the extrafusal fibers. Then any change in load (which had not been predicted in the alpha motor neuron firing program) would result in an action to restore the programmed rate of movement.

For example, suppose that the load is heavier than expected. Then the intrafusal fibers would shorten more rapidly, relative to the extrafusal fibers, than had been expected. This would lead to a sudden increase in the firing rate of spindle afferents (just as is seen in the voluntary isometric contraction in Fig. 13-50), which would, in turn, increase alpha motor neuron firing. Notice that a primitive version of this system might well send the same input to both extrafusal and intrafusal fibers, although more sophisticated functions would require a separate input for the intrafusal fibers.

Also, if this *is* the way things work, then Ia output during movement would provide a continuous commentary on the phase shift between intended and actual movement and hence would indicate to higher centers the extra energy cost of the movement at any given time.

AGONIST-ANTAGONIST INTERACTION AND RECIPROCAL INHIBITION

In terms of advancing a concise, clear, simple, and understandable description of motor control, the preceding development of ideas is disappointing, at best.

373

In spite of the large amount known about elements of the nervous system that are probably involved in motor control of length, it does not seem possible, at this point, to describe in any satisfying way how the elements interact, even in a simple movement.

In an attempt to delineate the reasons for our lack of understanding, here we add another complication—if only to glance over the abyss into the steaming caldron of the complexities of reality for but a brief moment before returning to the comfort and complacency of a simpler view.

What the preceding description lacked is any mention of the fact that **the most usual arrangement of muscle is a pairing of agonist and antagonist, with interaction** occurring in **both** the **mechanical** (anatomic) connections and the **neural** connections in the spinal cord.

There is a subtle irony in the possibility that physiologists, in trying to simplify their objects of study to manageable proportions by working with single muscles, may have prevented adequate analysis of the very mechanisms they seek to understand!

The neural interconnections between agonist and antagonist (previously shown in Fig. 13-37) give rise to **reciprocal inhibition,** i.e., the relaxation (inhibition) of the antagonist when the agonist contracts reflexly. At least one neural pathway is known that can account for reciprocal inhibition: Ia fibers from an agonist excite an interneuron, which in turn inhibits the alpha motor neuron of the antagonist. The inhibitory interneuron is also excited by pathways from higher centers (e.g., the pyramidal, rubrospinal, and vestibulospinal tracts), which suggests that this interaction can be extensively modified by higher centers.

Some of the mechanical and neural interactions of the agonist and antagonist are diagrammed in Fig. 13-51. But before you give up in complete disgust and frustration, please realize that at this point Fig. 13-51 is meant only to illustrate the following points. (1) The interrelationships are symmetrical: as the agonist acts on the antagonist, so does the antagonist on the agonist. (2) As more interactions are diagrammed (compared, say, with Fig. 13-48), more and more loops become apparent. For example, the agonist Ia fiber is involved in not only an *autogenic* loop (Ia_1, α_1, M_1, T_1, TD, $\dot{A}$, $\dot{L}_1$, Ia_1), but also loops through the antagonist (Ia_1, N_1, α_2, M_2, T_2, TD, $\dot{A}$, $\dot{L}_1$, Ia_1, as well as Ia_1, N_1, α_2, M_2, T_2, TD, $\dot{A}$, $\dot{L}_2$, Ia_2, N_2, α_1, M_1, T_1, TD, A, L_1, Ia_1). (3) If you care to take the trouble, Fig. 13-51 illustrates that **the connections of reciprocal inhibition also form negative feedback loops opposing changes in muscle length** (expressed in this system as changes in angle of the joint). (4) The complexity diagrammed is already so great that intuitive solutions are impossible. Only detailed quantitative analysis can assess the relative importance of each of these interacting loops. (5) This diagram, as complicated as it appears, has left out many factors known to be involved in muscle control, such as group II fibers, static control loops, other inputs to the neurons shown, etc.

Fig. 13-51. Some interactions during dynamic interaction of antagonistic muscles at a joint. Note that much has been left out and that shape of transfer functions shown is hypothetical. Symbols: M = muscle, T = tension, TD = tension difference, $\dot{A}$ = angular velocity, $\dot{L}$ = length velocity, F = firing rate, α = alpha motor neuron, N = inhibitory interneuron, + = excitatory synapse, − = inhibitory synapse.

375

For those masochists who are not satisfied until they have grasped the many enigmas of Fig. 13-51, the following description is offered. The figure diagrams dynamic changes so as to contrast with Fig. 13-48, but contains many simplifications solely for the purposes of clarity. For example, the complex relationship of alpha firing to muscle tension during dynamic changes is beyond the scope of this book and is left unspecified. The muscle tensions developed by the agonist and antagonist are mechanically added (algebraically) by the anatomic relationships around the joint. This gives rise to a tension difference TD, which determines the angular velocity $\dot{A}$ of the joint (assuming, for simplicity, that there is only a viscous, frictional load). The change in angular velocity in turn affects the velocity of length change $\dot{L}$ in the two muscles in opposite sign because of the anatomic mechanics (diagrammed at the top of the figure). The spindle primary dynamic response to length change occurs *only* when the muscle is being lengthened; this behavior can be readily seen in Fig. 13-22.

The negative feedback loops tending to maintain the position of the joint in spite of a disturbance can be seen by considering the following: Assume that a sudden increase in load imparts a positive $\dot{A}$ to the joint (moving in the direction of the agonist). In turn, this causes the agonist to shorten ($-L_1$), dropping the Ia_1 firing to zero and thus reducing the excitation of α_1 while increasing that of α_2 (by reduction of the inhibition of N_1). Both changes in motor neurons will tend to oppose the change: the antagonist will contract, and the agonist will reduce its activity (not adding force in the direction of the movement caused by the disturbance).

Now, it is of interest that *the loop described* provides only for a shutting off of excitation to α_1, but *cannot provide an inhibition to α_2 that is proportional to the disturbance to $\dot{A}$*. On the face of it, this would be undesirable from a control system viewpoint. However, note that *another loop does provide an inhibition to α_1 that is proportional to $\dot{A}$*: The antagonist spindle is activated by the disturbance, excites N_2, which in turn inhibits α_1, an inhibition that will be greater if $\dot{A}$ is greater. This provides an indication of the importance of reciprocal inhibition: **Only through this path can information on the magnitude of the velocity of change be transmitted to the agonist if the externally applied motion is such as to shorten the agonist and lengthen the antagonist.** Such information would be important if the agonist's motor neuron were receiving considerable excitation from other inputs (not shown) at the moment that the disturbance occurred.

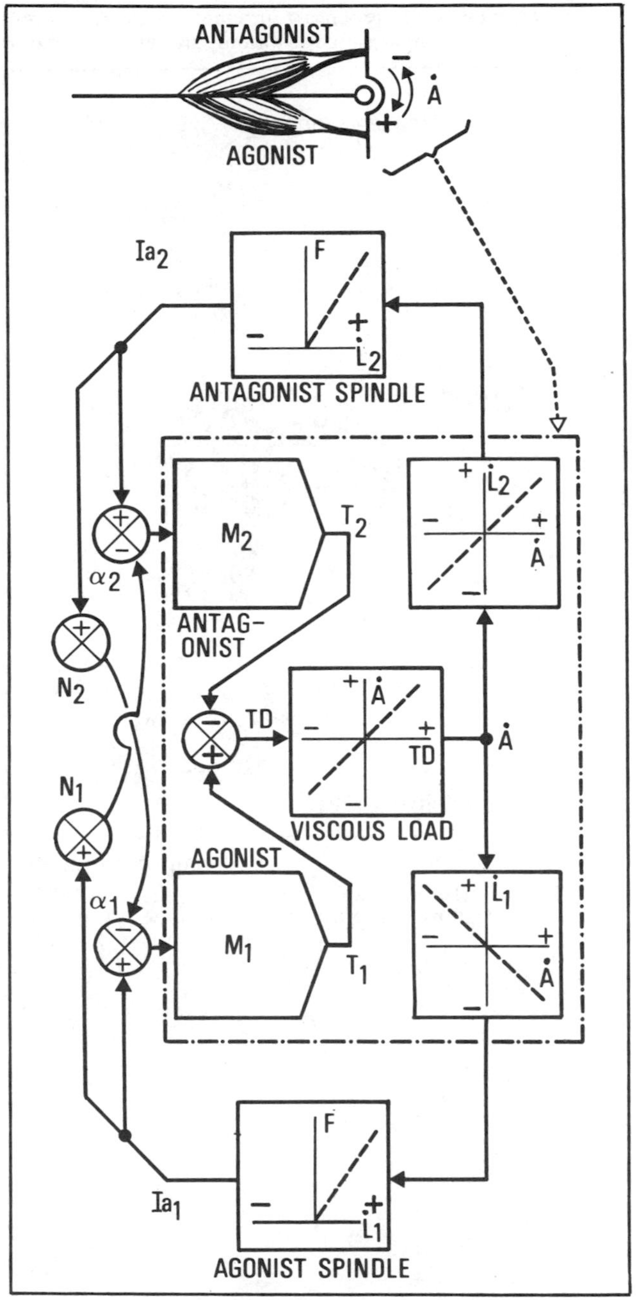

HINT

32. The classic method is to cut the dorsal roots, which leads to relaxation in the case of gamma rigidity, but has less effect on alpha rigidity.

375

376

Question: Now that you have studied Fig. 13-51, what was left out of the diagram? (Hint 33↓)
1

Question: Can you find some other loops involving Ia_1 besides those already delineated? (Hint 34↓)
2

If you are concerned that some of the loops specified contain linking elements (say, *A*) and thus are not independent, do not forget there can be considerable delays in various parts of the loops, so that a long path may not act back on the common element at the same time as a short loop, further adding to the complications!
3

4 Let us now go to something much easier to understand.

DETECTION OF MUSCLE TENSION—THE GOLGI TENDON ORGAN

Up to now, we were concerned mainly with detection of muscle length. The Golgi tendon organ and its afferent Ib fiber was mentioned only in passing (e.g., Figs. 13-20, 13-21, and 13-22; Tables 13-2 and 13-3; and Fig. 13-24).
5

The importance of the measurement of muscle tension should be apparent from a brief recollection of the muscle length-tension relationship and how it is affected by different loads (e.g., see Fig. 13-11), as well as from the fact that tendon organs are about as numerous as muscle spindles (Table 13-2). An enlargement of a Golgi tendon organ is shown in Fig. 13-52.
6

The following should be noted in Fig. 13-52. **The tendon organ is** placed at the musculotendinous boundary, **in series mechanically with the muscle.** A given organ can be affected by the tension of more than one muscle fiber (and hence the tendon organ can be activated by up to about 10 motor units [32, p. 136]). This is another example of convergence in the peripheral sensory system.
7

At times, tendon organs have been found in series with muscle spindles [15, p. 58], but the significance of this observation is unclear. (It does, however, give great possibilities for more loops for those servo-control theorists still struggling on the problems of the preceding section!)
8

The responses of the Golgi tendon organ to passive stretch (Fig. 13-21) and to active contraction of the muscle (Fig. 13-22) are just as you would expect for a transducer sensitive to increases in tension. Also the **tendon organ shows some sensitivity to the rate of tension change,** i.e., a dynamic as well as static responsiveness (Fig. 13-53) [15, p. 112].
9

The great sensitivity of the tendon organ to active tension should be noted. **The contraction of a single motor unit is sufficient to produce strong activation of a tendon organ** (Fig. 13-53). However, the tendon organ is *much less sensitive to passive tension*, because part
10

Fig. 13-52. Stained Golgi tendon organ teased from musculotendinous junction of the Achilles tendon in human, magnified ×147.

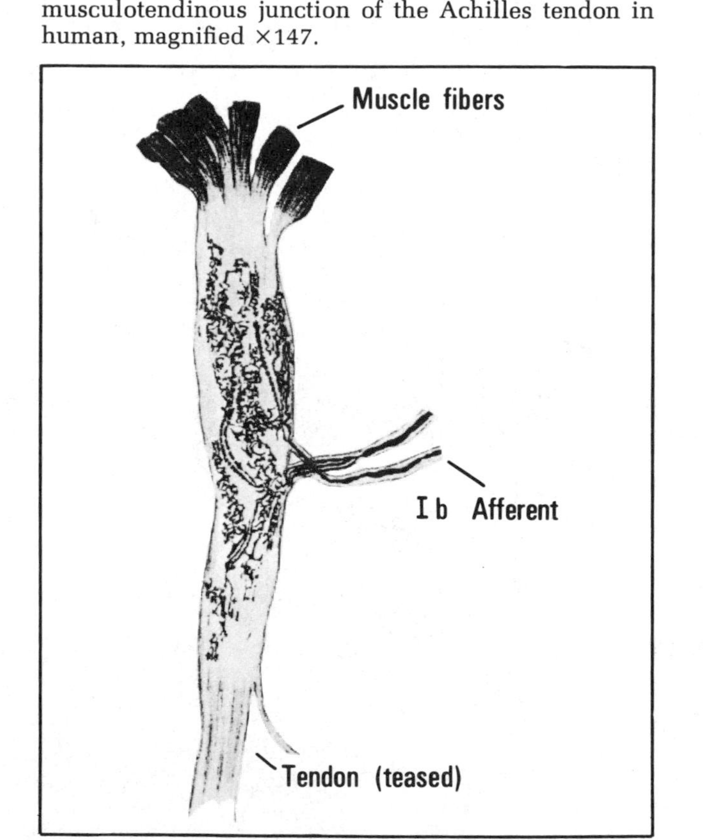

(From G. V. Ciaccio, Intorno alle piastre nervose finali né tendini de' vertebrati, *Mem. R. Accad. Sci., Bologna*, Ser. 4, 10: 301, 1890.)

376

377

of the tension is borne by the muscle connective tissue sheath rather than the muscle fibers (and tendon organs).

Some textbooks make the mistake of asserting that the tendon organs have a high threshold for tension and hence are not activated except by extreme force, which might damage the muscle. This view was based on the earliest studies, which found *high thresholds to passive stretch*. Since the tendon organ senses only the tension in the muscle fiber, it is easy to see how erroneous conclusions were reached.

Selective sensitivity of tendon organs to active tension (with relative insensitivity to passive tension) must have considerable consequences on the agonist-antagonist interactions (though this has not been studied yet). Consider that contractions of the agonist would stimulate agonist tendon receptors, while those in the antagonist would not be stimulated, even though both muscles had the same external tension.

The afferent discharge of tendon organs, like that of muscle spindles, apparently does not reach consciousness. You can obtain some indication of this by standing with your heels off the floor and then raising your left foot. By palpation you can find that your right gastrocnemius muscle has become much tenser (after all, it is now supporting *twice* the weight that it was when both feet were on the floor) *even though you are not conscious of the change*. A word of caution: In performing this experiment, you may find that you have to adopt some unusual postures which, if performed in a public place (such as a library), may cause others to question your judgment, if not your sanity.

SERVO-CONTROL OF MUSCLE TENSION

Ib fibers have many connections, but most interesting at this point is that which inhibits alpha motor neurons via an interneuron. Such a connection forms **a negative feedback loop,** which **tends to keep tension constant** (Fig. 13-54).

One function of tension feedback control would be in compensating for muscle fatigue, where a given motor neuron input to the muscle gave less tension output than previously.

Another possibility also has been suggested: The constant-tension loop may act to linearize the constant-length loop in the stretch reflex in the face of the length-tension variations in muscle strength with length [32, p. 441].

The tension and length-control loops are directly opposed to one another, even though they both are negative feedback loops, **because of the inverse nature of length and tension in muscle against most loads.** Thus, if a load change increases both length and tension, the tension control loop (Fig. 13-54) will *inhibit* the alpha motor neuron in order to keep the tension constant, whereas the length control loop (Fig. 13-48) will *excite* the alpha motor

Fig. 13-53. Firing pattern of single tendon organ upon stimulating single alpha motor neuron in split ventral root. Tension was about 18 g and lasted about 1 s. Note that velocity sensitivity is shown both at rise of tension and by rapid cessation of firing as tension starts to fall.

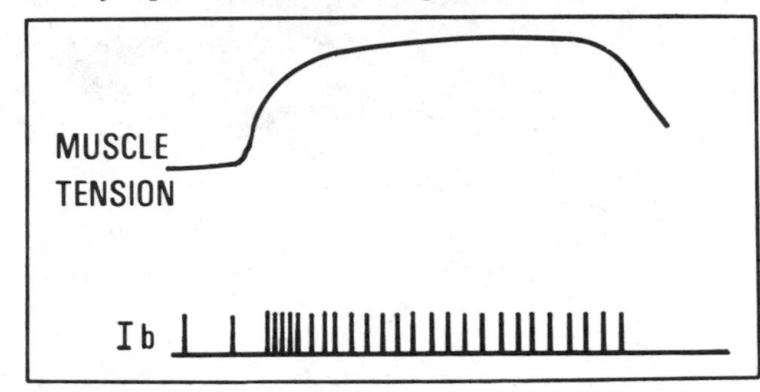

(Modified from J. Houk and E. Henneman, Responses of Golgi tendon organs to active contractions of the soleus muscle of the cat, *J. Neurophysiol.* 30:466, 1967.)

Fig. 13-54. Tension-control negative feedback loop of Golgi tendon organ (static response). Symbols: F = firing frequency, T = muscle tension, L = length, + = excitatory synapse, − = inhibitory synapse. Shapes of graphed functions are hypothetical. Many other connections have been left out. Note that load is not part of this feedback loop.

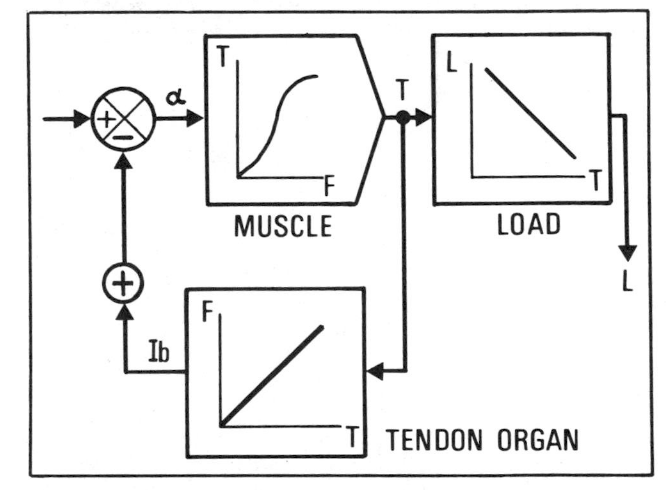

377

378

neuron to maintain length. Obviously, **it is not possible to simultaneously maintain both length and tension constant.** Thus, **in any accurate control of tension or length, one of the loops must be turned off when the other is turned on.** A possible neural mechanism for such an action is not hard to find.

PRESYNAPTIC INHIBITION AS A NEURAL SWITCH

The mechanisms of presynaptic inhibition are described in detail in "Inhibitory Systems, in Particular" in Chap. 11. Here (at last!) we have a place where presynaptic inhibition is not only present, but needed! Presynaptic inhibition is one method by which the monosynaptic length-control loop can be made inactive without also preventing the participation of the alpha motor neuron in other actions.

Consider the monosynaptic length-control loop. Basically it consists of only two elements, a sensory cell and a motor cell. If the loop were made inactive by postsynaptic inhibition of the alpha motor neuron, that inhibition would interfere with muscle control by other systems. But, **presynaptic inhibition can reduce or turn off the excitatory effects of the spindle afferents on the alpha motor neuron without changing the excitability of the motor neuron.**

QUESTION: What is another method by which the monosynaptic length-control loop can be made inactive without directly changing alpha motor neuron excitability? (Hint 35↓)

Among their many connections, **Ib fibers** are particularly **noted for their relatively large presynaptic inhibitory effects,** on both Ia and Ib fibers, as shown in Fig. 13-55. The significance of all these connections is, of course, unknown.

Presynaptic inhibition also can function to change the gain of other loops, since it affects the amount of transmitter released from a given presynaptic volley.

Especially unclear is the function of presynaptic inhibition of Ib fibers on Ib fibers, as diagrammed in Fig. 13-55. Possibly this is another form of lateral inhibition with respect to muscle tension, which conceivably might "distribute" the excitability throughout a muscle as the most excitable motor unit begins to fatigue, thus reducing the developed tension and hence the inhibition of that motor unit's tendon organs upon other motor neurons.

Thus we conclude that a mechanism is available by which the length control loop can be made inactive while the tension control loop is active. This mechanism involves presynaptic inhibition (by Golgi tendon organ Ib fibers) of the terminals of spindle Ia afferents.

Fig. 13-55. Distribution of ipsilateral presynaptic inhibitory effects arising from and terminating in muscle afferents. Width of arrow shows approximate strength of effects.

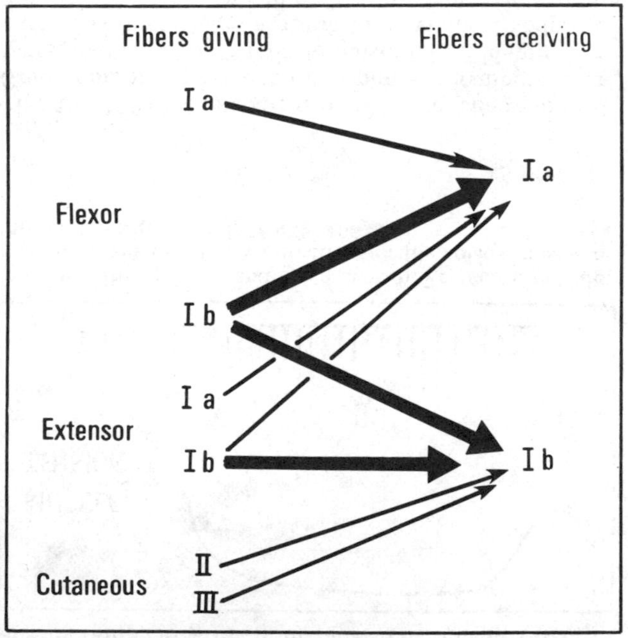

Fibers giving | Fibers receiving
Flexor: Ia, Ib → Ia
Extensor: Ia, Ib → Ib
Cutaneous: II, III → Ib

(From J. C. Eccles, R. F. Schmidt and W. D. Willis, Depolarization of central terminals of group Ib afferent fibers of muscle, *J. Neurophysiol.* 26:1, 1963.)

378

379

STATIC AND DYNAMIC ASPECTS OF JOINT RECEPTORS

1 Another receptor that undoubtedly plays an important role in posture and movement is the one that senses joint angle.

2 Ruffini endings are found in the joint capsule, whereas Golgi endings (like those in tendons) are found in ligaments. Free nerve endings are also found in the joint structures.

3 **The responses of receptors sensitive to joint angle have both a static and a dynamic aspect.** (So what else is new?)

4 The static aspect, signaling angle accurately for many hours (i.e., the receptors adapt slowly) is shown in Fig. 13-56. In that figure, **each receptor signals,** by its firing rate, **only a portion of the entire range of movement of the joint. Different receptors cover different angles.** So convergence of these sensory endings is necessary for the firing rate of higher-order cells to be proportional to joint angle over a larger range, as has been found in the thalamus [38, pp. 1407–1409].

5 Even though these cells are slowly adapting, **joint receptors also show sensitivity to the velocity of angular movement** (Fig. 13-57).

6 Thus, we again realize that the **CNS is supplied with information from the joint receptors on both static position and the dynamics of movement.**

Fig. 13-56. Firing patterns of single joint receptors (presumably Ruffini endings) to slow changes in knee angle. Note that each receptor has "preferred angle" over which it responds, and at other angles it is silent. This angle can be larger if there is force on the joint.

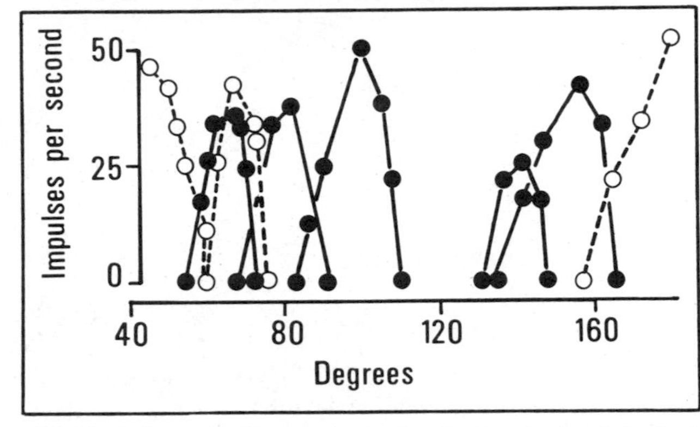

(Modified from S. Skoglund, Anatomical and physiological studies of knee joint innervation in the cat, *Acta Physiol. Scand.* 36 [Suppl. 124]:1, 1956.)

HINTS

33. A partial list includes (1) other inputs to the neurons, (2) the static aspects of Ia activity, (3) spindle secondaries and their polysynaptic connections, (4) gamma efferents, and (5) the changes in $TD \rightarrow \dot{A}$ resulting from load changes. In addition, you probably guessed some of these: (6) the feedback effect of muscle length in changing the responsiveness of muscle (tension) to alpha input; (7) the nonlinearities of some anatomic connections that do not algebraically add T_1 and T_2; (8) the inertial (mass) aspects of the load; (9) the capacitance (spring) aspects of the load, etc. What else?

34. Here is another: Ia_1, α_1, M_1, T_1, TD, $\dot{A}$, $\dot{L}_2$, Ia_2, α_2, M_2, T_2, TD, $\dot{A}$, $\dot{L}_1$, Ia_1. Still others can be found. The point is to completely convince you that quantitative analysis may be the only way of understanding all these interactions and that simple servo-control generalizations will hardly explain all the behavior of the real system.

379

380

The reflex connections and effects of joint afferents are unclear at present and are best left at that. 1

The predominant method of study has been to stimulate nerves innervating the joint capsule, but the effects have been quite variable. Sometimes the stimulation effects were so widespread as to seem to cause changes in overall spinal cord excitability. 2

QUESTION: If a nerve from a joint capsule is electrically stimulated at a high frequency with supramaximal shocks, state in words the "message" that is communicated to the CNS. (Hint 36↓) 3

In retrospect, it is hard to imagine what an experiment in which many joint afferent fibers are stimulated simultaneously can reveal: pathways and connections, perhaps, but certainly not normal patterns of activity. It is a credit more to the nervous system than to the experimenter that electrical stimulation has revealed as much as it has in the laboratory. As P. B. C. Matthews [32, p. 333] has said, "To some extent the spinal centres appear to be duly organised so that they can only emit 'sensible answers' even though the mixture of afferent inputs elicited by electrical stimulation presents them with a 'stupid question,' for the central integrative processes which lead to the pre-potence of some reflexes over others usually manage to ensure that only one kind of response is emitted at any one time." 4

One may presume that joint reflexes, if they exist, may be somewhat complex, since **the appropriate response at full flexion must be completely different from that at full extension!** Furthermore, the responses of a muscle that crosses two joints must be complicated with regard to the independent influences from the two joints. 5

Proprioceptive input from joints and cutaneous afferents is certainly important in positional control, not only because these fibers reach consciousness (as is tested in the usual neurological examination), but also because their absence (during a selective anesthetic block) can have a significant effect on control of movement. 6

Local anesthetic "ring" block around the base of the thumb can block joint afferents while not affecting extrinsic muscle afferents (the muscles being located in the forearm). **Such a block reduces or abolishes reflex increases in muscle activity** as a result of unexpected changes in load. This suggests that the joint and cutaneous afferents either directly participate in the control of movement or at least provide a *background excitability without which a stretch reflex cannot be elicited* [32, pp. 589–590]. 7

Fig. 13-57. Velocity sensitivity of single Ruffini joint ending, in terms of firing rate to 5° movement at different rates of movement. Note both the static and dynamic aspects of the response.

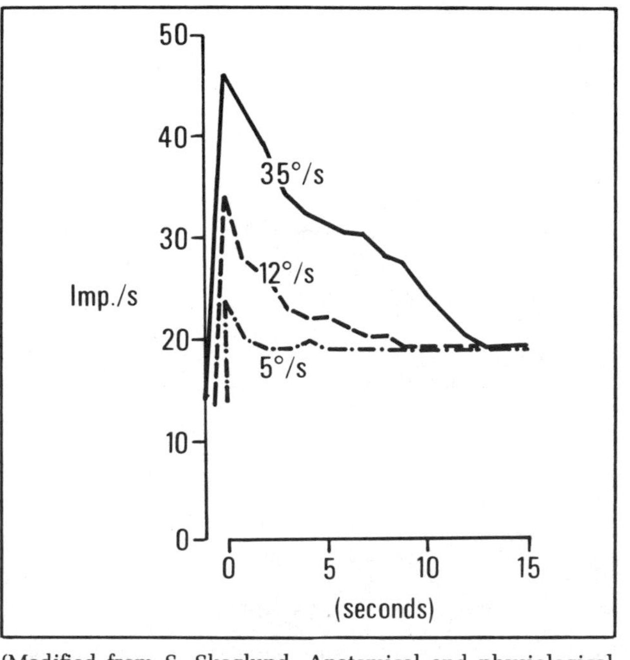

(Modified from S. Skoglund, Anatomical and physiological studies of knee joint innervation in the cat, *Acta Physiol. Scand.* 36 [Suppl. 124]:1, 1956.)

380

381

Table 13-5. Static and Dynamic Aspects of Motor Control

Abstract Term	Body Functions	Muscle Fiber Types	Length Receptors	Control of Muscle Spindle	Tension Receptors	Joint Position	Load
Static	Posture	Slow	II and Ia (static response)	γ-S	Ib (static response)	Receptor (static response)	Weight (steady)
Dynamic	Movement	Fast	Ia (dynamic response)	γ-D	Ib (dynamic response)	Receptor (dynamic response)	Δ weight; friction; inertia

Note that experiments on hand muscles (generally fast, dynamic muscles) may not provide generalizations applicable to tonic (slow) postural muscles. 1

Thus, it is certainly possible, though hardly proved, that control of limb movement may be related as much to joint afferents as to any muscle receptors! 2

The interaction of joint and muscle receptors is little understood, yet may be significant in conscious humans. For example, Ia afferents do not reach consciousness; yet input from muscle spindles can distort subjective judgment of joint angles by as much as 40° [32, p. 497]. 3

STATICS AND DYNAMICS REVISITED

At this point, we might do well to review the various static and dynamic aspects of motor control, to see how the distinction between posture and movement runs through most of the mechanisms described so far. Table 13-5 summarizes the distinctions presented. 4

You should *not* think that there are two separate systems. The separation between static and dynamic is made to emphasize the **different requirements of posture and movement** and to indicate that the **elements necessary for their control seem to be present** (even though the details of the interactions are, at times, very hazy). You certainly need to keep this sort of distinction in mind as we deal with more and more complex responses in the next sections. 5

HINT

35. By reducing the gamma efferent firing, so as to reduce or eliminate the spindle firing rate.

381

382

MORE COMPLEX REFLEX RESPONSES

With your background knowledge of various sensory aspects of motor control (just completed!), you are in a position to consider more complex reflex responses: those which probably involve more than one muscle and, finally, more than one limb. 1

The **clasp-knife response** can be observed in those experimental animals in whom the lengthening reaction (previously described) is seen. The clasp-knife response also is commonly observed in spastic human patients when the physician flexes and extends the patient's elbow. 2

The response, which is named by analogy to the way a pocketknife blade behaves when it is being opened and closed, **is observed as a sudden, rather complete loss of muscular** (stretch reflex) **resistance to a forced lengthening of the muscle.** At first, it was thought that this response manifested the inhibitory actions of Ib fibers on alpha motor neurons. But the all-or-nothing aspect of the loss of muscular activity would not be expected if the tension and length control loops were simultaneously active. Thus, we can only hypothesize that the length control servo-loop may be suddenly inactivated, either by the Ib presynaptic inhibitory effects on Ia pathways (previously mentioned) or by inhibition of fusimotor activity (as has been observed [32, pp. 447–448]). By either mechanism a sufficient Ib barrage might "open" the length control loop, which initially opposes the forcible extension of the muscle. But the basic underlying mechanism would be a switching from one control loop to another, rather than assuming (as did the older view) that the two opposing control loops (tension and length) "battle it out." 3

The **withdrawal reflex** can be observed in decerebrate, lightly anesthetized, and conscious animals and humans. When strong electric shocks are delivered to the skin (or to nerves to the skin) or to joint or muscle nerves (group III), there ensues a *coordinated withdrawal of the limb,* brought about by the combined actions of several muscles—the withdrawal reflex. 4

This response is remembered most easily if it is considered to be a response to "pain" (pinch, ischemia, hypertonicity, etc.), although it is not certain whether all the fibers that can elicit this response are nociceptive. 5

In passing, note that *no mention was made of the reflex effects of group III* (finely myelinated) *and group IV* (unmyelinated) fibers. The effects are rather nonspecific, are not easily classified as obvious reflexes, and, furthermore, are modified rather easily by the condition of the animal (depth of anesthesia, other inputs, length of experiment, etc.). However, since unmyelinated nerves outnumber myelinated nerves 2:1, it is well to remember that these fibers are present, and undoubtedly have a function, even though that function has not been adequately revealed by physiological experimentation! The responses to stimulation of small fibers (III and IV) are rather widespread in the nervous system, as shown by the multiple muscles involved in the withdrawal reflex, the frequent involvement of other limbs (as in the crossed 6

382

383

1 extensor reflex, described later), and prolonged effects in other spinal reflexes (as shown in Fig. 13-39, where the tendon jerk is inhibited by cutaneous stimulation). Fusimotor neuron excitability is readily affected by afferents from cutaneous stimuli and nociceptive tendon receptors [32, p. 451].

2 Often **crossed extensor reflexes** are observed at the same time as the withdrawal reflex. That is, at the time that there is flexion of the stimulated limb, **the contralateral limb will show extension.**

3 This reflex can be remembered on a teleological basis: it keeps an animal that is raising a limb off the ground from falling! The crossed extensor reflex is seen also in rotational tonic neck reflexes (observed in some disease states and in decerebrate animals with labyrinthine destruction). For example, when the neck is forcibly turned, there may be flexion of one limb and extension of the other.

4 Incidentally, the neck muscles are heavily innervated and are involved in a number of strong reflexes. This is not surprising since the orientation of the head is of fundamental importance in analysis of input from the major special senses located in the head: the visual, auditory, and vestibular systems. In addition, in four-legged animals, lateral and vertical movements of the head may cause extensive redistribution of load among the postural muscles (e.g., as in the postures of the male elk).

PUTTING THE PARTS TOGETHER—STEPPING EXPERIMENTS

5 Until now, we gave various descriptions of the elements (either anatomic or physiological) that *might* go into the complexities of simple, coordinated motor acts, such as walking.

HINTS

36. Statics: "The joint is in all positions at once." Dynamics: "The joint is moving rapidly in all directions at once."
37. See Table 13-6, numbers 3, 6, 7, 8.
38. A partial list includes (1) the extent (magnitude) of the reflex action of Ib fibers and of spindle afferents; (2) the action of spindle primaries as distinct from spindle secondaries; (3) any differences in the activity of γ-S and γ-D fibers; (4) the effects of such standard reflexes as reciprocal inhibition and the stretch reflex; (5) what initiates alpha firing, since spindles do not; (6) what terminates alpha firing; (7) the effects of load variations on the observations; (8) the effects of different patterns of locomotion (e.g., galloping) on the observations; (9) activity of joint receptors and the magnitudes of any reflex effects; etc. You probably thought of a couple of others!

383

384

Table 13-6. Experimental Observations during Stepping Movements of Cat[a]

Observation	Conclusion or Comment
1. Walking movements of forelimbs occur only if mesencephalic stimulation and limb motion occur (by hand or treadmill).	Higher-center activity alone is insufficient for cyclic stepping. Proprioceptive inputs alone are insufficient for cyclic stepping.
2. Rate of stepping is dependent on mesencephalic stimulus strength and treadmill speed.	Proprioceptive inputs are important in overall action, but interact with impulses from higher centers.
3. Deafferentated hindlimbs make walking "movements" if forelimbs are moving (intact afferents).	Cyclic alpha activity can be independent of gamma or spindle input (in hindlimb). (These movements are not identical with those when afferents are intact—afferents do have an effect.)
4. Tendon organs are best activated by contractions of own muscles, rather than by contractions of antagonists.	Tendon organs have higher threshold to passive (external) tension than to active (intramuscular) tension.
5. Spindle afferents discharge when their own muscle is contracting.	Fusimotor fibers are active during contraction (e.g., see Fig. 13-49).
6. Alpha fibers fire somewhat before spindle afferents.	Start of alpha firing is not dependent on stretch reflex from spindle (e.g., see Fig. 13-50).
7. Electromyographic activity in extensors occurs 10 ms before foot touches ground.	Start of alpha activity is not dependent on stretch reflex from loading.[b]
8. Spindle activity occurs even when alpha discharge in muscle has ended, as muscle is lengthened by activity of antagonist.	Spindle–alpha motor neuron reflex is inhibited by presynaptic inhibition for monosynaptic connections and by presynaptic or postsynaptic inhibition for polysynaptic ones. End of alpha firing may be due to Renshaw and/or Ib inhibitions.

[a] Except for 7, adapted from [54] and [55].
[b] Adapted from [13].

1 The real question is, can we begin to understand the interactions of these elements in walking?

2 A complete stepping cycle of a limb can be divided into two parts: the *swing phase,* during which the limb is off the ground and moving forward, and the *stance phase,* during which the limb is in contact with the ground and bearing weight.

3 **Each phase contains both flexions and extensions, some active and others passive.** The swing phase begins with hip and knee flexion and continues with a passive swing-through that results from inertia of the leg. In the later parts of the swing phase, the knee extends,

384

385

increasing the length of the stride. Hip extensors are active at the end of the swing phase to slow the moving leg (recapturing some of the inertial energy). In the stance phase, all extensors are active, even when the load of the body gives rise to a temporary flexion (yielding) of knee and ankle in the middle of the phase (while hip extension is maintained throughout the phase).

Yielding is more prominent in the gallop than in walking—the dynamics of dynamics!

Measurement of spindle and tendon afferents and of electromyographic activity in stepping movements has been accomplished in high decerebrate animals by stimulating the mesencephalon electrically to initiate and maintain treadmill walking [32, p. 569; 54, 55].

For technical reasons (to reach the dorsal and ventral roots), the body of the animal was supported by a metal frame; so the normal loads on the muscles could not be duplicated by the experiment. However, the technical achievements of this experiment are noteworthy.

The experimental observations and their significance are summarized in Table 13-6. (See also Granit [15, p. 259] and Matthews [32, p. 569].)

QUESTION: What is the evidence in this experiment (see Table 13-6) that alpha motor neurons can discharge without the excitation of spindle afferents? (Hint 37↑)

QUESTION: From Table 13-6, what aspects of the spinal cord control of motor activity (stepping) were not determined in these experiments? (Hint 38↑)

The observations in these experiments that alpha and gamma firing are not always identical are consistent with what was suggested by observations on the various forms of decerebrate rigidity. Other observations in awake animals also support this idea: (1) Local deafferentation leads to ataxic movements of that limb [32, p. 548] and (2) the action of a drug that seems to act peripherally to inhibit spindle discharges gives rise to ataxia in dog or cat [32, p. 549]. In both cases, fine control is lost, but gross movement is still possible.

While contemplating Table 13-6, bear in mind the description of the various possible forms of muscle action, as described at the beginning of this chapter. Many of these actions occur in stepping. Thus, **during the stance phase, there may be no time at which either length or tension is constant, even though information on these variables is important in control of the movement.**

385

386

Again, muscular activity of the hip flexors may not be continuous throughout the swing phase. In some cases, there is a burst of activity at the start of the phase (at the "resting length," at which maximum force is developed!), and then the leg swings through as a result of inertia, without added musclar force. So, the **requirements for some reflex or other may be varying from moment to moment.** And this somewhat alters the concept of reflex (if you had been thinking of it as the control of the nervous system by external events).

1

HIGHER CENTERS IN CONTROL OF MOVEMENT

Detailed discussion of this topic is beyond the scope of this book. But some points can be made that may help your future understanding of this difficult subject. Obviously, it is no surprise to hear that these higher centers are linked in a complex hierarchical system of interacting feed-back loops. These loops are even less well understood than the "simple" spinal control systems discussed in this chapter.

2

In computer jargon, the simpler reflexes of the spinal cord are the subroutines that are called by the routines of programs being controlled by higher centers in the nervous system. Where the routines and programs are generated and stored, called forth, or enacted—these questions are still glimmers in the eyes of young experimenters.

3

We can cut through these complexities with the following brave (foolhardy?) generalizations, which we hope you find useful.

4

1. **Postural and movement programs,** some innate (a foal stands and can gallop within hours of birth), others learned (as one might learn the complex coordinated movements of classical ballet), **are stored within the central nervous system.**
2. **These programs are called forth** in appropriate sequence **in response to sensory inputs** integrated in the reticular formation and in the thalamus, hypothalamus, limbic system, and cortex.
3. **Then these programs are modified and perfected** for conditions of load, orientation with respect to gravity, position of noninvolved body parts, etc., **by the cerebellum.** The cerebellum functions at least in part through spinal cord reflexes.
4. **Spinal cord reflexes** may be modulated to **provide continuous, rapid, local control** of the fine gradations of muscle function.

5

Finally, **what we described here is** not so much the pyramidal as the **extrapyramidal system of motor control.** What is the *pyramidal* system? Well, the primates (but not carni-vores such as the dog and cat—our main experimental subjects, note!) have developed a "bypass" motor control system that runs directly from the cortex to the spinal cord. In monkeys, apes, and humans, this system seems to be important in the *initiation* of volun-tary movement (although coordinated volitional movement can still occur after selective

6

386

387

lesions of the pyramidal tracts). This system also may be involved in the proven ability of conscious humans to learn to voluntarily control the firing of single motor units (when adequate electronic feedback is provided).

1

In this chapter we tried to show that **the elements of muscle control** (i.e., sensors and loops for length, tension, and joint position) **are known to exist at the spinal level. So they can be called into action or disconnected in tens of milliseconds,** as appropriate to the bodily activity of the moment. How the brain accomplishes this motor control is not known in detail at present, and for answers we need a much deeper understanding of the interactions of most parts of the brain. All the higher functions of the brain, including memory, learning, goal-seeking, etc., are ultimately directed toward motor system activity—on whose temporal and spatial sequence of muscle contractions the survival of the animal (and ultimately of the species) depends, in the macroscopic world of prey and predator.

2

COMMENT ON CLINICAL IMPLICATIONS

As we end this chapter, you may well ask why we did not describe and clarify the clinical implications of the physiological knowledge presented. In particular, why have the many diseases that involve derangements of the motor system, such as spasticity, palsy, ataxia, paresis, and paralysis, not been explicated? In reply, we can do no better than to quote (with slight modification) the excellent statement of P. B. C. Matthews [32, pp. 465–466]:

3

"The application of [physiological knowledge] to the elucidation of the derangements underlying human [disease] so far seems to have led to little more than a wide range of more or less controversial assertions. . . . Part of the confusion arises because the clinical literature grows too easily by feeding upon new physiological knowledge rather than by any new analysis of the diseased state itself. This is because improved understanding of the underlying normal physiological mechanisms readily lends itself to fresh speculation on the origin of well-known clinical syndromes, even when no new clinical observation has been made and the [significance of the] physiological work has not been fully digested.

"The real trouble arises quite simply from the paucity of methods currently available for studying human beings who naturally wish to suffer no permanent ill effects as the result of investigation; this equally limits the scope of the few methods which do exist. Most of the methods which are applicable to animals can no longer be employed with precision even if they can be used at all. The best hope is that suitable 'model syndromes' can be developed in animals each of which closely resembles some human malfunction. After a full analysis of a 'model' it might then be possible to decide on the basis of a few relatively simple confirmatory procedures whether the conclusions [drawn from] the model could be transferred more or less *in toto* to the human syndrome. . . . [Regrettably, at] present, there are . . . too few methods available to allow . . . a thorough correspondence to be established between a model syndrome in an animal and the real syndrome in man."

4

387

388

A FINAL WHIMSY

For purely nostalgic and whimsical reasons, you might like to gaze for one last time at Fig. 13-58/2-12 to realize how much your hard-earned knowledge has changed your perception of a simple diagram.
1

How much more meaning do you see in Fig. 13-58, compared with when you first saw it as Fig. 2-12? A "simple" process is no longer simple, but is rich in detail and implications—by such changed perceptions you can measure how far you have come. For your knowledge you had to work hard. Now, in addition, you must pay another price: Fig. 2-12 can never again seem simple to you—it is only *simplified,* and overly simplified at that!
2

Welcome to the club!
3

Fig. 13-58/2-12. A final look at what first seemed simple, in order to demonstrate the effects of sampling from the tree of knowledge upon perception.

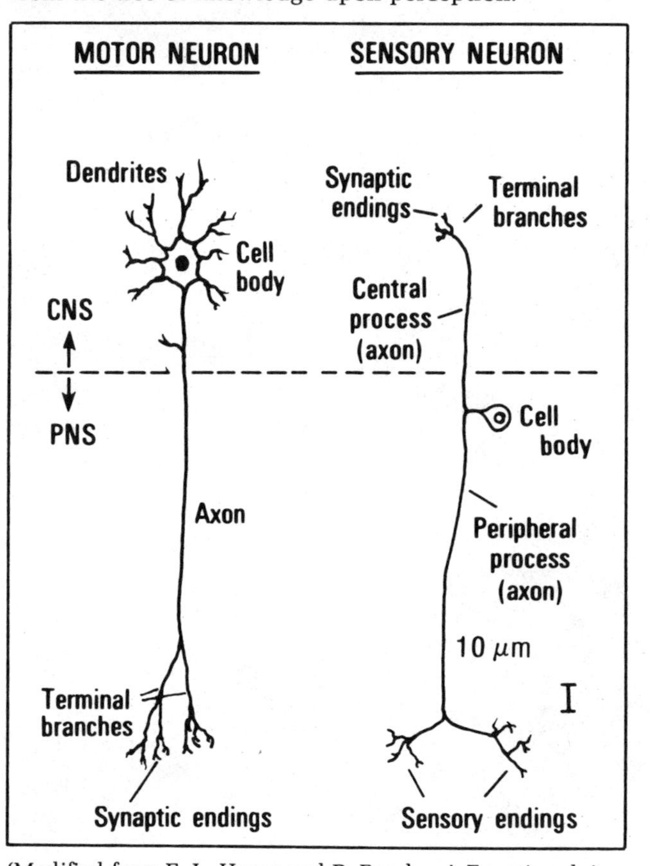

(Modified from E. L. House and B. Pansky, *A Functional Approach to Neuroanatomy* [2d Ed.]. New York: McGraw-Hill, 1967.)

388

389

14 Epilogue

390

This book, like all books, must end somewhere. We chose to do so at this point in the description of neurophysiology because, as you can guess from Chap. 13, *from here on, as one deals with "higher centers," the details become more numerous and the principles more difficult to extract.*

1

This is to be expected. *An area of scientific study* does not progress evenly—*it becomes first complex as discoveries are made and then simpler as principles are perceived* (more accurately: invented). Then the cycle repeats, at a new, higher level. At present, there is sufficient knowledge of the workings of the lower portions of the nervous system to describe principles. But it will be some time before the mass of detail known about the higher centers can be so described.

2

Furthermore, on the basis of numbers of cells alone, the higher centers are more complex and the interactions are of a different order from those described so far. The technical difficulties in dealing with a nervous system containing literally billions of cells (as in the mammalian CNS) and the (realistically) dim hope of understanding, within our lifetimes, the principles underlying such an unhomogeneous conglomeration of interactions have led some neurophysiologists to intensify their study of the less complex nervous systems of invertebrates.

3

(One is reminded of the story of the drunk who was asked why he was searching for his lost house keys under the porch light rather than out in the dark street where he dropped them. He replied, "Because the light is better here.")

4

While such studies may never explain the operation of the mammalian brain, they have been, and will continue to be, important in delineating the various *mechanisms possibly involved in neural functioning*, many of which are present in mammals.

5

Some of these examples have been mentioned in this book: the squid giant axon; the lateral eye of *Limulus;* invertebrate stretch receptors; ionic mechanisms in some invertebrate synapses; the electrical synapses in invertebrates; the glial cells of the leech; the motor control system of locusts; and the command cells of lobsters. If you study more neurophysiology, you will find still more examples. Thus the debt owed to invertebrate neurophysiology by those interested in the goal of understanding the mammalian brain is great indeed. It is to be expected that the debt will grow larger as the study of invertebrates progresses to detailed analysis of most of the cells in a limited (100-cell) nervous system, and as the neural bases of habituation, conditioning, and even learning (in these systems) are revealed. Indeed, it is often surprising how similar the neural mechanisms are for similar functions in species widely separated phylogenetically.

6

390

1 But can we expect that detailed, even exhaustive, study of simple systems will be sufficient to explain the mechanisms of the behavior of complex ones? Unequivocally, *no!*

2 On a purely technical level, the study of a complex system requires not only a much greater number of observations, but also a much greater degree of accuracy in the measurements to be made. For example, it may be that a 0.1 percent change in the threshold of all the cells in a small isolated group does not significantly affect them, but such a small threshold change conceivably could have a profound impact on the behavior of a large, complexly interconnected mass of cells.

3 Here we must approach the philosophical problem of *reductionism*, a problem present in all fields of science: If all the pieces of a given system were known in sufficient detail, would it be possible to predict the behavior of the whole? As you may recognize, this problem has occurred in physics, as well as in psychology, and will occur in neurophysiology someday. But the resolution of this problem does not truly lie in any particular field, but in the nature of scientific investigation itself.

4 The problem and its resolution are most easily understood in terms of the black-box approach.

5 A **black box,** by definition, is something whose behavior you can study, but whose inner mechanism is unobservable. That is, the black box represents a mechanism whose structure can be inferred only from observations of its inputs and outputs. In general, the unobservable inner mechanism of the box is the law that governs its behavior (the predictable relationship between input and output).

6 Some synonyms for black-box terms are shown in Table 14-1. The black-box view can help illuminate the relationship of science to medical practice (see Table 14-2). As you can see from Table 14-2, in this view the goals of medicine and of science are markedly different, even though medicine ultimately must rely on its scientific aspects in order to succeed.

7 What is important to realize is that **each field has its own black boxes.**

8 Moreover, if *some* of the fields of scientific inquiry are placed in the hierarchical order of Fig. 14-1, you can see that **the black boxes of one field are "opened" by the field "below."** That is, as *scientists attempt to answer the questions of their field, at times they are able to delineate some of the mechanisms operative in the system "above," at the same time accepting, without serious reservation, the dogma of the field "below."*

9 For example, the psychologist might hope to explain the behavior of social groups by understanding the behavior of an individual, without knowing in detail how the brain works. The physiological psychologist tries to explain the psychologist's ideas of motiva-

392

Table 14-1. Some Black-Box Synonyms

Input	Law	Output	Field
Input	Transfer function or input-output relationship	Output	Engineering
Stimulus	Law of behavior	Response	Psychology
Experimental manipulation	Predictable action	Experimental result	Scientific investigation
Values of constants and independent variables	Equations	Values of dependent variables	Computational mathetmatics
Cause of disease	Action of disease	Symptoms	Medicine
Present	Nature's "laws"	Future	Philosophy of "naive realism"
Garbage	Computer action	Garbage	Computer technology, misused

Table 14-2. Black-Box View of Scientific Investigation, Scientific Prediction, and Medical Practice

Activity	"Given"	Goal
Scientific investigation	Input and output	Discovery of law
Scientific prediction	Input and law	Prediction of output
Medical practice	Law and output, i.e., knowledge of disease and symptoms	Determination of input, i.e., disease (cause of symptoms)

392

tion or reward by means of brain mechanisms, without needing to understand the detailed mechanisms of the action potential. The neurophysiologist tries to understand the "pieces" that might fit together into a "brain," but does not question the basic tenets of the theories of macromolecular interactions. And so on.

For those planning a scientific career, the following comments on Fig. 14-1 might be helpful. In ascending the list, one moves more and more toward "the present problems of the human condition," whereas in descending the scale, one finds greater and greater rigor of scientific proof. In our experience, a scientist is attracted upward by an interest in human affairs and propelled downward by a need for certainty. The level at which each scientist settles depends on that individual's psychological needs, as well as the state of the field at that point in history (since rigor increases as a field evolves). The semiquantitative neurophysiologist of today would have been in organic chemistry in the first decades of this century and will have to be in psychology in next century's first decades if she or he is going to match the degree of compulsiveness of the field to the degree of compulsiveness in her or his personality.

An idea that is a philosophical hangover from 19th-century physics is that if all were understood and known at the atomic level, then everything else, including the future, would be predictable. Modern physicists are not so naive; nor should we be. But even though we renounce old-fashioned reductionism, we can foresee that scientists in all the fields in our proposed hierarchy will be essential for the task of connecting the ideas of each field to the others. That is, **the world is sufficiently complex that no available amount of knowledge in one field will be sufficient to predict all the behavior in the field "above."** Thus, we feel that the study of the mammalian CNS need not await the final analysis of simpler invertebrate neural networks.

In fact, often unexpected, inexplicable observations in the field "above" are powerful stimulants to investigators in fields "below."

Thus, we must conclude that **the study of the behavior of a box (the subject of a scientific field) must keep pace with the search for the mechanisms that govern its behavior.**

But we must keep in mind that no matter how accurate a measurement may be, its significance is limited by the nature of our perceptions. When the subject of one scientist's interest is within another's margin of error (and each is operating within the constraints noted), understanding of our surroundings is *necessarily* a step-by-step process in which we attain a solidly based, far-reaching view only after ascending many constructs, each resting on a hard foundation of experimentation and the integrated work of many scientists. Only in this way can the huge scope of the universe, from the subatomic to the supra-

393

Fig. 14.1. One hierarchical arrangement of some related scientific fields. (Note that the terms "above" and "below" in reference to these fields is for the purpose of description only and does not imply a moral judgment!)

Political "science"
Sociology
Social psychology
Psychology
Physiological psychology
Neurophysiology
Molecular biology
Biochemistry
Organic chemistry
Inorganic chemistry
Physical chemistry
Atomic physics
Theoretical physics
Mathematics

393

394

galactic, begin to be encompassed by the collective work of humans whose subjective, individual comprehensions of magnitude and complexity are so limited.

This view of the limitations of the human intellect is based on the following observations. First, subjective comprehension of magnitude seems limited to about one percent since, for example, the third digit rarely has any significance in prices, whether they are in dollars or millions of dollars. Second, subjective comprehension of complexity seems to be limited to three interacting variables, based on the frustrations of attempting to adjust several interacting knobs on a television set. If you think that such a lofty view of science and the universe should not be based on such apparently trivial observations, we can say only this: For some there can be no greater intellectual thrill than that of effortless flights of pure fancy in which one ascends to exhilarating heights supported by the slenderest of hypothetical threads. Such moments cannot be sustained long against the unremitting gravity of rigorous scientific logic. Yet, would you deny us this one last Icarian flight?

394

395

References

1. Aidley, D. J. *The Physiology of Excitable Cells.* Cambridge: Cambridge University Press, 1971.
2. Allweis, C. Control system diagrams in physiology, biology, and medicine. *Isr. J. Med. Sci.* [*Suppl.*] 7:1, 1971.
3. Brazier, M. A. B. The Historical Development of Neurophysiology. In J. Field, H. W. Magoun, and V. E. Hall (Eds.), *Handbook of Physiology: Section 1: Neurophysiology,* Vol. 1. Washington: American Physiological Society, 1959. Chap. 1.
4. Bullock, T. H. Signals and Neuronal Coding. In G. C. Quarton, T. Melnechuk, and F. O. Schmitt (Eds.), *The Neurosciences: A Study Program.* New York: Rockefeller University Press, 1967. Pp. 347–352.
5. Cole, K. S. *Membranes, Ions, and Impulses: A Chapter of Classical Biophysics.* Berkeley: University of California Press, 1968.
6. Cole, K. S. Some Aspects of Electrical Studies of the Squid Giant Axon Membrane. In W. J. Adelman, Jr. (Ed.), *Biophysics and Physiology of Excitable Membranes.* New York: Van Nostrand Reinhold, 1971. Chap. 4.
7. Cooke, I., and Lipkin, M., Jr. (Eds.). *Cellular Neurophysiology.* New York: Holt, Rinehart and Winston, 1972.
8. Davies, P. W. Introduction to the Problem of Excitation and Conduction. In V. B. Mountcastle (Ed.), *Medical Physiology*, 12th Ed. St. Louis: Mosby, 1968. Chap. 50.
9. Davies, P. W. Classical Electrophysiology. In V. B. Mountcastle (Ed.), *Medical Physiology*, 12th Ed. St. Louis: Mosby, 1968. Chap. 51.
10. Davies, P. W. Membrane Theory and Resting Potential. In V. B. Mountcastle (Ed.), *Medical Physiology*, 12th Ed. St. Louis: Mosby, 1968. Chap. 53.
11. Davies, P. W. Recovery Processes. In V. B. Mountcastle (Ed.), *Medical Physiology*, 12th ed. St. Louis: Mosby, 1968. Chap. 54.
12. Eccles, J. C. *The Physiology of Synapses.* New York: Springer-Verlag, 1964.
13. Engberg, I., and Lundberg, A. An electromyographic analysis of muscular activity in the hindlimb of the cat during unrestrained locomotion. *Acta Physiol. Scand.* 75:614, 1969.
14. Eyzaguirre, C. *Physiology of the Nervous System: An Introductory Text.* Chicago: Year Book Medical, 1969.
15. Granit, R. *The Basis of Motor Control: Integrating the Activity of Muscles, Alpha and Gamma Motoneurons and Their Leading Control Systems.* London: Academic Press, 1970.
16. Gray, J. A. B. Initiation of Impulses at Receptors. In J. Field, H. W. Magoun, and V. E. Hall (Eds.), *Handbook of Physiology: Section 1: Neurophysiology,* Vol. 1. Washington: American Physiological Society, 1959. Chap. 4.
17. Grundfest, H. Dynamics of the Cell Membrane as an Electrochemical System. In M. Lavallée, O. F. Schanne, and N. C. Hébert (Eds.), *Glass Microelectrodes.* New York: Wiley, 1969. Chap. 10.
18. Hodgkin, A. L. *The Conduction of the Nervous Impulse.* Springfield, Ill.: Thomas, 1964.
19. Hodgkin, A. L., and Huxley, A. F. Currents carried by sodium and potassium ions through the membrane of the giant axon of *Loligo. J. Physiol. (Lond.)* 116:449, 1952.
20. Hodgkin, A. L., and Huxley, A. F. The components of membrane conductance in the giant axon of *Loligo. J. Physiol. (Lond.)* 116:473, 1952.
21. Hodgkin, A. L., and Huxley, A. F. The dual effect of membrane potential on sodium conductance in the giant axon of *Loligo. J. Physiol. (Lond.)* 116:497, 1952.
22. Hodgkin, A. L., and Huxley, A. F. A quantitative description of membrane current and its application to conduction and excitation in nerve. *J. Physiol. (Lond.)* 117:500, 1952.
23. Hodgkin, A. L., Huxley, A. F., and Katz, B. Measurement of current-voltage relations in the membrane of the giant axon of *Loligo. J. Physiol. (Lond.)* 116:424, 1952.

396

24. Hodgkin, A. L., and Rushton, W. A. H. The electrical constants of a crustacean nerve fibre. *Proc. R. Soc. Lond. (Biol.)* 133:444, 1946.
25. Jewett, D. L. The role of tested multilevel textbooks in medical education. *Perspect. Biol. Med.* 15:450, 1972.
26. Jewett, D. L. Query-directed lecturing as a means of increasing teaching productivity in science education. *Perspect. Biol. Med.* 15:460, 1972.
27. Katz, B. *Nerve, Muscle, and Synapse.* New York: McGraw-Hill, 1966.
28. Kennedy, E. P. Some Recent Developments in the Biochemistry of Membranes. In G. C. Quarton, T. Melnechuk, and F. O. Schmitt (Eds.), *The Neurosciences: A Study Program.* New York: Rockefeller University Press, 1967. Pp. 271–280.
29. Lavallée, M., Schanne, O. F., and Hébert, N. D. (Eds.). *Glass Microelectrodes.* New York: Wiley, 1969.
30. Marshall, J. M. The Heart. In V. B. Mountcastle (Ed.), *Medical Physiology,* 12th Ed. St. Louis: Mosby, 1968. Chap. 4.
31. Marshall, J. M. Vertebrate Smooth Muscle. In V. B. Mountcastle (Ed.), *Medical Physiology,* 12th Ed. St. Louis: Mosby, 1968. Chap. 56.
32. Matthews, P. B. C. *Mammalian Muscle Receptors and Their Central Actions.* London: Edward Arnold, 1972.
33. Matthews, P. B. C. Personal communication, 1973.
34. McKusick, V. A. *Heritable Disorders of Connective Tissue,* 4th Ed. St. Louis: Mosby, 1972.
35. Meadows, D. H., Meadows, D. L., Randers, J., and Bahrens, W. M., III. *The Limits to Growth: A Report for the Club of Rome's Project in the Predicament of Mankind.* New York: Universe Books, 1972.
36. Mountcastle, V. B. Physiology of Sensory Receptors: Introduction to Sensory Processes. In V. B. Mountcastle (Ed.), *Medical Physiology*, 12th Ed. St. Louis: Mosby, 1968. Chap. 61.
37. Mountcastle, V. B., and Baldessarini, R. J. Synaptic Transmission. In V. B. Mountcastle (Ed.), *Medical Physiology*, 12th Ed. St. Louis: Mosby, 1968. Chap. 58.
38. Mountcastle, V. B., and Darian-Smith, I. Neural Mechanisms in Somesthesia. In V. B. Mountcastle (Ed.), *Medical Physiology*, 12th Ed. St. Louis: Mosby, 1968. Chap. 62.
39. Nastuk, W. L., and Mountcastle, V. B. Neuromuscular Transmission. In V. B. Mountcastle (Ed.), *Medical Physiology*, 12th Ed. St. Louis: Mosby, 1968. Chap. 57.
40. Ochs, S. *Elements of Neurophysiology*. New York: Wiley, 1965.
41. Ochs, S. Fast transport of materials in mammalian nerve fibers. *Science* 176:252, 1972.
42. Palti, Y. Analysis and Reconstruction of Axon Membrane Action Potential. In W. J. Adelman, Jr. (Ed.), *Biophysics and Physiology of Excitable Membranes.* New York: Van Nostrand Reinhold, 1971. Chap. 9.
43. Palti, Y. Digital Computer Reconstruction of Axon Membrane Action Potential. In W. J. Adelman, Jr. (Ed.), *Biophysics and Physiology of Excitable Membranes.* New York: Van Nostrand Reinhold, 1971. Chap. 10.
44. Partridge, L. D. Integration in the Central Nervous System. In J. H. V. Brown and D. S. Gunn (Eds.), *Engineering Principles in Physiology*, Vol. 1. New York: Academic Press, 1973. Chap. 4.
45. Patton, H. D. Special Properties of Nerve Trunks and Tracts. In T. C. Ruch, H. D. Patton, J. W. Woodbury, and A. L. Towe (Eds.), *Neurophysiology*, 2d Ed. Philadelphia: Saunders, 1965. Chap. 3.
46. Patton, H. D. Receptor Mechanism. In T. C. Ruch, H. D. Patton, J. W. Woodbury, and A. L. Towe (Eds.), *Neurophysiology*, 2d Ed. Philadelphia: Saunders, 1965. Chap. 4.
47. Patton, H. D. Spinal Reflexes and Synaptic Transmission. In T. C. Ruch, H. D. Patton, J. W. Woodbury, and A. L. Towe (Eds.), *Neurophysiology*, 2d Ed. Philadelphia: Saunders, 1965. Chap. 6.
48. Platt, J. R. *The Step to Man.* Huntington, N.Y.: Krieger, 1966.
49. Powers, W. T. Feedback: Beyond behaviorism. *Science* 179:351, 1973.
50. Quinn, P. J. *The Molecular Biology of Cell Membranes.* Baltimore: University Park Press, 1976.

396

397

51. Ratliff, F. *Mach Bands: Quantitative Studies on Neural Networks in the Retina*. San Francisco: Holden-Day, 1965.
52. Ruch, T. C. Neural Basis of Somatic Sensation. In T. C. Ruch, H. D. Patton, J. W. Woodbury, and A. L. Towe (Eds.), *Neurophysiology*, 2d Ed. Philadelphia: Saunders, 1965. Chap. 15.
53. Rushton, W. A. H. Peripheral Coding in the Nervous System. In W. A. Rosenblith (Ed.), *Sensory Communication: Contributions to the Symposium on Principles of Sensory Communication: July 19–August 1, 1959, Endicott House, M.I.T.* Cambridge: M.I.T. Press, and New York: Wiley, 1961. Pp. 169–181.
54. Severin, F. V., Orlovskii, G. N., and Shik, M. L. Work of the muscle receptors during controlled locomotion. *Biophysics* 12:575, 1967 (translation of *Biofizika* 12:502, 1967).
55. Shik, M. L., Orlovskii, G. N., and Severin, F. V. Organization of locomotor synergism. *Biophysics* 11:1011, 1966 (translation of *Biofizika* 11:879, 1966).
56. Sjodin, R. A. Ion Transport Across Excitable Cell Membranes. In W. J. Adelman, Jr. (Ed.), *Biophysics and Physiology of Excitable Membranes*. New York: Van Nostrand Reinhold, 1971. Chap. 3.
57. Stark, L. *Neurological Control Systems: Studies in Bioengineering*. New York: Plenum Press, 1968.
58. Stevens, S. S. *The Psychophysics of Sensory Function*. In W. A. Rosenblith (Ed.), *Sensory Communication: Contributions to the Symposium on Principles of Sensory Communication: July 19–August 1, 1959, Endicott House, M.I.T.* Cambridge: M.I.T. Press, and New York: Wiley, 1961. P. 13.
59. von Békésy, G. *Sensory Inhibition*. Princeton: Princeton University Press, 1967.
60. Walsh, E. G. *Physiology of the Nervous System*. London: Longmans, Green, 1957.
61. Werner, G. The Study of Sensation in Physiology: Psychophysical and Neurophysiologic Correlation. In V. B. Mountcastle (Ed.), *Medical Physiology*, 12th Ed. St. Louis: Mosby, 1968. Chap. 70.
62. Woodbury, J. W. The Cell Membrane: Ionic and Potential Gradients and Active Transport. In T. C. Ruch, H. D. Patton, J. W. Woodbury, and A. L. Towe (Eds.), *Neurophysiology*, 2d Ed. Philadelphia: Saunders, 1965. Chap. 1.
63. Woodbury, J. W. Action Potential: Properties of Excitable Membranes. In T. C. Ruch, H. D. Patton, J. W. Woodbury, and A. L. Towe (Eds.), *Neurophysiology*, 2d Ed. Philadelphia: Saunders, 1965. Chap. 2.
64. Woodbury, J. W., Gordon, A. M., and Conrad, J. T. Muscle. In T. C. Ruch, H. D. Patton, J. W. Woodbury, and A. L. Towe (Eds.), *Neurophysiology*, 2d Ed. Philadelphia: Saunders, 1965. Chap. 5.
65. Zierler, K. L. Mechanism of Muscle Contraction and Its Energetics. In V. B. Mountcastle (Ed.), *Medical Physiology*, 12th Ed. St. Louis: Mosby, 1968. Chap. 55.

397

399

Summary Tables, Figures, and Paragraphs

T refers to table page. F refers to figure page. Decimal refers to page and paragraph.

Decimal number following page number is the paragraph number, e.g., 131.7 signifies paragraph 7 on page 131. T refers to table page; F refers to figure page. Generally nouns form the main entries and modifiers the subentries.

402

402

404

404

405

405

407

407

408

408

411

411